T0190471

Communications
in Computer and Information Science   **2146**

Editorial Board Members

Joaquim Filipe⬤, *Polytechnic Institute of Setúbal, Setúbal, Portugal*
Ashish Ghosh⬤, *Indian Statistical Institute, Kolkata, India*
Lizhu Zhou, *Tsinghua University, Beijing, China*

Rationale

The CCIS series is devoted to the publication of proceedings of computer science conferences. Its aim is to efficiently disseminate original research results in informatics in printed and electronic form. While the focus is on publication of peer-reviewed full papers presenting mature work, inclusion of reviewed short papers reporting on work in progress is welcome, too. Besides globally relevant meetings with internationally representative program committees guaranteeing a strict peer-reviewing and paper selection process, conferences run by societies or of high regional or national relevance are also considered for publication.

Topics

The topical scope of CCIS spans the entire spectrum of informatics ranging from foundational topics in the theory of computing to information and communications science and technology and a broad variety of interdisciplinary application fields.

Information for Volume Editors and Authors

Publication in CCIS is free of charge. No royalties are paid, however, we offer registered conference participants temporary free access to the online version of the conference proceedings on SpringerLink (http://link.springer.com) by means of an http referrer from the conference website and/or a number of complimentary printed copies, as specified in the official acceptance email of the event.

CCIS proceedings can be published in time for distribution at conferences or as post-proceedings, and delivered in the form of printed books and/or electronically as USBs and/or e-content licenses for accessing proceedings at SpringerLink. Furthermore, CCIS proceedings are included in the CCIS electronic book series hosted in the SpringerLink digital library at http://link.springer.com/bookseries/7899. Conferences publishing in CCIS are allowed to use Online Conference Service (OCS) for managing the whole proceedings lifecycle (from submission and reviewing to preparing for publication) free of charge.

Publication process

The language of publication is exclusively English. Authors publishing in CCIS have to sign the Springer CCIS copyright transfer form, however, they are free to use their material published in CCIS for substantially changed, more elaborate subsequent publications elsewhere. For the preparation of the camera-ready papers/files, authors have to strictly adhere to the Springer CCIS Authors' Instructions and are strongly encouraged to use the CCIS LaTeX style files or templates.

Abstracting/Indexing

CCIS is abstracted/indexed in DBLP, Google Scholar, EI-Compendex, Mathematical Reviews, SCImago, Scopus. CCIS volumes are also submitted for the inclusion in ISI Proceedings.

How to start

To start the evaluation of your proposal for inclusion in the CCIS series, please send an e-mail to ccis@springer.com.

Kangshun Li · Yong Liu

Editors

Intelligence Computation and Applications

14th International Symposium, ISICA 2023
Guangzhou, China, November 18–19, 2023
Revised Selected Papers, Part I

 Springer

Editors
Kangshun Li 🔟
South China Agricultural University
Guangzhou, China

Yong Liu 🔟
The University of Aizu
Fukushima, Japan

ISSN 1865-0929   ISSN 1865-0937 (electronic)
Communications in Computer and Information Science
ISBN 978-981-97-4392-6   ISBN 978-981-97-4393-3 (eBook)
https://doi.org/10.1007/978-981-97-4393-3

© The Editor(s) (if applicable) and The Author(s), under exclusive license
to Springer Nature Singapore Pte Ltd. 2024

This work is subject to copyright. All rights are solely and exclusively licensed by the Publisher, whether the whole or part of the material is concerned, specifically the rights of translation, reprinting, reuse of illustrations, recitation, broadcasting, reproduction on microfilms or in any other physical way, and transmission or information storage and retrieval, electronic adaptation, computer software, or by similar or dissimilar methodology now known or hereafter developed.
The use of general descriptive names, registered names, trademarks, service marks, etc. in this publication does not imply, even in the absence of a specific statement, that such names are exempt from the relevant protective laws and regulations and therefore free for general use.
The publisher, the authors and the editors are safe to assume that the advice and information in this book are believed to be true and accurate at the date of publication. Neither the publisher nor the authors or the editors give a warranty, expressed or implied, with respect to the material contained herein or for any errors or omissions that may have been made. The publisher remains neutral with regard to jurisdictional claims in published maps and institutional affiliations.

This Springer imprint is published by the registered company Springer Nature Singapore Pte Ltd.
The registered company address is: 152 Beach Road, #21-01/04 Gateway East, Singapore 189721, Singapore

If disposing of this product, please recycle the paper.

Preface

The 14th International Symposium on Intelligence Computation and Applications (ISICA 2023) was held on November 18–19, 2023 in Guangzhou, China, and served as a forum to present the current work of researchers and software developers from around the world as well as to highlight activities in the Intelligence Computation and Applications areas. It aimed to bring together research scientists, application pioneers, and software developers to discuss problems and solutions and to identify current and new issues in this area. ISICA 2023 received a total of 178 papers, all of which underwent peer review, and it ultimately accepted 82 papers.

These two-volumes features the most up-to-date research, organized in the following five parts. Section 1 explores the frontiers of evolutionary intelligent optimization algorithms. Section 2 focuses on the exploration of computer vision. Section 3 presents machine learning and its applications. Section 4 discusses big data analysis and information security. Section 5 covers some new Intelligent applications of computers. One of ISICA's missions is to explore how complex systems can inherit simple evolutionary mechanisms, and how simple models can produce complex morphologies.

On behalf of the Organizing Committee, we would like to warmly thank the sponsors, South China Agricultural University, Guangdong Key Laboratory of Big Data Analysis and Processing at Sun Yat-sen University, and Guangdong Polytechnic Normal University, which helped in one way or another to achieve our goals for the conference. We wish to express our appreciation to Springer for publishing the proceedings of ISICA 2023. We also wish to acknowledge the dedication and commitment of both the staff at Springer's Beijing office and the CCIS editorial staff. We would like to thank the authors for submitting their work, as well as the Program Committee members and reviewers for their enthusiasm, time, and expertise. The invaluable help of active members from the Organizing Committee, including Yan Chen, Lixia Zhang, Lei Yang, Wenxiang Wang, Shumin Xie, Jiaxin Xu, Tian Feng, Zifeng Jiang, Jiayu Zhang, Zhensheng Yang, Tao Lai, Ruolin Ruan, Shuizhen He, Junjie Wang, Mingchen Xie, Weicong Chen, Zhihao Zhou, Juhong Wu, Zhidong Zeng, Tianjin Zhu, Wensen Mo, Wenbin Xiang, Hassan Jalil, and Al-Daba Saqr in setting up and maintaining the online submission systems by Easy Chair, assigning the papers to the reviewers, and preparing the camera-ready version of the proceedings is highly appreciated. We would like to thank them personally for their help in making ISICA 2023 a success.

April 2024

Kangshun Li
Yong Liu

Organization

Honorary Chairs

Yuping Chen   Wuhan University, China
Yuanxiang Li   Wuhan University, China
Wensheng Zhang   Institute of Automation, Chinese Academy of Sciences, China

General Chairs

Kangshun Li   South China Agricultural University, China
Witold Pedrycz   University of Alberta, Canada
Jian Yin   Sun Yat-sen University, China
Yu Tang   Guangdong Polytechnic Normal University, China

Program Chairs

Yong Liu   University of Aizu, Japan
Kangshun Li   South China Agricultural University, China
Zhiping Tan   Guangdong Polytechnic Normal University, China
Yiu-ming Cheung   Hong Kong Baptist University, China
Jing Liu   Xidian University, China
Hailin Liu   Guangdong University of Technology, China
Hui Wang   Shenzhen Institute of Information Technology, China
Feng Wang   Wuhan University, China
Xuesong Yan   China University of Geosciences, Wuhan, China
Wenyin Gong   China University of Geosciences, Wuhan, China
Xuewen Xia   Minnan Normal University, China
Xing Xu   Minnan Normal University, China
Yinglong Zhang   Minnan Normal University, China

Local Arrangement Chairs

Zhijian Wu   Wuhan University, China
Yan Chen   South China Agricultural University, China

Publicity Chairs

Shunmin Xie   South China Agricultural University, China
Feng Wang   Wuhan University, China
Wei Li   Jiangxi University of Science and Technology,
                              China
Lixia Zhang   South China Agricultural University, China
Lei Yang   South China Agricultural University, China

Program Committee

Ehsan Aliabadian   University of Calgary, Canada
Rafael Almeida   University of Calgary, Canada
Ehsan Amirian   University of Calgary, Canada
Nik Bessis   University of Derby, UK
Yiqiao Cai   Huaqiao University, China
Zhangxing Chen   University of Calgary, Canada
Iyogun Christopher   University of Calgary, Canada
Guangming Dai   University of Calgary, Canada
Lixin Ding   Wuhan University, China
Ciprian Dobre   University Politehnica of Bucharest, Romania
Xin Du   Fujian Normal University, China
Christian Esposito   University of Salerno, Italy
Zhun Fan   Shantou University, China
Massimo Ficco   University of Campania Luigi Vanvitelli, Italy
Razvan Gheorghe   University Politehnica of Bucharest, Romania
Maoguo Gong   Xidian University, China
Zhaolu Guo   Jiangxi University of Science and Technology,
                              China
Tomasz Hachaj   Pedagogical University of Krakow, Poland
Guoliang He   Wuhan University, China
Jun He   Aberystwyth University, UK
Han Huang   South China University of Technology, China
Xiaomin Huang   Sun Yat-sen University, China
Ying Huang   Gannan Normal University, China

Sanyou Zeng   China University of Geosciences, Wuhan, China
Changhe Li   China University of Geosciences, Wuhan, China
Ming Yang   China University of Geosciences, Wuhan, China
Dazhi Jiang   Shantou University, China
Zhiping Tan   Guangdong Polytechnic Normal University, China
Shuling Yang   South China University of Technology, China
Hui Li   Xi'an Jiaotong University, China

Contents – Part I

Frontiers of Evolutionary Intelligent Optimization Algorithms

An Improved NSGA-II Algorithm with Markov Networks 3
Yuyan Kong, Jintao Yao, Juan Wang, Peiquan Huang, and Zhenzhen Qiu

An Improved Particle Swarm Optimization Algorithm Combined with Bat
Algorithm 18
Hongyu Xiao, Nannan Zhao, Zihang Gao, and Xiaojun Cui

An Improved Whale Optimization Algorithm Combined with Bat
Algorithm and Its Applications 26
Xiaofeng Wang, Jian'ou Wang, and Chanjuan Lin

Fusion of Nonlinear Inertia Weight and Probability Mutation for Binary
Particle Swarm Optimization Algorithm 39
Jiayu Zhang and Kangshun Li

An Evolutionary Algorithm Based on Replication Analysis for Bi-objective
Feature Selection 49
Li Kangshun and Hassan Jalil

Improved Particle Swarm Algorithm Using Multiple Strategies 62
Yunfei Yi, Zhiyong Wang, and Yunying Shi

A Reference Vector Guided Evolutionary Algorithm with Diversity
and Convergence Enhancement Strategies for Many-Objective
Optimization 73
Lei Yang, Yuanye Zhang, and Jiale Cao

Research on Mine Emergency Evacuation Scheme Based on Dynamic
Multi-objective Evolutionary Algorithm 88
Furong Jing, Hui Liu, and Yanhui Zang

An Adaptive Dynamic Parameter Multi-objective Optimization Algorithm 101
Yu Lai and Lanlan Kang

Adaptive Elimination Particle Swarm Optimization Algorithm
for Logistics Scheduling 113
Kexin Lin, Wei Li, and Yuqi Ou

A Modified Two_Arch2 Based on Reference Points for Many-Objective
Optimization 125
  Shuai Wang, Dong Xiao, Futao Liao, Shaowei Zhang, Hui Wang,
  Wenjun Wang, and Min Hu

Floorplanning of VLSI by Mixed-Variable Optimization 137
  Jian Sun, Huabin Cheng, Jian Wu, Zhanyang Zhu, and Yu Chen

A Multi-population Hierarchical Differential Evolution for Feature
Selection 152
  Jian Guan, Fei Yu, and Zhenya Diao

Research on State-Owned Assets Portfolio Investment Strategy Based
on Improved Differential Evolution 165
  Dong Ji and Dandan Cui

A Particle Swarm Optimization Algorithm with Dynamic Population
Synergy 178
  Qianqian Dong, Wei Li, and Fufa He

Preference-Based Multi-objective Optimization Algorithms Under
the Union Mechanisms 192
  Yi Zhong and Lanlan Kang

Auto-Enhanced Population Diversity with Two Options 207
  Yangcong Ou, Ming Yang, and Jing Guan

Exploration of Computer Vision

Robot Global Relocation Algorithm Based on Deep Neural Network
and 3D Point Cloud 223
  Yan Chen, Zhengying Li, and Wenbin Qiu

Safety Zone and Its Utilization in Collision Avoidance Control of Industrial
Robot 231
  Yongcai Zhang, Yih Bing Chu, and Tian Jiang

Entropy of Interval Type-2 Fuzzy Sets and Its Application in Image
Segmentation 247
  Jianqiao Shen, Haijun Qian, and Huabei Nie

An Improved Algorithm for Facial Image Restoration Based on GAN 254
  Jibo Zhang, Jia Yuan, Dongbo Zhang, and Lu Xiang

A Study of PyTorch-Based Algorithms for Handwritten Digit Recognition 266
Kangshun Li, Mingchen Xie, and Xuhang Chen

A High-Quality Video Reconstruction Optimization System Based
on Compressed Sensing 277
*Yanjun Zhang, Yongqiang He, Jingbo Zhang, Zhihua Cui,
and Xingjuan Cai*

Conv and Efficient Multi-Scale Attention Module for YOLOv5 292
Xuan Guo and Weidong Huang

BRA-YOLO: Object Detection Algorithm with Bi-Level Routing
Attention for YOLOv5 302
Xing Huang and Weidong Huang

A Pest Detection Algorithm Based on Improved YOLO 312
Kangshun Li, Shuizhen He, and Jiancong Wang

Improving Interactive Differential Evolution for Cartoon Face Image
Combination 326
Bo Tang, Fei Yu, Qingrong Ou, Bang Liang, and Jian Guan

Mask Reconstruction Augmentation and Attention Aggregation for Stereo
Matching 340
Zhaokui Li, Zhongxin Yang, Jinen Zhang, and Jinrong He

Machine Learning and Its Applications

Construction of an Intelligent Salary Prediction Model and Analysis of BP
Neural Network Applications 357
Xuming Zhang, Ling Peng, and Ping Wang

Human Action Recognition Classification Based on 3D CNN Deep
Learning 369
Li Kangshun, Tianjin Zhu, and Hangchi Cheng

A KNN Algorithm Based on Mixed Normalization Factors 388
Hui Wang, Tie Cai, Yong Wei, and Jiahui Cai

Modified Carnivorous Plant Algorithm Based on Lévy Flight
for Optimizing the BP Model 395
Chen Ye, Peng Shao, and Shaoping Zhang

CR-IFSSL: Imbalanced Federated Semi-Supervised Learning with Class
Rebalancing 409
   Yutong Xie, Haiyan Liang, Xianmin Wang, Jing Li, Ziyu Cheng,
   Siming Huang, Feng Liu, and Li Guo

Kernel Fence GAN: Unsupervised Anomaly Detection Model Based
on Kernel Function 420
   Lu Niu and Shaobo Li

A News Recommendation Approach Based on the Fusion of Attention
Mechanism and User's Long and Short Term Preferences 429
   Yi Xiong

Research of the Three-Dimensional Spatial Orientation for Non-visible
Area Based on RSSI 443
   Huabei Nie, Jianqiao Shen, Haihua Zhu, Ani Dong, Yongcai Zhang,
   and Yi Niu

Collaborative Filtering Recommendation Algorithm Based on Improved
KMEANS 451
   Xuesong Zhou, Changrui Li, and Jia Shi

Emotion Analysis of Weibo Based on Long Short-Term Memory Neural
Network 463
   Li Kangshun, Weicong Chen, and Yishu Lei

Author Index 471

Contents – Part II

Machine Learning and Its Applications

Human Flow Prediction Model Based on Graph Convolutional Recurrent
Neural Network 3
  Hongwei Su and Maria Amelia E. Damian

Research on Text Classification Algorithm Based on Deep Learning 15
  Li Kangshun, Junjie Wang, and Wenbin Zhu

MobilenetV2-Based Network for Bamboo Classification
with Tri-Classification Dataset and Fog Removal Training 28
  Yan Chen, Dehao Shi, and Hongxing Peng

A Filter Similarity-Based Early Pruning Methods for Compressing CNNs 39
  Zifeng Jiang and Kangshun Li

Visualization of Convolutional Neural Networks Based on Gaussian Models ... 49
  Hui Wang and Tie Cai

A New Feature Selection Algorithm Based on Adversarial Learning
for Solving Classification Problems 56
  Xiao Jin, Bo Wei, Wentao Zha, and Jintao Lu

A Simulated Annealing BP Algorithm for Adaptive Temperature Setting 71
  Zi Teng, Zhixun Liang, Yuanxiang Li, and Yunfei Yi

Research on the Important Role of Computers in the Digital Transformation
of the Clothing Industry 95
  Ping Wang and Xuming Zhang

Multimedia Information Retrieval Method Based on Semantic Similarity 103
  Xuanyi Zong, Jingwen Zhao, Zhiqiang Chen, and Jinfeng He

Iterative Learning Control for Encoding-Decoding Method with Data
Dropout at Both Measurement and Actuator Sides 113
  Yongxian Chen and Yunshan Wei

A Domain Adaptive Segmentation Label Generation Algorithm
for Autonomous Driving Scenarios 127
  Kangshun Li and Tian Feng

Visualization Analysis of Convolutional Neural Network Processes   135
   Hui Wang, Tie Cai, Yong Wei, and Zeming Chen

Packet Performance Predictor Based on Graph Isomorphism Network
for Neural Architecture Search   142
   Yue Liu, Jiawang Li, Zitu Liu, and Wenjie Tian

Big Data Analysis and Information Security

Reversible Data Hiding Algorithm Based on Adaptive Predictor
and Non-uniform Payload Allocation   159
   Dan He

Research on Bayberry Traceability Platform Based on Blockchain   170
   Hongyu Xiao, Zihang Gao, Xiaojun Cui, and Nannan Zhao

Research on Satellite Navigation and Positioning Based on Laser Point
Cloud Data   179
   Yuming Sun and Hua Wang

Research and Application of System with Bayberry Blockchain Based
on Hyperledger Fabric   187
   Hongyu Xiao, Zihang Gao, Xiaojun Cui, and Nannan Zhao

A Multiparty Reversible Data Hiding Scheme in Encrypted Domain Based
on Hybrid Encryption   196
   Bing Chen, Lu Chai, Yong Wang, Jingkun Yu, and Wanhan Fang

Research on Smart Agriculture Big Data System Based on Spark
and Blockchain   208
   Yuming Sun and Hua Wang

Efficient Public Key Encryption Equality Test with Lightweight
Authorization on Outsourced Encrypted Datasets   214
   Chengyu Jiang, Sha Ma, and Hao Wang

Fake News Detection Model Incorporating News Text and User Propagation . . .   225
   Shuxin Yang, Jiahao Li, and Weidong Huang

Data Analysis of University Educational Administration Information
Based on Prefixspan Algorithm   240
   Yiying Xu, Yi Liu, and Haili Yu

The Application and Exploration of Big Data in College Student
Information Management 253
    Jun Zhang and Yuanbing Wang

Research on Precision Marketing Strategy of Guangdong Characteristic
Products Enabled by Big Data in Rural 262
    Hua Wang and Yuming Sun

Research and Application of Offline Log Analysis Method for E-commerce
Based on HHS 270
    Haoliang Wang, Kun Hu, Lili Wang, and Jingtong Shang

Research on Communication Power of Cross-Cultural Short Video Based
on Qualitative Comparative Analysis 278
    Wen Meng and Chao Yu

Multi-recipient Public-Key Authenticated Encryption with Keyword
Search 287
    Kejin He, Sha Ma, and Hao Wang

Analysis of Information Security Processing Technology Based
on Computer Big Data 297
    Hua Wang and Fuyu Zhu

Intelligent Application of Computer

Research on the Quality Evaluation and Optimization of Ideological
and Political Education in Universities Driven by Artificial Intelligence 309
    Yuanbing Wang and Jun Zhang

Research on the Routing Protocol Algorithm Driven by the Dedicated
Frequency Points of the Internet of Things to Build a Network 319
    Lingwei Wang and Hua Wang

Artificial Intelligence in Intelligent Clothing: Design and Implementation 326
    Ping Wang and Xuming Zhang

Edible Oil Price Forecasting: A Novel Approach with Group Temporal
Convolutional Network and BetaAdaptiveAdam 337
    Lei Yang, Huade Li, Rui Xu, Zexin Xu, and Jiale Cao

Design and Application of a Teaching Evaluation Model Based
on the Theory of Multiple Intelligences 351
    Luyan Lai

Study on TNM Classification Diagnosis of Colorectal Cancer Based
on Improved Self-supervised Contrast Learning 360
  Tao Lai and Kangshun Li

Construction and Quality Evaluation of Learning Motivation Model
from the Perspective of Course Ideology and Politics 372
  Luyan Lai and Yongdie Che

The 3D Display System of Art Works Based on VR Technology 384
  Huyuan Lu, Qiner Xu, Zhiqiang Chen, and Beixin Zhong

Petrochemical Commodity Price Prediction Model Based on Wavelet
Decomposition and Bayesian Optimization 394
  Lei Yang, Rui Xu, Huade Li, and Zexin Xu

Mutate Suspicious Statements to Locate Faults 407
  Guangsheng Zhan, Shi Cheng, and Jinbao Zhang

Iterative Learning Control with Variable Trajectory Length in the Presence
of Noise 419
  Yuangao Yan, Xixian Tan, and Yunshan Wei

Quality Control Model of Value Extraction of Residual Silk Reuse Based
on Improved Genetic Algorithm 430
  Qi Ji, Mingxing Li, and Chao Shen

Key Technology and Application Research Based on Computer Internet
of Things 442
  Lingwei Wang and Hua Wang

Research on the Innovative Application of Computer Aided Design
in Environmental Design 450
  Lei Wang

Research on Intelligent Clustering Scoring of English Text Based
on XGBOOST Algorithm 459
  *Zhaolian Zeng, Wanyi Yao, Jia Zeng, Jiawei Lei, Feiyun Chen,
  and Peihua Wen*

Machine Learning-Assisted Optimization of Direction-Finding Antenna
Arrays 476
  *Qing Zhang, Miao Gong, Gouqiong Li, Xinyu Ma, Yiheng Chen,
  Fei Zhao, and Sanyou Zeng*

Author Index 487

Frontiers of Evolutionary Intelligent Optimization Algorithms

An Improved NSGA-II Algorithm with Markov Networks

Yuyan Kong[1], Jintao Yao[2(✉)], Juan Wang[1], Peiquan Huang[1], and Zhenzhen Qiu[3]

[1] College of Computer and Informatics, Guangdong Polytechnic of Industry and Commerce,
Guangzhou 510510, China
[2] College of Mathematics and Informatics, South China Agricultural University,
Guangzhou 510642, China
justin_yjt@163.com
[3] Electromechanic Engineering College, Guangdong Engineering Polytechnic College,
Guangzhou 510520, Guangdong, China

Abstract. NSGA-II algorithm is one of the most representative multi-objective Evolutionary Algorithms. With the help of elite preserving strategy and fast non-dominated sorting method, NSGA-II can effectively maintain the diversity of population and reduce the computational complexity. It has been widely used to solve different problems. However, traditional crossover and mutation operators in NSGA-II have poor linkage learning ability, so it is not easy for NSGAII to identify and exploit the interaction between variables. To make matters worse, it will inevitably damage randomly good building blocks in solution and make the process of searching the optimal Pareto front extremely difficult. In this paper, we propose an improved NSGA-II based on Markov network that replaces crossover and mutation operators by building Markov networks of promising solutions and sampling the built model to generate new solutions. Markov network can describe, identify and maintain the interaction at the variable level abstractly and accurately, which can identify and protect the good building blocks. At the same time, reduction of the manual parameter setting such as crossover probability will direct the MN-NSGA-II intelligently search for Pareto optimal front. The experimental results also show that the MNNSGA-II is effective and has better global convergence than NSGA-II.

Keywords: MOPs · Multi-objective Evolutionary Algorithm · Markov Network

1 Introduction

Multi-objective Optimization Problems (MOPs) [1–4] usually consist of at least two conflicting optimization objectives. Due to the internal correlation among optimization objectives, an absolute optimal solution cannot be obtained. Instead, Pareto solution Set or Non-dominated solution set that approximates optimal Pareto front can only be searched in feasible objective space based on the proximity and diversity criteria [5]. As a population-based probabilistic search method, Evolutionary Algorithm (EA) [6] has been proved to be one of the most effective optimization algorithms for solving

© The Author(s), under exclusive license to Springer Nature Singapore Pte Ltd. 2024
K. Li and Y. Liu (Eds.): ISICA 2023, CCIS 2146, pp. 3–17, 2024.
https://doi.org/10.1007/978-981-97-4393-3_1

MOPs problems due to its potential parallelism and distributed characteristics. Therefore, Multi-objective Optimization Evolutionary Algorithm (MOEA) [7, 8] has incomparable advantages compared with traditional optimization methods [9–13] and become a hot research direction in the field of evolutionary algorithms. Among MOEA algorithms, NSGA-II (Non-dominated Sorting Genetic Algorithm II) algorithm [9] is one of the most famous and widely used. On the basis of NSGA algorithm [13], NSGA-II adds the fast non-dominated sorting method to construct non-dominated sets. In addition, NSGA-II algorithm also proposed the concept of congestion distance, and adopted a dynamic congestion mechanism to estimate density information to maintain the uniform distribution and diversity of solutions, which effectively improve the global search ability of NSGA-II algorithm.

The outstanding effect of NSGA-II algorithm in successfully solving various MOPs problems has attracted many scholars to NSGA-II algorithm. In order to further improve NSGA-II's performance or expand NSGA-II to solve more MOPs problems in the real world. Some modifications are made to reduce the computational complexity and weaken the global convergence for practicality. For example, the convergence speed and convergence rate are improved by using the prior information of the problem to heuristic the initial population, increase the selection pressure, simplify the calculation of congestion distance, parallel processing, reduce the scale of external Archive, and weaken the Pareto domination relationship (α-domination or ε-domination) [14]. More studies focus on modifying the crossover and mutation operations in NSGA-II algorithm to maintain the diversity of population during iteration, so as to improve the global convergence ability, such as adaptive crossover and mutation probability calculation, arithmetic crossover, local mapping crossover, Gaussian variation, inversion variation [15, 16]. Although these modifications to crossover and mutation operators can improve the performance of NSGA-II algorithm, they still follow genetic operators of traditional evolutionary algorithm without substantial changes, retain the randomness characteristics, and lack the linkage learning ability to identify and maintain the linkage (dependency) relationship between the variables in solution. In addition, manually setting many parameters such as cross probability also has great constraints on the performance and real application of algorithm.

In recent years, estimation of distribution algorithms (EDAs) [17], which are based on graph theory and probabilistic statistical theory, have been proposed. In order to overcome the difficulties from genetic operators in EAs, EDAs directly replace variation operator and crossover operator to generate new solutions by repeatedly constructing and sampling selected probabilistic graphic model (PGM) [18] for solving difficult problems. Since PGM can abstractly and accurately describe, identify and maintain the interaction of variables in solution, the excellent building blocks in solution can be effectively identified and protected during iteration. In addition, the manual setting of important parameters is reduced, which makes EDAs self-directedly and intelligently searching for promising solution space. Consequently, EDAs have become a relatively new kind of evolutionary algorithms for solving large-scale and highly complex difficult problems in the real world. The most representative EDAs include BOA [17] and MOA [19]. Approximate estimation and concise representation of complex probability distribution are the core of EDAs, which can succinctly and accurately describe the exact linkage information about fitness function and the structure of problem to be solved. Therefore, whether PGM selected is reasonable will directly affect the performance and efficiency of algorithm. MOA uses Markov network to simulate multivariable data and represent the dependency relationship between variables in solution. MOA realizes decomposition of solution by approximating Gibbs distribution of population, which uses Markov network to accurately learn linkage relationship between variables in a more natural way than Bayesian network. The bidirectional nature of variable interaction eliminates loop problem in Markov network, which can effectively reduce the construction complexity of Markov networks, and effectively solve the high-order difficult problems with complex multivariable symmetric interaction. Therefore, this paper proposes an improved NSGA-II algorithm based on Markov network (MN-NSGA-II), which generates new candidate solutions by sampling a Markov network learnt from a set of promising solutions during iteration, so as to obtain the better comprehensive performance.

2 Background

2.1 MOPs (Multi-objective Optimization Problems)

Definition 1. (*MOPs*). Let $f(x) = (f_1(x),...,f_n(x))^T$ be the objective function vector, $g(x) = (g_1(x),...g_p(x))^T$ and $h(x) = (h_1(x),...,h_q(x))^T$ be the constraint function vector respectively, where $x = (x_1,...,x_n)^T$ is a solution vector with n decision variables. Under the constraint conditions of $g_i(x) \leq 0$ $(i = 1,...,p)$ and $h_j(x) = 0$ $(j = 1,...,q)$, the minimization of all objective functions of $f(x)$ is MOPs as shown in formula (1).

$$
\begin{aligned}
minf(x) &= (f_1(x), \cdots, f_n(x))^T, n \geq 2 \\
s.t.\, g_i(x) &\leq 0, i = 1, \cdots, p \\
h_j(x) &= 0, j = 1, \cdots, q
\end{aligned}
\tag{1}
$$

Definition 2. (*Pareto Dominance*). Given two feasible solutions x_a and x_b, x_a dominates x_b if one at least of the following two conditions are met, which is denoted as $x_a \succ x_b$.

(1) For all $i \in 1, 2, ..., n$, if and only if $f_i(x_a) < f_i(x_b)$.

(2) For all $i \in 1, 2, ..., n$, if and only if $f_i(X_a) \leq f_i(x_b)$, and $f_j(x_a) < f_j(x_b)$ for at least one index $j \in 1,2,...,n$.

Definition 3. (*Pareto Set*). If a feasible solution x^* is not dominated by any other solutions, x^* is called a Pareto optimal solution. Therefore, all Pareto optimal solutions forms a Pareto Set (PS).

Definition 4. (*Pareto Front*). The Pareto front (PF) is the projection of Pareto set in the objective space.

2.2 NSGA-II Algorithm

As a well-known MOEA algorithm, NSGA takes the non-dominated front as the set of optimal solution. Compared with the traditional genetic algorithm, the major difference lies in employment of non-dominated sorting and fitness sharing strategies. First of all,

NSGA uses Pareto dominance and Pareto rank to identify and sort the non-dominated solutions. In order to obtain PF, NSGA specifies virtual fitness value for each solution and applies sharing niche technology to adjust the fitness value appropriately and make the distribution of solutions more uniform. Finally, NSGA selects the fitter solutions according to fitness value as new population. However, nondominated sorting inevitably leads NSGA to high computational complexity. Fortunately, NSGA-II proposes a fast non-dominated sorting method to construct nondominated set, which can effectively reduce the time complexity. In addition, NSGA-II also proposes the concept of crowding distance, which uses a dynamic congestion mechanism to estimate solution density information for maintaining the uniform distribution and diversity of population.

Fast Non-Dominated Sorting Method. Unlike non-dominated sorting method in NSGA algorithm, the fast non-dominated sorting method sets two parameters n_p and s_p to reduce computational complexity, where n_p represents the number of solutions dominating solution p and s_p represents the number of solutions dominated by p. In the process of fast non-dominated sorting, np and sp of each solution are obtained first by traversing the population. And then, n_p and s_p are used to sort and rank the population. Compared with the computational complexity $O(MN^3)$ of NSGA, NSGA-II reduces it to $O(MN^2)$, where M is objective number and N is population size.

Elite Preserving Strategy. In order to avoid the loss of excellent solutions during the iteration, NSGA-II algorithm executes the elite preserving strategy to directly transfer the excellent solutions in population to next generation. In other words, the nondominated solutions found in each generation move on to the next generations till some solutions dominate them.

Crowding Distance. NSGA algorithm uses shared niche technology to maintain population diversity, but shared radius σ_{share} is set manually, which easily leads algorithm to premature convergence. Therefore, the crowding distance is proposed in NSGA-II algorithm. In order to estimate the density of solutions surrounding a particular solution, the crowding distance is calculated, which is the average distance of two solutions on either side of the solution along each of the objectives. When two solutions have different crowding distances, the solution with the large crowded distance is considered to be present in a less crowded region. Obviously, the probability of converging to optimal Pareto front is greatly increased.

Procedure of NSGA-II Algorithm. The procedure of NSGA-II algorithm is described as follows:

Step 1: An initial parent population P_t of size N is generated randomly, and t is set to 1.
Step 2: The fast not-dominated sorting operator, selection operator, crossover operator and mutation operator are performed repetitively on P_t until the first offspring population Q_t is generated, and t is set to 2.
Step 3: The parent population P_t and the offspring population Q_t are merged into a new population R_t of size N, and the fast non-dominated sorting operator is performed on R_t. And then, the individuals of R_t are ranked into different fronts according to their Pareto levels.

Step 4: After performing repetitively, the fast not-dominated sorting operator, crowding distance calculating and elite preserving, N individuals are selected from R_t to create the next parent population P_{t+1}.

Step 5: Set $t=t+1$, and selection operator, crossover operator and mutation operator on P_t are performed to generate new candidate solutions. If the stopping criteria are not satisfied, the procedure transfers to Step 3.

3 The Modified NSGA-II with Markov Network

3.1 Markov Network

Markov network is an undirected probabilistic graph model, which can be formally represented as (G, Ψ), where G is the network structure and Ψ is the parameter set. Each node in G corresponds to a variable in the solution under given data set, and each edge in G corresponds to a conditional dependency relationship between a pair of random variables in the solution. For example, Fig. 1 shows a Markov network with six nodes. The local Markov property can characterize a Markov network as a neighborhood system such as a set of Markov Blankets, but the global Markov property can also characterize the Markov network as a set of cliques.

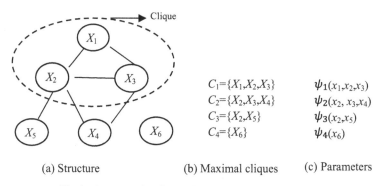

(a) Structure   (b) Maximal cliques   (c) Parameters

Fig.1. An example of a Markov network with six nodes.

Theorem 1. *Global Markov Property.* Given an undirected graph G, if all paths between node A and node B can be separated by a node in the node set C, then A and B are said to be conditionally independent given C. For a Markov network, its joint probability distribution, $p(x)$, can be formulated with the help of the global Markov property.

Definition 5. (*Clique*): Given an undirected graph G, a clique in G is a fully connected subset of the nodes. In Fig. 1, $\{X_1, X_2, X_3\}$ is a clique with three nodes.

Definition 6. (*Sub Clique*). Given an undirected graph G, a sub clique of a clique in G is a fully connected subset of nodes within this clique. In Fig. 1, $\{X_1, X_2\}$, $\{X_2, X_3\}$ and $\{X_1, X_3\}$ are all sub cliques of $\{X_1, X_2, X_3\}$.

Definition 7. (***Maximal Clique***). A clique is called maximal clique when it is not a sub clique of any other clique. In Fig. 1, $\{X_1, X_2, X_3\}$ not only is a clique, but also is a maximal clique with three nodes.

In order to formulate the joint probability distribution of a Markov network, its parameters in terms of cliques in G must be defined or learned firstly. In general, C (a set of cliques) consists of all cliques in the structure of Markov network, but only the maximal cliques are considered here. Therefore, a Markov network can also be viewed as a set of maximal cliques.

Definition 8. (***Clique Potential Function***). Given the structure G of a Markov network, its parameter $\Psi = \{\psi_1(c_1), \psi_2(c_2),..., \psi_m(c_m)\}$, is a set of positive potential functions defined on the set of maximal cliques $C = \{C_1, C_2,..., C_m\}$. c_i is used to denote the set of values taken by the set of variables in the maximal clique C_i. $\psi_i(c_i)$ is a clique potential function of C_i. Therefore, if a set of variables $X = \{X_1, X_2, ..., X_n\}$ is given, its joint probability distribution $p(x)$ for any Markov network is shown as Eq. (2) in term of clique potential function:

$$p(x) = \frac{1}{Z} \prod_{i=1}^{m} \psi_i(c_i) \tag{2}$$

where, $Z = \sum_{x \in \Omega} \prod_{i=1}^{m} \psi_i(c_i)$ is the partition function for ensuring $\sum_{x \in \Omega} p(x) = 1$. m is the number of maximal cliques in G, and Ω is the set of all possible solutions. For example, the Markov network in Fig. 1 has four maximal cliques shown as Fig. 1(b), and the corresponding parameters are shown as Fig. 1(c). Therefore, $p(x)$ of the structure G shown as Fig. 1(a) can be factorized in terms of clique potential functions shown as Eq. (3), where $x = \{x_1, x_2, x_3, x_4, x_5, x_6\}$ is a particular set of values of $X = \{X_1, X_2, X_3, X_4, X_5, X_6\}$.

$$p(x) = \frac{1}{Z} \psi_1(x_1, x_2, x_3) \psi_2(x_2, x_3, x_4) \psi_3(x_2, x_5) \psi_4(x_6) \tag{3}$$

3.2   Learning a Markov Network from a Set of Candidate Solutions

After a set of promising solutions is selected from population by means of corresponding selection method, MN-NSGA-II algorithm estimates the distribution of selected solutions to construct a Markov network and calculate the important parameters related to linkage among variables in solution. In general, this procedure is composed of two sub procedures: structure learning and parameter learning.

The main purpose of structure learning is to find the interactions between variables in solution. In other words, all of maximal cliques in a Markov network can be find out during this sub-procedure. In this paper, we extend chi-square statistic test method to learn the structure. The steps of structure learning are described as follows:

Step 1: Start with a completely undirected graph G.
Step 2: Perform Chi-Square statistic test with some threshold for each pair of variables x_i and x_j in G. If the test result is less than given threshold, x_i and x_j are independent of each other, and then the edge between x_i and x_j is removed subsequently from G.

Step 3: In order to further simplify the resulting structure of **step 2**, it is necessary to perform 1-order conditional independence test. The corresponding edge also will be removed if an independency relationship is·found during this test.

After the exact dependency structure is found out by statistically learning a Markov network, the parameter learning, which is more difficult, can be performed immediately. In this paper, we use the Markov fitness modeling (MFM) [19] approach to approximate the parameters in a Markov network. The steps of parameter learning are described as follows:

Step 1: Turn the joint probability distribution of a given Markov network into a Gibbs distribution. In essence, the structure learning is to decompose a Markov network into a set of maximal cliques, so the joint probability distribution of a Markov network can be defined in terms of the corresponding Gibbs distribution shown as Eq. (4) according to Hammersley-Clifford theorem [19].

$$p(x) = \frac{f(x)}{Z} = \frac{e^{-U(x)/T}}{Z} = \frac{f(x)}{\sum_{y \in \Omega} f(y)} = \frac{e^{-U(x)/T}}{\sum_{y \in \Omega} e^{-U(y)/T}} \tag{4}$$

where, $Z = \sum_{y \in \Omega} e^{-U(y)/T}$ is a normalizing constant, $f(x)$ is the fitness of a candidate solution x, T (setting $T = 1$ here) is a temperature parameter of Gibbs distribution and is used to control the shape of Gibbs distribution, and $U(x)$ is the energy of Gibbs distribution, which is the sum of clique potential functions according to all maximal cliques in G and is shown as Eq. (5):

$$U(x) = \sum_{i=1}^{m} u_i(c_i) \tag{5}$$

where, $u_i(c_i)$ represents a clique potential function over a maximal clique C_i. Therefore, the Eq. (4) can also be written as Eq. (6).

$$p(x) = \frac{e^{-\sum_{i=1}^{m} u_i(c_i)/T}}{Z} \tag{6}$$

Step 2: Relate the energy $U(x)$ to the fitness $f(x)$ of each candidate solution x. Generally, the probability of a candidate solution is proportional to its fitness as shown in Eq. (4), so the relationship between a candidate solution's fitness and an energy function is easily derived from Eq. (4), which is shown in Eq. (7).

$$U(x) = -\ln(f(x)) \tag{7}$$

Step 3: Define the energy $U(x)$ in terms of clique potential function. As stated earlier, the clique potential function represents the strength of linkage between the variables in a maximal clique. Here, we only consider the interactions among multiple variables. For each candidate solution x, its energy $U(x)$ can be represented as the Eq. (8):

$$U(x) = \sum_{i=1}^{m} u_i(c_i) = \sum_{i=1}^{m} \gamma_{C_i} x_{C_{i,1}} x_{C_{i,2}} \dots x_{C_{i,j}} \dots x_{C_{i,k}} \tag{8}$$

where, $\gamma = \{\gamma_{C_1}, \gamma_{C_2}, \cdots, \gamma_{C_m}\}$ is the parameter of a Markov network, and $x_{C_{i,j}}$ is the jth variable in a maximal clique C_i.

Step 4: Determine the parameters of a Markov network. Once a set of candidate solutions is formed, each candidate solution's energy $U(x)$ can be determined by its fitness as shown in Eq. (7). In terms of the Eq. (7) and the Eq. (8), an equation shown in (9) is formed.

$$-\ln(f(x)) = \sum_{i=1}^{m} \gamma_{C_i} x_{C_{i,1}} x_{C_{i,2}} \cdots x_{C_{i,j}} \cdots x_{C_{i,k}} \qquad (9)$$

Therefore, we can obtain a set of equations relating the parameter γ. In order to determine unknown parameters, we have to solve this set of equations by using singular value decomposition (SVD) method. For mathematical reason, the values $\{0,1\}$ for variable xi are replaced by $\{-1,1\}$ in the process of solving equations, which would be reset to $\{0,1\}$ when this step ends.

3.3 Sampling the Learned Markov Network Using Gibbs Sampler

Once the structure learning and parameter learning are completed, the Gibbs sampler is used to sample the learned Markov network to generate new candidate solutions. Concretely, the probabilities for all variables in solution are repeatedly sampled by Gibbs sampler to generate new values $\{0,1\}$ for them. The probability that variable xi is equal to 1 is shown as Eq. (10):

$$p(x_i = 1) = \frac{p(x_i^+)}{p(x_i^+) + p(x_i^-)} = \frac{e^{-U(x_i^+)/T}}{e^{-U(x_i^+)/T} + e^{-U(x_i^-)/T}} = \frac{1}{1 + e^{2W_i/T}} \qquad (10)$$

where, x_i^+ and x_i^- denote x having a particular $x_i = +1$ and$_i$ $x_i = -1$ respectively, $W_i = (U(x_i^+) - U(x_i^-))/2$ is an energy function for all maximal cliques containing x_i.

The steps of sampling the learned Markov network to generate a new candidate solution follow as:

Step 1: Set $x^o = \{x_1^o, x_2^o, \cdots, x_n^o\}$ and $T = 1$.
Step 2: Randomly select a variable x_i^o in x^o, compute in turn $U(x_i^+)$, $U(x_i^-)$, and W_i according to Eq. (8), and then compute $p(x_i = 1)$ according to Eq. (10). If $p(x_i = 1)$ is larger than given threshold, x_i^o is set to 1; otherwise, x_i^o is set to 0.
Step 3: Terminate the sampling if all variables have their new values; otherwise, turn to Step 2.

3.4 Procedure of MN-NSGA-II Algorithm

The comprehensive procedure of MN-NSGA-II algorithm is described as follows:

Step 1: An initial parent population P_t of size N is generated randomly, and t is set to 1.
Step 2: The fast not-dominated sorting operator and selection operator on P_t are performed repetitively to form a set of promising candidate solutions, which are used to construct a Markov network. The learned Markov network is sampled repeatedly to generate new candidate solutions until the first offspring population Q_t is formed, and t is set to 2;

Step 3: The parent population P_t and the offspring population Q_t are merged into a new population R_t of size N, and the fast non-dominated sorting operator is performed on R_t. And then, the candidate solutions in R_t are ranked into different fronts according to their Pareto levels.

Step 4: After performing repetitively, the fast not-dominated sorting operator, crowding distance calculating and elite preserving, N candidate solutions are selected from R_t to form the next generation parent population P_{t+1}.

Step 5: The selection operator over P_t is performed to form a set of promising candidate solutions, which will be used to learn and build a Markov network.

Step 6: The operators of structure learning and parameter learning are executed, which can find out all maximal cliques in a Markov network and their potential functions hidden in the set of selected promising candidate solutions.

Step 7: The learned Markov network is sampled repeatedly to generate new candidate solutions, which form the next generation offspring population Q_{t+1}.

Step 8: If the stopping criteria are not satisfied, t is set to $t+1$ and the procedure transfers to Step 3. Otherwise, the procedure is terminated.

4　Experiment Results

4.1　Experiment Setting

In order to verify the real performance of MN-NSGA-II algorithm, this paper conducts comparative tests on 9 functions [9] by using NSGA-II algorithm and MN-NSGA-II algorithm respectively. The initial parameter settings of NSGA-II algorithm and MN-NSGA-II algorithm are the same, that is, the population number is 100 and the number of iterations is 200. In addition, each test function is run 30 times respectively, and the algorithm performances such as the accuracy and diversity are analyzed using GD, IGD and HV metrics.

4.2　Experiment Results

The means and standard deviations of GD, IGD and HV on each test function are listed in Table 1 to Table 3. As shown in Table 1, MN-NSGA-II algorithm obtains the minimum GD on almost all of functions, while NSGA-II obtains the minimum GD only on few test functions. This result indicates that the convergence of MN-NSGA-II algorithm is closer to the optimal Pareto set than NSGA-II algorithm, and the global convergence ability of MN-NSGA-II is further improved comparing to NSGA-II algorithm.

In Table 2, IGD values obtained by MN-NSGA-II algorithm only on Laumanns and Viennet2 functions are slightly higher than those obtained by NSGA-II algorithm, which indicates that MN-NSGA-II algorithm can not only have better convergence, but also effectively ensure the better uniform distribution of solutions.

In Table 3, the maximum HV values obtained by the two algorithms on all test functions are very close, but the performance of MN-NSGA-II algorithm is slightly better, which is basically consistent with the IGD metric. All in all, the results in Table 3 further indicates that the comprehensive performance of MN-NSGA-II algorithm is more favorable.

Table 1. Means and Standard Deviations (SD) of GD metric.

Problems	NSGA-II		MN-NSGA-II	
	Mean	SD	Mean	SD
Tanaka	2.0230E-04	1.7225E-05	**1.6496E-04**	**1.9282E-06**
Fonseca	2.2193E-04	1.3926E-05	**2.1221E-04**	**2.4648E-05**
Schaffer	9.1608E-04	3.6832E-05	**9.1034E-04**	**1.1192E-05**
Laumanns	3.1000E-03	9.1432E-05	**3.004e-03**	**8.3696E-06**
Lis	8.7062E-04	3.8351E-05	**3.9089E-04**	**3.2456E-06**
Binh1	7.8590E-03	6.6532E-04	**7.7080E-03**	**9.6175E-05**
Viennet1	**9.4150E-03**	2.6050E-03	1.3100E-02	**1.3000E-03**
Viennet3	**4.5397E-04**	**2.7895E-05**	9.2904E-04	8.6282E-05

Table 2. Means and Standard Deviations (SD) of IGD metric.

Problems	NSGA-II		MN-NSGA-II	
	Mean	SD	Mean	SD
Tanaka	0.0338	2.6600E-04	**0.0040**	**2.6719E-05**
Fonseca	0.0060	**3.8998E-05**	**0.0056**	1.4422E-04
Schaffer	0.0222	1.1000E-03	**0.0198**	**7.1734E-04**
Laumanns	**0.0216**	6.0660E-04	0.0262	**5.9722E-05**
Lis	0.0047	9.1526E-05	**0.0040**	**4.6143E-05**
Binh1	0.2770	1.6200E-02	**0.2383**	**2.2000E-03**
Viennet1	0.1549	3.0000E-03	**0.1511**	**1.1000E-03**
Viennet3	0.0505	1.3000E-03	**0.0450**	**9.8081E-04**

Table 3. Means and Standard Deviations (SD) of HV metric.

Problems	NSGA-II		MN-NSGA-II	
	Mean	SD	Mean	SD
Tanaka	0.3841	9.8654E-04	**0.3972**	**3.1164E-05**
Fonseca	0.2231	**1.0781E-05**	**0.2235**	2.0852E-04
Schaffer	0.8583	1.8851E-04	**0.8587**	**9.0605E-05**
Laumanns	**0.8583**	7.3412E-05	0.8565	**1.2714E-05**
Lis	0.2488	8.5113E-05	**0.2535**	**5.3881E-05**

(*continued*)

Table 3. (*continued*)

Problems	NSGA-II		MN-NSGA-II	
	Mean	SD	Mean	SD
Binh1	0.8582	2.3630E-04	**0.8695**	**5.0095E-05**
Viennet1	0.3413	6.3399E-04	**0.3492**	**4.0129E-05**
Viennet3	**0.1830**	**3.0666E-05**	0.1854	4.3304E-05

To demonstrate some of cases graphically, this paper compares NSGA-II algorithm and MN-NSGA-II algorithm on several test functions. Figures 2, 3, 4, 5 and 6 show the Pareto fronts obtained by two algorithms on two-objective problems Tanaka, Schaffer, Lis and Binh1 after 200 generations. Figure 7 shows the real Pareto optimal fronts of threeobjective problems Viennet1 and Viennet3. Figures 8 and 9 illustrate that the Pareto fronts of the two algorithms on Viennet1 and Viennet3 after 200 generations.

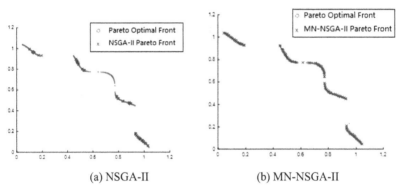

(a) NSGA-II   (b) MN-NSGA-II

Fig. 2. Pareto fronts in Tanaka problem.

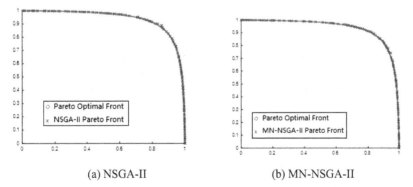

(a) NSGA-II   (b) MN-NSGA-II

Fig. 3. Pareto front in Fonseca problem.

The plots of final non-dominated solutions in Figs. 2, 3, 4, 5 and 6 clearly show that the Pareto fronts generated by MN-NSGA-II algorithm are more continuous and uniform than those by NSGA-II algorithm on all two-objective problems, and MN-NSGA-II algorithm closely converges to Pareto optimal fronts. Obviously, it is further proven that MN-NSGA-II algorithm has better comprehensive performance in accuracy, convergence and diversity.

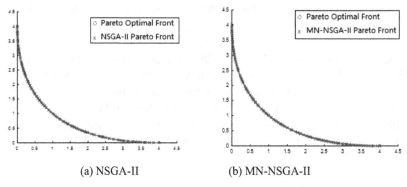

(a) NSGA-II   (b) MN-NSGA-II

Fig. 4. Pareto front in Schaffer problem.

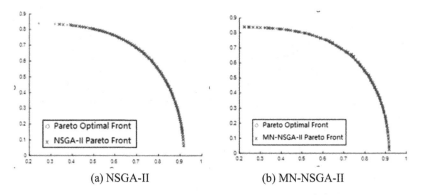

(a) NSGA-II   (b) MN-NSGA-II

Fig. 5. Pareto front in Lis problem.

Similarly, Figs. 8 and 9 show that the non-dominated solutions found by MN-NSGA-II algorithm are better than those by NSGA-II algorithm on three-objective problems Viennet1 and Viennet3. Intuitively speaking, the non-dominated solutions found by MN-NSGA-II algorithm spread nearly uniform on the Pareto optimal fronts. Therefore, it also can be concluded that MN-NSGA-II algorithm has more prominent advantages in solving higher dimensional multi-objective problems.

Summarily, the better experimental results on MN-NSGA-II algorithm are due to the linkage learning ability of Markov network, which can accurately discover and maintain excellent building blocks in solution. Furthermore, MN-NSGA-II algorithm is always extended in the most promising direction in the process of searching optimal Pareto front, which shows a good global convergence.

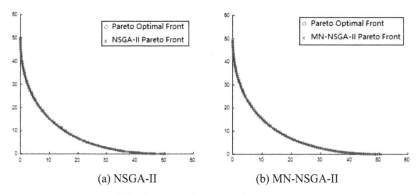

(a) NSGA-II   (b) MN-NSGA-II

Fig. 6. Pareto front in Binh1 problem.

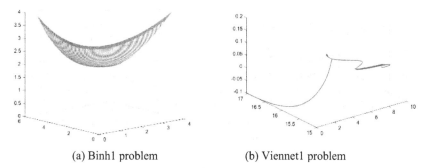

(a) Binh1 problem   (b) Viennet1 problem

Fig. 7. Pareto Optimal fronts on Binh1 problem and Viennet1 problem.

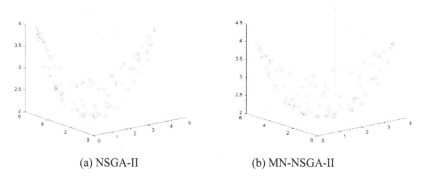

(a) NSGA-II   (b) MN-NSGA-II

Fig. 8. Pareto fronts in Viennet1 problem.

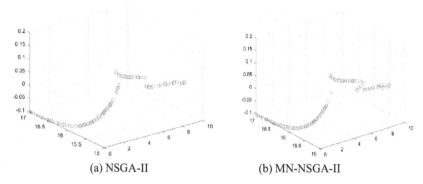

(a) NSGA-II   (b) MN-NSGA-II

Fig. 9. Pareto fronts in Viennet3 problem.

5 Conclusion

The linkage relationship between variables in solution is the key to successfully solving multi-objective optimization problems. However, the traditional genetic operators such as crossover and mutation in NSGA-II algorithm have serious randomness, which makes algorithm lack important linkage learning ability and cannot effectively discover and maintain the interaction relationship between variables. Therefore, success rate in discovering the optimal Pareto front is reduced. For the above reasons, we propose a modified NSGA-II based on Markov network, which discovers the linkage relationship between variables by learning Markov network, and generates new solutions by sampling the learned Markov network. Comparing with NSGA-II, MN-NSGA-II can achieve a certain improvement in overall performance without crossover operator and mutation operator, which has also been fully verified by experimental results.

Acknowledgements. This work is partially supported by Guangdong Province University Characteristic Innovation Project (2020KTSCX231) and National Natural Science Foundation of China (12171162).

References

1. Hyoungjin, K., Meng-sing, L.: Adaptive directional local search strategy for hybrid evolutionary Multiobjective optimization. Appl. Soft Comput. **19**, 290–311 (2014). https://doi.org/10.1016/j.asoc.2014.02.019
2. Lu, H., Yen, G.G.: Rank-density-based multiobjective genetic algorithm and benchmark test function study. IEEE Trans. Evol. Comput. **7**(4), 325–343 (2003)
3. Mei, Y., Tang, K., Yao, X.: Decomposition-based mimetic algorithm for multiobjective capacitated arc routing problem. IEEE Trans. Evol. Comput. **15**(2), 151–165 (2003)
4. Schutze, O., Lara, A., Coello, C.A.: On the influence of the number of objectives on the hardness of a multiobjective optimization problem. IEEE Trans. Evol. Comput. **15**(4), 444–455 (2011). https://doi.org/10.1109/TEVC.2010.2064321
5. Coello, C.A.: Evolutionary multi-objective optimization: a historical view of the field. IEEE Comput. Intell. Mag. **1**(1), 28–36 (2006)

6. Ngoc-Luong, H., Thi-Thanh-Nguyen, H., Wook-Ahn, C.: Entropy-based efficiency enhancement techniques for evolutionary algorithms. Inf. Sci. **188**, 100–120 (2012)
7. Zitzler, E., Deb, K., Thiele, L.: Comparison of multiobjective evolutionary algorithms: empirical results. Evol. Comput. **8**(2), 173–195 (2000)
8. Maoguo, G., Licheng, J., Dongdong, Y., Wenping, M.: Evolutionary multi-objective optimization algorithms. J. Softw. **20**(2), 271–289 (2009)
9. Deb, K., Pratap, A., Agarwal, S., Meyarivan, T.: A fast and elitist multiobjective genetic algorithm: NSGA-II. IEEE Trans. Evol. Comput. **6**(2), 182–197 (2002). https://doi.org/10.1109/4235.996017
10. Zitzler, E., Laumanns, M., Thiele, L.: SPEA2: improving the strength Pareto evolutionary algorithm for multiobjective optimization. In: Proceedings of Evolutionary Methods for Design, Optimization and Control, Barcelona, Spain, vol. 3242, pp. 95–100 (2002). https://doi.org/10.3929/ethz-a-004284029
11. Zhang, Q., Zhou, A., Jin, Y.: RM-MEDA: a regularity model-based multiobjective estimation of distribution algorithm. IEEE Trans. Evol. Comput. **12**(1), 41–63 (2008)
12. Qu, B.Y., Suganthan, P.N.: Multi-objective evolutionary algorithms based on the summation of normalized objectives and diversified selection. Inf. Sci. **180**(17), 317–318 (2010). https://doi.org/10.1016/j.ins.2010.05.013
13. Srinivas, N., Deb, K.: Multi-objective function optimization using non-dominated sorting genetic algorithms. Evol. Comput. **2**(3), 221–248 (1995)
14. Verma, S., Pant, M., Snasel, V.: A comprehensive review on NSGA-II for multiobjective combinatorial optimization problems. IEEE Access **9**, 57757–57791 (2021). https://doi.org/10.1109/ACCESS.2021.3070634
15. Ji, B., Sun, H., Yuan, X., Yuan, Y., Wang, X.: Coordinated optimized scheduling of locks and transshipment in inland waterway transportation using binary NSGA-II. Int. Trans. Oper. Res. **27**(3), 1501–1525 (2020). https://doi.org/10.1111/itor.12720
16. Yılmaz, Ö.F.: Operational strategies for seru production system a bi-objective optimisation model and solution methods. Int. J. Prod. Res, 58(11), 3195–3219 (2020). https://doi.org/10.1080/00207543.2019.1669841
17. Hauschild, M., Pelikan, M.: An introduction and survey of estimation of distribution algorithms. Swarm Evol. Comput. **1**(1), 111–128 (2011). https://doi.org/10.1016/j.swevo.2011.08.003
18. Larrañaga, P., Karshenas, H., Bielza, C., Santana, R.: A review on probabilistic graphical models in evolutionary computation. J. Heuristics **18**(5), 795–819 (2012). https://doi.org/10.1007/s10732-012-9208-4
19. Shakya, S., Santana, R., Lozano, J.A.: A Markovianity based optimisation algorithm. Genet. Program Evolvable Mach. **13**, 159–195 (2012). https://doi.org/10.1007/s10710-011-9149-y

An Improved Particle Swarm Optimization Algorithm Combined with Bat Algorithm

Hongyu Xiao[1], Nannan Zhao[1], Zihang Gao[1,2], and Xiaojun Cui[1(✉)]

[1] Wenzhou Vocational College of Science and Technology, Wenzhou 325006, China
cxjxhy@163.com
[2] Cangnan Industrial Research Institute of Modern Agriculture, Cangnan 325899, China

Abstract. In this paper, an improved Particle swarm optimization algorithm combined with Bat algorithm is proposed to solve global optimization problems. The inspiration of Bat algorithm comes from the Animal echolocation behavior of bats in nature. Because of its effectiveness, Bat algorithm has been studied by many scholars and has been applied to engineering optimization, economic scheduling, classification and other problems. However, when dealing with complex optimization problems, bat optimization algorithms tend to fall into local optimization. This paper combines the Bat algorithm (BA) with the comprehensive learning particle swarm algorithm (CLPSO), proposes an improved particle swarm algorithm(BCLPSO), and tests it on CEC2017 function. The experimental results show that the improved particle swarm optimization algorithm outperforms other comparative algorithms in terms of testing functions. In addition, the algorithm has been successfully applied to the mathematical modeling problem of multi disc clutch brake design.

Keywords: Bat optimization algorithm · Comprehensive learning particle swarm algorithm · CEC2017 benchmark function · multiple disk clutch brake design

1 Introduction

Optimization problem refers to finding the optimal solution under certain constraints, in order to achieve the optimal solution of the objective function. When dealing with complex optimization problems, methods such as formula calculation and enumeration often consume a lot of time but cannot find the optimal solution. Therefore, finding an effective method to handle complex optimization problems is currently the main research direction. In recent years, more and more people have begun to engage in the research of biomimetics. People have taken inspiration from the biological behavior of fish, birds and insects, and proposed a large number of Swarm intelligence optimization algorithms with adaptive evolution ability. These optimization algorithms solve optimization problems through coordination and cooperation between individuals and information sharing between groups. Because of their intelligent selection of optimal solutions, they are called swarm intelligence optimization algorithms. Some classical

© The Author(s), under exclusive license to Springer Nature Singapore Pte Ltd. 2024
K. Li and Y. Liu (Eds.): ISICA 2023, CCIS 2146, pp. 18–25, 2024.
https://doi.org/10.1007/978-981-97-4393-3_2

Swarm intelligence algorithms have been proposed and applied to engineering, medical care, agriculture and production scheduling. Particle swarm optimization (PSO) [1] algorithm developed by Kennedy and Eberhart in 1995. This algorithm is a population intelligence algorithm inspired by the predation of bird flocks. Particle swarm optimization algorithm and improved algorithm are applied to scheduling problems, engineering problems and other optimization problems. Zhang et al. [2] added orthogonal learning (OL) strategy to the Particle swarm optimization algorithm to improve the ability of the algorithm to jump out of the local optimum. The ant colony optimization (ACO) algorithm [4, 5] was proposed by Italian scholars Dorigo and Maniezzo in the 1990s. This algorithm is a population intelligence algorithm inspired by the ant in the process of feeding through pheromones to find the best path.

2 Bat Optimization Algorithm

The bat optimization algorithm (BA) [3] is a population intelligence algorithm inspired by the behavior of bats that prey on prey through Animal echolocation. The principle of Animal echolocation is that bats generate three-dimensional images according to the pulses they send out. The bat optimization algorithm mainly achieves this behavior through two control parameters: pulse loudness (A) and pulse emission rate (r). A and r are independent settings that control the flight status of each bat. These Animal echolocation algorithms have the following characteristics in bat optimization:

Each bat searches its surroundings by Animal echolocation;

In a given situation, bat i has a flight speed V_i, and each bat has a frequency at a certain position X_j. . The frequency range is [fmin, fmax]. The loudness A and pulse rate r change with the position of the prey;

The variation range of loudness A [A_{min}, A_0]; The variation range of pulse rate r [r_0, r_{max}];

The formula for bat position X_i^t and speed V_i^t is:

$$f_i = f_{min} + (f_{max} - f_{min}) * \beta \tag{1}$$

$$V_i^{t+1} = V_i^t + (X_i^t - X_*) * f_i \tag{2}$$

$$X_i^{t+1} = X_i^t + V_i^{t+1} \tag{3}$$

The parameter loudness A_i and pulse emission rate r_i:

$$A_i^{t+1} = \alpha A_i^t \tag{4}$$

$$r_i^{t+1} = r_i^0 \lceil 1 - \exp(-\gamma t) \rceil \tag{5}$$

$$A_i^t \to 0, r_i^t \to r_i^0, t \to \infty \tag{6}$$

$$X_i^t = X_* + \text{rand} * A_i^t \tag{7}$$

where α and γ are two constants;rand is a random number belonging to $[-1, 1]$.

3   Comprehensive Learning Particle Swarm Algorithm

Integrated learning particle swarm optimization (CLPSO) algorithm is a swarm intelligence algorithm proposed by Liang et al. [6] in 2006. It incorporates a comprehensive learning strategy (CLS) into the particle swarm optimization algorithm, which updates the velocity of the particle not by using the global optimal solution, but by using the individual best position pbest of other particles. The integrated learning strategy can maintain the diversity of the population and prevent premature convergence. The updated formula of speed and position in the integrated learning particle swarm algorithm is as follows:

$$v_{id} = w * v_{id} + c * r_{id}\left(\text{pbest}_{f_i(d),d} - x_{id}\right) \tag{8}$$

$$x_{id} = x_{id} + v_{id} \tag{9}$$

$$Pc_i = a + b * \frac{\exp\left(\frac{10*(i-1)}{N-1}\right)}{\exp(10) - 1} \tag{10}$$

where $f_i(d)$ represents the dimension value of the d dimension in the best position of the i th particle individual. $f_i = [f_i(1), f_i(2), \ldots, f_i(D)]$ is the learning sample vector set by the particle; pbest $_{f_i(d),d}$ represents the dimension value corresponding to the optimal position of a particle among all particles. The dimension of which particle is learned depends on the learning probability parameter Pc. For each dimension of the particle, we generate a random number. If this random number is greater than Pc, it will learn from the corresponding dimension in its own individual best position. Otherwise, it learns from the corresponding dimension in the optimal position of other particle individuals.

4 Bat Enhanced Comprehensive Learning Particle Swarm

The comprehensive learning particle swarm algorithm updates the speed of particles through the individual best position pbest of all particles, which can avoid the algorithm falling into the local optimal solution too early, but also makes it unable to perform local search near the global optimal solution. The improved algorithm in this paper first selects the three optimal solutions in the comprehensive learning particle swarm optimization algorithm as the global optimal solution of the Bat algorithm, and the optimal solution of each particle individual in the comprehensive learning particle swarm optimization algorithm is used as the position of the population individual of the Bat algorithm, and the optimal solution searched will replace the optimal solution in the comprehensive learning particle swarm optimization algorithm. The specific steps of the algorithm are described as follows:

1) We first initialize the particles and parameters; Calculate the fitness value of each particle
2) Update each particle using a comprehensive learning particle swarm algorithm
3) The three optimal solutions in the comprehensive learning particle swarm optimization algorithm are selected as the global optimal solution of the Bat algorithm, and the optimal solution of each particle individual in the comprehensive learning particle swarm optimization algorithm is used as the position of the population individual of the Bat algorithm. If the optimal solution searched is better than the optimal solution in the comprehensive learning particle swarm optimization algorithm, the optimal solution in the comprehensive learning particle swarm optimization algorithm is replaced.
4) Continuously cycle through steps (2) and (3) until the termination condition is met.

Pseudo code of BCLPSO

Set the parameters of BCLPSO such as the population size, the maximum number of iterations, the boundary of search and the dimensionality of the space.

Initialize the grey bat population $M_i(i = 1, ..., n)$

Create the initial population x; Calculate the objective function value of x: fit(x)

Record the best position of each individual particle pbest and Fitness of personal best positions fit(pbest); learning probability $Pc_i(i = 1,2, ..., N)$

l=1;

while l < T

    for i = 1:n

        if flag(i)> m

            generation method for particle f_i using Algorithm 1

            flag(i)=0

        end if

        Updating velocities and locations using Eqs. (8)–(9)

        Compute new Fitness of population x_i

        If $fit(x_i) < fit(pbest_i)$

        $pbest_i = x_i$; flag(i)=0

      Else

       flag(i)= flag(i)+1

      End if

    End for

    Update global optimal solution gbest

    Select the best three solutions x_a, x_b and x_c from fit(x) as X_*

    Calculate the fitness of each search agent

    For each search agent

        Update the position of the current search agent by equation(7)

    End for

    Select the best solutions M_i

    If fit(gbest)> $fit(M_i)$

      gbest $= M_i$

      $x_a = M_i$

    End if

  End while

  Return gbest

5   Testing Experiments

The benchmark function for this experiment comes from the function in CEC2017 [7]. Different types of benchmark functions can provide a more comprehensive evaluation of the performance of all algorithms. In order to obtain more fair results, all algorithms set the same parameters in the experiment: the number of particles and the maximum number of evaluations were set to 30 and 10000, respectively. For each benchmark function, all algorithms were independently tested 30 times. As shown in Fig. 1, the convergence Run chart graphs of four test functions, such as F1 and F4, are respectively displayed. It can be seen from the graphs that the improved algorithm in this paper has certain advantages in Rate of convergence and accuracy.

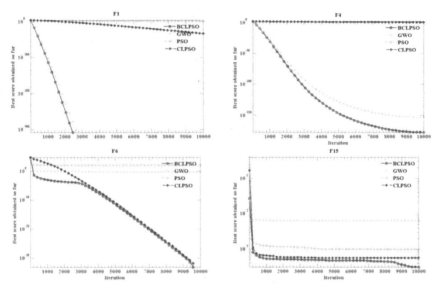

Fig. 1. Improved particle swarm optimization algorithm and other optimization algorithms convergence trend graph

6   Multiple Disk Clutch Brake Design Design Problems

The constraint processing method used in this paper is to eliminate this solution when the fitness of the solution is greater than zero, and when it violates the constraint, and then use recursive and iterative methods, each recursive call will generate a new point until a feasible solution is found [8]. The experiment was designed with five variables: $\vec{x} = (r_i, r_0, t, F, Z)$. The above five variables are considered discrete and their values range as follows:

$$r_i = (60 \sim 80)\,\text{mm}$$
$$r_0 = (90 \sim 110)\,\text{mm}$$
$$t = (1, 1.5, 2, 2.5, 3)$$
$$F = (600, 610, 620, \ldots, 980, 990, 1000)$$
$$Z = (2 \sim 20)$$

The constraint formula of minimum mass (f_1) and minimum stopping time (T) of the braking system is as follows:

$$f_1(\vec{x}) = \pi\left(x_2^2 - x_1^2\right)x_3(X_5 + 1)\rho$$

$$f_2(\vec{x}) = T = \frac{I_2\omega}{M_h + M_f}$$

The target constraints are as follows:

$$g_1(\vec{x}) = x_2 - x_1 - \Delta R > 0$$
$$g_2(\vec{x}) = L_{max} - (x_5 + 1)(x_3 + \delta) \geq 0$$
$$g_3(\vec{x}) = p_{max} - p_{rz} > 0$$
$$g_4(\vec{x}) = p_{max}V_{sr,max} - p_{rz}V_{sr} \geq 0$$
$$g_5(x) = V_{sr,max} - V_{sr} \geq 0$$
$$g_6(\vec{x}) = M_h - sM_s \geq 0$$
$$g_7(\vec{x}) = T \geq 0$$
$$g_8(\vec{x}) = T_{max} - T \geq 0$$
$$r_{i,min} \leq x_1 \leq r_{i,max}$$
$$r_{0,min} \leq x_2 \leq r_{0,max}$$
$$t_{min} \leq x_3 \leq t_{max}$$
$$0 \leq x_4 \leq F_{max}$$
$$2 \leq x_5 \leq Z_{max}$$

Some swarm intelligence algorithms have been used to calculate the design problem of multi disc clutch brakes, including artificial bee colony optimization algorithm (ABC) [9], water circulation algorithm (WCA) [10], and teaching and learning based optimization algorithm (TLBO) [11]. Table 1 lists the optimization results of different algorithms in the design of multi disc clutch brakes. It can be seen that the improved algorithm in this paper has achieved good optimization results, proving that the algorithm can also effectively improve the utilization rate of engineering materials and significantly reduce the consumption of engineering materials.

Table 1. Comparison of Optimization Results

Algorithm	Parameter					Optimal result
	r_i	r_0	t	F	Z	
BCLPSO	60.0000	110.0000	3.0000	630.0000	8.0000	0.313656
ABC [9]	80.0000	90.0000	3.0000	1000.0000	2.0000	0.313657
WCA[10]	70.0000	90.0000	1.0000	910.0000	3.0000	0.313656
TLBO[11]	70.0000	90.0000	1.0000	810.0000	3.0000	0.313657

7 Conclusion

In this paper, an improved Particle swarm optimization algorithm is proposed, and Bat algorithm is introduced into the integrated Particle swarm optimization algorithm. This algorithm can enhance the mutation ability of Particle swarm optimization algorithm, improve the local search ability of the algorithm, enable particles to learn from each other, and enhance the ability of the algorithm to search for the optimal solution. Compare the proposed BCLPSO algorithm with three population intelligent algorithms. The experimental results show that BCLPSO outperforms other comparison algorithms on different benchmark functions. In addition, the proposed algorithm is applied to the design problem of multi disc clutch brakes. The results indicate that BCLPSO has achieved good results in engineering application problems, indicating that the BCLPSO proposed in this paper can effectively solve constraint problems.

Acknowledgments. This work was supported by the Basic Agricultural Science and Technology Project of Wenzhou under Grant N20220003.

References

1. Kennedy, J., Eberhart, R.: Particle swarm optimization. In: IEEE International Conference on Neural Networks - Conference Proceedings, vol. 4, pp. 1942–1948 (1995)
2. Zhan, Z.H., Zhang, J., Liu, O.: Orthogonal learning particle swarm optimization. In: Proceedings of the 11th Annual Genetic and Evolutionary Computation Conference, GECCO-2009 (2009)
3. Yang, X.S.: A new metaheuristic Bat-inspired Algorithm. In: Studies in Computational Intelligence, pp. 65–74 (2010)
4. Stützle, T.G., Dorigo, M.: Ant colony optimization. IEEE Trans. Evol. Comput. **8**(4), 422–423 (2004)
5. Deng, W., Xu, J., Zhao, H.: An improved ant colony optimization algorithm based on hybrid strategies for scheduling problem. IEEE access **7**, 20281–20292 (2019)
6. Liang, J.J., et al.: Comprehensive learning particle swarm optimizer for global optimization of multimodal functions. IEEE Trans. Evol. Comput. **10**(3), 281–295 (2006)
7. Awad, N., Ali, M., Liang, J., Qu, B., Suganthan, P.: Problem definitions and evaluation criteria for the CEC 2017 special session and competition on single objective real-parameter numerical optimization. Technical Report (2016)
8. Coello, C.A.C.: Theoretical and numerical constraint-handling techniques used with evolutionary algorithms: a survey of the state of the art. Comput. Methods Appl. Mech. Eng. **191**(11–12), 1245–1287 (2002)
9. Karaboga, D., Basturk, B.: Artificial Bee Colony (ABC) optimization algorithm for solving constrained optimization problems. In: Melin, P., Castillo, O., Aguilar, L.T., Kacprzyk, J., Pedrycz, W. (eds.) IFSA 2007. LNCS (LNAI), vol. 4529, pp. 789–798. Springer, Heidelberg (2007). https://doi.org/10.1007/978-3-540-72950-1_77
10. Eskandar, H., Sadollah, A., Bahreininejad, A., Hamdi, M.: Water cycle algorithm – a novel metaheuristic optimization method for solving constrained engineering optimization problems. Comput. Struct. **110–111**, 151–166 (2012)
11. Rao, R.V., Savsani, V.J., Vakharia, D.P.: Teaching–learning-based optimization: a novel method for constrained mechanical design optimization problems. Comput. Aided Des.. Aided Des. **43**(3), 303–315 (2011)

An Improved Whale Optimization Algorithm Combined with Bat Algorithm and Its Applications

Xiaofeng Wang[1,2(✉)], Tian'ou Wang[2,3], and Chanjuan Lin[1]

[1] Wenzhou Vocational College of Science and Technology, Wenzhou, China
wangxiaofeng@wzvcst.edu.cn
[2] Cangnan Industrial Research Institute of Modern Agriculture, Cangnan, China
[3] Cangnan County Bureau of Agriculture and Rural Development, Cangnan, China

Abstract. The whale optimization algorithm(WOA) is inspired by the hunting behavior of humpback whales. Due to its good effect in searching for optimal solutions, it has been applied to various problems such as engineering optimization, economic dispatch, and classification. However, when the WOA optimizes some complex problems, it may ignore global problem processing due to the selection of local optimal solutions, resulting in poor algorithm performance. This paper proposes an improved whale optimization algorithm combined with the bat algorithm(WOACBA) to solve global optimization problems. The test results of the algorithm on the CEC2014 benchmark function verify that it is superior to other comparison algorithms on the test function. It is applied to the mathematical modeling of welded beam design, the results show that WOACBA achieved relatively good application results in solving engineering application problems.

Keywords: Bat Algorithm · Whale Optimization Algorithm · Global optimization problems · CEC2014

1 Introduction

Optimization problem refers to seeking the optimal solution to the performance of the established scheme under certain conditions, or to find the maximum or minimum value of the specified target performance index under the premise of multi-parameter comparison, so as to make the set target function index reach the optimal. The search for an effective technical way to optimize the solution of complex problems has become a frontier research direction of the current scientific research. Optimization technology is an applied technology that mathematically quantifies and solves computational problems in various fields, and is currently widely used in various disciplines such as sensors, video processing, image recognition and so on. It has blossomed in the fields of 5G, artificial intelligence, driverless, smart city, etc., and has generated great economic and social value. Optimization technology can reduce the extra loss in the process of technical operation through reasonable allocation and path selection, improve the system resource

© The Author(s), under exclusive license to Springer Nature Singapore Pte Ltd. 2024
K. Li and Y. Liu (Eds.): ISICA 2023, CCIS 2146, pp. 26–38, 2024.
https://doi.org/10.1007/978-981-97-4393-3_3

matching ability, and thus enhance the work efficiency. This efficiency enhancement is more obvious when applied on a large scale. With the advancement of digitalization, the cross-application of information technology and different disciplines such as management, economics and sociology has become a mainstream trend, which makes the optimization of complex combinatorial models more and more demanding. However, traditional optimization methods (e.g., Newton's method, simplex method, etc.) need to search all the data or space due to algorithmic bottlenecks, which is not only prone to "combinatorial explosion" of search, but also a waste of computing resources. The application of engineering models has many difficulties such as nonlinearity, complexity, irregularity, etc. In order to obtain the optimal solution quickly and form practical applications under this complex model, the researchers put forward higher requirements for the optimization algorithm.

In recent years, more and more scholars have begun to engage in the study of Bio mimicry. The activities of biological groups follow certain natural laws, which are the optimal solutions for the survival of such groups. Scholars summarize the experience of such natural laws and create a variety of intelligent optimization algorithms, which are used to solve many complex problems in various fields of production. Currently applied algorithms include: Genetic Algorithms(GA) that mimic the evolutionary mechanisms of organisms in nature [1], Ant Colony Algorithms(ACA) that mimic the behavior of ants discovering paths during the search for food [2, 3], Particle Swarm Optimization (PSO) that mimic the predatory behavior of bird flocks [4], Bat Algorithms(BA) that mimic the echolocation behavior of bats to perform global optimization [5] and Whale Optimization Algorithms(WOA) that mimic the chase behavior of humpback whales populations [6]. These algorithms have in common that they are formed and developed by simulating natural phenomena or mimicking the behavior of groups of organisms, possessing the characteristics of simplicity, reliability, and parallel processing basic algorithm.

1.1 Bat Algorithm (BA)

BA [5] is a swarm intelligence algorithm that mimics the behavior of bats in echolocation to hunt for prey. It detects prey and searches for paths by mimicking bats emitting acoustic pulses and bouncing off objects to produce 3D images for echolocation. The BA first initializes a set of random solutions, searches for optimality through iterations, and searches for local new solutions around the optimal solution, thus enhancing the possessive search. Specifically, bats emit strong pulses at a lower frequency during prey search and weak pulses at a higher frequency during hunting. The bat algorithm achieves this behavior through two main control parameters: impulse loudness (A) and impulse firing rate (r). Each bat has independent parameters A and r controlling the flight state. We make the following assumptions in the control of the BA:

Assumption 1: Bats search the environment by echolocation;

Assumption 2: In the given situation, each bat "i" has a flight speed "V_i" and each bat has a frequency at a certain position "X_i". The frequency range is [f_{min}, f_{max}]. The pulse loudness A and the pulse emission rate r change with the position of the prey;

Assumption 3: Loudness A varies in the range [A_{min}, A_0]; pulse rate r varies in the range [r_0, r_{max}].

Mathematical formulations of bat behavior can be derived as follows: the formulas for individual bat positions X_i^t and velocities V_i^t are:

$$f_i = f_{min} + (f_{max} - f_{min}) * \beta \tag{1}$$

$$V_i^{t+1} = V_i^t + (X_i^t - X_*) * f_i \tag{2}$$

$$X_i^{t+1} = X_i^t + V_i^{t+1} \tag{3}$$

Among them, the range of values of f_i is $[f_{min}, f_{max}]$. β is a value randomly selected from the uniform distribution range $[0, 1]$. X_* is the global optimal position generated by the population iteration, which can be found after comparing all the solutions provided by the bats.

In order to effectively balance the global exploration and local exploitation of the BA, this paper dynamically adjusts the parameters loudness A_i and pulse emission rate r_i associated with each bat according to the equations.

$$A_i^{t+1} = \alpha A_i^t \tag{4}$$

$$r_i^{t+1} = r_i^0 [1 - \exp(-\gamma t)] \tag{5}$$

Among them, α and γ are two constants.

In general, the loudness A_i decreases after the bat discovers its prey, and the rate of pulse emission r_i increases as the bat i approaches its prey, so we set $0 < \alpha < 1$, $\gamma > 0$, and

$$A_i^t \to 0, r_i^t \to r_i^0, \ t \to \infty \tag{6}$$

based on the value of the pulse emissivity r, we replace b with a new solution a:

$$X_i^t = X_* + rand * A_i^t \tag{7}$$

Among them, rand is randomly taken between $[-1, 1]$, and A_i^t denotes the loudness emitted by the ith bat when feeding on prey.

1.2 Whale Optimization Algorithm (WOA)

WOA is a swarm intelligence optimization algorithm proposed by Mirjalili et al. in 2016 [6]. The model of this algorithm originates from simulating three behavioral mechanisms such as searching and foraging, contracting and encircling and spiral updating in the cooperative feeding phase of whales, and the three mechanisms are independent of each other. When hunting, whales form unique bubbles along a spiral path to achieve prey capture. The captured prey is assumed to be the optimal solution and the location of the whale is the potential solution. The whale determines a position updating strategy (updating strategy i.e., the three behavioral mechanisms) by iterating random positions and position vectors toward the target travel path and keeps approaching the optimal solution.

When searching for the optimal solution, individuals of the whale population will randomly select whales in the optimal position to approach. To model this, we can update the position vector of the search agent X with the following equation:

$$\overrightarrow{D} = \left| \overrightarrow{C} \cdot \overrightarrow{X_{rand}} - \overrightarrow{X} \right| \tag{8}$$

$$\overrightarrow{X}(t+1) = \overrightarrow{X_{rand}} - \overrightarrow{A} \cdot \overrightarrow{D} \tag{9}$$

Among them, C and A are important coefficient factors, Xrand is the position vector of the randomly selected search agent from the current aggregate, and t denotes the current iteration. A and C can be computed by the following equations:

$$\overrightarrow{A} = 2 \overrightarrow{a} \cdot \overrightarrow{r} - \overrightarrow{a} \tag{10}$$

$$\overrightarrow{C} = 2 \overrightarrow{r} \tag{11}$$

Among them, r is a random vector in [0,1] and a is a variable that decreases from 2 to 0 as the iteration progresses.

During the process of encircling the prey, the search agent targets the currently found optimal candidate solution and updates its state. This performance can be expressed as follows:

$$\overrightarrow{D} = \left| \overrightarrow{C} \cdot \overrightarrow{X^*}(t) - \overrightarrow{X}(t) \right| \tag{12}$$

$$\overrightarrow{X}(t+1) = \overrightarrow{X^*}(t) - \overrightarrow{A} \cdot \overrightarrow{D} \tag{13}$$

Among them, X* is the position vector of the target.

In the bubble net attack phase, the search agent moves towards the target in a spiral motion. The model is shown below:

$$\vec{X}(t+1) = \overrightarrow{D'} \cdot e^{bl} \cdot \cos(2\pi l) + \overrightarrow{X^*}(t) \tag{14}$$

$$D' = \left| \overrightarrow{X^*}(t) - \vec{X}(t) \right| \tag{15}$$

Among them, D' is the distance from the search agent to the target, b is a constant, and l is a random number in [−1, 1].

In WOA, the probability that a search agent chooses to update its position in a contraction inclusion model or a spiral model is half of its probability. The mathematical model is as follows:

$$\vec{X}(t+1) = \begin{cases} \overrightarrow{X^*}(t) - \vec{A} \cdot \vec{D} & \text{if } p < 0.5 \\ \overrightarrow{D'} \cdot e^{bl} \cdot \cos(2\pi l) + \overrightarrow{X^*}(t) & \text{if } p \geq 0.5 \end{cases} \tag{16}$$

Among them, p is a random number in [0,1].

2 Improved WOA Combined with the BA (WOACBA)

The WOA borrows from humpback whale hunting to search for an objective value, quickly discovering a locally optimal solution and allowing other individuals in the group to quickly search in this region. However, the algorithm cannot perform a partial search in the neighborhood of the globally optimal solution when it is obtained. The improved algorithm in this paper first specifies the optimal solution in the WOA, and then searches for the global optimal solution near this local optimal solution through the BA, which is not only avoids falling into the local optimum, but also conducive to searching in the vicinity of the local solution. The specific steps of the algorithm are described as follows:

Step1:Initialize each particle and parameter of the whale optimization algorithm, and calculate the adaptation value of each individual;

Step2:Execute the whale optimization algorithm, and calculate and update the position of each particle using Eq. 8–16;

Step3:Select the optimal solution in the whale optimization algorithm as the global optimal solution of the bat algorithm. If the searched optimal solution is better than the optimal solution in the whale optimization algorithm, execute Eqs. 1–7 and replace the optimal solution in the whale optimization algorithm according to Eq. 7.

Step4:Loop the obtained optimal solution continuously for steps2, Step3:until the termination condition is satisfied.

The pseudo-code for the WOACBA is shown below:

Algorithm Pseudo code of the WOACOA

Set the parameters of the WOACBA such as the population size, the maximum number of iterations, the boundary of search and the dimensionality of the space.

Initialize the bat population $M_i(i = 1, ..., n)$

Create the initial population x; Calculate the objective function value of x: fit(x)

Record the best position of each individual particle pbest and Fitness of personal best positions fit(pbest); learning probability $Pc_i(i = 1,2, ..., N)$

l=1;

while l < T

for i = 1:n

    Updating velocities and locations using Eqs. (8)–(16)

    Compute new Fitness of population x_i

    If $fit(x_i) < fit(pbest_i)$

      $pbest_i = x_i$; flag(i)=0

        Else

          flag(i)= flag(i)+1

        End if

    if flag(i)> m

      generation method for particle f_i using Bat Algorithm

      flag(i)=0

    end if

    End for

For each search agent

    Update the position of the current search agent by equation(7)

End for

Select the best solutions M_i

If fit(gbest)> $fit(M_i)$

  $gbest = M_i$

  $x_a = M_i$

End if

End while

Return gbest

---

3 Simulation Experiment and Analysis

3.1 Benchmark Experiments

The benchmark functions for this experiment are from the functions in CEC2014[7]. Different types of benchmark functions can evaluate the performance of existing algorithms from various aspects. In order to obtain fairer results, all algorithms are set with the same parameters in the experiments: the number of particles and the maximum number of iterations are set to 30 and 3000. For each benchmark function, all algorithms are tested independently for 30 times. This experiment tests the functions F3(x), F4(x), F6(x), F12(x), F13(x), F18(x), F21(x) and F23(x) as Table 1, and compares the convergence trends of the WOACBA and the WOA, the BA, the PSO and the GA.

Table 1. Test Functions and Formula Algorithms

Functions	Formula
F3:Rotated Discus Function	$f_3(x) = 10^6 x_1^2 + \sum_{i=2}^{D} x_i^2;$ $F_3(x) = f_3(M(x - o_3)) + F_3^*$
F4:Shifted and Rotated Rosenbrock's Function	$f_4(x) = \sum_{i=1}^{D-1} \left(100(x_i^2 - x_{i+1})^2 + (x_i - 1)^2 \right);$ $F_4(x) = f_4(2.048(x - o_4)/100 + 1) + F_4^*$
F6:Shifted and Rotated Weierstrass Function	$f_6(x) = \sum_{i=1}^{D} \left(\sum_{k=0}^{kmax} \left[a^k cos(2\pi b^k(x_i + 0.5)) \right] \right) -$ $D \left(\sum_{k=0}^{kmax} \left[a^k cos(2\pi b^k * 0.5) \right] \right), a = 0.5, b = 3, kmax = 20;$ $F_6(x) = f_6(M(x - o_6)/100) + F_6^*$
F10:Shifted and Rotated Katsuura Function	$f_{10}(x) = \frac{10}{D^2} \prod_{i=1}^{D} \left(1 + i \sum_{j=1}^{32} \frac{|2^j x_i - round(2^j x_i)|}{2^j} \right)^{\frac{10}{D^{1.2}}} - \frac{10}{D^2};$ $F_{12}(x) = f_{10}(5(x - o_{12})/100) + F_{12}^*$
F13:Expanded Griewank's plus Rosenbrock's Function	$f_{11}(x) = \left| \sum_{i=1}^{D} x_i - D \right|^{1/4} + (0.5 \sum_{i=1}^{D} x_i^2 + \sum_{i=1}^{D} x_i)/D + 0.5;$ $F_{13}(x) = f_{10}(5(x - o_{12})/100) + F_{13}^*$
F18:Hybrid Function 2	$F(x) = g_1(M_1 Z_1) + g_2(M_2 Z_2) + \cdots + g_N(M_N Z_N) + F^*(x); P_i =$ [0.3,0.3,0.4]: used to control the percentage of $g_i(x)$ N = 3: dimension for each basic function; g_2: Bent Cigar Function f2; g_{12}: HGBat Function f12; g_8: Rastrigin's Function f8

(continued)

Table 1. (*continued*)

Functions	Formula
F18:Hybrid Function 5	$F(x) = g_1(M_1Z_1) + g_2(M_2Z_2) + \cdots + g_N(M_NZ_N) + F^*(x); P_i =$ [0.1,0.2,0.2,0.2,0.3]: used to control the percentage of $g_i(x)$ N = 5: dimension for each basic function; g1: Scaffer's F6 Function:f14 g2: HGBat Function f12 g3: Rosenbrock's Function f4 g4: Modified Schwefel's Function f9 g5: High Conditioned Elliptic Function f1
F18:Hybrid Function 6	$F(x) = \sum_{i=1}^{N} \{\omega_i * [\lambda_i g_i(x) + bias_i]\} + F^*; P_i =$ [0.1,0.2,0.2,0.2,0.3]: used to control the percentage of $g_i(x)$ N=5 : number of basic functions $\sigma = [10,20,30,40,50]$: new shifted optimum postion for each $g_i(x)$,define the global and local optimaof on f1.Fun Bias=[100,200,300,400] : defines which optimum is global optimum i: used to control each gi(x)'s coverage range, a small i give a narrow range for that gi(x) g1: Rotated Rosenbrock's Function F4' g2: High Conditioned Elliptic Function F1' g3 Rotated Bent Cigar Function F2' g4: Rotated Discus Function F3' g5: High Conditioned Elliptic Function F1'

It can be seen that the WOACBA has relatively high convergence speed and convergence accuracy on the three functions F3, F4, F13, and has the same convergence speed and convergence accuracy with the WOA on the five functions of F6, F12, F18, F21 and F23 from Fig. 1. At the same time, the convergence curve of WOACBA has a relatively smooth convergence curve, with fewer cases of falling into local extremes, and better ability to find optimization. Therefore, WOACBA shows better optimization performance in the CEC-2014 test function set than typical algorithms such as the WOA, the BA, the PSO and the GA, which show more significant solution superiority.

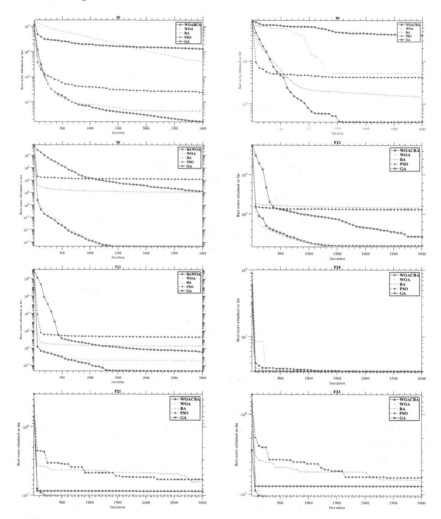

Fig. 1. Convergence trend of WOACBA compared to other optimization algorithms

The mean and standard deviation of the different algorithms tested to form the optimum on the 8 tested functions are shown in Table 2. It can be seen that compared with the other four algorithms, the WOACBA searches for the optimal solution on the remaining seven functions, except on function F12 where the WOACBA ranks second. The experiment proves that WOACBA has strong search ability on CEC2014 benchmark functions.

Table 2. Comparison data of different algorithms on CEC2014 function.

Fun	Item	WOACBA	WOA	BA	PSO	GA
F3	AVG	1.7E +01	4.3E+03	4.3E+01	2.4E+02	1.3E+04
	STD	7.5E+00	5.7E+ 03	9.5E+00	6.8E+01	3.6E+03
	Rank	1	4	2	3	5
F4	AVG	3.8E-01	5.2E+00	1.5E+00	4.2E+00	4.0E+01
	STD	4.6E-01	1.1E+01	1.4E-01	1.1E-01	1.2E+01
	Rank	1	4	2	3	5
F6	AVG	4.7E-04	5.2E-04	1.1E+01	1.2E+02	1.3E+01
	STD	1.7E-04	1.6E-04	1.7E+00	9.7E+00	5.0E+ 0
	Rank	1	2	3	5	4
F12	AVG	8.3E-05	6.3E-05	5.9E+00	3.7E+00	1.3E-03
	STD	3.0E-05	1.0E-05	3.2E+00	4.3E-01	2.9E-03
	Rank	2	1	5	4	3
F13	AVG	2.3E-03	3.7E-02	1.9E+00	1.9E+01	3.6E-01
	STD	9.9E-04	7.2E-02	1.0E-01	2.4E+00	4.1E-01
	Rank	1	2	4	5	3
	p-value					
F18	AVG	3.0E+00	3.0E+00	3.0E+00	3.0E+00	3.0E+ 0
	STD	2.6E-07	8.7E-06	8.0E-03	1.6E-02	0.0E+00
	Rank	2	3	4	5	1
F21	AVG	− 1.0E+01	− 1.0E+01	− 1.0E+01	− 1.0E+01	− 1.0E +01
	STD	5.1E-04	3.0E-03	3.4E+00	9.5E-01	3.3E+00
	Rank	1	2	4	5	3
F23	AVG	− 1.1E+01	− 1.1E+01	− 6.6E+00	− 6.2E+00	− 7.9E+00
	STD	6.5E-04	1.2E-03	3.7E+00	4.6E-01	3.7E+00
	Rank	1	2	4	5	3

3.2 Application of Engineering Models

WOACBA can be applied to optimization of engineering models. In this paper, the Welded beam design problem (WBD) is used as an example to seek the minimum value of the manufacturing cost of the model based on the optimization constraints. In this case, the optimization constraints are bucking load (P_C), shear stress (τ), bending stress in the beam (θ), and deflection rate (δ). The manufacturing cost of the model is controlled by four optimization parameters, which include the height of the rod(t), thickness of the weld(h), thickness of the rod(b), and the length of the rod(l). The mathematical model of the welded beam design problem is described as shown below.

Consider

$$\vec{x} = [x_1 x_2 x_3 x_4] = [h\ l\ t\ b] \tag{17}$$

Objective

$$f(\vec{x})_{min} = 1.10471 x_2 x_1^2 + 0.04811 x_3 x_4 (14.0 + x_2) \tag{18}$$

Subject to

$$g_1(\vec{x}) = \tau(\vec{x}) - \tau_{max} \le 0 \tag{19}$$

$$g_2(\vec{x}) = \sigma(\vec{x}) - \sigma_{max} \le 0 \tag{20}$$

$$(\vec{x}) = \delta(\vec{x}) - \delta_{max} \le 0 \tag{21}$$

$$g_4(\vec{x}) = x_1 - x_4 \le 0 \tag{22}$$

Thickness of the weld (h), thickness of the bar (b), and length of the bar (l). The mathematical model of this problem can be described as:

$$g_5(\vec{x}) = P - P_C(\vec{x}) \le 0 \tag{23}$$

$$g_6(\vec{x}) = 0.125 - x_1 \le 0 \tag{24}$$

$$g_7(\vec{x}) = 1.10471 x_1^2 + 0.04811 x_3 x_4 (14.0 + x_2) - 5.0 \le 0 \tag{25}$$

Variable ranges:

$$0.1 \le x_1 \le 2$$

$$0.1 \le x_2 \le 10$$

$$0.1 \le x_3 \le 10$$

$$0.1 \le x_4 \le 2$$

where:

$$\tau(\vec{x}) = \sqrt{(\tau')^2 + 2\tau'\tau'' \frac{x_2}{2R} + (\tau'')^2} \tag{26}$$

$$\tau' = \frac{P}{\sqrt{2} x_1 x_2} \tag{27}$$

$$\tau'' = \frac{MR}{J} \tag{28}$$

$$M = P(L + \frac{x_2}{2}) \tag{29}$$

$$R = \sqrt{\frac{x_2^2}{4} + (\frac{x_1 + x_3}{2})^2} \tag{30}$$

$$J = 2\left\{\sqrt{2}x_1x_2\left[\frac{x_2^2}{4} + (\frac{x_1 + x_3}{2})^2\right]\right\} \tag{31}$$

$$\sigma(\overrightarrow{x}) = \frac{6PL}{x_4x_3^2} \tag{32}$$

$$\delta(\overrightarrow{x}) = \frac{6PL^3}{Ex_3^2x_4} \tag{33}$$

$$P_C(\overrightarrow{x}) = \frac{4.013E\sqrt{\frac{x_3^2x_4^6}{36}}}{L^2}(1 - \frac{x_3}{2L}\sqrt{\frac{E}{4G}}) \tag{34}$$

$P = 6000\text{lb}, L = 14\text{in}..\delta_{max} = 0.25 \text{ in}.., E = 30 \times 1^6\text{psi}, G = 12 \times 10^6\text{psi}, \tau_{max} = 13600\text{psi}, \sigma_{max} = 30000\text{psi}$

This problem has received attention from many researchers. Among them, Kaveh et al. used RO [8] for this mathematical model. Abualigah et al. [9] optimized this mathematical model by using HS and the optimal manufacturing cost for HS is 2.3807, WOA also optimized this Model and the optimal manufacturing cost obtained is 1.730499, as shown in Table 3.

Table 3. Comparison of WOACBA with literature for the WB case

Technique	Best variables				Best cost
	h	l	t	b	
WOACBA	**0.20547**	**3.2572**	**9.03877**	**0.20572**	**1.69572**
WOA [6]	0.205396	3.484293	9.037426	0.206276	1.730499
RO[8]	0.203687	3.528467	9.004233	0.207241	1.735344
HS [9]	0.2442	6.2231	8.2915	0.2433	2.3807
IHS[10]	0.20573	3.47049	9.03662	0.20573	1.7248

The WOACBA is used to optimize this mathematical model and the results obtained are compared with other methods. As shown in Table 1, the minimum manufacturing cost optimized by the WOACBA is 1.69572, which indicates that with the four parameter values set to 0.20574, 3.2572, 9.03877, and 0.20572, the model can achieve a manufacturing cost of 1.69572, which is a minimum cost smaller than the results obtained by the other algorithm. Therefore, the WOACBA proposed in this paper can better optimize this mathematical model and obtain the minimum value of manufacturing cost of welded beam design.

4 Conclusion

In this paper, the WOACBA is proposed by introducing the BA into the WOA, which can enhance the variation ability of the WOA, improve the local search ability of the algorithm, and enhance the ability of the algorithm to search for the optimal solution through the mutual learning among particles. The proposed WOACBA is compared with four swarm intelligence algorithms respectively. The experimental results show that WOACBA exhibits more significant solution superiority compared with other typical algorithms on different benchmark functions. In addition, the proposed algorithm is applied to the least-cost solution of welded beam design, and the results show that WOACBA achieves better results in engineering application problems, which indicates that the WOACBA proposed in this paper can solve the constraint problem effectively.

References

1. Gen, M., Lin, L.: Genetic Algorithms and Their Applications, pp. 635–674. Springer Handbook of Engineering Statistics. Springer, London (2023)
2. Wu, L., Huang, X., Cui, J., et al.: Modified adaptive ant colony optimization algorithm and its application for solving path planning of mobile robot. Expert Syst. Appl. **215**, 119410 (2023)
3. Deng, W., Xu, J., Zhao, H.: An improved ant colony optimization algorithm based on hybrid strategies for scheduling problem. IEEE Access **7**, 20281–20292 (2019). https://doi.org/10.1109/ACCESS.2019.2897580. Article no. 8635465
4. Zhang, Y., Kong, X.: A particle swarm optimization algorithm with empirical balance strategy. Chaos, Solitons Fractals X **10**, 100089 (2023)
5. Shehab, M., Abu-Hashem, M.A., Shambour, M.K.Y., et al.: A comprehensive review of bat inspired algorithm: variants, applications, and hybridization. Arch. Comput. Methods Eng. **30**(2), 765–797 (2023)
6. Chakraborty, S., Saha, A.K., Sharma, S., et al.: A hybrid whale optimization algorithm for global optimization. J. Ambient. Intell. Humaniz. Comput.Intell. Humaniz. Comput. **14**(1), 431–467 (2023)
7. Liang, J., Qu, B.Y., Suganthan, P.: Problem definitions and evaluation criteria for the CEC 2014 special session and competition on single objective real-parameter numerical optimization. Computational Intelligence Laboratory (2014)
8. Kaveh, A., Ardebili, S.R.: A comparative study of the optimum tuned mass damper for high-rise structures considering soil-structure interaction. Periodica Polytech. Civ. Eng. **65**(4), 1036–1049 (2021)
9. Abualigah, L., Elaziz, M.A., Khasawneh, A.M., et al.: Meta-heuristic optimization algorithms for solving real-world mechanical engineering design problems: a comprehensive survey, applications, comparative analysis, and results. Neural Comput. Appl., 1–30 (2022)
10. Luo, J., He, F., Yong, J.: An efficient and robust bat algorithm with fusion of opposition-based learning and whale optimization algorithm. Intell. Data Anal. **24**(3), 581–606 (2020)

Fusion of Nonlinear Inertia Weight and Probability Mutation for Binary Particle Swarm Optimization Algorithm

Jiayu Zhang and Kangshun Li[(✉)]

College of Mathematics and Information, South China Agricultural University, Guangzhou, Guangdong, China
likangshun@sina.com

Abstract. With the rapid development of Internet information technology, vast amounts of textual data are constantly emerging. However, raw data is often unstructured and lacks readability, posing significant challenges for data mining. Feature selection, as a crucial step in machine learning, is essential for enhancing model performance. In this paper, we propose an Improved Binary Particle Swarm Optimization (IBPSO) algorithm that addresses the limited search capability of traditional swarm intelligence algorithms in high-dimensional optimization problems. By incorporating a dynamic nonlinear decreasing inertia weight update strategy and a local probability crossover mutation strategy, IBPSO effectively enhances algorithm performance while overcoming the limitations of conventional approaches. The proposed algorithm demonstrates superior performance through testing on benchmark functions.

Keywords: Binary Particle Swarm Optimization · Feature Selection · High-Dimensional

1 Introduction

Since the concept of swarm intelligence algorithms was introduced in the 1990s, it has attracted the attention and research of researchers in various fields worldwide [1]. With the subsequent proposals of algorithms such as Particle Swarm Optimization (PSO), Ant Colony Optimization (ACO), Bee Colony Algorithm, Frog Leap Algorithm, etc., swarm intelligence algorithms have become a hot research topic. Particle Swarm Optimization (PSO) is a heuristic search algorithm based on swarm intelligence. It simulates the foraging behavior of bird flocks and finds the optimal solution to a problem through collaboration and information sharing among particles. PSO is characterized by its simplicity in implementation, low number of adjustable parameters, fast convergence speed, and strong environmental adaptability [2].

To address the issue of traditional PSO algorithms easily getting trapped in local optima when dealing with high-dimensional data, the literature proposed an enhanced particle swarm optimization algorithm based on multiple populations and velocities [3].

© The Author(s), under exclusive license to Springer Nature Singapore Pte Ltd. 2024
K. Li and Y. Liu (Eds.): ISICA 2023, CCIS 2146, pp. 39–48, 2024.
https://doi.org/10.1007/978-981-97-4393-3_4

This algorithm utilizes multiple populations to achieve information sharing, quickly discovers global and local information, prevents particles from getting trapped in local optima, and demonstrates superior performance compared to other improved PSO algorithms on multiple benchmark functions. In this literature [4], a multi-objective particle swarm optimization algorithm integrating grid technology and multi-strategies was proposed. By randomly selecting convergence and distribution evaluation index strategies, combined with grid technology, the diversity and convergence of the population are enhanced. Simulation results show that the proposed algorithm exhibits better convergence. The literature introduced two new particle swarm optimization models that utilize convolutional neural networks and long short-term memory to predict the inertia weight during the movement of the particle swarm, improving the convergence performance of the particle swarm [5].

Compared to foreign countries, the research on clustering technology in China started relatively late, but after long-term research, it has also achieved fruitful results and has been applied in multiple fields. The literature introduces a particle swarm optimization algorithm with a crossover strategy, integrating genetic algorithms' crossover and mutation mechanisms [6]. This algorithm updates particles through crossover and mutation, effectively addressing PSO's issues with local optima and low accuracy. Addressing the limitations of single-objective optimization and ignoring feature interactions, the literature suggests a feature selection algorithm combining particle swarm optimization and neighborhood rough set theory [7]. This algorithm selects highly correlated features randomly, suitable for various data types without domain knowledge. Experimental comparisons on 14 datasets verify its effectiveness. The literature presents a hybrid strategy PSO algorithm integrating five PSO variants and optimizing mutation choices through an adaptive selection strategy. On benchmark test functions, it outperforms other algorithms in solution accuracy [8]. The literature introduces a dynamic subpopulation approach for PSO, enhancing its global search and improving text document clustering results, effectively avoiding local optima [9]. The literature proposes a novel hybrid genetic bat algorithm for text classification, reducing feature space dimensionality and extracting effective features [10].

2 Related Work

The standard Particle Swarm Optimization (PSO) algorithm was designed to solve continuous function optimization problems, relying on searching and optimizing solutions within a continuous solution space. However, many practical applications and engineering problems, such as combinatorial optimization and the Traveling Salesman Problem, involve discrete spatial characteristics. To effectively address discrete space optimization problems, researchers adjusted and developed the standard PSO, introducing the Binary Particle Swarm Optimization (BPSO) algorithm tailored for discrete optimization tasks [11]. BPSO addresses discrete optimization by mapping the concepts of position and velocity from continuous PSO into a binary space. In BPSO, the position vector of particles is represented as a binary encoding, typically using combinations of integers 0 and 1. The update rules of the algorithm are also adjusted to accommodate the discrete search space. The position updates of particles in BPSO are represented by probability values in the real number interval [0,1]. The Sigmoid function assists in determining whether a particle should take 0 or 1 in each dimension. Therefore, the Sigmoid function is used in BPSO to convert the particle's velocity vector into position updates. The formulas for particle velocity update, the Sigmoid mapping function, and particle position update are as follows:

$$v_{id}^{t+1} = \omega \times v_{id}^{t} + c1 \times r1 \times \left(pbest_{id}^{t} - x_{id}^{t}\right) + c2 \times r2 \times \left(gbest_{d}^{t} - x_{id}^{t}\right) \qquad (1)$$

$$s\left(v_{id}^{t+1}\right) = \frac{1}{1+e^{-v_{id}'+1}} \qquad (2)$$

$$x_{id}^{i+1} = \begin{cases} 1, & \text{if } rand() < s(v_{id}^{t+1}) \\ 0, & \text{otherwise} \end{cases} \qquad (3)$$

In this context, the constant t represents the current iteration number, w is the inertia weight value, $c1$, $c2$ are learning factors, and $r1$ and $r2$ are random numbers uniformly distributed within the interval [0,1]. v_{id}^{t} Represents the velocity of the i particle in the d dimension during the t iteration, while x_{id}^{t} represents its position. Both the velocity and position vectors are bounded within a fixed range. $pbest_{id}^{t}$ Denotes the individual particle's best solution history at the t iteration, and $gbest_{d}^{t}$ represents the global best solution of the population at the t iteration.

3 Improved Binary Particle Swarm Optimization

The fundamental principle of the standard Particle Swarm Optimization (PSO) algorithm lies in the collaboration and information sharing among particles to seek the optimal solution. While PSO has demonstrated good optimization capabilities in many fields, its global search capability becomes relatively weak when faced with large-scale and high-dimensional optimization problems. This can lead to the algorithm easily getting trapped in local optima, especially when multiple local optima exist in the search space, making it challenging for the particles to find the global optimal solution.

To overcome these limitations, researchers have proposed numerous strategies to enhance the PSO algorithm and adapt it to more complex optimization challenges. These strategies include dynamic weight adjustment, which adjusts the inertia weight based on the search status of the algorithm to achieve a balance between exploration and exploitation; multi-swarm particle swarm optimization, which enhances search efficiency through information exchange among multiple swarms; local search enhancement, which combines the PSO framework with local search to help the algorithm escape from local optima; and hybrid particle swarm optimization, which combines the characteristics of PSO with other algorithms (such as genetic algorithms, simulated annealing algorithms, etc.) to form more powerful hybrid optimization methods. To improve the algorithm's ability to search for optimal solutions in the search space, this paper proposes an improved binary particle swarm optimization algorithm (IBPSO) to better address convergence issues.

3.1 Dynamic Nonlinear Decreasing Inertia Weight Updating Strategy

In the Particle Swarm Optimization (PSO) algorithm, the magnitude of the inertia weight, denoted as ω, has a significant impact on the search capability of the algorithm. Therefore, by adjusting the value of ω, it is possible to control the global search ability and local search ability. According to the velocity update formula of the standard Particle Swarm Optimization algorithm, it is shown as follows:

$$v_i^{t+1} = w \cdot v_i^t + c_1 \cdot r_1 \cdot \left(P_{best_i}^t - x_i^t\right) + c_2 \cdot r_2 \cdot \left(G_{best}^t - x_i^t\right) \tag{4}$$

According to the exploration and optimization characteristics of swarm intelligence algorithms, the focus should be on global search of the solution space during the early stages of the search to locate potential optimal solutions. In contrast, later stages should concentrate on local search to fully explore the optimal solution for the optimization problem. However, in the traditional PSO algorithm, the inertia weight w is set as a fixed value, failing to consider the different requirements of the algorithm for the search space range during different search stages. To address this issue, this paper sets w as a dynamic weight, creating a dynamic inertia weight to meet the varying search capabilities required at different stages.

The implementation of the dynamic nonlinear inertia weight update based on the above concept is as follows:

$$w = \begin{cases} w_{max} & , 0 < t < 0.4 \cdot T \\ w_{max} - (w_{max} - w_{min}) \times \left(\frac{t}{T}\right)^2 & , 0.4 \cdot T \le t \le T \end{cases} \tag{5}$$

Among them, w_{max} represents the maximum value of the inertia weight w set initially for the algorithm; w_{min} represents the minimum value of the inertia weight w set initially for the algorithm; t represents the current number of iterations during the search; and T represents the maximum number of iterations.

3.2 Partial Probability Cross Variation

When the standard Particle Swarm Optimization (PSO) algorithm is applied to solve high-dimensional data optimization problems, the search speed of particles in the later stages of the search process can rapidly decrease or almost approach zero. As a result, the best fitness value of the population hardly improves during each search iteration. This phenomenon is caused by the population getting trapped in a local optimum. To address this issue, this paper introduces a crossover mutation strategy that tracks changes in extreme values. The specific implementation is as follows: a marker variable "tag" is introduced to track whether the historical best fitness of the population remains unchanged or whether the change in fitness value is consistently below a certain threshold. This is used to determine if the population may be trapped in a local optimum. If such a phenomenon is detected, the crossover mutation mechanism of the population is triggered. By applying crossover mutation methods to individual particles, new individuals are generated, introducing more diversity to help escape the local optimum situation.

However, the phenomenon of the population's best fitness value not being updated for a long time or periodically during the search iteration does not necessarily mean that the algorithm has fallen into a local search optimal solution. The reason for this situation may be that the Particle Swarm Optimization algorithm has already found or approached the global optimal solution. If conventional crossover mutation operations are used to change the position of these solutions, it may result in losing the already found high-quality solutions, achieving the opposite effect of the algorithm's search goal. Based on Darwin's biological evolution theory of "natural selection": there is widespread variation among individuals within the same population, and those individuals with advantageous traits that can adapt to the environment will survive and reproduce, while those that cannot adapt will be eliminated [12]. This paper proposes a local probabilistic crossover mutation strategy. The basic idea is as follows:

(1) Sorting and grouping particles based on fitness values: Calculate the fitness values of all particles in the population according to the objective function of the algorithm, and sort them in ascending order. Based on the sorting results, select the top 50% of particles with lower fitness values as the high-quality solution group, and the remaining particles as the low-quality solution group.

(2) Local probabilistic genetic crossover operation for particles in the high-quality solution group: Use a probabilistic method to select a certain number of high-quality solution particles, and perform crossover operations on them in pairs. This generates an equal number of new particles for population updating. The specific formula is as follows:

$$x_{new} = p \cdot x_{old1} + (1 - p) \cdot x_{old2} \tag{6}$$

x_{new} represents the generation of new offspring of the population, x_{old1} and x_{old2} represent the non-repetitive individuals of the previous generation of the population, and p represents a random real number in the interval $(0,1)$.

(3) Mutation operation for all particles in the low-quality solution group: All particles in the low-quality solution group undergo mutation operations. Through mutation, new genetic individuals are generated, increasing the diversity of the population (Table 1).

Table 1. Tagged variable algorithm.

Algorithm 1: Set and track the marker variable **tag**
1.If $G_{best} \leq fitness_{max}^{current}$ **then**
2.   **tag** = **tag**+1;
3.Else
4.   **tag** Reset to 0;
5.End

As shown in Algorithm 1, the fitness value of the best individual in the current search iteration, $fitness_{max}^{current}$, is compared with the best fitness value of the population, $G_{best}^{fitness}$. If no better solution is generated in this iteration compared to $G_{best}^{fitness}$, the marker variable is incremented by one; otherwise, it is reset to zero (Table 2).

Table 2. Local probability crossover mutation algorithm.

Algorithm 2: Local Probability Crossover Mutation
if tag $\geq 0.05 \cdot t_{max}$ **then**
Sorting all particles in the population based on their fitness values;
for i = 1 to $\dfrac{size_{pop}}{2}$ **do**
With a probability of p, perform crossover operations on the selected particles generate new particles;
for i = $\dfrac{size_{pop}}{2}$ to size **do**
Randomly select one or several attributes of dimension on chromosome number i;
Replace the corresponding positional information of chromosome number i;
End

4 Algorithm Experiments and Result Analysis

To better test the global search capability and convergence performance of the IBPSO algorithm, this paper selects five benchmark test functions for evaluation. These test functions typically possess multiple local optima and a single global optimum, making them suitable for assessing the algorithm's global search ability and convergence speed. Simulation experiments are conducted on two unimodal functions (Sphere and Rosenbrock) and three multimodal functions (Rastrigin, Griewank, and Ackley) to evaluate the comprehensive performance of the algorithm. Additionally, three other algorithms are chosen for comparison, including the standard binary particle swarm optimization (BPSO), adaptive mutation binary particle swarm optimization (AMBPSO), and inertia weight linear decay binary particle swarm optimization (IWDBPSO) (Table 3).

The testing parameters of the simulation experiments in this section have been adjusted and optimized as follows: the population size N is set to 40, 50, 60, and 80, respectively, to evaluate the impact of different population sizes on the algorithm's performance. To match the different population sizes, the corresponding maximum number of iterations, MaxIter, is also set to 400, 600, 800, and 1000, respectively. Additionally, the dimension of the unknown search space is set to 300 dimensions, aiming to test the algorithm's search capability in high-dimensional spaces. To reduce experimental

Table 3. Benchmarking function.

Function	Expression	range	extremum
Sphere	$f_1(x) = \sum\limits_{i=1}^{D} x_i^2$	$[-100,100]$	$\{0\}^D$
Rosenbrock	$f_2(x) = \sum\limits_{i=1}^{D-1} \left(100(x_i^2 - x_{i+1})^2 + (x_i - 1)^2\right)$	$[-10,10]$	$\{0\}^D$
Rastrigin	$f_3(x) = \sum\limits_{i=1}^{D} (x_i^2 - 10\cos(2\pi x_i) + 10)$	$[-5.12,5.12]$	$\{0\}^D$
Girewank	$f_4(x) = \sum\limits_{i=1}^{D} \frac{x_i^2}{4000} - \prod\limits_{i=1}^{D} \cos(\frac{i}{\sqrt{i}}) + 1$	$[-600,600]$	$\{0\}^D$
Ackley	$-20\exp\left(-0.2\sqrt{\frac{\sum_{i=1}^{n} x_i^2}{n}}\right) - \exp\left(\frac{1}{n}\sum\limits_{i=1}^{n}\cos(2\pi x_i)\right) + 20 + e$	$[-32,32]$	$\{0\}^D$

errors and obtain more reliable results, each group of experiments will be independently repeated 30 times. Such repeated experiments can eliminate the influence of random factors, making the results more stable and reliable. By collecting data from multiple experiments, the performance of the algorithm can be more comprehensively evaluated, providing a solid basis for subsequent research.

Additionally, following the original parameter settings, the experimental parameters in BPSO are set as follows: both learning factors $c1$ and $c2$ are set to 2, and the inertia weight w is set to 0.9. For the parameter settings in AMBPSO, both learning factors $c1$ and $c2$ are also set to 2, and w linearly increases within the range of [0.4, 0.9]. In IWDBPSO, the learning factors $c1$ and $c2$ are set to 2, and w linearly decreases within the range of [0.4, 0.9]. Comparison will be made based on the mean of the function values.

The results are shown in the following table (Tables 4, 5, 6, 7, and 8):

Table 4. Sphere function.

Sphere	BPSO	AMBPSO	IWDBPSO	IBPSO
N = 40 MaxIter = 400	140.53	126.3	127.8	118.8
N = 50 MaxIter = 600	131.74	122.3	123.9	117.8
N = 60 MaxIter = 800	126.67	120.4	121.6	114.1
N = 80 MaxIter = 1000	122.86	118.9	119.8	112.5

Through the analysis of experimental results, it can be observed that the standard BPSO algorithm does not perform well on multiple test functions. The main issues lie in the premature convergence of the BPSO algorithm and its tendency to get trapped

Table 5. Rosenbrock function.

Rosenbrock	BPSO	AMBPSO	IWDBPSO	IBPSO
N = 40 MaxIter = 400	13019.66	11885.07	11575.1	11247.5
N = 50 MaxIter = 600	12398.27	11311.67	11068.5	10881.9
N = 60 MaxIter = 800	12081.93	11012.6	10732.6	10552.3
N = 80 MaxIter = 1000	11845.36	10839.13	10160.1	10037.2

Table 6. Rastrigin function.

Rastrigin	BPSO	AMBPSO	IWDBPSO	IBPSO
N = 40 MaxIter = 400	3266.9	3107.7	3095.2	2951.7
N = 50 MaxIter = 600	3019.5	2897.9	2861.6	2743.2
N = 60 MaxIter = 800	2912.7	2694.3	2588.3	2467.1
N = 80 MaxIter = 1000	2470.9	2286.3	2194.4	2065.3

Table 7. Girewank function.

Girewank	BPSO	AMBPSO	IWDBPSO	IBPSO
N = 40 MaxIter = 400	0.45	0.44	0.31	0.22
N = 50 MaxIter = 600	0.38	0.33	0.27	0.15
N = 60 MaxIter = 800	0.34	0.28	0.25	0.18
N = 80 MaxIter = 1000	0.33	0.25	0.20	0.13

Table 8. Ackley function.

Ackley	BPSO	AMBPSO	IWDBPSO	IBPSO
N = 40 MaxIter = 400	4.93	4.81	4.72	4.28
N = 50 MaxIter = 600	4.77	4.52	4.35	3.81
N = 60 MaxIter = 800	4.62	4.39	4.08	3.64
N = 80 MaxIter = 1000	4.54	4.23	3.89	3.63

in local optima. Although the AMBPSO algorithm has made some improvements by introducing an adaptive inertia weight, due to the increasing trend of its inertia weight, although the global ability is enhanced in the early stage, the local search ability of the algorithm becomes insufficient in the later stage of search, and it is also prone to falling into local optimal solutions. The IWDBPSO algorithm effectively enhances the local search ability of the algorithm by adopting an optimization strategy of linearly decreasing inertia weight, but the large change of inertia weight in the early stage of search iteration weakens the global search ability to some extent. In addition, when dealing with high-dimensional feature selection problems, the search ability of the BPSO algorithm often appears insufficient, leading to reduced population diversity and thus premature convergence issues.

In this paper, an improved binary particle swarm optimization algorithm (IBPSO) is proposed. In the early stage of search, IBPSO uses a larger inertia weight to enhance the global search ability of particles. In the later stage of search, a nonlinearly decreasing inertia weight with the number of iterations is adopted, so that the entire population gradually turns to local search. Furthermore, to address the phenomenon of long-term unchanged population historical optimal extrema, IBPSO introduces a local probability mutation strategy to generate new individuals through local mutation. This strategy not only retains the high-quality solutions in the population but also increases population diversity by eliminating inferior solutions. Experimental results show that compared to the previous comparison algorithms, the proposed IBPSO algorithm has significantly improved performance in terms of solution accuracy and search ability.

Acknowledgement. This work was supported by Key Realm R&D Program of GuangDong Province with Grant No. 2019B020219003.

References

1. Kennedy, J., Eberhart R.: Particle swarm optimization. In: IEEE International Conference on Neural Networks, Perth, Australia, Proceedings IEEE, pp. 1942–1948 (1995)
2. Esmin, A., Coelho, R.A., Matwin, S.: A review on particle swarm optimization algorithm and its variants to clustering high-dimensional data. Artif. Intell. Rev. **44**(1), 23–45 (2015)
3. Ning, D.W.J.: Enhanced particle swarm optimization with multi-swarm and multi-velocity for optimizing high-dimensional problems. Appl. Intell. Int. J. Artif. Intell. Neural Netw. Complex Probl. Solving Technol. **49**(2) (2019)
4. Zou, K., Liu, Y., Wang, S.,et al.: A Multiobjective particle swarm optimization algorithm based on grid technique and multistrategy. J. Math. (2021). https://doi.org/10.1155/2021/162 6457
5. Pawan, Y.V.R.N., Prakash, K.B., Chowdhury, S., et al.: Particle swarm optimization performance improvement using deep learning techniques. Multimedia Tools Appl. (2022)
6. Wang, B., Wang, G., Wang, Y., et al.: A K-means clustering method with feature learning for unbalanced vehicle fault diagnosis. Smart Resilient Transp. **3**(2), 162–176 (2021)
7. Liang, S., Liu, Z., You, D., et al.: PSO-NRS: an online group feature selection algorithm based on PSO multi-objective optimization. Appl. Intell. **53**(12), 15095–15111 (2023)
8. Pang, J., Li, X., Han, S.: PSO with mixed strategy for global optimization. Complexity **2023** (2023)
9. Selvaraj, S., Choi, E.: Dynamic sub-swarm approach of PSO algorithms for text document clustering. Sensors **22**(24), 9653 (2022)
10. Eliguzel, N., Cetinkaya, C., Dereli, T.: A novel approach for text categorization by applying hybrid genetic bat algorithm through feature extraction and feature selection methods. Expert Syst. Appl. **202**, 117433 (2022)

11. Kennedy, J., Eberhart, R.C., Shi, Y., et al.: Swarm Intelligence. The Morgan Kaufmann Series in Evolutionary Computation (2001)
12. Darwin, C.R.: The origin of species by means of natural selection: or, the preservation of favored races in the struggle for life. G. Richards (1917). https://doi.org/10.5962/bhl.title.59991

An Evolutionary Algorithm Based on Replication Analysis for Bi-objective Feature Selection

Li Kangshun[1,2] and Hassan Jalil[2(✉)]

[1] School of Computer Science, Guangdong University of Science and Technology,
Dongguan 523000, China
[2] College of Mathematics and Informatics, South China Agricultural University,
Guangzhou 510642, China
hassanjalil722@yahoo.com

Abstract. Feature selection is a complicated optimization problem with significant practical applications. Nevertheless, its strength lies in significantly reducing the size of datasets and increasing classification efficiency. Several evolutionary algorithms (EAs) have been used to solve feature selection problems. However, most EAs are unsuitable to deal with real-world problems with multiple objectives. The multi-objective feature selection problems normally consist of two objectives: minimizing the number of feature selections and minimizing the classification errors. In this paper, a replication analysis method based on the evolutionary algorithm (RAEA) is proposed to classify bi-objective selected features. The proposed method has improved the framework of dominance-based EA from two viewpoints: firstly, modify the reproduction procedure to improve the features of offspring, and secondly, the replication analysis technique has been proposed to filter out unnecessary solutions. We conducted some experiments with the proposed method and compared it with five traditional MOEAs. The experimental results show that RAEA performs better results on most datasets, indicating that RAEA not only performs best in the optimization process but also performs better results in generalization and classification.

Keywords: Classification · replication analysis · multi-objective evolutionary algorithm · feature selection · multi-objective optimization

1 Introduction

Feature selection plays an important role in the system classification [1] such as neural networks (NNs) and machine learning, whose main purpose is to increase the classification accuracy by selecting small subsets of related features from the original larger feature set [2, 3]. A multi-objective feature selection problem [4, 5], frequently has two main goals: improving classification performance and reducing irrelevant features that are theoretically different [6]. Many traditional evolutionary algorithms [7] are proposed for the solution of feature selection [8] and also to optimize real-world problems [9].

© The Author(s), under exclusive license to Springer Nature Singapore Pte Ltd. 2024
K. Li and Y. Liu (Eds.): ISICA 2023, CCIS 2146, pp. 49–61, 2024.
https://doi.org/10.1007/978-981-97-4393-3_5

EAs do not require domain information and suppositions related to search space as compared with state-of-the-art search methods. As a population-based stochastic search approach, EAs are mainly suitable to solve the MOPs because of their intrinsic parallelism allows them to identify a Pareto non-dominated solution set in one run rather than several independent runs as in case of traditional approaches. These EAs are also known as multi-objective evolutionary algorithms (MOEAs) [10].

Based on environmental selection techniques, MOEAs are divided into four types: Pareto dominance-based MOEAs, decomposition-based MOEAs, performance indicator-based MOEAs, and decision-variable analysis-based EAs. In comparison, the dominance-based MOEAs are still incompetent due to checking for non-dominance in the population. Thus, dominance-based MOEAs show a serious degradation in performance. In most dominance-based systems, the reproduction method and diversity maintenance are not efficient to solve the bi-objective feature selection problems that must be updated.

There is a key factor to reproduce the best offspring, which affects the convergence and diversity of an evolutionary algorithm. Many studies have been conducted on the PSO reproduction method and multi-objective evolutionary feature selection based on DE, such as local search mechanisms based on differential learning [18].

One more significant challenge in feature selection is the frequent replication that occurs in either the objective or decision space by evolution. The replication occurs in a decision space called decision vector replication. Its solution is very simple: eliminate the replication and keep the single, unique solution. It is useless to keep the similar multiple solutions. So, the replication that occurs in objective space is called objective vector replication. It has the most complex effects on classification performance, and its optimizations are difficult to handle. There are two possible solutions. Even with the same classification performance (such as objective value) on training data, one might get relatively different performance on test data due to keeping the test data unknown while training [19].

We propose a replication analysis-based evolutionary algorithm named RAEA for bi-objective feature selection in classification to solve the above-mentioned problems. A replication analysis method is presented at the beginning of selection to determine which replicated solutions should be excluded from the objective space. The key idea to estimate the dissimilarities of decision space is to use the distance between solutions since it removes the lower dissimilarities. With this, the population diversity is preserved well in an objective space while keeping the possible optimal solutions in test data.

2   Research Background

2.1   Related Works

The MOP with m objectives optimization at the same time is formulated as:

$$\begin{aligned} minimize \ \ y &= F(x) = (f_1(x), f_2(x), \ldots, f_m(x))^T \\ subject \ to \ \ x &\in \Omega; \end{aligned} \tag{1}$$

where $x \in \Omega$ is a decision vector in D-dimensional decision space Ω, and y is the objective vector in objective space. If the number of objectives $m = 2$, then MOP turns into the bi-objective optimization problem (BOP), which is its initial case. In this paper, two objectives need to be optimized: the number of selected features and the classification error while employing the selected features. Furthermore, those two objectives' values need to convert the ratios between 0 and 1. Assume that $f_1(x)$ selects k features for a solution, then $f_1(x) = k/D$, signifying the ratio of selected features, whereas $f_2(x)$ signifies the error rate of classification.

In this paper, an individual is represented as a binary code 0 or 1 with genes, whereas value 0 specifies that the corresponding features are not selected and value 1 specifies that they are selected. According to a recent study [2], evolutionary feature selection methods are categorized into filter and wrapper methods. Filter methods [21] are independent of the classifier and ignore the classification performance of feature selection. In contrast, the wrapper methods [6, 22] contain the classifier as a "black box" to estimate the fitness solution, such as classification performances. In the proposed RAEA method, we select a wrapper method for selected features.

2.2 Main Objectives

Evolutionary algorithms are frequently used for multi-objective feature selection [4, 6, 8, 22]. However, many studies only used EAs and did not take into account the characteristics of feature selection [2]. One important aspect is solution replication, which happens often in both the objective and decision spaces of feature selection. In either the decision or objective space, replicated solutions are uncommon in a continuous optimization problem. By solution replication is a common and difficult problem for combinatorial or discrete optimization problems such as feature selection, especially when there are several ineffective features or few training instances, as stated above, the replication can occur in two conditions: objective vector replication and decision vector replication. The prior is simply solved by keeping unique points in the decision space. But others are very difficult to handle for the given reasons.

Firstly, if all replicated solutions are kept in an objective space, then it will make the diversity of the population in an objective space worse because objective vectors overlap. Consider the worst condition in which the maps to all solutions have the same objective vectors; then no diversity will be in objective space. Secondly, the best diversity will occur with many unique points in the objective space when we keep a single solution for every repeated objective vector. Though it is hard to choose the replicated solution to remove when training or optimizing, it is because these replicated solutions are unique in the decision space or may have different classification accuracies in hidden test data. These two conditions are contradictory. When we remove replicated solutions in objective space, it can be best for optimizing the process but may worsen the classifications. So, it is necessary to discover the cooperation between optimizing processes (such as increasing the diversity of the population in objectives) and classifications (such as keeping the best possible solutions to the test data), which needs to be considered in the replication analysis.

3 Proposed Algorithm

First, we define the RAEA's general framework.

Algorithm 1 *RAEA* (N, D, τ)

Input: population size N; decision space dimensionally D; termination criterion τ;

Output: population Pop;

1: if $(3 * N) < D$ **then**
2:    randomly sample $Pop = (x^1, \ldots, x^N)$ from the decision space *s.t.* $\sum_j x_j^i \leq (3 * N), j = 1, \ldots, D$;

3: else
4:    randomly sample $Pop = (x^1, \ldots, x^N)$ from the decision space *s.t.* $\sum_j x_j^i \leq D, j = 1, \ldots, D$;

5: end if
6: set the objective function evaluation number $t = N$;
7: while $t < \tau$ **do**
8:    run $S = Reproduction$ (Pop;
9:    $t = t + N$
10:   get union population $Pop = Pop \cup S$;
11:   remove all the replicated solutions in the decision space;
12:   run $Pop = Replication\_Analysis$ (Pop) in **Algorithm 2**;
13:   rank Pop into different fronts via nondominated sorting;
14:   find k *s.t.* $num(k) \geq |Pop|$ & $num(k - 1) < |Pop|$;
15:   set α and β as the indexes of the first $(k-1)$ fronts and the kth front's solutions respectively;
16:   run $op = Diversity\_Select$ (Pop, α, β, N);

17: **end while**

3.1 General Framework

The main architecture of RAEA, as indicated in Algorithm 1, may be separated into three sections: initialization, reproduction, and environmental selection. RAEA's startup approach is designed for large-scale feature selection. As indicated in Line 4, RAEA still adopts the traditional initialization technique, which randomly selects features from the decision space without regard to constraints, when the total number of features is no more than three larger than the population size. However, when the number of features become very larger, RAEA randomly samples the initial solutions from the decision space with the limitation that the selected features for all solutions are not beyond three times the population size, defined as in Line 2. There are two objectives to set these limitations: First, according to a recent study [3], using smaller number of features for initialization may be beneficial for the convergence of feature selection in some scopes; second, because covering the whole decision space with a significantly smaller population is difficult, so reducing the exploration range to three times the population size is a reasonable trade-off choice.

RAEA's termination criterion is based on the evaluations of objective functions, such as Eq. (1), that increase by every N generation. In the while loop section (Lines 7 to 17), Line 8 shows the reproduction method, whereas Lines 10 to 16 show the environmental selection method. The RAEA's selection operation is mostly separated into three sections: the replication analysis (Line 12), the non-dominated sorting (Lines 13 to 15), and the diversity maintenance (Line 16). The second section of selection is not different from state-of-the-art MOEAs based on dominance [11, 12, 19], whereas the technique num (k) signifies the k fronts' solutions (Line 14). Furthermore, a non-dominated sorting technique is mentioned in the paper [11]. Furthermore, entire replicated solutions in decision spaces are removed at the start of the environmental selection (Line 11). In term of computational complexity, the most time-consuming aspect of the multi-objective feature selection method is evaluating the objective values, such as training classification models with a specific feature subset. The proposed method does not require additional objective function evaluation, i.e., the use of additional local search because its computational complexity contains the appropriate levels.

Algorithm 2 *Replication_Analysis (Pop)*

Input: population *Pop*;
Output: population *Pop*;
1: generate a $|Pop|*|Pop|$ matrix *Dist*, while *Dist(i, j)* denotes Manhattan distance between *Pop(i)* and *Pop(j)*;
2: get objective vectors *Obj* of solutions in *Pop*, and find all the unique objective vectors *Obj^u*;
3: generate an empty removal set *R*;
4: **For** $i = 1, ..., |Obj^u|$ **do**
5:   find indexes j *s.t.* $Obj(j) == Obj^u(i)$;
6:   **if** $|j|>1$ **then**
7:     set t as the number of selected features and D as the number of total features;
8:     generate a $|j|$ size dissimilarity set *Diss*;
9:     **for** $k = 1,...,|j|$ **do**
10:       $Diss(k)=min(Diss(j(k),j))$;
11:    **end for**
12:    $Diss = (Diss/2)/t$;
13:    set threshold $\delta=0.8 - 0.6 * (t-1)/(D-1)$, as noted in the text below;
14:    find indexes k **s.t.** $Diss(k)< \delta$;
15:      $R = R \cup j (randi(k, |k|-1))$;
16: **end if**
17: **end for**
18: remove solutions *Pop(R)* from *Pop*;

3.2 Replication Analysis

Algorithm 2 is the key concept of replication analysis, which include the use of Manhattan distances between solutions in decision space for approximation of their different degrees to remove the redundant replicated solution in objective space. In Line 5, variable j denotes the set indexes according to the solutions s.t. (referred to as subject to) and the state in which they contain same objective vectors to discover all solutions with repeated objective vectors. From Lines 8 to 12, measuring the solution's different degrees by the minimum Manhattan distance to all solutions in decision space, the value must be between 0 and 1. The Manhattan distance between two decision vectors essentially exposes the difference in selected features. In Line 13, set the removal threshold (δ) according to the total features (D) and selected features (t). The value δ differs from 0.20 to 0.80, as the t value changed from D to 1, which makes a more comprehensive removal criterion. After that, from Lines 14 to 15, these replicated solutions are separated into two sections according to the different degrees: remote solutions with variation are equal to or higher than the removal thresholds, and cluster solutions with differences are lower than the removal threshold. Among them, it reserves just a single clustered solution and whole remote solution, whereas the technique *randi* (k, $|k|$-1) defines the randomly selection of the ($|k|$-1) members from array k (Line 15). Therefore, a large δ value will probably make replicated solutions disappear. Moreover, the set δ is negatively associated with t (Line 13). This needs a high difference to be solved with a few selected features to keep.

The replication analysis process is showed in Fig. 1, where $(x^1,...,x^4)$ represents four different solutions for the selected feature problems, with all eight features in the decision space expected to contain similar objective vectors and map similar points in the objective space. In Fig. 1, each column signifies the feature and each row signifies the solution vectors, whereas the selected features are represented with a value 1. In Algorithm 2 from Line 8 to Line 12, the dissimilarity *Diss* for each solution are measured as 0.25, 0.25, 0.75, and 0.25 correspondingly, whereas the existing threshold $\delta = 0.80 - 0.60$ (4-1)/8-1) is almost equivalent to 0.543. In Algorithm 2, from Lines 14 to 18, x^3 belongs to remote groups and is kept reserved directly, whereas the other three solutions belonging to the clustered groups can be reserved randomly. The result shows that it keeps only two replicated solutions out of four based on the proposed replication analysis process.

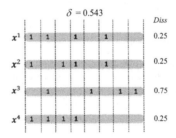

Fig. 1. Proposed replication analysis method

4 Experimental Setup

4.1 Comparison Algorithms

In this paper, we compared RAEA with traditional MOEAs such as MOEA/D [13], PMEA [23], MP-DEA [14], NSGA-II [11], and RVEA [24]. Among them, NSGA-II adopts same crowding distance measuring techniques and non-dominated sorting as RAEA. The classic MOEA based on decomposition is MOEA/D, whose convergence is based on the combination function's fitness rather than the Pareto dominance connection. The recently proposed MP-DEA is an improved form of MOEA/D that adaptively regulates the weight vector's distribution to handle the complex Pareto front. RVEA chooses the dominance relation for a good balance of diversity and convergence. The other proposed PMEA is MOEA based on an indicator that makes use of polar-metric usage [25], such as environmental selection criterion.

For comparison algorithms, we used variation operators that are bitwise mutation approaches and single-point crossover. Set the crossover rate to 1, and set the mutation rate to $1/D$, where D is denoted as the decision variable. We made a comparison of MOEA based on hypervolume (HypE) and knee points-driven MOEA (KnEA) [17] with the proposed method RAEA, whereas experiment results show that RAEA performs better on given datasets. We do not present those performances in details in this paper because of the limited number of pages.

4.2 Classification Datasets

In our experiments, we used 10 different real-world datasets for classification-based learning, and their repository of UCI machine learning is available online [26]. Table 1 shows that the number of classes changed from 2 (Leukemia) to 26 (ISOLET5). Furthermore, the features ranged from 7070 (such as Leukemia) to 256 (Semeion), covering both simple and difficult issues. Furthermore, some datasets contain binary classification issues, while others have multi-class classification issues. If the instance changes, Leukemia has a lower instance (72), while Madelon has a higher instance (2600). In this paper, we used both complex and comprehensive data sets to test the simplification and efficiency of comparison algorithms to handle the selected features in classification.

Table 1. Classification data sets in descending order of feature number

No.	Datasets	Instances	Features	Classes
1	Leukemia	72	7070	2
2	Brain1	90	5920	5
3	Leukemia1	72	5327	3
4	SRBCT	83	2308	4
5	Multiple_Features	2000	649	10
6	ISOLET5	1559	617	26
7	Madelon_Train_Validation	2600	500	2
8	LSVT_Voice_Rehabilitation	126	310	2
9	Arrhythmia	452	278	16
10	Semeion_Handwritten_Digit	1593	256	10

4.3 Performance Indicator

In this paper, a hypervolume indicator is used [15]. Firstly, for hypervolume, normalize the objective values achieved by algorithms using nadir points and ideal points to scale [0, 1] in the objective direction. So, set the nadir point and ideal point to (1.1, 1.1) and (0, 0), and also set the hypervolume reference point to (1, 1). In general, the higher the hypervolume values, the greater the performance of an algorithm.

Furthermore, apply a 0.05 level (Wilcoxon test) to measure RAEA performance, which is different from comparison algorithms. Adopt the Freidman test for comparative rank performances among the comparison algorithms. We require a little change in the hypervolume values because hypervolume can be suitable for the Freidman test. For example, higher HV performances can correspond to lower Freidman ranking values, making it easier for experiment analysis and comparison, As a result, if hv signifies the normalized HV value between (0 and 1), we will use (1 - hv) for the Freidman test rather than directly using hv.

4.4 Parameter Settings

In this paper, MATLAB is used for the experiments based on PlatEMO [27]. In each dataset, the comparison algorithms are independently run 30 times, whereas initial random seeds are set same for all the algorithms but changed by different data sets and runs. Therefore, all the performances can be reproduced at any time by other researchers. The algorithm's parameter settings are organized in the corresponding literatures [11, 13, 14, 16, 17]. The population size N is adaptively set to the number of features, but not beyond 200 to balance efficiency and diversity. The termination criterion has been set to 100 times the population size for each dataset, which means that the maximal evolutionary generation is almost 100.

For the classification, the data sets are randomly divided in this paper into the test and training subsets with approximately 30% and 70% proportions, respectively, according to the stratified splitting method [2, 3, 5]. Note that this division is fixed to the specified data sets by algorithms on the all the 30 runs to understand the statistical performance, which varies by different data sets. During the training procedure, apply the 10-fold cross-validation to calculate classification's errors in training sets to avoid the features selection bias [19, 28], and set the KNN closest neighbor to 5 to balance the efficiency and accuracy.

5 Experimental Performance

5.1 Analyze the Experimental Results

We used the candidate solutions to calculate the algorithm's performance, which must depend on information obtained by training data in experiments. This is due to keeping the test data invisible to the decision-makers. We separately choose the candidate solution by using two cases of training data. In case 1, select the non-dominated Pareto fronts from the training data to measure the performances on the training and test data. In case 2, we choose solutions with better classification accuracy for each size of the selected feature subset based on training data. If there are numerous better solutions for feature selection, we randomly select only one of them. Note that the candidate solution uses only the non-dominated to additionally measure the values of hypervolume in both cases. Therefore, both cases essentially get the same performance of hypervolume on training data but may have different performances on test data. In all the test data tables, we used **Test** to signify the calculated performances by case 1 and **Test *** to signify the calculated performances by case 2.

In Table 2, the detailed performance of hypervolume on test data sets was calculated by case 1, including the metric values of mean and standard deviation obtained by each algorithm. To make it easier for readers, we also compute the general win ratio of RAEA by comparing it to the given algorithms in Table 3 and the average ranking of Freidman in Table 4 according to hypervolume metrics terms.

The results clearly show that RAEA outperforms all of the training datasets and the majority of the test datasets in Table 2. Table 2 shows that only a few RAEA losses occur in less-dimensional datasets, with no meaningful change. In Table 3 (**Test** ∗ rows), the performance calculated by case 2 is even better with the significant winning ratios and small loss ratios. Case 2 selects the most reliable candidate solutions.

Table 2. Performance of MEAN HV on test data, with the best output emphasized in blue.

No.	RAEA	NSGA-II	MOEA/D	MP-DEA	RVEA	PMEA
1	9.3605e-01	5.4403e-01	5.8426e-01	5.3443e-01	5.3278e-01	5.3950e-01
	± 4.75e-02	± 1.57e-02	± 2.95e-02	± 1.39e-02	± 1.92e-02	± 1.55e-02
2	7.9995e-01	4.8159e-01	5.3751e-01	4.7436e-01	4.6724e-01	4.8284e-01
	± 4.75e-02	± 3.09e-03	± 4.65e-03	± 8.93e-03	± 3.35e-03	± 5.93e-03
3	9.5013e-01	5.4235e-01	5.9347e-01	5.2952e-01	5.2777e-01	5.3994e-01
	± 2.64e-02	± 1.98e-02	± 2.61e-02	± 2.20e-02	± 1.69e-02	± 1.68e-02
4	9.4252e-01	2.9477e-01	3.5521e-01	2.6969e-01	2.6993e-01	2.8592e-01
	± 3.54e-02	± 2.00e-02	± 1.94e-03	± 1.71e-03	± 2.51e-03	± 2.36e-03
5	9.7180e-01	7.8710e-01	9.3305e-01	9.8346e-01	9.7460e-01	7.7089e-01
	± 2.24e-03	± 9.65e-03	± 5.61e-03	± 1.13e-02	± 1.52e-02	± 9.67e-0
6	8.6684e-01	6.9466e-01	8.0031e-01	6.7897e-01	6.8588e-01	6.7531e-01
	± 8.11e-03	± 1.48e-02	± 1.40e-02	± 1.30e-02	± 1.24e-02	± 1.09e-02
7	8.7691e-01	5.8170e-01	8.2269e-01	5.6826e-01	5.5723e-01	5.6539e-01
	± 1.71e-02	± 1.47e-02	± 3.38e-02	± 1.73e-02	± 1.69e-02	± 1.63e-02
8	9.1730e-01	8.3162e-01	8.9962e-01	9.3204e-01	8.0134e-01	7.9704e-01
	± 2.07e-02	± 3.19e-02	± 2.37e-02	± 3.32e-02	± 2.17e-02	± 2.55e-02
9	6.9650e-01	6.5983e-01	6.8286e-01	5.8987e-01	5.5498e-01	5.5338e-01
	± 1.34e-02	± 1.94e-02	± 1.18e-02	± 1.92e-02	± 2.39e-02	± 6.15e-02
10	8.2027e-01	7.1894e-01	7.9948e-01	7.6159e-01	7.2447e-01	7.1078e-01
	± 9.54e-03	± 1.25e-02	± 1.01e-02	± 1.24e-02	± 1.63e-02	± 1.22e-02

Table 3. RAEA's win, loss, and draw rations as compared to the other algorithms on all datasets, where the bracketed values show better outputs or no substantial differences.

Metric	Data	Win	Loss	Draw
HV	Train	100% (93%)	0% (0%)	0% (7%)
	Test	96% (75%)	4% (0%)	0% (25%)
	Test∗	98% (88%)	2% (0%)	0% (12%)

Table 4. Average Freidman's rankings on all datasets

Metric	Data	RAEA	NSGA-II	MOEA/D	MP-DEA	RVEA	PMEA
HV	Train	1.22	3.55	3.03	3.79	5.03	4.39
	Test	1.68	3.60	2.69	3.73	4.78	4.42
	Test*	1.54	3.73	2.77	3.76	4.66	4.48

5.2 Non-dominated Fronts Distribution Analyzation

Figure 2 shows the distribution of the first non-dominated front with average performances of hypervolume obtained by the algorithms. Column (col) 1 of Fig. 2 displays the non-dominated solutions on training data, whereas col 2 displays the non-dominated solutions on test data. Take note that the solutions in col 2 are derived from col 1, and remove those dominated in test data but might be non-dominated in training data. Then we select the three data sets (e.g., LSVT voice, Multiple Features, and ISOLET5). Since their non-dominated front distribution is achieved by different algorithms, they are easily distinguished graphically. On training and test data, RAEA's non-dominated solution is typically more diverse (e.g., reserve most Pareto optimal solutions) and well converged (e.g., with few selected features and minor classification errors) than other algorithms. Note that few non-dominated solutions of col 1, as shown in Fig. 2 change their position or even vanish in col 2, such as containing different classification errors or being dominated by other solutions that are very mutual in classification problems, including selected features, because of variances among test and training data.

6 Conclusions

In this paper, we propose a new method to solve the replication analysis-based bi-objective feature selection problem. Our proposed RAEA method is compared with five traditional algorithms using 10 benchmark classification data sets, which contain many ranges of features from 7070 to 256. The results prove that RAEA performs better than comparison algorithms by using hypervolume and selecting a small number of features successfully while improving classification accuracy in many data sets.

In the future, we will propose the idea of further examining the sensitivities of RAEA parameters and researching RAEA performance on further classification data sets, especially in multi-modal evolutionary environments.

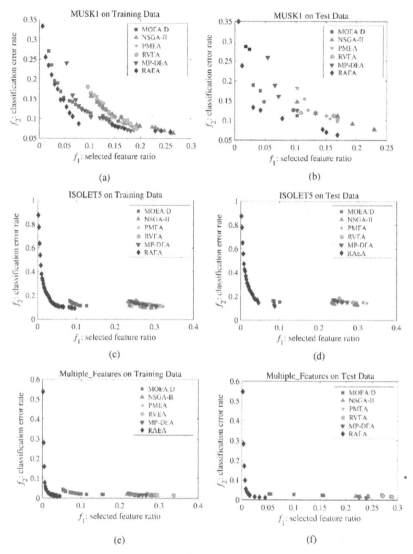

Fig. 2. Non-dominated front distributions obtained by the algorithms in objective space: (a) MUSK1 on Training Data; (c) ISOLET5 on Training Data; (e) Multi-Features on Training Data; (b) MUSK1 on Test Data; (d) ISOLET5 on Test Data; (f) Multi-Features on Test Data.

References

1. Tang, J., Alelyani, S., Liu, H.: Feature selection for classification: a review. Data Classif. Algorithms Appl. **37** (2014)
2. Xue, B., Zhang, M., Browne, W.N., Yao, X.: A survey on evolutionary computation approaches to feature selection. IEEE Trans. Evol. Comput. **20**(4), 606–626 (2015)

3. Xue, B., Zhang, M., Browne, W.N.: Particle swarm optimization for feature selection in classification: novel initialization and updating mechanisms. Appl. Soft Comput. 1(18), 261–276 (2014)

4. Mukhopadhyay, A., Maulik, U.: An SVM-wrapped multiobjective evolutionary feature selection approach for identifying cancer-microRNA markers. IEEE Trans. Nanobiosci. 12(4), 275–281 (2013)

5. Nguyen, H.B., Xue, B., Liu, I., Andreae, P., Zhang, M.: New mechanism for archive maintenance in PSO-based multi-objective feature selection. Soft. Comput. 20(10), 3927–3946 (2016)

6. Nguyen, B.H., Xue, B., Andreae, P., Ishibuchi, H., Zhang, M.: Multiple references points-based decomposition for multiobjective feature selection in classification: static and dynamic mechanisms. IEEE Trans. Evol. Comput. 24(1), 170–184 (2019)

7. Holland, J.H.: Adaptation in Natural and Artificial Systems: An Introductory Analysis with Applications to Biology, Control, and Artificial Intelligence. MIT Press, 29 April 1992

8. De La Iglesia, B.: Evolutionary computation for feature selection in classification problems. Wiley Interdiscip. Rev. Data Min. Knowl. Discov. 3(6), 381–407 (2013)

9. Cheng, R., Rodemann, T., Fischer, M., Olhofer, M., Jin, Y.: Evolutionary many-objective optimization of hybrid electric vehicle control: From general optimization to preference articulation. IEEE Trans. Emerg. Top. Comput. Intell. 1(2), 97–111 (2017)

10. Zhou, A., Qu, B.Y., Li, H., Zhao, S.Z., Suganthan, P.N., Zhang, Q.: Multiobjective evolutionary algorithms: a survey of the state of the art. Swarm Evol. Comput. 1(1), 32–49 (2011)

11. Deb, K., Pratap, A., Agarwal, S., Meyarivan, T.A.: A fast and elitist multiobjective genetic algorithm: NSGA-II. IEEE Trans. Evol. Comput. 6(2), 182–197 (2002)

12. Deb, K., Jain, H.: An evolutionary many-objective optimization algorithm using reference-point-based nondominated sorting approach, part I: solving problems with box constraints. IEEE Trans. Evol. Comput. 18(4), 577–601 (2013)

13. Zhang, Q., Li, H.: MOEA/D: A multiobjective evolutionary algorithm based on decomposition. IEEE Trans. Evol. Comput. 11(6), 712–731 (2007)

14. Elarbi, M., Bechikh, S., Coello, C.A., Makhlouf, M., Said, L.B.: Approximating complex Pareto fronts with predefined normal-boundary intersection directions. IEEE Trans. Evol. Comput. 24(5), 809–823 (2019)

15. While, L., Hingston, P., Barone, L., Huband, S.: A faster algorithm for calculating hypervolume. IEEE Trans. Evol. Comput. 10(1), 29–38 (2006)

16. Bader, J., Zitzler, E.: HypE: an algorithm for fast hypervolume-based many-objective optimization. Evol. Comput. 19(1), 45–76 (2011)

17. Zhang, X., Tian, Y., Jin, Y.: A knee point-driven evolutionary algorithm for many-objective optimization. IEEE Trans. Evol. Comput. 19(6), 761–776 (2014)

18. Zhang, Y., Gong, D.W., Sun, X.Y., Guo, Y.N.: A PSO-based multi-objective multi-label feature selection method in classification. Sci. Rep. 7(1), 1–2 (2017)

19. Tran, B., Xue, B., Zhang, M., Nguyen, S.: Investigation on particle swarm optimization for feature selection on high-dimensional data: Local search and selection bias. Connect. Sci. 28(3), 270–294 (2016)

20. Srinivas, N., Deb, K.: Multiobjective optimization using nondominated sorting in genetic algorithms. Evol. Comput. 2(3), 221–248 (1994)

21. Lazar, C., et al.: A survey on filter techniques for feature selection in gene expression microarray analysis. IEEE/ACM Trans. Comput. Biol. Bioinf. 9(4), 1106–1119 (2012)

22. Vignolo, L.D., Milone, D.H., Scharcanski, J.: Feature selection for face recognition based on multi-objective evolutionary wrappers. Expert Syst. Appl. 40(13), 5077–5084 (2013)

23. Xu, H., Zeng, W., Zeng, X., Yen, G.G.: A polar-metric-based evolutionary algorithm. IEEE Trans. Cybern. 51(7), 3429–3440 (2020)

24. Cheng, R., Jin, Y., Olhofer, M., Sendhoff, B.: A reference vector-guided evolutionary algorithm for many-objective optimization. IEEE Trans. Evol. Comput. **20**(5), 773–791 (2016)
25. He, Z., Yen, G.G.: Visualization and performance metric in many-objective optimization. IEEE Trans. Evol. Comput. **20**(3), 386–402 (2015)
26. Dua, D., Graff, C.: UCI machine learning repository (2017). http://archive.ics.uci.edu/ml
27. Tian, Y., Cheng, R., Zhang, X., Jin, Y.: PlatEMO: A MATLAB platform for evolutionary multi-objective optimization [educational forum]. IEEE Comput. Intell. Mag. **12**(4), 73–87 (2017)
28. Kohavi, R., John, G.H.: Wrappers for feature subset selection. Artif. Intell. **97**(1–2), 273–324 (1997)

Improved Particle Swarm Algorithm Using Multiple Strategies

Yunfei Yi[1,2], Zhiyong Wang[2], and Yunying Shi[1(✉)]

[1] School of Big Data and Computer, Hechi University, Yizhou 546300, Guangxi, China
Syy62311@163.com
[2] College of Computer Science and Engineering, Guangxi Normal University, Guilin 541001, Guangxi, China

Abstract. In order to address the issues of premature convergence and low search efficiency in the basic particle swarm algorithm, this paper analyzes the improved particle swarm optimization algorithms proposed by previous researchers. Based on this analysis, a modified algorithm for particle swarm optimization is proposed. The enhanced algorithm adjusts the inertia weight parameter in a nonlinear manner and dynamically changes the self-learning factor using chaos, allowing it to continuously change during iterations and achieve better optimization results. Additionally, to prevent the algorithm from getting trapped in local optima, a mutation operation is introduced into the improved algorithm. Finally, several classical test functions are used to conduct experiments, and the results are compared with the improved particle swarm optimization algorithms described in related literature. The experimental results demonstrate that the newly proposed algorithm achieves a certain degree of improvement in optimization accuracy compared to the basic particle swarm algorithm. This indicates that the modified algorithm effectively avoids premature convergence and possesses high search efficiency.

Keywords: Nonlinear Inertial Weights · Chaos · Mutation Operation

1 Introduction

Particle Swarm Optimization (PSO) algorithm is a swarm intelligence evolutionary algorithm proposed by social psychologist J. Kennedy and electrical engineer R.C. Eberhart in 1995, inspired by the foraging behavior of bird flocks [1, 2]. Due to its strong global optimization ability, fast convergence speed, and minimal parameter settings, this algorithm has attracted significant attention from researchers. It has been widely applied in various fields such as neural network optimization, multi-objective algorithm optimization, load scheduling, and fuzzy control [3–6].

Like other optimization algorithms, the original version of the Particle Swarm Optimization (PSO) algorithm has several limitations. Therefore, many scholars have made numerous improvements to it. In particular, the inertia weight [7] is a critical parameter in the PSO algorithm as it affects the particle flight direction and the ability to balance algorithm development. To enhance algorithm performance, researchers have proposed

© The Author(s), under exclusive license to Springer Nature Singapore Pte Ltd. 2024
K. Li and Y. Liu (Eds.): ISICA 2023, CCIS 2146, pp. 62–72, 2024.
https://doi.org/10.1007/978-981-97-4393-3_6

methods such as linear variation [8], nonlinear variation [9], chaotic variation [10], and adaptive variation [11, 12] to improve the inertia weight. Similarly, the learning factor is also an important factor that influences algorithm performance, as its range determines the degree of particle learning from neighboring particles and information exchange. To better utilize the learning factor, scholars have proposed operations such as nonlinear variable factors [13] and adaptive variation [14–16]. Furthermore, some scholars have balanced the algorithm's development capabilities by modifying the update equations, for instance, reducing equation parameters [17], introducing acceleration factors [18], compression factors [19], or considering population diversity [20, 21]. All these improvement strategies offer choices and insights for optimizing the Particle Swarm Optimization algorithm.

The aforementioned improvements are targeted at specific aspects and do not effectively balance the local and global development capabilities of the algorithm. This paper not only focuses on local improvements but also takes into account the overall development capability of the algorithm. In order to achieve better optimization results, we make improvements to multiple factors. Firstly, we use a nonlinear variation approach to improve the inertia weight, making it more realistic and enhancing the algorithm's convergence speed and stability. Secondly, we adopt a chaotic approach to improve the learning factor, increasing population randomness and improving the efficiency of global search. Lastly, through mutation operations, the algorithm is able to break out of local optima and find better solutions, further optimizing the algorithm's global development capability. The comprehensive application of these improvement measures allows the algorithm to be more effective in solving practical problems, with better robustness and adaptability.

2 Algorithm Introduction

2.1 Algorithm Introduction

The standard Particle Swarm Optimization (PSO) algorithm is a swarm intelligence optimization algorithm based on the foraging behavior of birds. In this algorithm, each particle is likened to a bird, and the fitness value is compared to the amount of food. Each particle searches for food in the search space, guided by the fitness function that determines the size of the fitness value. The direction of flight for each particle is determined based on the magnitude of the fitness value. In each iteration, particles update themselves based on two optimal positions: the individual best position (pbest) and the global best position (gbest). The algorithm stops when the termination condition is met, and the globally best position found at that time represents the optimal solution.

Suppose there is a swarm composed of N particles flying at certain velocities in a search space of D dimensions. $x_i^t = (x_{i1}^t, x_{i2}^t, \ldots, x_{ij}^t, \ldots, x_{iD}^t)^T$ Represents the position of particle i at the t-th iteration, $x_{ij}^t \in [L_D, U_D]$. L_D And U_D represent the lower and upper bounds of the search space respectively. $v_i^t = (v_{i1}^t, v_{i2}^t, \ldots, v_{ij}^t, \ldots, v_{iD}^t)^T$ Represents the velocity of particle i at the t-th iteration, $v_{ij}^t \in [v_{min}, v_{max}]$. v_{min} And v_{max} represent the lower and upper bounds of the velocity respectively. Individual optimal position-$p_i^t = (p_{i1}^t, p_{i2}^t, \ldots, p_{ij}^t, \ldots, p_{iD}^t)^T$, $1 \leq j \leq D$, $1 \leq i \leq N$. Global optimal position-$p_g^t = (p_{g1}^t, p_{g2}^t, \ldots, p_{gj}^t, \ldots, p_{gD}^t)^T$, $1 \leq j \leq D$.

The particle swarm iteration process is controlled by the following two formulas:

$$v_{ij}^{t+1} = wv_{ij}^t + c1r_1\left(p_{ij}^t - x_{ij}^t\right) + c2r_2\left(p_{gj}^t - x_{ij}^t\right) \tag{1}$$

$$x_{ij}^{t+1} = x_{ij}^t + v_{ij}^{t+1} \tag{2}$$

where r_1 and r_2 are random numbers between $(0,1)$. $c1$ and $c2$ are learning factors, usually $c1 = c2 = 2$. w is called the inertia weight. The particle continuously updates the speed and position according to formulas (1) and (2) until the maximum number of iterations is satisfied or the optimization accuracy is satisfied.

3   Improved PSO Algorithm

Each parameter of PSO has a very important influence on the convergence speed and optimization accuracy of the algorithm. The inertial weight controls the particle flight speed and direction, and has the ability of balancing the local search and global search of the algorithm. The learning factor controls how much influence each particle has on the individual and global optimal solutions during the search process. Setting the appropriate parameters can give full play to the influence of the algorithm and improve the convergence efficiency and solving accuracy of the algorithm.

3.1   Improvement of Inertia Weight

The traditional particle swarm optimization (PSO) adjusts the inertia weight parameters by linear decrement, but the search ability of the algorithm is easily reduced and the algorithm falls into local optimal. For this reason, this paper adopts the nonlinear decreasing method to control the change of weight parameters, and the specific formula is as follows:

$$w = w_{min} + (w_{max} - w_{min}) * \left(1 - \sqrt{1 - \left(\frac{t}{iter} - 1\right)^2}\right) \tag{3}$$

where t is the current number of iterations, iter is the maximum number of iterations, w_{max} is the maximum value of the weight coefficient, and w_{min} is the minimum value of the weight coefficient. In this paper, $w_{max} = 0.9$, $w_{min} = 0.4$. Different from the traditional linear decrement method, the improved nonlinear decrement method maintains a relatively gentle rate of change in the late iteration period, which is conducive to the particle swarm to conduct local search stably with a small rate of change at the end of the search, and is convenient to find the optimal solution.

3.2 Improvement of Self-learning Factors

In the process of particle iteration, the learning factors are divided into self-learning factors c1 and social learning factors c2. The value range of c1 and c2 determines the learning degree of self-cognition and social cognition of particles, and reflects the information exchange degree of particles between groups. In this paper, chaos strategy [22] is introduced into the improvement of self-learning factors, and the specific operations are as follows:

$$\begin{cases} r = 4.0 * r * (1 - r) \\ c1 = r * end + (\frac{I}{iter}) * (start - end) \end{cases} \tag{4}$$

where r_0 is a random number in $(0,1)$, and $r_0 \notin \{0, 0.25, 0.5, 0.75, 1\}$, the value of $start$ is 1.25, and the value of end is 2.50.

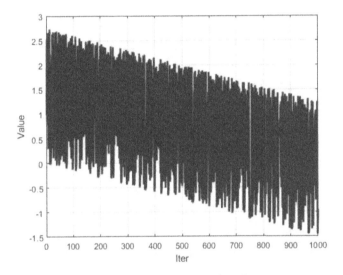

Fig. 1. Values of c1 during iteration

From the Fig. 1, we can clearly see that the value range of c1 shows a decreasing trend and has the random characteristics of chaos. Moreover, in the middle and late iterations, the probability of c1 taking a negative value increases, which is set to make the particle get rid of the influence of the historical optimal, search the solution space in a larger range, and jump out of the local optimal.

3.3 Improved Gaussian Variation

In view of the problem that traditional particle swarm optimization is easy to fall into local optimal, we refer to the variation ideas of some algorithms [23, 24], introduce Gaussian variation into particle swarm optimization, and improve it to enhance the ability of particles to jump out of local optimal. The specific improvement formula is as follows:

$$\begin{cases} x^{new} = (1 + \varphi * Gaussion(\sigma)) * x^{old} \\ \varphi = \frac{iter-l}{iter} \end{cases} \quad (5)$$

where x^{new} represents the new particle generated after the mutation, x^{old} is the old particle when the number of iterations is t, φ is the parameter controlling the mutation step, $Gaussion(\sigma)$ is a random variable satisfying the Gaussian distribution.

The specific mutation operation is to select particles for mutation with a certain probability P, and compare the fitness values of x^{old} and x^{new}. If the fitness after mutation is greater, the new particle is used to replace the old particle. Otherwise, no operation is performed. Among them, the value of probability P greatly affects the process of the algorithm. Therefore, in this paper, probability P is set as dynamic adaptive, and the formula is as follows:

$$P = (p_{min} - p_{min}) * \left(1 - \sqrt{1 - \left(\frac{t}{iter} - 1\right)^2}\right) + p_{max} \quad (6)$$

Among them, $p_{min} = 0.5$, $p_{max} = 0.9$. From the formula, we can see that the value of P is constantly increasing, which has the advantage that the update of the algorithm itself can be satisfied in the early stage of iteration without too much intervention. In the later stage of iteration, the algorithm basically converges to the global optimal position, and then we can mutate it to increase the population diversity. Helps algorithms get out of local optimality.

3.4 Algorithm Flow

The specific flow of the improved algorithm is as follows:

Step 1: Set the population number N and particle dimension D, and initialize the particle position and velocity.

Step 2: Calculate the fitness of each particle. If the fitness of the current particle is better than the historical individual optimal pbest, update the value of pbest. If a particle in the population has a better fitness than the globally optimal particle, the value of gbest is updated.

Step 3: Update the weight values and c1 values according to the weight function and chaotic self-learning factor proposed in this paper, and bring the calculated values to equations (1) and (2) to update the particle position and velocity.

Step 4: Select the particle for mutation operation according to a certain probability. If the fitness value of the particle after mutation is higher, update the particle; otherwise, no operation is carried out.

Step 5: If the number of iterations or the optimization accuracy is satisfied, stop the iteration and output the result; otherwise, go to step 2.

4   Simulation Experiment and Result Analysis

In order to verify the feasibility of the algorithm proposed in this paper, this section tests the benchmark functions of common standards. The test results are compared with the particle swarm optimization algorithm PSO, the particle swarm optimization algorithm with improved inertia weights (IWPSO) [9], a nonlinear dynamic adaptive inertial weight PSO algorithm (PSODAIW) [11], the nonlinear variable factor based particle swarm optimization algorithm (BPSO) [13], and the adaptive particle swarm optimization algorithm with perturbation acceleration factor (PAFPSO) [18], and it is further demonstrated that the proposed algorithm has a good effect.

Among the six standard test functions selected in this paper, f_1, f_2 and f_3 are single-peak test functions, while f_4, f_5 and f_6 are multi-peak test functions, which have multiple local minimums and are easy to fall into local optimality. The test functions are shown in Table 1.

Table 1. Test function table

Function Number	Test function	Functional formula
f_1	Sphere	$f(x) = \sum_{i=1}^{n} x_i^2$
f_2	Schwefel's 1.2	$f(x) = \sum_{i=1}^{n} \left(\sum_{j=1}^{i} x_j \right)^2$
f_3	Rosenbrock	$f(x) = \sum_{i=1}^{n} [100\left(x_{i+1} - x_i^2\right)^2 + (x_i - 1)^2]$
f_4	Rastrigin	$f(x) = \sum_{i=1}^{n} [x_i^2 - 10\cos(2\pi x_i) + 10]$
f_5	Griewank	$f(x) = 1 + \frac{1}{4000} \sum_{i=1}^{n} x_i^2 - \prod_{i=1}^{n} \cos(\frac{x_i}{\sqrt{i}})$
f_6	Ackley	$f(x) = -20\exp\left(-0.02\sqrt{n^{-1} \sum_{i=1}^{n} x_i^2}\right) - \exp\left(n^{-1} \sum_{i=1}^{n} \cos(2\pi x_i)\right) + 20$

4.1   Experimental Environment and Parameter Settings

In order to test the performance of the proposed algorithm, the selected operating system is Windows11, 64-bit; The processor is 12th Gen Intel(R) Core(TM) i7-12700 2.10 GHz; The hardware is 16G memory; The simulation software is MATLABR2019a.

The specific parameter settings of various algorithms are shown in Table 2. The number of iterations of each algorithm is set to 1000. The independent variable value range, test dimension and optimal function value of the five test functions are shown in Table 3.

Table 2. Parameter Settings of various algorithms

Algorithm	Population	w	C1	C2	N
PSO	30	0.4–0.9	2	2	1000
IWPSO	30	0.4–0.9	2	2	1000
PAFPSO	30	0.4–0.9	1.25–2.75	0.50–2.25	1000
BPSO	30	0.4–0.45	1.6–2.2	1.6–2.2	1000
PSODAIW	30	0.4–0.9	2	2	1000
Improved	30	0.4–0.9	−1.50–2.75	2	1000

Table 3. Test function parameter Settings

Function	Type	Dimension	Data Range	Theoretical
f_1	Unimodal	30	$[-100,100]$	0
f_2	Unimodal	30	$[-100,100]$	0
f_3	Unimodal	30	$[-30,30]$	0
f_4	Multimodal	30	$[-5.12,5.12]$	0
f_5	Multimodal	30	$[-600,600]$	0
f_6	Multimodal	30	$[-32,32]$	0

4.2 Simulation Results

In the experiment, each set of test functions was run 30 times each with a different algorithm to eliminate errors. The average value, minimum value and variance of the test function are calculated. The test results are shown in Table 4. The iterative process of particles in different test functions is shown in Fig. 2.

Table 4. Experimental results

Function	Algorithm	Optimal	Mean	Variance
f_1	PSO	4.17E-7	7.02E-6	4.11E-6
	IWPSO	3.27E-3	3.32E-2	2.83E-2
	PAFPSO	2.71E-9	1.07E-4	1.26E-4
	BPSO	4.55E-13	8.60E-11	1.23E-10

(continued)

Table 4. (*continued*)

Function	Algorithm	Optimal	Mean	Variance
	PSODAIW	3.30E-18	7.75E-14	1.54E-13
	Improved	**8.96E-39**	**1.31E-35**	**2.12E-35**
f_2	PSO	18.9	27.2	7.43
	IWPSO	45.0	59.4	14.9
	PAFPSO	2.99	10.5	4.36
	BPSO	8.35E-1	1.89	8.90E-1
	PSODAIW	4.42	10.3	3.88
	Improved	**1.94E-7**	**3.65E-4**	**3.41E-4**
f_3	PSO	23.5	49.5	30.5
	IWPSO	76.8	112	34.1
	PAFPSO	26.9	44.3	19.5
	BPSO	7.47	31.9	23.58
	PSODAIW	23.7	65.6	21.1
	Improved	21.8	**23.3**	**0.77**
f_4	PSO	47.5	98.2	44.08
	IWPSO	93.6	13.5	32.94
	PAFPSO	55.8	66.7	11.91
	BPSO	20.9	50.5	15.12
	PSODAIW	47.8	59.7	10.97
	Improved	**0**	**8.36**	**6.12**
f_5	PSO	1.26E-7	5.42E-3	6.68E-3
	IWPSO	9.39E-4	2.26E-2	1.57E-2
	PAFPSO	8.38E-7	9.37E-3	1.08E-2
	BPSO	1.63E-12	1.23E-2	1.26E-2
	PSODAIW	4.15	5.27	1.21
	Improved	**0**	**0**	**0**
f_6	PSO	1.42E-3	2.74E-3	1.21E-3
	IWPSO	6.35E-2	5.82E-1	6.59E-1
	PAFPSO	8.79E-4	3.14E-1	5.96E-1
	BPSO	6.34E-7	4.63E-1	9.26E-1
	PSODAIW	2.79E-10	6.85E-1	8.39E-1
	Improved	**7.99E-15**	**1.51E-14**	**4.49E-15**

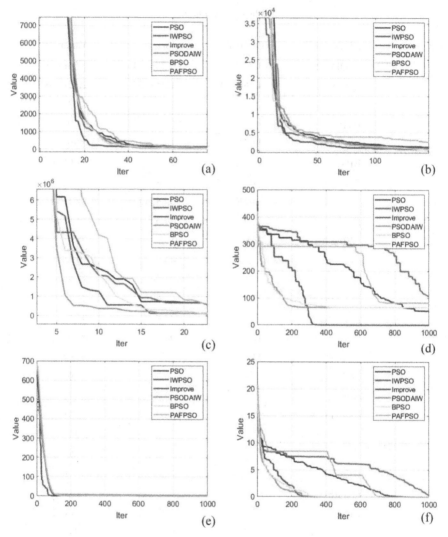

Fig. 2. Schematic diagram of particle iteration process in different test functions. (a) Sphere test procedure, (b)S'hwefel's test procedure, (c) Rosenbrock test procedure, (d) Rastrigin test procedure, (e) Griewank test procedure, (f) Ackley test procedure

4.3   Analysis of Experimental Results

By analyzing the experimental results in Table 4 (the bold ones indicate the best results in the test function), we can clearly see that compared with the traditional particle swarm optimization algorithm and other improved algorithms, although the optimization effect of multi-strategy PSO algorithm in f_2 function is not as high as the optimization accuracy of BPSO algorithm, its stability is better than those of them. And on most test functions, the improved algorithm has higher optimization precision and stronger stability. This shows that the chaotic learning factor and mutation strategy proposed by us have high search rate and the particles have the ability to jump out of the local optimal. However, by analyzing the particle iteration process diagram in Fig. 2, we can find that although the proposed algorithm performs well in the optimization accuracy, it still has shortcomings in the convergence speed and still needs to be further improved.

In general, the improved particle swarm optimization algorithm based on multiple strategies has certain effects, is better than other algorithms, and has better stability. However, satisfactory results have not been achieved on a small number of test functions, and the iteration speed has not reached the ideal state. These problems still need to be solved, which is also the next research direction.

5 Conclusion

Particle swarm optimization (PSO) is a new evolutionary algorithm. In this paper, an improved PSO based on multiple strategies is proposed to solve the problems of prematurity and low search efficiency of the basic PSO. By changing the value of inertia weight and learning factor, the global search and local exploration ability of the algorithm can be dynamically adjusted in the iterative process. In order to prevent the particles from falling into the local optimal in the iterative process, the variation operation is introduced to make the particles jump out of the local optimal. The experimental results show that the improved algorithm has higher optimization accuracy than other algorithms in the benchmark test function. Moreover, many experiments eliminate random factors, which further proves the effectiveness and stability of the algorithm.

Acknowledgments. This work is supported by the Guangxi Natural Science Foundation Joint Funding Project 2020GXNSFAA159172 and 2021GXNSFBA220023, in part by the Research Basic Ability Improvement Project for Young and Middle-aged Teachers of Guangxi Universities 2021KY0604, 2022KY0606 and 2023KY0633, in part by the Special Project of Guangxi Collaborative Innovation Center of Modern Sericulture and Silk 2023GXCSSC01 and Hechi University level research project 2023XJPT012.

References

1. Kennedy, J., Eberhart, R.C.: Particle swarm optimization. In: Proceedings of the IEEE International Conference on Neural Networks. IEEE Service Center, Piscataway, pp. 1942–1948 (1995)
2. Shi, Y., Eberhart, R.C.: Empirical study of particle swarm optimization. In: Proceedings of Congress on Evolutionary Computation, Washing DC, pp. 1945–1950 (1999)
3. Chen, Y., Chen, Z.: A prediction model of wall shear stress for ultra-high-pressure water-jet nozzle based on hybrid BP neural network. Eng. Appl. Comput. Fluid Mech. **16**(1), 1902–1920 (2022)
4. Zhang, X., Zheng, X., Cheng, R., et al.: A competitive mechanism based multi-objective particle swarm optimizer with fast convergence. Inf. Sci. **427**, 63–76 (2018)
5. Alzubi, I., Al-Masri, H.M.K., Abuelrub, A.: Modified particle swarm optimization algorithms for solving economic load dispatch. In: Proceedings of the 2022 3rd International Conference on Smart Grid and Renewable Energy (SGRE), pp. 1–5. IEEE (2022)

6. Pozna, C., Precup, R.E., Horváth, E., et al.: Hybrid particle filter–particle swarm optimization algorithm and application to fuzzy controlled servo systems. IEEE Trans. Fuzzy Syst. **30**(10), 4286–4297 (2022)
7. Bansal, J.C., Singh, P.K., Saraswat, M., et al.: Inertia weight strategies in particle swarm optimization. In: 2011 Third World Congress on Nature and Biologically Inspired Computing, pp. 633–640. IEEE (2011)
8. Wang, F., Zhang, H., Li, K., et al.: A hybrid particle swarm optimization algorithm using adaptive learning strategy. Inf. Sci. **436**, 162–177 (2018)
9. Qian, J., Zhang, J., Yao, D., et al.: A particle swarm optimization algorithm based on improved inertia weights. Comput. Digit. Eng. **50**(08), 1667–1670 (2022)
10. Chen, K., Zhou, F., Liu, A.: Chaotic dynamic weight particle swarm optimization for numerical function optimization. Knowl.-Based Syst. **139**, 23–40 (2018)
11. Wang, S., Liu, G.: A nonlinear dynamic adaptive inertial weight PSO algorithm. Comput. Simul. **38**(04), 249–253+451 (2021)
12. Zdiri, S., Chrouta, J., Zaafouri, A.: Cooperative multi-swarm particle swarm optimization based on adaptive and time-varying inertia weights. In: Proceedings of the 2021 IEEE 2nd International Conference on Signal, Control and Communication (SCC), pp. 200–207 (2021)
13. Xiao, Z., Long, Y., Guo, J., et al.: Particle swarm optimization based on nonlinear variable factor. Electron. Measur. Technol. **43**(05), 67–70 (2020)
14. JuanTong, Q., Zhao, W., Li, M.: Particle swarm optimization algorithm based on adaptive dynamic change. Microelectron. Comput. **36**(02), 6–10+15 (2019)
15. Yang, X., Li, H., Liu, Z.: Adaptive comprehensive learning particle swarm optimization with spatial weighting for global optimization. Multimedia Tools Appl. **81**(25), 36397–36436 (2022)
16. Lv, J., Shi, X.: Particle swarm optimization algorithm based on factor selection strategy. In: Proceedings of the 2019 IEEE 4th Advanced Information Technology, Electronic and Automation Control Conference (IAEAC), vol. 1, pp. 1606–1611 (2019)
17. El-Sherbiny, M.M.: Particle swarm inspired optimization algorithm without velocity equation. Egypt. Inform. J. **12**(1), 1–8 (2011)
18. Jiang, J., Tian, M., Wang, X., et al.: An adaptive particle swarm optimization algorithm with perturbation acceleration factor is proposed. J. Xidian Univ. **39**(04), 74–80 (2012)
19. Tang, J., Zheng, S., Wang, Z., et al.: A particle swarm optimization algorithm for improving learning factor and compression factor. Yunnan Water Power **38**(06), 77–79 (2022)
20. Huang, D., Yang, J., Yu, J.: A particle swarm optimization with fitness-distance balance strategy. In: Proceedings of the 18th International Conference on Computational Intelligence and Security (CIS 2022), pp. 336–340 (2022)
21. Hayashida, T., Nishizaki, I., Sekizaki, S., et al.: Improvement of particle swarm optimization focusing on diversity of the particle swarm. In: Proceedings of the 2020 IEEE International Conference on Systems, Man, and Cybernetics (SMC), pp. 191–197 (2020)
22. Liu, H., Zhang, X.W., Tu, L.P.: A modified particle swarm optimization using adaptive strategy. Expert Syst. Appl. **152**, 113353 (2020)
23. Anwaar, A., Ashraf, A., Bangyal, W.H.K., et al.: Genetic algorithms: brief review on genetic algorithms for global optimization problems. In: 2022 Human-Centered Cognitive Systems (HCCS), pp. 1–6 (2022)
24. Adsawinnawanawa, E., Kruatrachue, B., Siriboon, K.: Enhance particle's exploration of particle swarm optimization with individual particle mutation. In: Proceedings of the 2019 7th International Electrical Engineering Congress (iEECON), pp. 1–4 (2019)

A Reference Vector Guided Evolutionary Algorithm with Diversity and Convergence Enhancement Strategies for Many-Objective Optimization

Lei Yang$^{(\boxtimes)}$, Yuanye Zhang, and Jiale Cao

School of Mathematics and Informatics, South China Agricultural University, Guangzhou 510642, China
yanglei_s@scau.edu.cn

Abstract. Maintaining the balance between convergence and diversity is a key issue in evolutionary multi-objective optimization and a challenge in many-objective scenarios. Reference-vector-guided selection is an exemplary method for decomposition-based many-objective evolutionary algorithms (MaOEAs). Aiming at solving or alleviating the defects reference vector guided selection confronts, this paper proposes a reference vector guided evolutionary algorithm with diversity and convergence enhancement strategies (RVEA-DCES) for many-objective optimization. RVEA-DCES introduces two new strategies namely adaptive sparse region filling and convergence-only selection for diversity and convergence enhancement. The former is to improve diversity by adaptively adding solutions into sparse regions while the latter is to prevent the elimination of solutions with prominent convergence performance. Experimental results on WFG test suite up to 15 objectives indicate that RVEA-DCES is highly competitive in comparison with five state-of-the-art MaOEAs.

Keywords: Many-objective optimization · Reference vector guided selection · Diversity and convergence enhancement

1 Introduction

Multi-objective optimization problems (MOPs) can be formulated as follows:

$$min \quad F(x) = \{f_1(\boldsymbol{x}), f_2(\boldsymbol{x}), ..., f_m(\boldsymbol{x})\}$$
$$s.t. \quad \boldsymbol{x} \in \Omega \tag{1}$$

where $\Omega \subseteq R^n$ is the decision space, $\boldsymbol{x} = (x_1, x_2, ..., x_n)$ is the decision vector and $F(\boldsymbol{x}) \in R^m$ is the corresponding objective vector consisting of m objective values. If more than three objectives are considered, these problems are termed as many-objective optimization problems (MaOPs) [1]. Since these objectives often conflict with each other, finding a solution that optimizes all the objectives simultaneously is impossible. Hence a set of solutions which achieve the best

© The Author(s), under exclusive license to Springer Nature Singapore Pte Ltd. 2024
K. Li and Y. Liu (Eds.): ISICA 2023, CCIS 2146, pp. 73–87, 2024.
https://doi.org/10.1007/978-981-97-4393-3_7

trade-off among all objectives are what multi-objective optimization seeks. These solutions are known as Pareto optimal solutions and any improvement of a Pareto optimal solution in one objective must lead to a deterioration in at least one other objective. The set of all Pareto optimal solutions in decision space and objective space are Pareto Set (PS) and Pareto Front (PF) respectively.

Evolutionary algorithms (EAs) have proven to be an efficient way solving MOPs since they are able to obtain a set of solutions in a single run owing to the population-based nature. Thus, research area of multi-objective evolutionary algorithms (MOEAs) has attracted much attention and numerous MOEAs have been developed over the past three decades. These MOEAs can be mainly categorized into three categories, i.e., dominance-based MOEAs, decomposition-based MOEAs and indicator-based MOEAs.

Many MOEAs that perform well on MOPs with two or three objectives encounter difficulties in many-objective scenarios [2]. As the number of objective increases, the proportion of non-dominated solutions increases dramatically. Affected by this, dominance-based MOEAs loss selection pressure as the major selection criteria Pareto dominance fails to compare solutions. Indicator-based MOEAs integrate convergence and diversity into a scalar performance indicator to evaluate solutions. However, exponentially increased computation cost in many-objective scenarios makes them time-consuming and impractical. Decomposition-based MOEAs crystallised in [3] share a general framework. They decompose an MOP into a number of single-objective subproblems using scalarization approaches then optimize them simultaneously and cooperatively. Decomposition-based MOEAs represented by MOEA/D [3] are capable of dealing with MaOPs since no additional difficulties emerge, but their unstable performance and low search efficiency are not satisfying [4]. Another decomposition strategy has been proposed recently which decomposes the objective space into a set of subspaces using reference vectors [5]. This is equivalent to dividing an MOP into a set of multi-objective subproblems and solving them separately but simultaneously. Based on this, some MOEAs are proposed to deal with MaOPs [6,7] and achieve competitive results. Reference vector guided evolutionary algorithm (RVEA) [8] is one of the most representative algorithms on this branch. We will give a brief introduction about objective space decomposition and reference vector guided selection in Sect. 2.

There are some improvement researches on the basis of RVEA. In [9], reference vectors are adjusted according to the distribution of solutions after reference vector guided selection to make sure that most reference vectors are associated with solutions. In [10], a new search engine based on differential evolution (DE) is combined with RVEA to achieve further performance enhancement. In [11], modified inverted generational distance (IGD^+) indicator is combined with reference vector guided selection to form a two-stage selection strategy, which can balance convergence and diversity well when solving MaOPs.

Different from the above researches, we focus on other problems RVEA or reference vector guided selection strategy confronts. RVEA selects only one elite solution from each subspace, which often results in insufficient number of elite solutions to reach population size as there may exist some empty subspaces containing no solution especially in the early evolutionary stage. Although this problem was mentioned in [8,12], it was either unresolved or resloved sloppily. While in [9], a secondary selection criterion based on the dominance relationship is adopted so that a sufficient number of solutions can survive and be passed to the next generation. Further on this, if there exist empty subspaces, there will naturally be crowded subspaces. Selecting only one solution from each of these crowded subspaces probably lose some well-performed solutions. Last but not least, even if without empty subspaces, the distribution of elite solutions is hardly as uniform as preset reference vectors. There may be some sparse regions with fewer solutions and the existence of empty subspaces will worsen this.

To solve or alleviate the above issues and improve the performance of reference vector guided selection-based MaOEAs, this paper proposes some feasible methods. The main contributions are summarized as follows:

1. Two strategies for diversity and convergence enhancement namely adaptive sparse region filling and convergence-only selection are proposed, which will be conducted after reference vector guided selection. For the former, an adaptive approach is applied to remaining solutions to detect sparse regions of current elite population. Some solutions will be added into elite population EP_1 to fill these sparse regions later. For the latter, remaining solutions with outstanding convergence will be retained into elite population EP_2.
2. Together with a new elite retention strategy, solutions of EP_1 and EP_2 will be added into final elite population methodically to fill any vacancies of it brought by empty subspaces while improving its convergence and diversity.

The remainder of this paper is organized as follows. Section 2 introduces some background knowledge as preliminaries. We then elaborate details of the proposed RVEA-CDES in Sect. 3. Comparative studies are carried out in Sect. 4 followed by discussions about experimental results. Finally, we conclude this paper and present some future research issues.

2 Preliminaries

In this section, we introduce the reference vector set used in this study and briefly review the design of reference vector guided selection strategy.

2.1 Reference Vector

Generating a uniformly distributed reference vector set is essential for most decomposition-based MOEAs. We use Riesz s-Energy method [13] to generate reference points in this study. Originateing from the coordinate origin and connecting to these reference points, we can get the reference vector set. It is worth noting that converting these obtained reference vectors into unit vectors is convenient for subsequent calculation, so we conduct unitization beforehand by dividing each reference vector by its norm.

2.2   Reference Vector Guided Selection

Objective space decomposition is a prerequisite for reference vector guided selection. Commonly used objective space decomposition strategy proposed by [5] divides the objective space into some subspaces using an uniformly distributed reference vector set $V = \{v_0, v_1, ..., v_N\}$. Specifically, each solution will be allocated into a subspace according to the acute angles between its objective vector and all reference vectors. Subspaces are defined as $\Omega^i = \{s| \langle s, v_i \rangle \leq \langle s, v_j \rangle, \forall j \in 1, 2, ..., N\}$, where $i = 1, 2, ..., N$ and $\langle s, v_i \rangle$ represents the acute angle between solution s and i-th reference vector v_i. This angle can be calculated as:

$$\langle s, v_i \rangle = arccos \frac{s \cdot v_i}{\|s\| \|v_i\|} \tag{2}$$

According to the definition of subspace, each solution will associate a reference vector v_i if and only if it has the smallest vector angle between v_i compared with other reference vectors. Afterwards, all solutions associated with v_i constitute Ω^i. Apart from acute angle, perpendicular distance from solution to reference vector is alternative for objective space decomposition. Either way, they divide objective space according to the similarity between solution and reference vector.

Since solutions distribute in different subspaces after decomposition, solving the original MOP using solutions of each subspace constitutes a series of different multi-objective subproblems, and the diversity of reference vectors ensures the diversity of obtained solutions from each subproblem. Different strategies can be applied to solve these subproblems simultaneously. RVEA proposes a reference vector guided selection strategy. Specifically, it designs a scalarization approach to convert these multi-objective subproblems into single-objective subproblems, then selects one solution from each subspace that is optimal to the corresponding subproblem.

3   Proposed RVEA-DCES

3.1   Main Framework

The main framework of the proposed RVEA-DCES is presented in Algorithm 1. A set of N reference vectors are predefined to provide search directions. The upper bound of population size N' is set larger than N. After initializing the population with random individuals, commonly used genetic operations including selection, crossover and mutation are conducted to generate offspring population. Specifically, binary tournament selection is employed to select parents for reproduction. Simulated binary crossover (SBX) [14] and polynomial mutation (PM) [15] are employed to generate offspring solutions. Following the widely used elitism strategy, offspring population is combined with parent population to undergo environmental selection, during which elite solutions are selected to constitute an elite population for next iteration. Repeat the above procedures until termination condition, i.e., maximal function evaluations is met.

Algorithm 1 Main framework of the proposed RVEA-DCES

Input: upper bound of population size N', maximal function evaluations FEs, a set of reference vectors $V = \{v_0, v_1, ..., v_N\}$
Output: final population P
1: **Initialization:** create the initial population P with N random individuals, set $evaluations = 0$
2: **while** $evaluations < FEs$ **do**
3:     $Q \leftarrow$ offspring-generation(P)
4:     $FE \leftarrow |P|$
5:     $U \leftarrow P \cup Q$
6:     $P \leftarrow$ environmental selection(U, N, N')
7:     $evaluations \leftarrow evaluations + FE$
8: **end while**
9: **return** P

3.2 Angle-Penalized Distance

RVEA designs a scalarization approach Angle-Penalized Distance (APD) for many-objective optimization to balance convergence and diversity in each subspace. We utilize this approach in reference vector guided selection stage and APD is defined as:

$$d_{i,j} = (1 + P(\theta_{i,j})) \cdot \left\| f_i' \right\| \tag{3}$$

where $f_i' = f_i - z^{min}$ is the translated objective vector of solution i (z^{min} is the ideal point consisting of minimal objective values of current population) and $\left\| f_i' \right\|$ represents the Euclidean distance from f_i' to the ideal point. $P(\theta_{i,j})$ is a penalty function related to $\theta_{i,j}$, the acute angle between f_i' and v_j to measure the consistency between the direction of i and its associated reference vector:

$$P(\theta_{i,j}) = m \cdot \left(\frac{evaluations}{FEs}\right)^2 \cdot \frac{\theta_{i,j}}{\gamma_{v_j}}$$

$$\gamma_{v_j} = min \langle v_i, v_j \rangle, \quad i \in 1, 2, ..., N, i \neq j \tag{4}$$

where m is the number of objectives and γ_{v_j} is the smallest acute angle between v_j and other reference vectors, which is used to normalize angles in the subspace specified by v_j in case of dense or sparse distribution of reference vectors. It's obvious that a smaller APD indicates a better overall performance, that is to say this solution balances convergence and diversity better.

3.3 Environmental Selection

Algorithm 2 provides the pseudocode of environmental selection. Line 1 to Line 19 are reference vector guided selection referring to RVEA. Specifically, objective value translation is executed first to set ideal point the origin of coordinate system and guarantee that all the translated objective values are inside the first quadrant. After that, objective space is divided into a set of subspaces by associating each solution with its nearest reference vector, which has already been introduced in Sect. 2. Based on the results of population partition, APDs of all solutions are calculated and each solution with the smallest APD among its own subspace will be selected into population P for next generation.

Algorithm 2 Environmental Selection Strategy of the Proposed RVEA-DCES

Input: combined population U, the number of reference vectors N, upper bound of population size N'

Output: final elite population P for next generation

1: $P, EP_1, EP_2 \leftarrow \emptyset$
2: Calculate ideal point z^{min} of U
3: **for** $i = 1$ to $|U|$ **do**
4:     $f_i' \leftarrow f_i - z^{min}$
5: **end for**
6: **for** $i = 1$ to N **do**
7:     $P_i \leftarrow \emptyset$
8: **end for**
9: **for** $i = 1$ to $|U|$ **do**
10:     $k \leftarrow argmin_{j \in 1,2,\ldots,N} \left\langle f_i', v_j \right\rangle$
11:     $P_k \leftarrow P_k \cup U_i$
12: **end for**
13: **for** $k = 1$ to N **do**
14:     **for** $i = 1$ to $|P_k|$ **do**
15:         $d_{i,k} = (1 + P(\theta_{i,k})) \cdot \left\| f_i' \right\|$
16:     **end for**
17:     $s \leftarrow argmin_{i \in 1,2,\ldots,|P_k|} d_{i,k}$
18:     $P \leftarrow P_{k_s}$
19: **end for**
20: $R \leftarrow \complement_U P$
21: $EP_1 \leftarrow$ adaptive sparse region filling(P, R)
22: $EP_2 \leftarrow$ convergence-only selection(P, R)
23: elite rentention(P, R, EP_1, EP_2, N')
24: **return** P

According to the analysis in Sect. 1, selecting only one solution from each subspace often results in insufficient population size of P because of the existence of empty subspaces. To better solve this problem while improving diversity and convergence of P, adaptive sparse region filling strategy and convergence-only selection will be executed sequentially after reference vector guided selection.

Adaptive Sparse Region Filling. Although uniformly distributed reference vectors provide uniform search directions, actual distribution of elite solutions selected from each subspace may not be the same, Fig. 1 (left part) is an illustration. In this two-objective minimization problem, v_0, v_1, ..., v_5 are five uniformly distributed reference vectors and after reference vector guided selection, solutions represented by solid circles are selected into P for next generation as they are elite solutions in subspaces specified by v_1 to v_4 separately, while solutions represented by hollow circles are abandoned. The convergence performance of solution A and B are similar to elite solutions, and their absence lead to the emergence of two sparse regions surrounded by dotted lines. The absense of A and B are caused by their closeness to angle bisector of two reference vectors, which makes them far from both reference vectors.

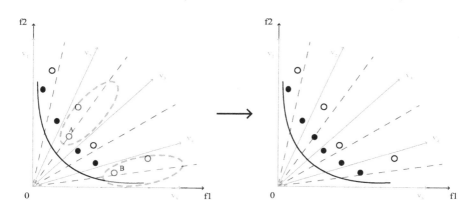

Fig. 1. Elite population before/after adaptive sparse region filling.

If we can figure out a way to retain A and B into population P, the distribution will undoubtfully become better, hence to improve diversity (right part of Fig. 1). This is the purpose of adaptive sparse region filling strategy and Algorithm 3 describes the details of it.

According to Algorithm 3, each solution R_i in R will be associated with a solution P_j in P first, satisfying that P_j has the smallest angle with R. Record this smallest angle as $\theta(R_i)$. It's obvious that the large $\theta(R_i)$ is, the better can R_i fill the sparse regions of P. Thus we put solutions in R with larger $\theta(R_i)$ into elite population EP_1 to retain them into P in subsequent procedures.

To limit the amount of solutions added into P, an angle threshold is set to $(m-1)^2 * \frac{\pi/2}{2(N-1)}$. This threshold is set intuitively according to the situation in two-dimensional space and extended to higher dimensionalities.

Convergence-Only Selection. Solutions that best balance convergence and diversity in their own subspace are selected during reference vector guided selection and adaptive sparse region filling strategy gives remaining solutions a second

---

Algorithm 3 Adaptive sparse region filling

Input: population P and remaining combined population Rx
Output: elite population EP_1
1: **for** $i = 1$ to $|R|$ **do**
2:     $\theta(R_i) \leftarrow \infty$
3:     **for** $j = 1$ to $|P|$ **do**
4:         Calculate $\alpha = \left\langle f'_{R_i}, f'_{P_j} \right\rangle$
5:         **if** $\alpha < \theta(R_i)$ **then**
6:             $\theta(R_i) \leftarrow \alpha$
7:         **end if**
8:     **end for**
9: **end for**
10: $\rho \leftarrow argmax_{i \in 1,2,\ldots,|R|} \theta(R_i)$
11: **while** $\theta(R_\rho) \geq (m-1)^2 * \frac{\pi/2}{2(N-1)}$ **do**
12:     $EP_1 \leftarrow EP_1 \cup R_\rho$
13: **end while**
14: **return** EP_1

---

chance to survive regardless of their convergence. Nevertheless, some solutions with prominent convergence may still be discarded and a possible case is shown in Fig. 2. Solution A, B and C densely distribute in the subspace specified by v_2 and the ascending order of their APDs is A, C, B. APDs of B and C are just a little bit larger than A. Since only one solution can survive as an elite, B and C will be eliminated despite they are more convergent than A. Retaining B and C will improve the convergence and benefit subsequent evolutionary process apparently as they may produce promising offspring solutions. This is what convergence-only selection aims and Algorithm 4 presents the procedure of it.

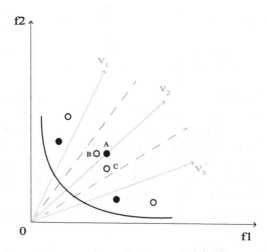

Fig. 2. A scenario where convergence-only selection works.

Algorithm 4 Convergence-only selection

Input: population P and remaining combined population R
Output: elite population EP_2
1: **for** $i = 1$ to $|R|$ **do**
2:    Determine the subspace R_i, locates and elite solution es of this subspace selected in reference vector guided selection.
3:    **if** $\left\| f'_{R_i} \right\| < \left\| f'_{es} \right\|$ **then**
4:       $EP_2 \leftarrow EP_2 \cup R_i$
5:    **end if**
6: **end for**
7: **return** EP_2

As can be seen from Algorithm 4, convergence-only selection is simple to implement as calculations it needs have already been done during reference vector guided selection. By comparing convergence only between remaining solutions in R and elite solutions of their own subspaces, solutions in R with outstanding convergence (e.g., B and C in Fig. 3) will be put into elite population EP_2.

Elite Retention. After conducting adaptive sparse region filling strategy and convergence-only selection, two elite population EP_1 and EP_2 are obtained. Finally, we will put solutions of them into final elite population P. A new elite retention strategy will be conducted here. First, put solutions in both EP_1 and EP_2 into P as they can improve diversity and convergence of P simultaneously. After that, remove these solutions from EP_1 and EP_2 and update the associated smallest angles of remainging solutions in EP_1, i.e., $\theta(EP_{1_i})$ of EP_{1_i} since P has more solutions now and $\theta(EP_{1_i})$ may decrease. Update procedure is presented in Algorithm 5. Then, selecting solutions from EP_1 and EP_2 in turn into P until the size of P reaches N' or EP_1 and EP_2 are empty. It is worth noting that the update procedure should be conducted once adding new solutions into P. As population size of EP_1 and EP_2 are both unknown, population size of final elite population P is not sure and dynamic. We then set an upper bound for it and details will be discussed in Subsect. 4.3.

Algorithm 5 UpdateAssociation

Input: elite population EP_1, solution set S new added into P
Output: updated elite population EP_1
1: **for** $i = 1$ to $|EP_1|$ **do**
2:    **for** $j = 1$ to $|S|$ **do**
3:       Calculate $\alpha = \left\langle f'_{EP_{1_i}}, f'_{S_j} \right\rangle$
4:       **if** $\alpha < \theta(EP_{1_i})$ **then**
5:          $\theta(EP_{1_i}) \leftarrow \alpha$
6:       **end if**
7:    **end for**
8:    **if** $\theta(EP_{1_i}) < (m-1)^2 * \frac{\pi/2}{2(N-1)}$ **then**
9:       Remove EP_{1_i} from EP_1
10:    **end if**
11: **end for**
12: **return** EP_1

---

4   Comparative Studies

To compare the proposed RVEA-DCES with 5 state-of-the-art MaOEAs, i.e., RVEA [8], onebyoneEA [16], MaOEA-IGD [17], MaOEA-IT [18] and hpaEA [19], comprehensive experiments are conducted on WFG test suite [20]. Each problem is tested on $3, 5, 8, 10, 15$ objectives with 20 independent runs. To have statistically sound conclusions, Wilcoxon rank sum test is adopted to test signif icance of differences among the results of RVEA-CDES and its competitors at a significance level of 0.05. According to the results of Wilcoxon rank sum test, "$-$" means that RVEA-DCES significantly outperforms the compared algorithm, "$+$" indicates that RVEA-DCES is significantly surpassed by the compared algorithm and "$\approx$" represents that there is no statistically significant difference.

4.1   Benchmark Test Problems

In the comprehensive experiments, WFG1-9 are taken as benchmark test problems. The number of decision variables of theses test problems is set to $n = K + L$, where $K = m - 1$ is the number of position-related working parameters and $L = 10$ is the number of distance-related working parameters. These working parameters will be mapped into $m - 1$ underlying position parameters and 1 underlying distance parameter via transformation functions.

4.2   Performance Indicators

A widely used performance indicator HV [21] is employed to evaluate the comprehensive performance of final non-dominated solution set obtained by each MaOEA. HV represents the volume of the region which is dominated by final solution set but dominates the reference point. HV concerns convergence and diversity simultaneously and a larger HV indicates a better overall performance.

4.3   Parameter Settings

In this subsection, we present the general parameter settings of experiments.

- *Population Size*: Some of the compared MaOEAs use reference vectors, their population size is then set the same to the number of reference vectors. Population size of other MaOEAs is set the same to these decomposition-based MaOEAs while population size of RVEA-DCES should be larger for elite retention. As introduced in Subsect. 3.3, population size of final elite population is dynamic and we set an upper bound for it according to the number of reference vectors and objectives. The detailed settings are summarized in Table 1. In general, more objectives corresponds to larger population size since more solutions are needed to search higher dimensional objective space.
- *Termination Condition*: All experimental runs terminate when the function evaluations reach the maximal value FEs. FEs are set the same for different benchmark problems but vary according to m. Specifically, 50000 for

3-objective, 100000 for 5-objective, 150000 for 8-objective, 210000 for 10-objective and 240000 for 15-objective. It is worth mentioning that although RVEA-DCES holds a larger population size, the same FEs ensure the fairness of comparative experiments to some extent since larger population size corresponds to less iterations.

– *General Parameter Settings for Genetic Operators*: As mentioned above, simulated binary crossover (SBX) and polynomial mutation (PM) are used to generate offspring population. For SBX, the distribution index η_c and crossover probability p_c are set to 20 and 1.0 respectively. For PM, the distribution index is set to $\eta_m = 20$ and mutation probability $p_m = 1/n$.

Table 1. Settings of the number of reference vectors (N) and population size (upper bound for RVEA-DCES) for compared MaOEAs

M	N	Other MaOEAs	RVEA-DCES
3	91	91	100
5	210	210	250
8	240	240	300
10	300	300	350
15	400	400	450

4.4 Results of WFG Test Suites

The average and standard deviation of HV on WFG1-9 over 20 independent runs are presented in Table 2. It can be observed from Table 2 that RVEA-DCES obtained 18 best results out of 45 test instances, making it the best one on WFG test suite. The comprehensive performance of other algorithms lagged behind RVEA-DCES obviously except RVEA and hpaEA, which obtained 10 and 9 best results.

WFG1 is designed to investigate an MOEA's ability coping with flat bias and mixed PF geometries. The performance of RVEA-DCES on WFG1 was poor and only better than MaOEA-IGD and MaOEA-IT. Obvious performance degradation compared to RVEA reveals that diversity and convergence enhancement strategies may not work on minority test problems. WFG2 has a disconnected PF while WFG3 is difficult with a degenerate PF. Without specialized approach handling irregular PFs, RVEA-DCES still obtained competitive results especially on WFG2. Irregular PFs will invalidate some reference vectors as subspaces specified by them are always empty. Thus, reference vector guided selection is bound to obtain insufficient number of elite solutions. Considering that solutions probably distribute crowdedly near PF segments, convergence-only selection can improve diversity of elite population by adding more solutions

Table 2. Comparison of HV average and standard deviation on WFG test suite

Problem	M	RVEA-DCES	RVEA	onebyoneEA	MaOEA-IGD	MaOEA-IT	hpaEA
*WFG1	3	6.8187e-1 (3.35e-2)	9.2044e-1 (2.02e-2)+	9.0225e-1 (7.04e-3)+	1.7500e-1 (1.38e-2) -	1.7916e-1 (5.37e-2) -	6.1614e-1 (1.45e-1) -
	5	6.3458e-1 (3.66e-2)	9.8741e-1 (2.23e-2)+	9.8488e-1 (1.96e-2)+	3.8389e-1 (1.82e-1) -	2.3309e-1 (8.72e-2) -	9.2958e-1 (6.10e-2)+
	8	7.7776e-1 (8.94e-2)	9.8335e-1 (3.49e-2)+	9.9645e-1 (1.03e-3)+	4.5089e-1 (2.14e-1) -	1.7328e-1 (9.88e-2) -	9.8433e-1 (3.65e-2)+
	10	8.1141e-1 (7.66e-3)	9.9714e-1 (5.89e-4)+	9.9791e-1 (6.01e-4)+	4.2429e-1 (1.75e-1) -	1.3887e-1 (8.00e-2) -	9.9874e-1 (9.80e-4)+
	15	9.4630e-1 (9.64e-3)	9.9835e-1 (6.20e-4)+	9.9887e-1 (1.69e-4)+	9.5206e-1 (3.66e-2)≈	5.4786e-2 (9.01e-3) -	9.9425e-1 (2.43e-2)+
	3	9.2631e-1 (2.23e-3)	9.2404e-1 (1.62e-3)≈	8.9226e-1 (8.15e-3) -	5.7862e-1 (8.28e-2) -	5.9047e-1 (5.06e-2) -	9.2533e-1 (6.29e-3)≈
	5	9.7458e-1 (5.94e-3)	9.8911e-1 (1.68e-3)≈	9.7803e-1 (5.24e-3)≈	9.3649e-1 (2.36e-2) -	6.0085e-1 (3.25e-2) -	9.8917e-1 (1.52e-3)≈
WFG2	8	9.4885e-1 (1.05e-2)	9.8083e-1 (4.51e-3)+	9.9011e-1 (2.73e-3)+	9.8637e-1 (1.40e-2)+	6.1130e-1 (3.67e-2) -	9.8370e-1 (2.77e-3)+
	10	9.4797e-1 (5.24e-3)	9.8560e-1 (4.06e-3)+	9.9382e-1 (1.83e-3)+	9.8282e-1 (3.95e-2)+	5.9518e-1 (4.54e-2) -	9.9164e-1 (1.44e-3)+
	15	9.7373e-1 (6.20e-3)	9.6871e-1 (7.13e-3) -	9.9486e-1 (1.09e-3)+	9.8213e-1 (1.10e-2)+	5.8966e-1 (1.09e-1) -	9.8958e-1 (2.31e-3)+
	3	3.0335e-1 (1.90e-2)	3.3870e-1 (9.34e-3)+	2.5353e-1 (1.84e-2) -	8.2609e-2 (2.39e-2) -	1.3404e-1 (1.93e-2) -	3.6690e-1 (1.49e-2)+
	5	1.8288e-1 (1.49e-2)	1.4649e-1 (2.04e-2) -	5.9976e-2 (1.11e-2) -	8.3393e-2 (2.87e-3) -	1.7099e-4 (6.69e-4) -	1.4331e-1 (2.08e-2) -
WFG3	8	0.0000e+0(0.00e+0)	0.0000e+0(0.00e+0) -	0.0000e+0 (0.00e+0)-	4.1938e-2 (1.92e-2)+	0.0000e+0 (0.00e+0)≈	7.1207e-3 (1.18e-2)+
	10	0.0000e+0(0.00e+0)	0.0000e+0(0.00e+0)≈	0.0000e+0(0.00e+0)≈	9.5502e-3 (1.64e-2)+	0.0000e+0 (0.00e+0)≈	0.0000e+0 (0.00e+0)≈
	15	0.0000e+0(0.00e+0)	0.0000e+0(0.00e+0)≈	0.0000e+0(0.00e+0)≈	0.0000e+0(0.00e+0)≈	0.0000e+0(0.00e+0)≈	0.0000e+0(0.00e+0)≈
	3	5.5166e-1 (1.64e-3)	5.4842e-1 (1.45e-3) -	5.0617e-1 (1.14e-2) -	1.1083e-1 (3.59e-2) -	2.5380e-1 (6.79e-2) -	5.5311e-1 (3.56e-3)+
	5	8.0125e-1 (2.48e-3)	7.9709e-1 (2.12e-3) -	6.7897e-1 (1.31e-2) -	1.1098e-1 (3.59e-2) -	3.2526e-1 (8.19e-2) -	7.6079e-1 (1.89e-2) -
WFG4	8	9.2425e-1 (1.85e-3)	8.9809e-1 (4.40e-3) -	7.6893e-1 (8.54e-3) -	1.0299e-1 (2.95e-2) -	3.3976e-1 (3.66e-2) -	7.2611e-1 (3.10e-2) -
	10	8.5383e-1 (1.19e-2)	9.5142e-1 (3.16e-3)+	8.0994e-1 (9.85e-3) -	1.0318e-1 (3.00e-2) -	2.8472e-1 (5.96e-2) -	7.5754e-1 (3.61e-2) -
	15	9.7684e-1 (4.87e-3)	9.7462e-1 (4.19e-3)≈	8.7663e-1 (8.54e-3) -	1.4727e-1 (5.24e-2) -	3.7286e-1 (7.58e-2) -	7.1136e-1 (2.37e-2) -
	3	5.1768e-1 (8.59e-4)	5.1604e-1 (4.28e-4)≈	4.7039e-1 (1.62e-2) -	3.8626e-1 (1.46e-1) -	2.2308e-1 (5.64e-2) -	5.1735e-1 (1.18e-3)≈
	5	7.5918e-1 (2.07e-2)	7.5864e-1 (1.15e-3)≈	6.4878e-1 (1.22e-2) -	5.3804e-1 (1.97e-1) -	2.7780e-1 (3.11e-2) -	7.5893e-1 (2.39e-3)≈
WFG5	8	8.7004e-1 (6.69e-4)	8.5813e-1 (1.50e-3) -	7.2225e-1 (1.33e-2) -	3.2925e-1 (2.61e-1) -	2.6617e-1 (2.20e-2) -	8.3182e-1 (1.15e-2) -
	10	8.1664e-1 (4.37e-3)	9.0122e-1 (7.96e-4)+	7.6274e-1 (1.22e-2) -	1.8254e-1 (2.33e-1) -	2.3130e-1 (1.20e-2) -	8.5240e-1 (1.54e-2)+
	15	8.1045e-1 (1.20e-2)	9.1589e-1 (3.78e-4)+	8.2086e-1 (1.08e-2)+	2.9497e-1 (3.24e-1) -	2.9092e-1 (2.61e-2) -	7.4595e-1 (1.38e-2) -
	3	4.9587e-1 (5.61e-3)	4.9539e-1 (1.25e-2) -	4.2868e-1 (2.54e-2) -	1.6952e-1 (1.28e-1) -	1.7019e-1 (2.83e-2) -	5.0387e-1 (1.12e-2)+
	5	7.3974e-1 (1.14e-2)	7.3721e-1 (1.23e-2) -	5.8526e-1 (2.03e-2) -	1.6244e-1 (6.47e-2) -	2.7465e-1 (2.43e-2) -	7.3192e-1 (1.23e-2) -
WFG6	8	8.5282e-1 (1.29e-2)	8.3284e-1 (1.26e-2) -	6.3253e-1 (2.21e-2) -	3.2803e-1 (1.98e-1) -	2.6338e-1 (2.28e-2) -	7.9509e-1 (2.08e-2) -
	10	8.6057e-1 (1.54e-2)	8.7281e-1 (2.08e-2)+	6.6452e-1 (2.20e-2) -	2.9536e-1 (1.86e-1) -	2.0214e-1 (2.09e-2) -	8.2436e-1 (1.82e-2) -
	15	8.6283e-1 (2.43e-2)	7.3981e-1 (5.79e-2) -	6.9891e-1 (2.29e-2) -	4.3803e-1 (2.37e-1) -	3.2097e-1 (5.56e-2) -	7.2783e-1 (1.65e-2) -
	3	5.5421e-1 (8.78e-4)	5.4987e-1 (1.06e-3) -	4.6335e-1 (1.43e-2) -	2.0889e-1 (8.66e-2) -	2.5496e-1 (1.96e-2) -	5.5625e-1 (9.27e-4)≈
	5	8.0197e-1 (9.05e-4)	8.0261e-1 (9.67e-4)≈	6.2798e-1 (1.52e-2) -	2.4814e-1 (2.33e-2) -	3.3457e-1 (2.76e-2) -	8.0033e-1 (2.66e-3) -
WFG7	8	9.2069e-1 (2.08e-3)	8.9611e-1 (5.73e-3) -	7.3382e-1 (1.20e-2) -	2.0516e-1 (6.84e-2) -	3.6428e-1 (3.89e-2) -	8.3268e-1 (2.70e-2) -
	10	9.5349e-1 (1.82e-3)	9.5279e-1 (3.03e-3)≈	7.7837e-1 (1.54e-2) -	2.3119e-1 (6.67e-2) -	2.7336e-1 (3.78e-2) -	8.6208e-1 (2.12e-2) -
	15	9.8301e-1 (1.83e-3)	9.8272e-1 (3.46e-3)≈	8.7134e-1 (9.51e-3) -	2.2080e-1 (6.29e-2) -	3.7847e-1 (3.82e-2) -	8.0735e-1 (1.53e-2) -
	3	4.6675e-1 (1.79e-3)	4.6471e-1 (2.02e-3)+	3.8878e-1 (1.07e-2) -	1.8511e-2 (3.56e-2) -	1.7923e-1 (1.20e-2) -	4.6951e-1 (5.07e-3)+
	5	6.7960e-1 (2.60e-3)	6.8897e-1 (2.11e-3)+	5.3629e-1 (1.06e-2) -	5.6656e-2 (5.38e-2) -	2.7749e-1 (2.23e-2) -	6.8835e-1 (2.65e-3)+
WFG8	8	7.5241e-1 (6.26e-2)	7.3059e-1 (9.06e-2) -	5.3307e-1 (5.61e-2) -	2.7305e-1 (1.85e-1) -	2.4934e-1 (2.91e-2) -	7.4475e-1 (5.66e-2) -
	10	7.6046e-1 (6.38e-2)	7.5879e-1 (7.32e-2)≈	5.9881e-1 (6.73e-2) -	2.2867e-1 (6.92e-2) -	1.5329e-1 (2.76e-2) -	8.1729e-1 (1.95e-2)+
	15	8.9911e-1 (5.60e-2)	7.2683e-1 (1.19e-1) -	6.3369e-1 (1.02e-1) -	2.0408e-1 (6.80e-2) -	3.5057e-1 (2.64e-2) -	8.4212e-1 (1.44e-2) -
	3	5.3465e-1 (3.08e-3)	5.3048e-1 (3.23e-3) -	7.5809e-1 (3.61e-3) -	1.9175e-1 (3.44e-2) -	1.7827e-1 (2.56e-2) -	5.2952e-1 (3.88e-3) -
	5	7.6197e-1 (4.30e-3)	7.5402e-1 (5.66e-3) -	6.4458e-1 (1.45e-2) -	3.2094e-1 (1.80e-1) -	2.7032e-1 (1.80e-2) -	7.5325e-1 (4.23e-3) -
WFG9	8	7.6210e-1 (1.13e-2)	8.4113e-1 (1.02e-2)+	7.1503e-1 (4.13e-2) -	3.6965e-1 (2.56e-1) -	2.8716e-1 (2.41e-2) -	8.1043e-1 (1.70e-2)+
	10	8.0049e-1 (1.44e-2)	8.8436e-1 (1.08e-2)+	7.5785e-1 (2.75e-2) -	3.1435e-1 (2.65e-1) -	2.1301e-1 (3.15e-2) -	8.4467e-1 (1.25e-2)+
	15	8.7354e-1 (1.50e-2)	8.5204e-1 (6.18e-2) -	8.0247e-1 (3.93e-2) -	2.2858e-1 (1.65e-1) -	3.6756e-1 (1.92e-2) -	7.4690e-1 (1.04e-2) -
+/ − / ≈			16/17/12	9/33/3	5/38/2	0/43/2	17/21/7

near these segments into it. WFG4-9 have the same concave PFs but different problem properties. RVEA-DCES showed a clear advantage over most competitors on WFG4-9, indicating its comprehensive and prominent ability dealing with concave MaOPs. It is worth noting that the performance of RVEA on WFG4-9 was also competitive. This is owing to the consistency between the distribution of reference vectors and the shape of concave PFs. On the basis of that,

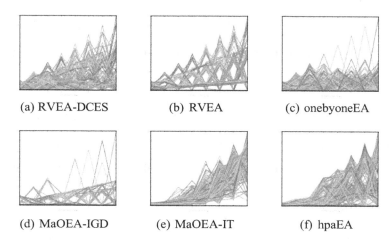

(a) RVEA-DCES   (b) RVEA   (c) onebyoneEA

(d) MaOEA-IGD   (e) MaOEA-IT   (f) hpaEA

Fig. 3. Parallel coordinates of final non-dominated solution sets obtained by each algorithm on 8-objective WFG4.

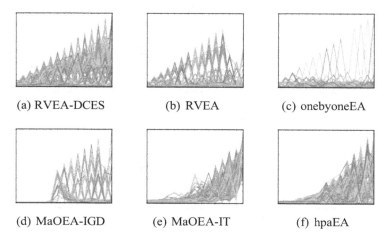

(a) RVEA-DCES   (b) RVEA   (c) onebyoneEA

(d) MaOEA-IGD   (e) MaOEA-IT   (f) hpaEA

Fig. 4. Parallel coordinates of final non-dominated solution sets obtained by each algorithm on 15-objective WFG8.

diversity and convergence enhancement strategies improve the performance of RVEA-CDES in most cases.

WFG4 is a multimodal problem which may trap MOEAs into local optima. Its 8-objective parallel coordinates are presented in Fig. 3. RVEA-DCES, RVEA and onebyoeEA were capable of covering the whole range of PF while MaOEA-IGD failed to cover the fourth objective. Besides, the range of approximate PFs of MaOEA-IT and hpaEA were smaller than ture PFs. The final non-dominated solution set of RVEA-DCES distributed denser compared to other rivals especially in the middle part of PF, thus it obtained the best HV on this test instance. This can be attributed to the adaptive sparse region filling strategy which enables exploration of sparse regions. Meanwhile, solutions retained by convergence-only selection strategy also marked the distribution denser.

Similar observation occured on 15-objective WFG8 which has non-separable property. Figure 4 shows that only RVEA-DCES covered the whole PF well. RVEA failed to cover the thirteenth to fifteenth objective while onebyoneEA failed to cover the eleventh objective, not to mention the terrible distribution of MaOEA-IGD, MaOEA-IT and hpaEA. Solution distributions of most compared algorithms degenerated evidently as the dimensionality increased, but RVEA-DCES was not affected obviously. This indicates that RVEA-DCES can maintain the balance between convergence and diversity well in many-objective scenarios.

5   Conclusion

To solve or alleviate the defects reference vector guided selection confronts, this paper proposes a reference vector guided evolutionary algorithm with diversity and convergence enhancement strategies for many-objective optimization. To fill any vacancies that may exist in final elite population, two diversity and convergence enhancement strategies together with an elite retention strategy are conducted after reference vector guided selection to obtain two elite population EP_1 and EP_2 and add solutions of them into final elite population methodically, which improves the diversity and convergence of it simultaneously.

To evaluate the performance of RVEA-DCES, comparative studies are conducted on WFG test suite up to 15 objectives. Experimental results against five state-of-the-art MaOEAs demonstrated the best overall performance of RVEA-DCES, making it a preferable choice for many-objective optimization.

Handling MOPs with irregular PF is a common problem for decomposition-based evolutionary algorithms. A widely used method is adaptive reference vector adjustment. In the future, we will add adaptive reference vector adjustment strategy into RVEA-DCES to enhance its ability handling MOPs with irregular PF.

References

1. Farina, M., Amato, P.: On the optimal solution definition for many-criteria optimization problems. In: Annual Meeting of the North American Fuzzy Information Processing Society Proceedings. NAFIPS-FLINT 2002 (Cat. No. 02TH8622), pp. 233–238. IEEE (2002)
2. Ishibuchi, H., Tsukamoto, N., Nojima, Y.: Evolutionary many-objective optimization: a short review. In: IEEE Congress on Evolutionary Computation (IEEE World Congress on Computational Intelligence), pp. 2419–2426. IEEE (2008)
3. Zhang, Q., Li, H.: MOEA/D: a multiobjective evolutionary algorithm based on decomposition. IEEE Trans. Evol. Comput. **11**(6), 712–731 (2007)
4. Xu, Y., Zhang, H., Zeng, X., Nojima, Y.: An adaptive convergence enhanced evolutionary algorithm for many-objective optimization problems. Swarm Evol. Comput. **75**, 101180 (2022)

5. Liu, H.L., Gu, F., Zhang, Q.: Decomposition of a multiobjective optimization problem into a number of simple multiobjective subproblems. IEEE Trans. Evol. Comput. **18**(3), 450–455 (2013)
6. Bai, H., Zheng, J., Yu, G., Yang, S., Zou, J.: A pareto-based many-objective evolutionary algorithm using space partitioning selection and angle-based truncation. Inf. Sci. **478**, 186–207 (2019)
7. Qiu, W., Zhu, J., Wu, G., Fan, M., Suganthan, P.N.: Evolutionary many-objective algorithm based on fractional dominance relation and improved objective space decomposition strategy. Swarm Evol. Comput. **60**, 100776 (2021)
8. Cheng, R., Jin, Y., Olhofer, M., Sendhoff, B.: A reference vector guided evolutionary algorithm for many-objective optimization. IEEE Trans. Evol. Comput. **20**(5), 773–791 (2016)
9. Liu, Q., Jin, Y., Heiderich, M., Rodemann, T.: Adaptation of reference vectors for evolutionary many-objective optimization of problems with irregular pareto fronts. In: 2019 IEEE Congress on Evolutionary Computation (CEC), pp. 1726–1733 (2019)
10. Lin, J., Zheng, S., Long, Y.: Improved reference vector guided differential evolution algorithm for many-objective optimization. In: Proceedings of the 2020 5th International Conference on Mathematics and Artificial Intelligence, pp. 43–49 (2020)
11. Li, F., Shang, Z., Shen, H., Liu, Y., Huang, P.Q.: Combining modified inverted generational distance indicator with reference-vector-guided selection for many-objective optimization. Appl. Intell. **53**(10), 12149–12162 (2023)
12. Qiu, W., Zhu, J., Yu, H., Fan, M., Huo, L.: An adaptive reference vector adjustment strategy and improved angle-penalized value method for RVEA. Complexity **2021** (2021)
13. Blank, J., Deb, K., Dhebar, Y., Bandaru, S., Seada, H.: Generating well-spaced points on a unit simplex for evolutionary many-objective optimization. IEEE Trans. Evol. Comput. **25**(1), 48–60 (2020)
14. Deb, K., Agrawal, R.B., et al.: Simulated binary crossover for continuous search space. Complex Syst. **9**(2), 115–148 (1995)
15. Deb, K., Goyal, M., et al.: A combined genetic adaptive search (GeneAS) for engineering design. Comput. Sci. Inform. **26**, 30–45 (1996)
16. Liu, Y., Gong, D., Sun, J., Jin, Y.: A many-objective evolutionary algorithm using a one-by-one selection strategy. IEEE Trans. Cybern. **47**(9), 2689–2702 (2017)
17. Sun, Y., Yen, G.G., Yi, Z.: IGD indicator-based evolutionary algorithm for many-objective optimization problems. IEEE Trans. Evol. Comput. **23**(2), 173–187 (2018)
18. Sun, Y., Xue, B., Zhang, M., Yen, G.G.: A new two-stage evolutionary algorithm for many-objective optimization. IEEE Trans. Evol. Comput. **23**(5), 748–761 (2018)
19. Chen, H., Tian, Y., Pedrycz, W., Wu, G., Wang, R., Wang, L.: Hyperplane assisted evolutionary algorithm for many-objective optimization problems. IEEE Trans. Cybern. **50**(7), 3367–3380 (2019)
20. Huband, S., Hingston, P., Barone, L., While, L.: A review of multiobjective test problems and a scalable test problem toolkit. IEEE Trans. Evol. Comput. **10**(5), 477–506 (2006)
21. While, L., Hingston, P., Barone, L., Huband, S.: A faster algorithm for calculating hypervolume. IEEE Trans. Evol. Comput. **10**(1), 29–38 (2006)

Research on Mine Emergency Evacuation Scheme Based on Dynamic Multi-objective Evolutionary Algorithm

Furong Jing, Hui Liu[(✉)], and Yanhui Zang

Electronic Information School, Foshan Polytechnic, Foshan 528137, China
jephirus@fspt.edu.cn

Abstract. The production safety in coal mines has always been highly valued by the state and society, and the emergency evacuation plan for coal mines is of great significance for reducing casualties and ensuring life safety. Therefore, this paper first establishes a mine emergency evacuation model, and then proposes a dynamic single-objective evolutionary algorithm to optimize the model. This algorithm can adapt to the optimization of emergency evacuation plans in dynamic environments such as secondary disasters and equipment faults, and has important application value.

Keywords: production safety · coal mines · dynamic single-objective evolutionary algorithm

1 Introduction

Safe production is the guarantee of the production activities in various industries. In many production safety accidents, the number of casualties caused by the direct cause of the accident will not be too large, but after the accident which brought indirect causes of death such as delayed evacuation, transfer, treatment, oxygen and food shortages, are far greater. Therefore, it is very important to do a good job of emergency evacuation plan and evacuation scheduling before the accident. Coal mining industry is a very risky industry, once an emergency occurs, it is necessary to use the limited transportation resources in the first time to evacuate the underground staff, otherwise the underground personnel will face explosion, collapse, hypoxia and other dangers.

To this end, scholars have also done a lot of research on emergency evacuation. Rahman et al. [1] studied the effect of passenger subjective awareness on railway emergency evacuation. Soltanzadeh et al. [2] studied the emergency exit setting and optimization of high-rise buildings, and Mao et al. [3] also simulated and optimized the mixed evacuation model in the emergency state of high-rise buildings. Jiang et al. [4] used the ant colony algorithm to study the emergency evacuation plan of commercial streets. Wang et al. [5] used Anylogic to simulate the emergency evacuation model of the subway, and analyzed the influence of personality characteristics on the evacuation model. Wang et al. [6] studied the effects of natural light on road-finding behavior during emergency

© The Author(s), under exclusive license to Springer Nature Singapore Pte Ltd. 2024
K. Li and Y. Liu (Eds.): ISICA 2023, CCIS 2146, pp. 88–100, 2024.
https://doi.org/10.1007/978-981-97-4393-3_8

evacuation of underground space. Yang et al. [7] proposed an urban flood emergency evacuation model based on Agent, exploring the chain effect of flood on human behavior and emergency evacuation, and conducted a simulation study with the "7.20" rainstorm event in Zhengzhou, Henan province. Zhu et al. [8] studied the emergency evacuation model through the simulation experiment, and showed that the dormitory interpersonal relationship has an important influence on the emergency evacuation behavior of college students. Vecliuc et al. [9] have studied the effects of synergistic behavior on emergency evacuation. Hu et al. [10] studied the effects of obstacles and luggage on emergency evacuation of bus carriage passengers. Adjiski et al. [11] studied the downhole fire emergency evacuation process using the Monte Carlo method with uncertain parameters.

However, there are still few research on emergency evacuation for coal mining, mainly due to the complex modeling of mines and dynamic problems such as secondary disasters and equipment failures in the optimization process. Therefore, this paper takes transport as the research object to simplify mine evacuation model, and then proposes a dynamic multi-objective evolution algorithm to optimize the mine evacuation model to adapt to the dynamic change of the mine evacuation environment. The research of this paper provides a new idea for the study of mine evacuation and has positive significance to ensure the safe production of mines.

2 Mine Emergency Evacuation Method Based on Dynamic Multi-target Evolutionary Algorithm

In this paper, the mine emergency evacuation structure model is constructed by combining the mainstream mine structure, and then the mine evacuation optimization model is constructed based on the structure model; and then a dynamic single target evolution algorithm considering the cost of the state migration is proposed to answer emergency situations in emergency evacuation; and finally the proposed algorithm is applied to the mine emergency evacuation model.

2.1 Mine Emergency Evacuation Model

After a lot of research, we drew the structure model map of the mainstream mine, as shown in Fig. 1. The mine mainly includes transportation roadway, shaft, coal seam operation area and other areas, among which, the shaft is divided into main shaft and auxiliary shaft, they are equipped with elevators, and at the bottom of the two Wells are also equipped with a underground parking lot. In the process of emergency evacuation, the personnel in the operation area first take the mine car to the underground parking lot through the roadway, and then take the elevator to rise to the ground. In the optimization of emergency evacuation scheme, the scheduling of vehicle capacity and elevator capacity is the key to improve the evacuation efficiency, which is a complex combination optimization problem due to the numerous vehicles and operation points.

The following in-depth analysis of the mine emergency evacuation optimization model is conducted from four aspects: optimization target, decision variable, objective function and constraint condition.

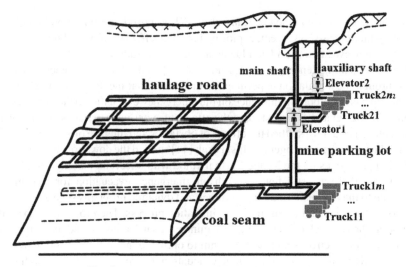

Fig. 1. Structure diagram of the mainstream mine

Optimization Objectives. In the optimization model of mine emergency evacuation, time is life. How to evacuate all the people underground to the ground in the shortest time is our only goal, and other factors such as economic cost and labor cost are not factors to consider. Therefore, the only optimization goal of this problem is the minimum total time for emergency evacuation.

Decision Variables and Their Coding. In the optimization scheduling, the schedulable resources can be used as the decision variable for this optimization problem. In the mine emergency evacuation optimization problem, the vehicle is the most important dispatching resource, so we take the corresponding vehicle in the operation area as the decision variable of the optimization problem. The vehicle involves picking up people at the work point and transporting them to the elevator in the underground parking lot. Therefore, we divided the vehicles into two groups according to the elevator of the destination, and then assigned all the operation points to each vehicle. The encoding scheme of decision variables is shown in Fig. 2. Among them, each chromosome consists of n operational segments, and each segment consisting of vehicle number and elevator number.

operation area 1		operation area 2		···	operation area n	
truck no.	elevator no.	truck no.	elevator no.	...	truck no.	elevator no.

Fig. 2. Decision variable coding scheme of mine emergency evacuation model

However, the only evacuation plan can not be determined only based on the vehicle number and elevator number of each operation point, because it also involves the transportation sequence and route splicing of each operation point. To this end, we used the DE algorithm to optimize the transport route for each vehicle, and subsequently calculated the delivery time of each vehicle using the DE-optimized route distance.

Objective Function. The above analysis of the optimization goal of mine emergency evacuation is the minimum total time of evacuation, but the calculation of the whole evacuation process is a complex process, many papers use RePast, Netlogo and other group behavior simulation software for simulation calculation. This paper presents a new method of directly calculating the emergency evacuation time to improve the computational efficiency.

We have noticed that the emergency evacuation time is mainly spent in the vehicle delivery time and the elevator delivery time, but the two also overlap, so the calculation process is quite complicated. To this end, we propose the calculation model of mine emergency evacuation time. As shown in Fig. 3, there are two elevators in the figure, and the vehicles have two groups according to the elevator, so the whole evacuation process can be divided into two groups according to the elevator, and the evacuation time of the whole mine is equal to the evacuation time with longer time. We took the second group as an example to analyze the evacuation time calculation of each group. After the emergency evacuation order was obtained, all vehicles started evacuation from the parking lot at the same time. Due to the different tasks of each vehicle, their delivery time was different, so we took the vehicle delivery time with the longest t_{T2} time in the vehicle as the vehicle delivery time. For the elevator, we take the return time of the first returning vehicle as the start time of the elevator, while the waiting time of the elevator t_{n2} is defined. Since the vehicle capacity is much greater than the elevator, we assume that the elevator is running at full load until all personnel are completely evacuated, and the running time of the elevator is. The total evacuation time $t_{n2} + t_{E2}$ of the second group was. The waiting time of the elevator t_{n2} is equal to the time of the fastest return vehicle for the first time, and the elevator delivery time t_{E2} is equal to the number of all working points in the group divided by the total capacity of the elevator (round up to an integer) and multiplied by the single round trip time of the elevator. Then the total evacuation time calculation formula of the model is shown in Eq. (1):

$$
\begin{cases}
T(X) = \max[t_{E1} + t_{n1}, t_{E2} + t_{n2}] \\
t_{E1} = (\sum_{j=1}^{N_p} p_j|_{x_{2j}==1})/c_1 \cdot t_{e1} \\
t_{E2} = (\sum_{j=1}^{N_p} p_j|_{x_{2j}==2})/c_1 \cdot t_{e2} \\
t_{n1} = \min(S_{11,1}/v, S_{12,1}/v, ..., S_{1n_1,1}/v) \\
t_{n2} = \min(S_{21,1}/v, S_{22,1}/v, ..., S_{2n_2,1}/v)
\end{cases}
\tag{1}
$$

Among them, n_1, n_2 are the number of vehicles in the first group and second groups, respectively, N_{S1k} is the number of work points assigned to the k-th vehicle in the first group, $S_{1k,i}$ is the round trip distance of the i-th work point of the k first vehicle in the first group, N_{S2k} is the number of operation points allocated to the k-th vehicle in the second group, $S_{2k,i}$ is the round trip distance of the i-th work point of the k-th vehicle in the

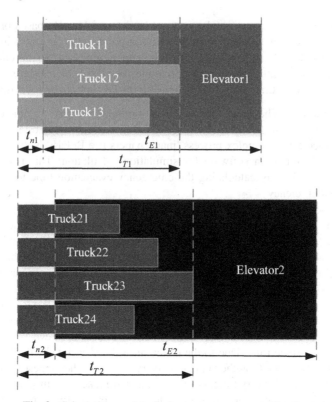

Fig. 3. Calculation model of mine emergency evacuation time

second group, v is the average speed of the vehicle; t_{E1} and t_{E2} are the operating hours of Elevator 1 and Elevator 2, respectively, c_1 and c_2 are the capacity of two elevators (maximum number of passengers delivered once), t_{e1} and t_{e2} are the time required by the two elevators for a single round trip; t_{n1} and t_{n2} are the waiting times for Elevators 1 and Elevator 2, respectively.

Restrictive Condition. During an emergency mine evacuation, many limitations must be met, including vehicle occupancy constraints, vehicle number constraints, elevator number constraints, elevator occupancy constraints, and so on. The restrictions are arranged as:

$$\begin{cases} P_{T,k} \leq C_T, P_{T,k} \in N \\ +n_1 + n_2 \leq N_T, n_1 \in N+, n_2 \in N \\ +P_{E1,k} \leq C_{E1}, P_{E1,k} \in N \\ +P_{E2,k} \leq C_{E2}, P_{E2,k} \in N \end{cases} \tag{2}$$

2.2 Dynamic Multi-objective Evolutionary Algorithm Considering Migration Costs

Adjustment of Optimization Objectives Considering Secondary Disasters and Equipment Failures. In the mine emergency evacuation, often accompanied by secondary disasters or equipment failure, which brings more uncertainty to the evacuation process, such as channel blockage, vehicle failure, or elevator failure will affect the evacuation efficiency. Therefore, it is of great practical significance to make appropriate adjustments to improve evacuation efficiency based on the existing evacuation program when secondary disasters or equipment failures occur. We view the different evacuation scenarios as different system states and evaluate the variability of the evacuation scenarios using the Euclidean distance between the two evacuation scenarios, which is also denoted as the migration cost of the system。 Based on this, this paper proposes a dynamic multi-objective evolutionary algorithm considering migration cost (Dynamic multi-objective evolutionary algorithm considering migration costs, DMOEA-CMC), The algorithm takes the evacuation time of the current evacuation scenario and the relocation cost of the two scenarios before and after as the optimization objectives, and the two optimization objectives are shown in Eq. (3):

$$\text{minimize } F(X) = (T(X), D(X, X_0))^T \tag{3}$$

Among others, X_0 is an evacuation program before a secondary disaster or equipment failure occurs, X is an existing evacuation scenario, $D(X, X_0)$ is the Euclidean distance between the two evacuation scenarios:

$$D(X, X_0) = \sqrt{\sum_{i=1}^{D} (X_i - X_{0,i})^2} \tag{4}$$

Algorithmic Design. The optimization of mine emergency evacuation considering secondary disasters and equipment failures requires the use of multi-objective evolutionary algorithms, and in this paper, we choose the NSGA-II algorithm framework based on fast non-dominated sorting as the basis for algorithm design. For this scenario, we also need to consider constraint handling and dynamic environment detection and response mechanisms. For constraints processing, this paper proposes a progressive penalty function method of punishment, that is, adding a penalty function and a penalty coefficient adjustment function to the objective function, as shown in Eq. (5):

$$F(X) = (T(X), D(X, X_0))^T + r(t)P(X) \tag{5}$$

Among others, $P(X)$ are the penalty function, $r(t)$ the penalty coefficient adjustment function, and $0 < r(t+1) < r(t) < ... < r(1)$. In this way, the algorithm starts with no immediate elimination of individuals that violate the constraints, which enriches the diversity of the population as evolution proceeds, the penalty coefficient of $r(t)$ is increasing and the algorithm starts to phase out the individuals that violate the constraints until finally, all individuals satisfy the constraints.

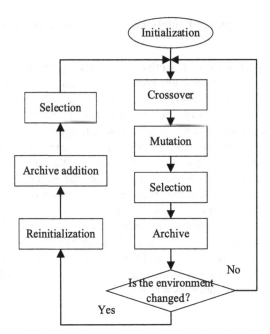

Fig. 4. Flowchart of dynamic multi-objective evolutionary algorithm for optimal scheduling of secondary disasters and equipment failures

For the dynamic environment problems caused by secondary disasters and equipment failures, since there are already multiple secondary disaster detection devices in the mine and equipment failures can be fed back in time, there is no need to do dynamic environment detection at the algorithmic layer, and the algorithmic level only needs to design the dynamic environment response mechanism. This paper adopts the dynamic response mechanism combining reset and archiving, the algorithm flow is shown in Fig. 4, when a secondary disaster or equipment failure brings environmental changes, on the one hand, the algorithm population should be re-initialized to enrich the algorithm's population diversity in the new environment, on the other hand, the algorithm will also add some of the archived elite individuals to the new population to improve the algorithm's convergence speed, and then select a certain number of individuals from them to be used as the new population of the evolutionary algorithm. The method enriches the diversity of the algorithm population while retaining the useful experience of the pre-search, and thus has a faster response speed and better global optimization capability for dynamic environments.

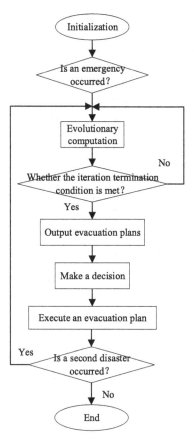

Fig. 5. Flow chart of mine emergency evacuation model optimized by a dynamic multi-objective evolutionary algorithm

2.3 Optimizing Mine Emergency Evacuation Models Using Dynamic Multi-objective Evolutionary Algorithms

The workflow of optimizing the mine emergency evacuation model using the DMOEA-CMC algorithm is shown in Fig. 5. When an emergency event is detected, the system enables the evolutionary algorithm to start optimizing the emergency evacuation scheme, and after a certain number of iterations and meeting the end conditions of the iteration, the algorithm outputs a set of evacuation scheme solutions, and the system then makes a decision based on the user's preference and selects one of the solutions from the solution set as the execution of the evacuation scheme. During the execution of the evacuation plan, the system will also continue to detect whether a secondary disaster or equipment failure occurs, and once a secondary disaster or equipment failure occurs, the system will again call the dynamic multi-objective evolutionary algorithm to optimize a new evacuation plan to adapt to the environmental changes brought about by the secondary disaster or equipment failure.

3 Simulation Experiment and Analysis

To test the effectiveness of the proposed DMOEA-CMC in the mine emergency evacuation optimization model, we carried out simulation tests in the MATLAB simulation platform. We choose two classical multi-objective evolutionary algorithms, MOEA/D and NSGA-II, and two constrained multi-objective evolutionary algorithms, PPS and RVEA, as the comparison algorithms, with the population size set to 100, and the maximum number of adaptive value evaluations set to 10,000. The simulation experiments are set up in two groups, namely, the secondary disaster impact experiments and the equipment failure impact experiments, and each group is set up with a different time of disasters or failures, to observe the effect of disaster or failure on emergency evacuation optimization modeling in mines. To facilitate the observation of the impact of the disaster or failure time on the emergency evacuation time.

3.1 Emergency Evacuation Optimization Experiment with Secondary Hazards

In this set of experiments, we first simulated the evacuation times of the five optimization algorithms corresponding to the evacuation scenarios without secondary disasters, and then simulated and tested six sets of tests in which the secondary disaster occurred 20, 40, 60, 80, 100, and 120 min after the primary disaster, and the recorded data are shown in Fig. 6.

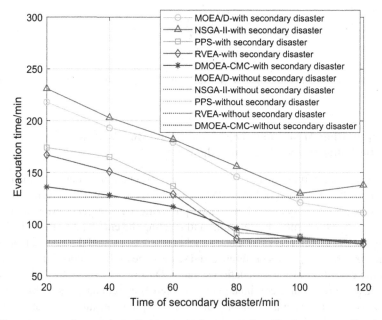

Fig. 6. Time impact of secondary disasters on the evacuation plans corresponding to six algorithms.

Firstly, it can be seen from the graph that the earlier the secondary disaster occurs, the longer the evacuation time required for all algorithms. This also reflects that the earlier the secondary disaster occurs, the greater the impact on the evacuation plan. Secondly, among the five algorithms, the optimization and adjustment effect of DMOEA CMC for early secondary disasters is significantly better than the other four comparative algorithms. For secondary disasters that occur in the later stage, its optimization effect is relatively similar to PPS and RVEA. Finally, compare the situation where no secondary disasters occurred, Secondary disasters that occurred 80 min ago have a significant impact on evacuation time.

To further observe the convergence performance of DMOEA-CMC in the impact of secondary disasters, we also recorded the convergence curves of DMOEA-CMC and five comparative algorithms when the second disaster occurred at 40 min. As shown in Fig. 7, it can be seen that the six algorithms before the occurrence of the second disaster can converge quickly. But after the second disaster, DMOEA – CMC can quickly converge again and achieve shorter evacuation adjustment plans. This indicates that the migration cost penalty strategy and dynamic response strategy of DMOEA CMC have played a positive role.

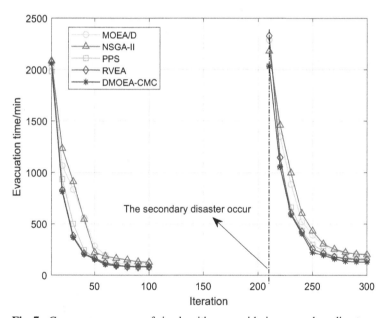

Fig. 7. Convergence curves of six algorithms considering secondary disasters.

3.2 Emergency Evacuation Optimization Experiment with Equipment Failure

In this experiment, we investigate the impact of equipment failures on emergency evacuation plans corresponding to various algorithms. Firstly, we simulated the evacuation time of five optimization algorithms corresponding to evacuation plans without equipment failure. Then we simulated and tested six sets of tests with equipment failures occurring 20, 40, 60, 80, 100, and 120 min after a disaster. The recorded data is shown in Fig. 8.

From the figure, it can be seen that firstly, the earlier the equipment failure occurs, the longer the evacuation time required for all algorithms. That is to say, the earlier the equipment failure occurs, the greater the impact on the evacuation plan. Secondly, the optimization and adjustment effect of DMOEA-CMC for early device failures is significantly better than the other four comparative algorithms. Its optimization effect is similar to PPS and RVEA for equipment failures that occur in the later stage, but significantly better than MOEA/D and NSGA-II. Finally, compared to the situation where no equipment failure occurred, a secondary disaster that occurred 80 min ago had a significant impact on evacuation time.

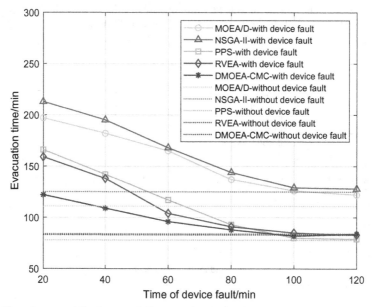

Fig. 8. Time Impact of Equipment Failure on the Evacuation Schemes Corresponding to Six Algorithms

To further observe the convergence performance of DMOEA-CMC under device failure, we also recorded the convergence curves of DMOEA-CMC and five comparative algorithms when the equipment failure occurred at 40 min. As shown in Fig. 9, it can be seen from the figure that the six algorithms before the equipment failure can converge quickly. However, after a device failure occurs, DMOEA-CMC can quickly converge again and achieve faster evacuation adjustment plans. Once again, it proves that the migration cost penalty strategy and dynamic response strategy of DMOEA CMC have played a positive role.

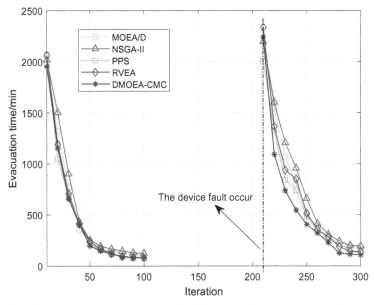

Fig. 9. Convergence curves of six algorithms considering equipment failures

4 Conclusion

This paper proposes a dynamic multi-objective evolutionary algorithm considering state transition costs to address the problems of secondary disasters and equipment failures in emergency evacuation of mines. This algorithm can not only quickly optimize the emergency evacuation plan for mines, but also optimize and adjust new evacuation plans again in the event of secondary disasters or equipment failures. In simulation experiments, the proposed DMOEA-CMC has faster convergence speed and achieves shorter evacuation plans than the four comparison algorithms in the event of secondary disasters or equipment failures. This study has good reference value for the practical application of emergency evacuation in mines.

Acknowledgement. This work is supported by the Guangdong Provincial Department of Education Project with the Grant No. 2022KTSCX320, the Foshan Polytechnic School Level Project with the Grant No. KY2022G04, and also supported by the Internet of Things Application Technology Innovation Team with the Grant No. 2021KCXTD074.

References

1. Rahman, N.A., Johari, M.S.M., Dias, C.: Exploratory study on self-awareness and self-preparedness of Malaysian rail passengers for emergency evacuations. Transp. Eng. **7**, 1–7 (2022)
2. Soltanzadeh, A., Mazaherian, H., Heidari, S., et al.: Placing egress components and smoke shafts in the core structure of residential high-rise buildings for emergency evacuation. Space Ontol. Int. J. **10**(1), 27–45 (2021)

3. Mao, H., Zhu, L.: Simulation and optimization of hybrid evacuation model for high-rise building under emergency condition. Int. J. Model. Simul. **14**(3), 1–14 (2023)
4. Xue, J., Huan, Y., Peihong, Z.: Emergency evacuation decision optimization for commercial pedestrian streets based on ant colony algorithm. China Saf. Sci. J. **31**(10), 144–151 (2021)
5. Wang, H., Xu, T., Li, F.: A novel emergency evacuation model of subway station passengers considering personality traits. Sustainability **13**(18), 1–15 (2021)
6. Wang, D., Liang, S., Chen, B., et al.: Investigation on the impacts of natural lighting on occupants' wayfinding behavior during emergency evacuation in underground space. Energy Build. **255**, 1–13 (2022)
7. Yang, Y., Yin, J., Wang, D., et al.: ABM-based emergency evacuation modelling during urban pluvial floods: a "7.20" pluvial flood event study in Zhengzhou, Henan Province. Chine. Sci. Earth Sci. Engl. Ed. **66**(2), 282–291 (2023)
8. Zhu, X., Xiangfei, L.I., Feng, X.: Simulation study on emergency evacuation behavior of college students in considering the influence of dormitory interpersonal relationship. J. Syst. Sci. Inf. Technol. **11**(1), 124–138 (2023)
9. Vecliuc, D., Leon, F., Bădică, C.: The effect of collaborative behaviors in emergency evacuation. Simul. Model. Pract. Theory **118**, 102554–102575 (2022).
10. Xu, H.H., Guo, R.Y.: Passenger emergency evacuation from bus carriage: results from realistic data and modeling simulations. J. Manag. Sci. **7**(4), 530–549 (2022)
11. Adjiski, V., Zubíek, V., Despodov, Z.: Monte Carlo simulation of uncertain parameters to evaluate the evacuation process in an underground mine fire emergency. J. South. Afr. Inst. Min. Metall. **119**(11), 907–917 (2019)

An Adaptive Dynamic Parameter
Multi-objective Optimization Algorithm

Yu Lai[1]([✉]) and Lanlan Kang[1,2]([✉])

[1] School of Information Engineering, JiangXi University of Science and Technology,
Ganzhou 341000, Jiangxi, China
lyckyt@outlook.com
[2] GanNan University of Science and Technology, Ganzhou 341000, Jiangxi, China
victorykll@163.com

Abstract. In order to improve the convergence speed of the multi-objective optimization algorithm while obtaining good distribution and diversity, an adaptive dynamic parameter multi-objective optimization algorithm is proposed (ADPMO). The new algorithm consists of three main strategies. Firstly, a new mutation method based on individual competition mechanism integrated with k-means clustering is proposed, which updates the velocity and position information of the individuals that have failed to compete in each cluster, for improve the diversity of the solution set and avoid premature convergence. Secondly, an adaptively dynamical parameters strategy is proposed. In the process of speed updating, parameters that change dynamically with the number of population iterations, for enhance the convergence speed and convergence of the algorithm. At last, a cross-mutation strategy is introduced, for making the population out of collapse state and increase the diversity of the optimal solution set. Compared with other state-and-art multi-objective algorithms on the two types of benchmark functions, it is verified that the individuals can more converge faster to the real Pareto frontier with good distribution.

Keywords: Parameter dynamics · Individual competition mechanism · Multiobjective optimization

1 Introduction

In many real-world problems, we often face the task of optimizing multiple objectives simultaneously. These problems are known as multi-objective optimization problems (MOPs) [1–5]. MOPs are widely used in engineering design, economics, bioinformatics, and other fields. However, the solutions to MOPs often involve trade-offs between objectives, making the task of finding the optimal solution complex and challenging.

Research on multi-objective optimization problems began in the 1960s, and since then, many methods have been developed to solve these problems. These methods can be roughly divided into two categories: deterministic methods and stochastic search methods. Deterministic methods, such as the weighted sum method, goal programming

© The Author(s), under exclusive license to Springer Nature Singapore Pte Ltd. 2024
K. Li and Y. Liu (Eds.): ISICA 2023, CCIS 2146, pp. 101–112, 2024.
https://doi.org/10.1007/978-981-97-4393-3_9

method, and constraint method, usually assume that the objective functions and constraints of the problem are known and can be solved mathematically. However, these methods are often insufficient when dealing with complex, nonlinear, non-convex, or uncertain problems.

Typically, the set of optimal solutions is called the Pareto optimal solution set, or the Pareto frontier (PF). Since many optimization problems in practical engineering problems are NP-hard [6], traditional mathematical programming methods become extremely difficult. In contrast, stochastic search methods, such as genetic algorithms, particle swarm optimization, and ant colony optimization, do not need to understand the objective functions and constraints of the problem in detail, but only need the ability to evaluate the quality of solutions. This makes them able to handle more complex and less understandable problems. Researchers have proposed a variety of multi-objective optimization algorithms.

(1) Dominance-based optimization algorithms: NSGA-II (Non-dominated Sorting Genetic Algorithm-II) [7], SPEA2 (Strength Pareto Evolutionary Algorithm-2) [8], PESA-II (Pareto Envelope-based Selection Algorithm-II) [9];
(2) Decomposition-based optimization algorithms: MOEA/D (Multi-Objective Evolutionary Algorithm based on Decomposition) [10], MOEA/D-m2m (Decomposition of a Multi-objective Number of Simple Multi-objective Subproblems) [11] and RVEA (Reference Vector Guided Evolutionary Algorithm) [12];
(3) Performance indicator-based optimization algorithms: IBEA (Indicator-Based Evolutionary Algorithm) [13], SMS-EMOA (S-Metric Selection Evolutionary Multi-Objective Algorithm) [14] and HypE (Hypervolume-based Evolutionary Algorithm) [15]. These algorithms have been successfully applied to solve various complex MOPs.

However, despite their excellent performance in many applications, they still face some challenges. For example, they may get stuck in local optima and fail to find the global optimum. In addition, the performance of these algorithms is often affected by their parameter settings.

Due to the increasing number and complexity of multi-objective optimization problems, on the basis of these classic algorithms, in order to solve the problem of excessive population grouping pressure and the population easily collapsing and stagnating in the objective space, Wang [16] proposed a multi-objective optimization algorithm based on the game mechanism of multi-attribute elite individuals, introducing the game mechanism of K-means clustering into multi-objective optimization problems. To solve the problem of difficult multi-objective optimization of workflow in cloud computing environment, Luo [17] proposed a flower differential pollination workflow multi-objective scheduling optimization algorithm. To improve the overall performance of the multi-objective firefly algorithm to better solve complex MOP problems, Xie [18] proposed a multi-objective firefly algorithm (MOFA-MCS) based on multiple cooperation strategies. To improve the uneven distribution of solutions in discontinuous irregular multi-objective problems with evenly distributed reference points, Qi [19] proposed a decomposition multi-objective optimization algorithm with adaptive weight adjustment (MOEAD-AWA).

Multi-objective optimization is an active and rapidly developing research field. Despite many significant achievements, there are still many challenges and opportunities. The loss of population diversity, rapid convergence, low convergence, easy to fall into local optima, easy to collapse in outer space and other problems have always been the focus of this paper, but these problems have not been well solved in the above algorithms. Therefore, the purpose of this paper is to propose a new multi-objective optimization algorithm to solve these problems.

The paper is organized as follows, with Sect. 1 presenting the background of this work, including some brief overviews of multi-objective optimization algorithms in recent literature. Section 2 presents details of the specifics of the relevant work, and the further analysis of the multi-objective optimization problem is summarized. Section 3 describes our proposed algorithm, including the problems solved and the strategies used. Section 4 gives the experimental design based on two types of benchmark functions and the comparative study and further analysis of experimental results. Finally, the discussion and results are given in Sect. 5.

2 Related Works

2.1 Multi-objective Optimization Problems Analysis

Taking the minimization problem as an example, consider the following form of multi-objective optimization problem.

$$\begin{aligned} \min F(X) &= (f_1(X), f_2(X), ..., f_m(X)) \\ s.t.\ X &\in \Omega \subseteq R^n \end{aligned} \tag{1}$$

Where: $X = (x_1, x_2, \cdots, x_n)$ is an n-dimensional decision vector in the decision space Ω, $F : \Omega \rightarrow \Theta \subseteq R^n$ is an m-dimensional goal vector which is a mapping from the n-dimensional decision space to the m-dimensional goal space Θ.

Definition 1. (Pareto domination) for any decision vector, if the following conditions are met:

$$\begin{cases} f_i(u) \leq f_i(v),\ \forall i \in 1, 2, \cdots, m \\ f_i(u) < f_i(v),\ \exists j \in 1, 2, \cdots, m \end{cases} \tag{2}$$

is said to dominate v by u, denoted $u \succ v$, where u is the nondominated solution and v is the dominated solution.

Definition 2 (Pareto optimal solution). The solution $x \in \Omega$ is said to be Pareto optimal if and only if there are no other solutions x' in Ω such that the vector $F(x') = (f_1(x'), f_2(x'), \cdots, f_m(x'))$ dominates the vector $F(x) = (f_1(x), f_2(x), \cdots, f_m(x))$.

Definition 3 (Pareto optimal solution set). For a given multi-objective optimization problem, the set of all Pareto optimal solutions is called Pareto optimal solution set.

Definition 4 (Pareto optimal frontier). The set of Pareto optimal solutions mapped to the objective space is called Pareto optimal frontier, denoted as PF.

The complexity of multi-objective optimization problems is manifested in that there is not a single solution in the decision space, so that all the objectives can be optimized, but a Pareto optimal set that balances all the objectives can be obtained. Therefore, we prefer to get an optimal set with better convergence and diversity. However, the PFs of many practical engineering problems is irregular, Based on the existing research, ADPMO algorithm tries to adopt different strategies to optimize the population with high diversity while maintaining convergence.

2.2 Population Renewal Formula

In 1995, Dr. Kennedy and Dr. Eberhart [20] proposed the Particle swarm optimization (PSO) algorithm. In the standard particle swarm algorithm [3], in order to find the optimal solution to the minimum optimization problem, the position and velocity of the particles are updated with the formula in Eq. (3).

$$
\begin{cases}
V_i(t+1) = \omega V_i(t) + c_1 R_1(t)(pbest_i(t) - X_i(t)) \\
+ c_2 R_2(t)(gbest(t) - X_i(t)) \\
X_i(t+1) = X_i(t) + V_i(t+1)
\end{cases}
\tag{3}
$$

Where, ω is the inertia factor, c_1, c_2 are the learning rates, Pbest is the individual optimal position, and Gbest is the global optimal position, R_1, R_2 represent random factors between [0, 1], X_i is the position vector, V_i is the flight speed.

The particle swarm algorithm process is as follows:

Algorithm: Particle Swarm Algorithm

---
1. Limit the maximum number of iterations, the initial population size, and the velocity and position of the particles, The initial position of the particle is set to Pbest, and then select Gbest from the Pbest of all particles.
2. Calculate the fitness value of each particle.
3. Compare the fitness value of the current particle and the previous generation Pbest, if better then update the Pbest, if not keep the Pbest unchanged. better then keep Pbest unchanged.
4. Select the Gbest from the contemporary Pbest, then compare it with the fitness value of the previous generation Gbest, and update the Gbest if it is better. compare it with the fitness value of the previous Gbest, and update the Gbest if it is better, and keep the Gbest unchanged if it is not good.
5. Update according to the updating formula of particle swarm velocity and position.
6. See if the maximum number of iterations is reached or a better global optimal solution cannot be found, otherwise start from 2. Continue the loop .

Through the above formula and algorithm flow, individuals can be guided by their own optimal position and global optimal position to approach the optimal solution of the optimization problem step by step. However, too fast convergence will cause the population to fall into local extremes.

2.3 Levy Flights

Levy flights [21] is a class of non-Gaussian stochastic processes, commonly used to step composition to describe the mathematical form of continuous random step composition trajectories such as human trip splitting, foraging search trajectories of organisms, etc., which is also one of the best strategies in stochastic wandering models. Levy flights trajectory is a Markov stochastic process, and the step size of the walk satisfies a heavy-tailed Levy distribution, as shown in Eq. (4).

$$L(s) \sim |s|^{-1-\beta}, 0 < \beta \leq 2 \tag{4}$$

The random step s of levy flights can be obtained from Eq. (5):

$$s = u/|v|^{1/\beta} \tag{5}$$

where, u, v both obey normal distribution, $u \sim N(0, \sigma_u^2)$, $v \sim N(0, \sigma_v^2)$, σ_u and σ_v, satisfy Eq. (6):

$$\begin{cases} \sigma_u = \left\{ \dfrac{\Gamma(1+\beta)\sin(\pi\frac{\beta}{2})}{\Gamma(\frac{1+\beta}{2})\beta * 2^{(\beta-1)/2}} \right\}^{1/\beta} \\ \sigma_v = 1 \end{cases} \tag{6}$$

where Γ is the standard Gamma function.

In the ADPMO algorithm, the introduction of the Lévy flight [18] brought a new impetus to the optimization process. This strategy empowers some particles to be able to deeply explore the region neighboring the current optimal solution in the same region, thus improving the accuracy of the local search. Meanwhile, some other particles can explore extensively in regions far away from the current optimal solution, thus ensuring the global search capability of the algorithm and avoiding falling into local optimality. It is worth noting that Levy Flight introduces large-scale jumps and multiple sharp changes in direction in the ADPMO algorithm, which not only expands the search breadth and enhances the diversity of the population, but also effectively solves the problem that it is difficult for the multi-objective algorithm to approximate the global Pareto front when dealing with multi-objective optimization problems (MOPs) with multiple local Pareto fronts.

3 Proposed Algorithm

3.1 Pseudo-Code for the Algorithm

Algorithm: **ADPMO**

Input:

     A MOP;

     A stop criterion;

     Pop size;

     Probability of Crossover and Mutation;

     K-means clustered subgroups k;

     Binary crossover and polynomial variance distribution indices;

     Boundary of X and V;

Output:

     The solution sets obtained in each environment;

     The objective function values obtained in each environment;

1. initialize population P(t) and compute the objective function values of its members
2. While stop condition satisfied
3.   k-means(P(t), k) and get_pacesetter(P(t), k);;
4.   competition(P(t), k);
5.   P`(t)=update_pop.f_v and x(P(t), pacesetter, k);
6.   levy(P`(t), k);
7.   sort(P`(t)) // non_domination_sort, crowding_distance_sort and compute the objective function values
8.   E(t)=choice_halfpop(P`(t));
9.   tournament_selection(E(t));
10.   E`(t)=cross_mutation and combine //combine offspring and parent
11.   sort and select pop individuals from the E`(t);
12.   P(t+1) = E`(t);
13. End

3.2 K-Means Clustering and Individual Competition

Combining the non-domination relationship with the optimal particle selection strategy of particle swarm algorithm, after k-means clustering [22] and grouping, we comprehensively evaluate the non-domination level and congestion degree of the subgroup, and replace Gbest with the pacesetter randomly selected from the optimal subgroup within each cluster. Replace Pbest with the individual who wins the competition within each cluster, In order to avoid populations falling into local extremes, finally update the speed and position information only for the individual who loses the competition, while neither the winning individual nor the pacesetter participates in this process.

$$\begin{cases} V_{f,i}(t+1) = \omega V_{f,i}(t) + c_1(X_{w,i}(t) - X_{f,i}(t)) \\ \quad + c_2(P_{k,i}(t) - X_{f,i}(t)) \\ X_{f,i}(t+1) = X_{f,i}(t) + V_{f,i}(t+1) \end{cases} \tag{7}$$

where i is the current population iteration number and k is the cluster grouping index, P_k is a random particle in the highest particle of the non-dominated sorting rank in subpopulation k, which is referred to as a pacesetter in this paper, w is the competition winning particle identifier, and f is the game losing particle identifier.

The competitive mechanism is specified as comparing the non-dominance rank of two individuals, and if equal then comparing the degree of crowding, with the individual with a high non-dominance rank and a low degree of crowding winning.

The value of the number of clustering groups k was experimented by wang [16] and it was determined that the algorithm IGD [23] curve performs better when k = 7 than when k is equal to other values and the final IGD value is also lower.

3.3 Proposed Adaptive Dynamic Parameters

In this paper, for the particle swarm optimization algorithm's speed update formula of the inertia weight ω and learning factor c_1, c_2, the change of these two types of parameters, the influence of the population in the target area of the search ability of the law, designed a dynamic adjustment of these two types of parameters.

Since larger inertia weights favor the ability of particles to explore new regions, they also favor the global search of the population. Smaller inertia weights favor the local search ability of particles, which in turn favors the local search of the population. At first, the population tends to search globally, but later will gradually enter the local search. According to this characteristic, the formula for the dynamic adjustment of W is:

$$\omega(r) = [\omega_{max} - \omega_{min}] * e^{\frac{-(r+1)^2}{(0.3*r_{max})^2}} + \omega_{min} \tag{8}$$

Where, r represents the number of iterations, r_{max} represents the maximum number of iterations, generally take $\omega_{min} = 0.4$, $\omega_{max} = 0.9$.

For learning factors C1 and C2, C1 weighs "own experience" and C2 weighs "social experience". The larger C favors searching in one's own domain as guided by one's own experience. The larger C2 favors the global search guided by "social experience". Combining these characteristics, In the early stage of the algorithm, a larger C and a smaller C2 are needed, while in the later stage, a smaller C and a larger C2 are needed. Dynamic adjustment formula is:

$$\begin{cases} c_1 = 1 + \frac{r_{max}-r-1}{r_{max}} \\ c_2 = 1 + \frac{r+1}{r_{max}} \end{cases} \tag{9}$$

3.4 Mutation

After screening half of the individuals in the population. Based on the level of non-dominance and congestion. Using the binary tournament, simulated binary crossover, and polynomial variation strategies in NSGA II [7].

$$v'_k = v_k + \delta * (u_k - l_k)$$

$$\delta = \begin{cases} [2u + (1 - 2u)(1 - \delta_1)^{\frac{1}{\eta_m+1}} - 1] & \text{if } u \leq 0.5 \\ 1 - [2(1 - u) + 2(u - 0.5)(1 - \delta_2)^{\eta_m+1}]^{\frac{1}{\eta_m+1}} & \text{if } u > 0.5 \end{cases} \tag{10}$$

In the formula, $\delta_{1,2} = (v_k - l_k)/(u_k - l_k)$, u is a random number in the interval [0, 1], η_m is the distribution index selected by the user, v_k represents the population of the previous generation, v_k' represents the newly generated population after polynomial variation, u_k represents upper: the upper bound of the variable, l_k represents lower: the lower bound of the variable [24].

After using the levy flight strategy and the variation strategy, the algorithm can solve the problem that it is easy to fall into the local optimum or collapse in the objective space, so that the algorithm can obtain a more uniform and better convergence of the optimization solution set, and when the population falls into the local extremes, using the above strategy can also make some of the individuals get out of the local extremes, and become the new navigator, which guides the population to get out of the local extremes, and improves the algorithm's degree of convergence.

Similarly, the simulated binary crossover and polynomial mutation of the screened population of size N/2 will result in an elite subpopulation of size N, which can help the original parent population to get out of the local extremes to a certain extent, and get out of the stagnation of the population in the target space.

4   Experiment

4.1   Experimental Design

In order to verify the effectiveness of ADPMO algorithm, ADPMO algorithm is applied to ZDT [23] test function series and DTLZ [23] test function series, and ADPMO algorithm is compared with NSGA II [7], SPEA2 [8], MOEA/D [10] and MOPSO [25] algorithm. In order to evaluate algorithm the optimal solution set of convergence and diversity, this article uses the IGD (Inverted Generational Distance) [23] to the convergence of the algorithm and distribution were compared and evaluated. The IGD index mainly evaluates the convergence performance and distribution performance of the algorithm by calculating the minimum distance sum between each reference point on the real Pareto front plane and the individual external space map obtained by the algorithm. The smaller the IGD value, the better the comprehensive performance of the algorithm.

With the exception of the ADPMO algorithm, all algorithms are run using PlatEMO [26], using the platform's default parameter values, which are completely consistent with the original text of the algorithm. The inertia weight w, learning factors c1 and c2 of ADPMO algorithm are set as dynamic changes, the crossover probability Pc is 0.7, the variation probability Pm is 1/number of decision variables, the denominator is the number of decision variables, the analog binary crossover and polynomial variation distribution index n1 and n2 are 2 and 5 respectively, and the number of k-means clustering groups of ADPMO algorithm is set to 7.

Table 1. Test function properties

Test function	Number of Decision Variables	Number of Objectives	Number of real PF samples
ZDT1	30	2	500
ZDT2	30	2	500
ZDT3	30	2	136
ZDT4	10	2	200
ZDT6	10	2	2992
DTLZ1	12	3	2500
DTLZ2	12	3	40000

4.2 Experimental Result

Table 2. IGD performance indicator

Test function	ADPMO	NSGAII	SPEA2	MOPSO	MOEA/D
ZDT1	**2.3070E−03**	2.6478E−03	2.9580E−03	1.5230E−01	1.6696E−02
ZDT2	**2.0770E−03**	2.8802E−03	3.3832E−03	8.2168E−01	3.5879E−02
ZDT3	4.0532E−03	3.3592E−03	**2.6856E−03**	1.6108E−01	4.5290E−02
ZDT4	**2.7730E−03**	1.7066E−02	6.8338E−03	7.9449E+00	2.5361E−02
ZDT6	**1.0840E−03**	5.9573E−03	6.1994E−03	2.3591E−03	6.0520E−03
DTLZ1	**3.0836E−01**	5.7869E−01	6.0552E−01	6.4380E+01	3.6358E−01
DTLZ2	3.5338E−02	4.0181E−02	3.0952E−02	3.7014E−02	**2.9045E−02**

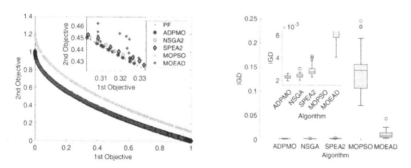

Fig. 1. Function ZDT1 test results and box-chart

Running ADPMO with 4 classical algorithms on ZDT1-4, ZDT6, DTLZ1-2 functions, the results of the run are visualized as shown in Figs. 1, 2, 3, 4, 5, 6 and 7. It

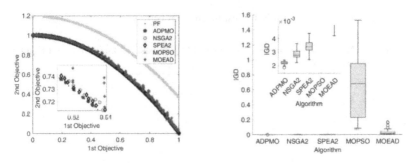

Fig. 2. Function ZDT2 test results and box-chart

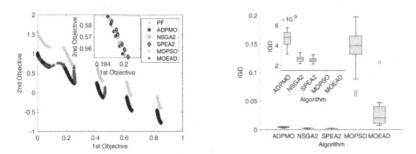

Fig. 3. Function ZDT3 test results and box-chart

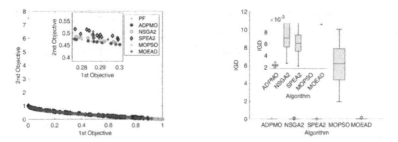

Fig. 4. Function ZDT4 test results and box-chart

can be seen that the ADPMO algorithm has good convergence and diversity, performing slightly worse than the other algorithms on the ZDT3 and DTLZ2 test functions, and performing particularly well on ZDT4. From this, we can conclude that the improved competition mechanism and dynamic parameter strategy are very good at stopping the decrease in population diversity, and the particle screening strategy before the bidding race is also very effective at maintaining the asymptotic convergence of the population, reducing the amount of computation on the objective function and increasing the convergence accuracy. Overall, the multiple strategies proposed in this paper improve the convergence of the algorithm and the diversity of the optimized solution set to some extent.

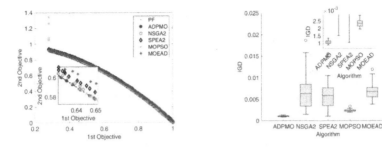

Fig. 5. Function ZDT6 test results and box-chart

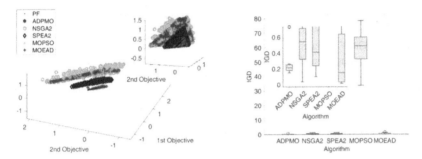

Fig. 6. Function DTLZ1 test results and box-chart

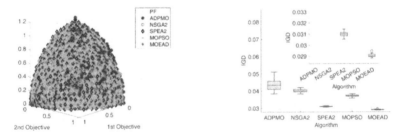

Fig. 7. Function DTLZ2 test results and box-chart

References

1. Zhang, J., Xing, L.: A survey of multiobjective evolutionary algorithms. In: 2017 IEEE International Conference on Computational Science and Engineering (CSE) and IEEE International Conference on Embedded and Ubiquitous Computing (EUC), pp. 93–100 (2017)
2. Ma, H., Shen, S., Yu, M., Yang, Z., Fei, M., Zhou, H.: Multi-population techniques in nature inspired optimization algorithms: a comprehensive survey. Swarm Evol. Comput. **44**, 365–387 (2019)
3. Feng, Q., Li, Q., Quan, W., Pei, X.: Overview of multiobjective particle swarm optimization algorithm. Chin. J. Eng. **43**, 745–753 (2021)

4. Xiao, R., Li, G., Chen, Z.: Research progress and prospect of evolutionary many-objective optimization. Control Decis. **38**, 1761–1788 (2023)
5. Falcón-Cardona, J.G., Coello, C.A.C.: Indicator-based multi-objective evolutionary algorithms: a comprehensive survey. ACM Comput. Surv. **53**, 1–35 (2021)
6. Liu, Y.: Exact algorithms for NP hard problems based on conflicts (2019)
7. Deb, K., Pratap, A., Agarwal, S., Meyarivan, T.: A fast and elitist multiobjective genetic algorithm: NSGA-II. IEEE Trans. Evol. Comput. **6**, 182–197 (2002)
8. Zitzler, E., Laumanns, M., Thiele, L.: SPEA2: Improving the strength pareto evolutionary algorithm. ETH Zurich (2001)
9. Corne, D.W., Jerram, N.R., Knowles, J.D., Oates, M.J.: PESA-II: region-based selection in evolutionary multiobjective optimization. In: Proceedings of the 3rd Annual Conference on Genetic and Evolutionary Computation, pp. 283–290. Morgan Kaufmann Publishers Inc., San Francisco (2001)
10. Zhang, Q., Li, H.: MOEA/D: a multiobjective evolutionary algorithm based on decomposition. IEEE Trans. Evol. Comput. **11**, 712–731 (2007)
11. Liu, H.-L., Gu, F., Zhang, Q.: Decomposition of a multiobjective optimization problem into a number of simple multiobjective subproblems. IEEE Trans. Evol. Comput. **18**, 450–455 (2014)
12. Cheng, R., Jin, Y., Olhofer, M., Sendhoff, B.: A reference vector guided evolutionary algorithm for many-objective optimization. IEEE Trans. Evol. Comput. **20**, 773–791 (2016)
13. Zitzler, E., Künzli, S.: Indicator-based selection in multiobjective search. In: Yao, X., et al. (eds.) PPSN 2004. LNCS, vol. 3242, pp. 832–842. Springer, Heidelberg (2004). https://doi.org/10.1007/978-3-540-30217-9_84
14. Beume, N., Naujoks, B., Emmerich, M.: SMS-EMOA: multiobjective selection based on dominated hypervolume. Eur. J. Oper. Res. **181**, 1653–1669 (2007)
15. Bader, J., Zitzler, E.: HypE: an algorithm for fast hypervolume-based many-objective optimization. Evol. Comput. **19**, 45–76 (2011)
16. Wang, X., Ji, W., Zhou, G., Yang, J.: Multi-objective optimization algorithm based on multi-index elite individual game mechanism. J. Syst. Simul. **35**, 494–514 (2023)
17. Luo, Z., Zhu, Z.: A multi-objective workflow scheduling algorithm based on flower pollination cloud environment. Acta Electroncia Sinica **49**, 470 (2021)
18. Xie, C., Zhang, F.: Multi-objective firefly algorithm based on multiply cooperative strategies. Acta Electroncia Sinica **47**, 2359 (2019)
19. Qi, Y., Ma, X., Liu, F., Jiao, L., Sun, J., Wu, J.: MOEA/D with adaptive weight adjustment. Evol. Comput. **22**, 231–264 (2014). https://doi.org/10.1162/EVCO_a_00109
20. Kennedy, J., Eberhart, R.: Particle swarm optimization. In: Proceedings of ICNN 1995 - International Conference on Neural Networks, vol. 4, pp. 1942–1948 (1995)
21. Zheng, J., Zhan, H.: Development of Lévy flight and its application in intelligent optimization algorithm. Comput. Sci. **48**, 190–206 (2021)
22. Wang, S., Liu, C., Xing, S.: Review on K-means clustering algorithm. J. East China Jiaotong Univ. **39**, 119–126 (2022)
23. Wang, L., Ren, Y., Qiu, Q., Qiu, F.: Survey on performance indicators for multi objective evolutionary algorithms. Chin. J. Comput. **44**, 1590–1619 (2021)
24. Deb, K., Goyal, M.: A combined genetic adaptive search (GeneAS) for engineering design. Department of Mechanical Engineering Indian Institute of Technology (1996)
25. Coello Coello, C.A., Lechuga, M.S.: MOPSO: a proposal for multiple objective particle swarm optimization. In: Proceedings of the 2002 Congress on Evolutionary Computation, CEC 2002 (Cat. No. 02TH8600), vol. 2, pp. 1051–1056 (2002)
26. Tian, Y., Cheng, R., Zhang, X., Jin, Y.: PlatEMO: a MATLAB platform for evolutionary multi-objective optimization (2017)

Adaptive Elimination Particle Swarm Optimization Algorithm for Logistics Scheduling

Kexin Lin, Wei Li[(⊠)], and Yuqi Ou

School of Information Engineering, JiangXi University of Science and Technology,
Ganzhou 341000, Jiangxi, China
Liwei@mail.jxust.edu.cn

Abstract. The particle swarm optimization algorithm sets few parameters, is easy to operate, and has now been successfully applied to various optimization problems. For the logistics scheduling problem, in the face of complex, large-scale search solution space will consume many resources. The standard particle swarm optimization can falling into local optimum easily when facing more complex problems. In this paper, three optimization strategies are designed to address the above problems, and the adaptive elimination particle swarm optimization (AELPSO) algorithm is proposed to optimize the logistics scheduling problem with a path planning scheme based on the principle of greedy strategy. The inertia weight linear decreasing method and asynchronous change adjusting learning factor method are introduced to reasonably balance the exploration and exploitation ability of the particle swarm. Combined with the idea of natural selection, the search efficiency of particles is enhanced. In the experimental part of this paper, AELPSO is used to solve the logistics scheduling problem on eight data sets and compared and analyzed. The experimental results show that the AELPSO algorithm can significantly improve the quality of the solution, and at the same time, it can obtain a better path planning scheme for the logistics scheduling problem.

Keywords: Logistics scheduling problem · Particle swarm optimization · Greed strategy · Adaptive learning

1 Introduction

Intelligent optimization algorithms in the evolutionary category mainly imitate and learn the rules and ideas observed in nature. These algorithms leverage behaviors such as heredity, selection, and mutation, as well as the principle of survival of the fittest, to undergo continuous evolution. Swarm intelligence optimization algorithm refers to the computational intelligence that searches by learning the behavioral rules of biological populations.

Swarm intelligence optimization algorithms are inspired by swarming organisms such as ants, birds, bees, and other animals and are proposed to solve distributed problems. The particle swarm optimization algorithm was proposed in 1995 by Eberhart and Kennedy [1]. The foraging behavior and characteristics of bird flocks inspired the particle swarm optimization algorithm. During the feeding process of a community of birds,

© The Author(s), under exclusive license to Springer Nature Singapore Pte Ltd. 2024
K. Li and Y. Liu (Eds.): ISICA 2023, CCIS 2146, pp. 113–124, 2024.
https://doi.org/10.1007/978-981-97-4393-3_10

each bird will communicate information with other individual birds to share personal feeding experiences and flight experiences.

The particle swarm optimization algorithm simulates the bird feeding process, abstracts the bird searching for the wealthiest food into the process of particle searching for the optimal solution in the solution space. The fundamental particle swarm optimization algorithm is an iterative optimization technique. It begins with the initial population, which consists of a group of particles with random positions and velocities. The algorithm then iteratively searches for the optimal value by updating functions. In each iteration, each particle updates its position information. It does so by considering individual and group history information, position, and velocity vectors. This iterative process allows the particles to approach the optimal solution gradually.

Capacitated Vehicle Routing Problem (CVRP) is a typical NP-hard problem; The current logistics scheduling problems are significant and characterized by numerous constraints. Additionally, the solutions to these problems are becoming increasingly complex. Moreover, the time and space complexity of solutions obtained using traditional algorithms increases exponentially. The particle swarm optimization algorithm proposed in this paper aims to optimize the resolution of the logistics distribution vehicle scheduling problem. It does so by reasonably and scientifically determining the scheduling task to reduce the cost of logistics distribution as much as possible while improving the quality of service. Furthermore, this approach plays a positive role in promoting the improvement of the economic benefits of enterprises and accelerating the development of the logistics industry [2].

The position vector and velocity vector of the particle swarm optimization algorithm are encoded and decoded based on actual numbers. This process constructs the priority matrix for delivering customer points using vehicles. The assignment of customer points to vehicles relies on this priority matrix and the current residual loads of each vehicle volume. Part of the dimensions is related to customer points and the function of delivery vehicles. Using the particle swarm optimization algorithm to address the logistics scheduling problem involves several steps. First, the particle coding problem and the path planning problem are solved. Then, the iterative optimization search commences according to the process outlined in the particle swarm optimization algorithm.

The rest of the paper is organized as follows: Sect. 2 details related work such as elementary particle swarm optimization algorithms and logistics scheduling. The Sect. 3 is a brief description of the adaptive elimination particle swarm optimization algorithm (AELPSO) proposed in this paper. Section 4 is the experiments, where the A-n32-k5 is used, the experimental results are compared for the application problem and other data sets, and the results are recorded. The Sect. 5 summarizes the work of this research and looks forward to future work.

2   Related Work

2.1   Elementary Particle Swarm Optimization Algorithm

The particle swarm optimization algorithm was proposed in 1995 by Eberhart and Kennedy. A particle represents a solving operator in the primary particle swarm optimization algorithm. The speed and direction of the particle searching the solution space are determined by the particle's history information and the swarm's history information. Suppose the search space of the problem is *d-dimensional* and the population size of the algorithm is *m* [3]. The position vector of particle i's *k-th* generation in the solution space can be expressed as:

$$(x_{i1}^{(k)}, x_{i2}^{(k)}, x_{i3}^{(k)}, \ldots, x_{id}^{(k)}), i = 1, 2, \ldots, m \tag{1}$$

the position vector is the position of a particle in the particle population in the solution space and represents a feasible solution to the problem to be solved. The velocity vector of a particle is a vector with the same dimension as the position vector of the particle. The velocity vector of the k-th generation of particle i in the solution space can be:

$$(v_{i1}^{(k)}, v_{i2}^{(k)}, v_{i3}^{(k)}, \ldots, v_{id}^{(k)}), i = 1, 2, \ldots, m \tag{2}$$

the velocity vector is the motion of each particle in the particle population in each dimension of the solution space, reflecting not only the direction but also the speed. The basic particle swarm optimization algorithm updates the state of each particle position in each generation. The D-th dimension neighborhood function of the i-th particle in the k+ 1st generation is calculated as follows:

$$v_{id}^{(k+1)} = \omega * v_{id}^{(k)} + c_1 r_1 (pbest_i - x_{id}^{(k)}) + c_2 r_2 (gbest - x_{id}^{(k)}) \tag{3}$$

$$x_{id}^{(k+1)} = x_{id}^{(k)} + v_{id}^{(k+1)} \tag{4}$$

where ω is the inertia weight, which acts as a partial inheritance of the particle's velocity, r_1 and r_2 are random numbers between [0, 1], i.e., the perturbation factors of particle velocity update. c_1 and c_2 are learning factors, which represent the influence of the particle's "individual learning" ability and "group learning" ability on the velocity, respectively [4].

The search for the particle in the solution space needs to satisfy certain boundary conditions; the position of the particle has to be in the solution space. If the value of a certain dimension d of the particle i exceeds the range of the solution space, i.e., $x_{id}^{(k)} > x_{id\ max}$ or $x_{id}^{(k)} < x_{id\ min}$, At this point, the value is readjusted to be within the range of the solution space.

Similarly, when the neighborhood function generates new particle velocity vectors, the constraints of the velocity vector condition $\left| v_{id}^{(k+1)} \right| < V_{max}$ need to be satisfied to prevent the particles from having too large a velocity in a given dimension, which would lead to too fast convergence [5].

The specific process of the fundamental particle swarm optimization algorithm is described below: In the first step, the parameters of the particle swarm optimization algorithm need to be determined. These parameters include the maximum number of iterations, the size of the particle swarm, the inertia weights, the learning factor, the boundaries of the solution space, the maximum velocity, and the dimensionality of the particle vector. Initialize the population and initialize the position vector and velocity vector of each individual, the values of each dimension of the position vector and velocity vector are random and are within the constraints of the solution space range. Then, evaluate each particle by the fitness function fitness and record the optimal fitness value for each individual and the optimal fitness value for the population. Each particle updates the velocity vector and position vector and makes the values of each dimension of the velocity vector and position vector satisfy the constraints. Eventually, the number of iterations plus one, if the number of iterations reaches the maximum number of iterations or other termination conditions, then the algorithm finds the end of the optimal solution to get the search for the optimal solution [6]. If the number of iterations does not reach the maximum number of iterations or other termination conditions are not met, the process returns to the initialization step.

2.2 Logistics Dispatch

CVRP is generally defined as the following: The task is organizing appropriate travel routes for a series of loading and unloading points. These routes ensure vehicles pass through the points in an orderly manner while adhering to certain constraints. These constraints may include demands for goods, delivery quantities, delivery and shipment times, vehicle capacity limitations, mileage limitations, time constraints, etc. The objective is to achieve a specific goal for the problem at hand, such as minimizing distance, costs, time, or vehicle number [7]. Logistics scheduling problems can be categorized in different ways according to different perspectives. A logistics scheduling problem can be identified from five aspects: The number of customer nodes, the number of warehouses, the presence or absence of customer service time constraints, vehicle load limitations, and customer demand.

To determine the logistics scheduling problem studied in this paper, the necessary premises and assumptions are made: Firstly, the number of distribution centers is set to 1. Secondly, all distributed goods are of the same kind, and their density is not considered. Thirdly, both the load limit of the distribution trucks and the metrics of the goods are standardized to unit weights. The logistics scheduling problem discussed in this paper is a static logistics planning problem, i.e., the locations of all customer points, distribution demands, and distribution centers are known and determined. Each distribution vehicle is loaded with goods from the distribution center. It visits each customer point in the sequence according to its respective planning path. After visiting the last customer point, the process returns to the distribution center. The vehicles arrive in a straight line from one node to another [8, 9]. Each customer point is served by only one distribution vehicle and will not be distributed by more than one vehicle, each loaded with a portion of the goods of a particular customer point. All distribution trucks in the distribution center are of the same model, with the same load limits and unit freight rates. Each distribution vehicle is subject to strict load limits, and the weight of the goods loaded at departure from the distribution center cannot exceed the determined load limit. There is also no single delivery mileage limit for each distribution vehicle and no limit on the number of customer locations served by each distribution vehicle. There is no time limit imposed on the distribution vehicle at the customer location. This includes both the time required for loading or unloading goods and the time spent traveling on the road. Secondly, the customer location only requires delivery and does not involve any pickups. Develop a mathematical model of the logistics scheduling problem [8, 10]. Objective function:

$$minZ = \sum_{i=1}^{N+1} \sum_{j=1}^{N+1} \sum_{k=1}^{K} d_{ij}x_{ijk} \tag{5}$$

Constraints:

$$\sum_{i=1}^{N} q_i y_{ki} \leq Q, k = 1, 2, 3, \ldots, K \tag{6}$$

$$\sum_{k=1}^{K} y_{ki} = 1, i = 1, 2, 3, \ldots, N \tag{7}$$

$$\sum_{i=1}^{N+1} x_{ijk} = y_{kj}, j = 1, 2, 3, \ldots, N, k = 1, 2, 3, \ldots, K \tag{8}$$

$$\sum_{j=1}^{N+1} x_{ijk} = y_{kj}, i = 1, 2, 3, \ldots, N, k = 1, 2, 3, \ldots, K \tag{9}$$

$$x_{ijk} = 0 \text{ or } 1, i, j = 1, 2, 3, \ldots, N+1, k = 1, 2, 3, \ldots, K \tag{10}$$

$$y_{kj} = 0 \text{ or } 1, i = 1, 2, \ldots, N, k = 1, 2, 3, \ldots, K \tag{11}$$

Among others:

$$x_{ijk} = \begin{cases} 1, i \leq k \leq j \\ 0, k < i, k > jj \end{cases} \tag{12}$$

$$y_{ijk} = \begin{cases} 1, i \in k \\ 0, i \notin k \end{cases} \tag{13}$$

3 Proposed Algorithm

The proposed approach consists of several components: First, a particle swarm optimization algorithm is employed to address Logistics Scheduling Problems. Second, an asynchronous change-adjusting learning factor method, known as AELPSO, is introduced, incorporating natural selection ideas. Lastly, an optimized path planning scheme based on a greedy strategy is implemented. The PSO process incorporating natural selection is shown in Fig. 1 below:

3.1 Linearly Decreasing Inertia Weights

Inertia weight directly affects the proportion of inheritance to the historical speed, a high inertia weight makes the algorithm exploratory, and the opposite makes the algorithm exploitable. The linearly decreasing inertia weight method means that as the number

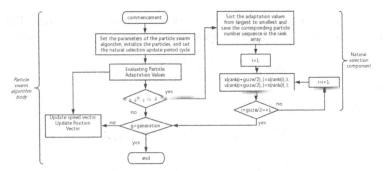

Fig. 1. Flowchart of the particle swarm optimization algorithm framework.

of iterations increases, the inertia weight decreases. Linearly from ω_{max} to ω_{min}, at generation g, the expression of inertia weight is:

$$\omega(g) = \omega_{max} - g * \frac{\omega_{max} - \omega_{min}}{generation} \tag{14}$$

In the particle swarm optimization algorithm, the term "generation" represents the maximum number of iterations. The strategy of variable inertia weights aims to enhance the early convergence of the standard particle swarm optimization algorithm. The mitigate particle oscillation near the global optimal solution during the late iteration.

3.2 Asynchronous Change Strategy

The particles exhibit characteristics of both "individual learning" and "social learning." To capitalize on this, the improvement strategy involves adjusting the variable learning factor. Specifically, during the early iteration period, the particle population is endowed with a more significant learning factor for "individual learning," enhancing their self-learning capability. As a result, during the late iteration phase, the "social learning" learning factor within the particle population becomes more prominent while their self-learning ability diminishes. Consequently, the particle population strongly emphasizes social learning but weaker self-learning. This adjustment facilitates a more robust local search capability towards the latter stages of iteration, which aids in converging towards the optimum solution. It is favorable to converge to the global optimal solution. The mathematical expression of the asynchronous change strategy is as follows:

$$c_1(g) = -g * \frac{c_{1\,max} - c_{1min}}{generation} \tag{15}$$

$$c_2(g) = c_{1\,max} + g * \frac{c_{2\,max} - c_{2min}}{generation} \tag{16}$$

The learning factor associated with individual learning decreases linearly with the number of iterations, and the learning factor associated with social learning increases linearly with the number of iterations. It makes the balance of exploratory and exploitative abilities of the particle population in the pre-exploration and post-exploratory stages much more reasonable.

3.3 Integration of the Idea of Natural Selection

During the process of natural selection, over time, individuals within a population evolve towards the most suitable traits. Unsuitable characteristics gradually diminish, and unfit individuals are eventually eliminated. Applying the idea of natural selection to the particle swarm optimization algorithm is to select a part of particles to be eliminated and replaced every few iteration cycles: Half of the particles with low fitness values are eliminated. Their position vectors and velocity vectors will be replaced by half of the particles with good fitness values but retaining their individual optimal memories $pbest_i$ and optimal positions. This elimination of particles to update particles allows the particles optimizing in the worse region to update to the parts of the particles with better population optimization. The updating of their respective velocities is, in turn, influenced by the individual optimal posts before updating their positions.

3.4 Path Planning with Greedy Strategies

Calculate the distance between each distribution customer point and the distribution center of each vehicle, and then arrange the priority according to the order of the distance between each node from small to large, and construct the successor node priority matrix. Then, according to the successor node priority matrix, starting from the distribution center, the nodes that have not been selected and have the highest priority are selected as successor nodes until all the nodes have been selected. Then, the path of the distribution vehicle is obtained according to the order of the selected nodes. Then, the distance summation operation is performed so that the path planning scheme of a vehicle is obtained.

The distribution vehicle planning path task is to start from the distribution center, pass through, and only pass through all the customer points in its customer point set once each, and then return to the distribution center. Improved path planning scheme based on greedy strategy: Add the distribution center to the set of customer points, then traverse each node, take each node as the starting point of the Hamiltonian circuit, calculate the path length of the Hamiltonian circuit starting from each node as the rule of the greedy strategy described in the previous section, and then select the Hamiltonian circuit with the shortest path length as the distribution path of the vehicle.

Algorithm 1: Optimized path planning scheme combined with AELPSO

Step1: Construct the distance matrix dist2() in the order of the serial numbers of the nodes in dist(i,:), (i = 1,2,3,...,vk) (including distribution center 1) from smallest to largest.

Step2: Construct the set S ={ task(i,:)},

Step3: Add the distribution center serial number 1 to the set S, such that S' = S, j = size(S), k - 1,

Step4: Select a node o in the set as the starting point and insert c into the first position of the one-dimensional array temp route(k,:),

Step5: $c_1 = c_2$,

Step6: Selecting $c_2 \in S$ makes dist2(c_1, c_2) = min(dist2$(c_1,:)$),

Step7: Remove c_2 from the set S' by inserting it into the end position of the one-dimensional array temp route(k,:),

Step8: Modify dists2$(:,c_2)$ = Inf, $c_1 = c_2$,

Step9: If set S' ≠ Ø, return to Step6,

Step10: Remove c from the set S by inserting c into the end position of the one-dimensional array temp route(k,:),

Step11: If the set S = Ø, the construction of temp route is complete, otherwise k = k+1, return to Step4,

Step12: Initialize cost = zeros(1,j), $k_1 = 1, k_2 = 1$,

Step13: do

   cost(k_1) = cost(k_1) + dist2(k_1,k_1+1),

   $k_2 = k_2 + 1$,

   While (temp route(k_1, k_2+1) ~= temp route$(k_1,1)$),

Step14: Choose c such that cost(c) = min(cost), then the vehicle's path length is cost(c), and the path is temp route(c,:).

3.5 AELPSO of Solution Capability

The standard particle swarm optimization algorithm uses fixed parameters. The AELPSO in this paper has a more vital ability to explore the particle population in the pre-particle population to enhance the diversity of particles. With the increase in the number of iterations, the particle population has a more vital ability to develop, enhance the ability to search for the optimal locally, and promote the convergence of the particle swarm as a whole. The range of inertia weight is set to [0.4, 0.9], and the content of learning factor change is developed to [0.5, 4.0]. According to the convergence diagram of the algorithm, it can be seen that the AELPSO has increased the number of generations for the convergence of the algorithm compared to the standard particle swarm optimization algorithm, indicating that the AELPSO can effectively avoid the premature convergence of the particle swarm optimization algorithm. The optimized algorithm has more pre-exploration ability, and the algorithm's search for a solution is better. The optimal solution that appeared during testing of the A-n32-k5 data set by the standard particle swarm optimization algorithm is 832, while the optimal solution that appeared using the AELPSO for solving this data set is 810. According to its logistics scheduling scheme, it is known that the optimal solution produced by the AELPSO is better than that produced by the standard particle swarm optimization algorithm, and the optimized particle swarm optimization algorithm is able to find the solution better than the optimal solution of the standard particle swarm optimization algorithm many times. The average value of the standard particle swarm optimization algorithm for solving the A-n32-k5 dataset 10 times is 839. In contrast, the AELPSO solves the value less than 840, indicating that the introduction of the inertia weight linear decreasing method, asynchronous change adjusting learning factor method, and the idea of natural selection can make the particle swarm optimization algorithm solve the logistics scheduling problem with better results.

4 Experimental Study

4.1 Experimental Setting

The logistics scheduling problem dataset studied in this paper is derived from the classic CVRP problem example, taking the A-n32-k5.vrp dataset as an example. Firstly, the parameters of the algorithm are set to the classical values, such as the learning factor c_1 and c_2 set to the same value of 2, and inertia weight is set to the average value of 0.65 between 0.4 and 0.9. Based on the computer configuration, we will set the population size to 40. The number of iterations will be varied from 200 to 2000, increasing by increments of 200 in each step. For each iteration count, the algorithm will be run ten times to compute statistical values. The results will be evaluated comprehensively, taking into account both the optimization effectiveness of the algorithm and its running time. The maximum number of iterations for the algorithm will be set at 1200 times.

Set the maximum number of iterations of the algorithm to 1200, the value of the learning factor c_1 and c_2 is 2, and the inertia weight is 0.4, 0.5, 0.6,..., 0.9, respectively, and run the algorithm ten times for each case to calculate the statistical values, and set the inertia weight to 0.7 according to the data obtained by comprehensively considering the definition of inertia weight to make the particle population's exploratory ability and search ability is in a more balanced and relatively strong state. In this paper, the learning factor is set to 2 according to the classical value, and the value of the velocity boundary is taken as 20% of the maximum boundary value $\max(x_{max}, y_{max})$ for v_{max}.

4.2 Application Examples of Logistics Scheduling

For example, in the data set A-n32-k5, a delivery truck with the delivery task customer point number (27, 8, 14, 18, 20, 32, 22) is added to the distribution center. The greedy strategy is applied to each node as a starting point to construct the length of the Hamiltonian loop as follows Table 1.

Table 1. Length of loops starting at different nodes.

Starting point number	Loop	Length
1	1-27-8-14-22-32-20-18-1	160. 5585
27	27-8-14-22-32-20-18-1-27	160. 5585

(*continued*)

Table 1. (*continued*)

Starting point number	Loop	Length
8	8-14-22-32-20-18-27-1-8	160. 7042
14	14-8-27-1-22-32-20-18-14	156. 2816
18	18-20-32-22-14-8-27-1-18	160. 5585
20	20-18-32-22-14-8-27-1-20	162. 0679
32	32-20-18-22-14-8-27-1-32	163. 6861
22	22-32-20-18-14-8-27-1-22	156. 2816

According to Table 1, the length of the loop constructed with customer point 14 as the starting point is 156.2816, while the size of the loop constructed with the distribution center as the starting point is 160.5585. Therefore, the optimized path planning algorithm can obtain a distribution scheme with shorter vehicle distribution distances for the same customer point allocation scheme [10]. The reduction of the distribution distance of each vehicle leads to the reduction of the total distribution distance. This reduction has a cascading effect, influencing both the position of the group optimal particles and the position of the individual optimal particles. Consequently, these changes impact the updating of the particles, as well as the process of finding the global optimal solution. For example, in the data set A-n32-k5, the optimal solution using the optimized algorithm is 797.8290, and its logistics scheduling scheme is as follows: the length of the loop constructed with the customer point 14 as the starting point is 156.2816, while the length of the loop constructed with the starting point of the distribution center is 160.5585. Therefore, the optimized path planning algorithm, which can be used in the case of the same customer point distribution scheme, can get the distribution scheme with a shorter vehicle distribution distance [8]. The reduction of the distribution distance of each vehicle results in a decrease in the total distribution distance. This decrease has an impact on both the position of the group optimal particles and the position of the individual optimal particles. Subsequently, these changes influence the updating of the particles and the search for the global optimal solution. For example, in the data set A-n32-k5, the optimal solution using the optimized algorithm is 797.8290. Using the AELPSO to solve the eight data sets described in Table 1, the situation is as follows in Table 2.

As shown in Table 2, the particle swarm optimization algorithm with AELPSO body and path planning scheme has made big progress in the quality of solution compared with the standard particle swarm optimization algorithm, and the five successive runs of the algorithm to solve the eight data sets respectively, the best solution is within 5% of its global optimal solution, and the worst solution is within 10% of its global optimal solution. From the standard deviation, the stability of the algorithm's solutions is good, and no solutions with large deviations are produced. The optimized path planning scheme has the potential to significantly enhance the solution's quality. However, this improvement comes with a drawback. The scheme of finding the shortest path per particle per vehicle

Table 2. Optimizing the particle swarm optimization algorithm for the dataset case.

Dataset	Optimized particle swarm optimization algorithm for solving							
Title	Optimal solution	Average	Minimum	Gap	Maximum	Gap	Variance	Time
A-n32-k5	784	802	797	1.66%	803	2.42%	2	68.8 s
A-n36-k5	799	828	824	3.13%	839	5.01%	6	84.2 s
A-n39-k5	822	864	857	4.26%	879	6.93%	10	90.0 s
A-n44-k7	937	996	983	4.91%	1016	8.43%	15	102.3 s
A-n53-k7	1010	1073	1063	5.25%	1082	7.13%	7	124.9 s
A-n60-k9	1408	1431	1423	1.07%	1438	2.13%	5	154.3 s
A-n69-k9	1168	1226	1203	3.00%	1266	8.39%	25	168.1 s
A-n80-k10	1764	1884	1872	6.12%	1896	7.48%	10	201.1 s

necessitates more cycles compared to the original algorithm. Consequently, the time complexity of the optimized algorithm experiences a dramatic increase. Furthermore, as the number of nodes and vehicles increases, the algorithm's runtime also increases proportionally.

4.3 Discussion of AELPSO Performance

Particle swarm optimization algorithms for solving specific logistics scheduling problems include particle decoding algorithms, path planning algorithms, and particle swarm optimization algorithms. The position vectors and velocity vectors of AELPSO are encoded and interpreted based on real numbers, where some of the dimensions are related to the customer points, and some measurements are related to the delivery vehicles. The encoding and decoding scheme of AELPSO involves constructing priority matrices for the assigned vehicles regarding the customer points. Based on these priority matrices and the current remaining load capacity of each vehicle volume, the algorithm assigns customer points to vehicles. The path planning algorithm based on the greedy strategy eliminates the limitation of the initial path planning algorithm. This initial algorithm constructs the Hamiltonian loop starting from the distribution center. Applying the greedy strategy to construct with different nodes as the start and end points may produce different results. AELPSO and path planning schemes have made significant progress in terms of solution quality compared to standard particle swarm optimization algorithms. The optimized path planning scheme can significantly improve the quality of the solution. However, the original algorithm for finding the shortest path for each vehicle for each particle necessitates more cycles compared to the original algorithmic scheme of loops. Consequently, the time complexity of the optimized algorithm experiences a significant increase. Moreover, as the number of nodes increases, the number of times the algorithm runs also increases. Additionally, with each run of the algorithm, the time complexity further escalates. The more vehicles there are, the more time the algorithm takes to run once, which is one of the disadvantages of the improved algorithm.

5 Conclusion

This paper provides a detailed description of the designing situation and defines the model of the logistics scheduling problem that is required to be solved in this paper. An actual number encoding and decoding particle scheme is employed to accomplish the task of allocating customer points in the logistics scheduling problem. Additionally, a path planning scheme based on the greedy strategy is introduced to derive insights into planning distribution routes for delivery vehicles in the logistics design experiment. Following experiments, the parameters of the algorithm are set to solve the dataset and analyze it using the standard particle swarm optimization algorithm. Further, the AELPSO and optimized path planning algorithm are proposed, and the data set is solved and analyzed. AELPSO with inertia weight linear decreasing method, asynchronous change adjusting learning factor method, natural selection idea, and optimized path planning scheme based on greedy strategy are introduced and applied to solve the data set. Solving the data set using the standard particle swarm algorithm is more effective, while solving the data set using the AELPSO can further reduce the gap with the optimal solution. However, this improved greedy strategy leads to a significant increase in the algorithm's time complexity compared to the original standard particle swarm algorithm. Therefore, future work will focus on reducing the algorithm's time complexity.

References

1. Kennedy, J.: Parameter selection in particle swarm optimization. In: Proceedings of the IEEE International Conference on Neural Networks, Australia (1998)
2. Lee, C.Y., Lee, Z.J., Lin, S.W., et al.: An enhanced ant colony optimization (EACO) applied to capacitated vehicle routing problem. Appl. Intell. **32**(1), 88–95 (2010)
3. Li, W., Meng, X., Huang, Y., et al.: Multi population cooperative particle swarm optimization with a mixed mutation strategy. Inf. Sci. **529** (2020)
4. Cong, W.U., Jianhui, Y., School of Computer Science and Technology, et al.: Vehicle routing problem of logistics distribution based on improved particle swarm optimization algorithm. Comput. Eng. Appl. (2015)
5. Li, W., Chen, Y., Cai, Q., et al.: Dual-stage hybrid learning particle swarm optimization algorithm for global optimization problems. Complex Syst. Model. Simul. **2**(4), 288–306 (2022)
6. Nickabadi, A., Ebadzadeh, M.M., Safabakhsh, R.: A novel particle swarm optimization algorithm with adaptive inertia weight. Appl. Soft Comput. **11**(4), 3658–3670 (2011)
7. Li, W., Meng, X., Huang, Y., et al.: An efficient particle swarm optimization with multidimensional mean learning. Int. J. Pattern Recognit Artif Intell. **3**, 35 (2021)
8. Sadok, A.: A genetic local search algorithm for the capacitated vehicle routing problem. Association of Computer, Communication and Education for National Triumph Social and Welfare Society (ACCENTS), (48) (2020)
9. Ai, T.J., Kachitvichyanukul, V.: Particle swarm optimization and two solution representations for solving the capacitated vehicle routing problem. Comput. Ind. Eng. **56**(1), 380–387 (2009)
10. Ai, T.J., Kachitvichyanukul, V.: A particle swarm optimization for the capacitated vehicle routing problem. Int. J. Log. SCM Syst. **2**, 50–55 (2007)

A Modified Two_Arch2 Based on Reference Points for Many-Objective Optimization

Shuai Wang[1], Dong Xiao[1], Futao Liao[1], Shaowei Zhang[1], Hui Wang[1(✉)], Wenjun Wang[2], and Min Hu[1]

[1] School of Information Engineering, Nanchang Institute of Technology, Nanchang 330099, China
huiwang@whu.edu.cn
[2] School of Business Administration, Nanchang Institute of Technology, Nanchang 330099, China

Abstract. Many-objective optimization problems (MaOPs) refer to those multi-objective problems (MOPs) having more than three objectives. For MaOPs, the performance of most multi-objective evolutionary algorithms (MOEAs) often deteriorates because it is hardly to achieve a balance between convergence and population diversity. To address this issue, this paper proposes a modified Two_Arch2 based on reference points (called Two_Arch2-RP). Firstly, a simplified reference point based dominance (RP-dominance) and cosine distance are used to update the convergence archive (CA). Then, the niche-preserving operation based reference points is combined with Pareto dominance to maintain the diversity archive (DA). To test the optimization capability of the proposed Two_Arch2-RP, seven DTLZ benchmark problems with 3, 5, 8, and 15 objectives are used. Experimental results show that Two_Arch2-RP obtains superior performance when compared with five other state-of-the-art algorithms.

Keywords: Evolutionary algorithm · many-objective optimization · reference points · two-archive

1 Introduction

Multi-objective optimization problems (MOPs) [1–3] widely exist in real world applications, such as engineering design [4], software engineering [5], and industrial scheduling [6]. In the past decades, evolutionary algorithms (EAs) have achieved good performance in solving MOPs with 2 or 3 objectives [2,7]. Then, many different multi-objective EAs (MOEAs) were proposed. When the number of objectives exceeds 3, the MOPs are called many-objective optimization problems (MaOPs). For MaOPs, the optimization performance of most MOEAs is seriously affected. The main reason is that most solutions in the population become incomparable. Balancing population diversity and convergence is a challenging task.

© The Author(s), under exclusive license to Springer Nature Singapore Pte Ltd. 2024
K. Li and Y. Liu (Eds.): ISICA 2023, CCIS 2146, pp. 125–136, 2024.
https://doi.org/10.1007/978-981-97-4393-3_11

In order to solve MaOPs, various many-objective evolutionary algorithms (MaOEAs) have been proposed [8–11]. Generally, these MaOEAs can be classified into three categories: Pareto-based, decomposition-based, and indicator-based. In addition, some hybrid MaOEAs were proposed. In [12], Wang et al. [12] suggested an improved two-archive algorithm (called Two_Arch2) for MaOPs. Two_Arch2 is based on both Pareto dominance and indicator, which combines the advantages of Pareto and indicator-based MOEAs. In Two_Arch2, two archives namely convergence archives (CA) and diversity archives (DA) are used. For CA, a convergence indicator ($I_{\epsilon+}$) [13] is employed to choose offspring and improve the convergence. For DA, the Pareto dominance is used as the principle for enhancing diversity.

By the suggestions of [12], though Two_Arch2 achieves excellent performance on many MaOPs, it has some deficiencies. Two_Arch2 shows good convergence but poor diversity. In addition, the update of DA archives involves a high time cost, and it is also necessary to set the number of CA archives and the parameter p for calculating the distance [14]. To address these issues, this article proposes a modified Two_Arch2 based on reference points (Two_Arch2-RP). The main contributions of this work can be summarized as follows:

- A simplified reference point based-dominance (RP-dominance) [15] and cosine distance are used to replace the indicator $I_{\epsilon+}$ to update CA.
- The niche-preserving operation based reference points is combined with the Pareto dominance to update DA.
- The proposed Two_Arch2-RP can effectively maintain diversity by the utilizing of reference points. Finally, it can make a good balance between diversity and convergence.

The rest of this paper is organized as follows. Section 2 briefly introduces the two-archive algorithm and its variants. Section 3 describes the proposed approach. Experimental results and analysis are given in Sect. 4. Finally, this paper is concluded in Sect. 5.

2   Related Work

To address the challenges posed by problems with a large number of objectives, Praditwong and Yao [16] devised a new MOEA (called Two_Arch), which employs a two-archive approach. In Two_Arch, two archives namely convergence archives (CA) and diversity archives (DA) are utilized to promote both convergence and diversity. Throughout the search process, non-dominated solutions are employed for updating CA and DA. Specifically, if a non-dominated solution dominates any solution in either CA or DA, it is added to CA. If a non-dominated solution cannot dominate any solution in CA and DA, it is added to DA. The detailed steps of update CA and DA can be found in [16].

Since Two_Arch relies only on Pareto dominance and CA lacks a mechanism to maintain diversity, the algorithm performs poorly for MaOPs. To alleviate these problems, Wang et al. [12] proposed an improved version of the algorithm (called Two_Arch2). Firstly, they adopted the quality indicator $I_{\epsilon+}$ [13] based on the indicator-based evolutionary algorithm (IBEA) as the selection rule for CA. Then, the Euclidean distance does not allow a better assessment of the solution distance in high-dimensional spaces. When the diversity archive (DA) overflows, non-dominated solutions are removed based on the fractional distance (L_p-norm, $p < 1$).

As mentioned before, though Two_Arch2 has obtained good performance on many MaOPs, it still has some disadvantages. To improve the performance of Two_Arch2, several modified versions of Two_Arch2 were proposed. Lei et al. [17] designed two update strategies based on the aggregation framework and integrated them into Two_Arch2. In [14], a new two-archive algorithm based on a multi-indicator (SRA3) was proposed. In SRA3, good solutions can be efficiently chosen in environment selection based on performance indicators. An adaptive parameter strategy is used for parental selection without setting additional parameters.

3   Proposed Approach

In this paper, an improved Two_Arch2 algorithm based on reference points (Two_Arch2-RP) is proposed. In Two_Arch2-RP, there are two main modifications. Firstly, a simplified RP-dominance and and cosine distance are used to replace the indicator $I_{\epsilon+}$ to update CA. Then, the niche-preserving operation based reference points is combined with the Pareto dominance to update DA. The detailed Two_Arch2-RP is described as follows.

3.1   Convergence Archive

The update of CA in Two_Arch2 is based on the indicator $I_{\epsilon+}$ taken from IBEA [13]. However, the $I_{\epsilon+}$ is overly biased towards the convergent solutions. Reference point-based dominance (RP-dominance) [15] is a dominance relation based on a set of well-distributed reference points. It can guide the candidate solutions toward convergence along the direction of the reference points. The uniform reference points can guarantee the distribution of solutions. To further improve the convergence and promote a better distribution synchronously, we use a simplified version of RP-dominance (SRP) and cosine distance to update CA in this paper.

Simplified RP-dominance (SRP): Supposing that v and u are two solutions. All solutions are associated with the reference point of the closest Euclidean distance. Each reference point may associate with multiple solutions, which constitute a subregion. The subregion index of these two solutions are $I(v)$ and $I(u)$. Let $d(v)$ and $d(u)$ represent the distance between v and u in the closest reference point of the projection point to the origin. If u dominates v (denoted $u \prec_{SRP} v$), the following conditions hold true:

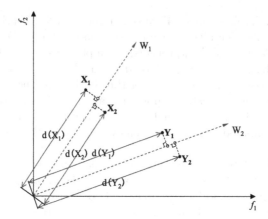

Fig. 1. Illustration of the simplified RP-dominance.

Algorithm 1: Update CA

---

1    $CA_t = CA_t \cup$ Offspring;
2    $(F_1, F_2, F_3, \dots) \leftarrow$ Non-dominated sorting is conducted on CA by the simplified RP-dominance;
3    $i \leftarrow 1$, $CA_{t+1} \leftarrow \emptyset$;
4   **while** $\mid CA_{t+1} \mid + \mid F_i \mid \le N$ **do**
5       $CA_{t+1} = CA_{t+1} \cup F_i$;
6       $i \leftarrow i + 1$;
7   **end**
8   Calculate the cosine distance of all solutions in F_i;
9   **while** CA_{t+1} *is not full* **do**
10   Choose a solution $X^* \in F_i$ with the smallest cosine distance;
11   Move X^* to CA_{t+1};
12   **end**

---

1) v Pareto dominates u.
2) u and v are Pareto equivalent, $I(v) = I(u)$ and $d(v) < d(u)$.

Unlike the original RP-dominance, we only establish dominance relations in the same subregion because CA mainly focuses on convergence. Figure 1 presents an example of the simplified RP-dominance (SRP). As seen, four non-dominated solutions are associated according to the vertical distance from the reference points. The objective space is divided into red and green subregions. $d(X_1) > d(X_2)$ in the W_1 subregion, and $d(Y_1) < d(Y_2)$ in the W_2 subregion. Therefore, solutions X_2 and Y_1 with better convergence are preferred within the same region. At this time, X_2 is said to dominate X_1 in modified RP-dominance (denoted $X_2 \prec_{SPRP} X_1$) and Y_1 is said to dominate Y_2 in the simplified RP-dominance (denoted $Y_1 \prec_{SRP} Y_2$).

Algorithm 1 describes the update method of CA. Firstly, the offspring is added to CA (Line 1). Then, the non-dominated sorting is conducted on CA by the simplified RP-dominance (Line 2). For Lines 4-7, some non-dominated solutions are added to CA_{t+1}, and the last non-dominated set F_i is obtained. The cosine distance of all solutions in F_i is calculated (Line 8). When CA is not

---

Algorithm 2: Update DA

---

1   $DA_t = DA_t \cup$ Offspring;
2   $(F_1, F_2, F_3, \ldots) \leftarrow$ Non-dominated sorting is conducted on DA by the Pareto dominance;
3   $i \leftarrow 1, DA_{t+1} \leftarrow \emptyset$;
4   **while** $| DA_{t+1} | + | F_i | \leq N$ **do**
5   $\quad DA_{t+1} = DA_{t+1} \cup F_i$;
6   $\quad i \leftarrow i + 1$;
7   **end**
8   $Q = DA_{t+1} \cup F_i$;
9   Calculate the PBI value of the population Q;
10 All solutions in Q are associated with reference points according to the PBI value;
11 **while** DA_{t+1} *is not full* **do**
12 $\quad$ The niche-preserving operation taken from NSGA-III is used to choose a solution $\breve{X}$
        from F_i;
13 $\quad$ Move $\breve{X}$ to DA_{t+1};
14 **end**

---

full, solutions with the smallest cosine distance are added to CA_{t+1} until the
size of CA_{t+1} is equal to N (Lines 9-12).

3.2   Diversity Archive

In Two_Arch2, the fractional distance (L_p-norm, $p < 1$) is used to replace the
Euclidean distance to gauge the similarity. However, the diversity maintenance
based on the L_p-norm distance cannot work well for MaOPs. To emphasize the
diversity, this paper uses a new diversity maintenance method based on reference
points taken from NSGA-III [18].

In the update of DA, the Pareto dominance is still employed to choose non-
dominated solutions. All solutions before the critical layer are added to DA.
Solutions in DA and the critical layer are associated with reference points in
terms of the penalized boundary intersection (PBI) method. Then, the niche-
preserving operation taken from NSGA-III is used as the selection principle to
choose non-dominated solutions from the critical layer. Unlike NSGA-III, we
use the PBI method to associate solutions with reference points. The PBI is a
reference vector-based decomposition method used in MOEA/D [19]. In PBI, $d1$
and $d2$ control convergence and diversity. So, the convergence and diversity can
be balanced by a preset penalty factor θ, which is usually set to 5. In this paper,
we also adopt the suggested value.

Algorithm 2 describes the update method of DA. Similarly, DA is firstly
combined with offspring (Line 1). Non-dominated sorting is conducted on DA
by the Pareto dominance. All solutions before the critical layer are added DA
(Line 2-7). Then, solutions in the current DA (DA_{t+1}) and the critical layer (F_i)
are stored in a temporary population Q. The PBI value of Q is calculated. All
solutions in Q are associated with reference points according to the PBI value
(Line 8-10). By the niche-preserving operation, solutions are chosen from F_i and
added to DA_{t+1} until DA_{t+1} is full (Lines 11-14).

Algorithm 3: Proposed Approach (Two_Arch2-RP)

---

1 Randomly initialize the population P and reference point $W_i = (W_1, W_2, \ldots, W_N)$;
2 $(F_1, F_2, F_3, \ldots) \leftarrow$ Non-dominated sorting is conducted on the population P by the Pareto dominance;
3 Set CA $= P$ and DA$=F_1$;
4 **while** $FEs \leq MAXFES$ **do**
5   Reproduction (crossover between CA and DA and mutation on CA only);
6   Update CA according to Algorithm 1;
7   Update DA according to Algorithm 2;
8 **end**

3.3   Framework of Proposed Approach

In Two_Arch2-RP, we use the same framework of Two_Arch2 including population initialization and genetic operations. The main differences between Two_Arch2-RP and Two_Arch2 focus on the update of CA and DA. In Two_Arch2-RP, the simplified RP-dominance is employed to generate nondominated solutions, and the cosine distance is used as the selection principle to update CA. To maintain the diversity of DA, the niche-preserving operation based reference points is utilized.

The main framework of Two_Arch2-RP is listed in Algorithm 3. At the beginning, the population P and the reference points are initialized (Line 1). Then, CA and DA are initialized based on the idea of Two_Arch2 (Lines 2-3). The reproduction used for generating offspring consists of crossover and mutation, which keep the same with Two_Arch2 (Line 5). The two archives CA and DA are updated according to Algorithm 1 and Algorithm 2, respectively (Lines 6-7).

4   Experimental Study

4.1   Benchmark Problems and Parameter Settings

In order to verify the performance of Two_Arch2-RP, seven well-known MaOPs are chosen from the DTLZ benchmark set [20]. For each problem, the number of objectives M is set to 3, 5, 8, and 15, respectively. The number of the corresponding decision variables D is set to 12, 14, 17, and 24, respectively. Performance of Two_Arch2-RP is compared with five other state-of-the-art MaOEAs. The involved algorithms are listed as below.

– NSGA-III [18].
– MOEA/D [19].
– KnEA [21].
– RVEA [22].
– Two_Arch2 [12].
– Proposed approach Two_Arch2-RP.

In order to have a fair comparison, the maximum number of function evaluations ($MAXFEs$) and population size (N) use the same settings. When M is set to 3, 5, 8, and 15, $MAXFEs$ is set to 30000, 50000, 80000, and 120000, respectively. And the corresponding population size N is set to 91, 126, 156, and 240, respectively.

For each problem, each algorithm is run 20 times. All algorithms and default settings used for comparisons were implemented on the PlatEMO platform [23]. To measure the performance of different MaOEAs, two popular performance indicator, HV [24] and IGD [13], are used. Both indicators are usually used to evaluate the convergence and diversity of MaOEAs. A smaller IGD value means a higher quality solution, and a larger HV means a better performance of the algorithm. Moreover, the Wilcoxon rank sum test with the significance level 0.05 is used to analyze the HV and IGD results. The symbols "$+/-/=$" summarize the comparison results between our approach and other compared algorithms, where "$+$", "$-$", and "$\approx$" represent that Two_Arch2-RP is significantly worse than, significantly better than, and statistically similar to the compared algorithm, respectively.

4.2   Results and Discussions

Table 1 shows the HV values obtained by MOEA/D, NSGA-III, RVEA, KnEA, Two_Arch2, and Two_Arch2-RP. It can be seen that Two_Arch2-RP outperforms MOEA/D, NSGA-III, RVEA, KnEA, and Two_Arch2 on 13, 17, 12, 23, and 20 test instances, respectively. Two_Arch2-RP performs better than other algorithms on DTLZ4 and DTLZ7 with 8 and 15 objectives. On DTLZ5 and DTLZ6, the performance of MOEA/D is slightly better than Two_Arch2-RP, but Two_Arch2-RP is not worse than MOEA/D on DTLZ3, DTLZ4, and DTLZ7. On all 28 test instances, NSGA-III, RVEA, KnEA, and Two_Arch2 achieve better HV values than Two_Arch2-RP on no more than 4 out of 28 test instances. It means that Two_Arch2-RP is not worse than NSGA-III, RVEA, KnEA, and Two_Arch2 on at least 24 test instances. In contrast with Two_Arch2, our approach Two_Arch2-RP obtains better HV values on 20 test instances. It shows that the proposed strategies can effectively improve the performance of Two_Arch2.

Table 1. HV values of different MaOEAs on the DTLZ benchmark set.

Problem	M	MOEA/D	NSGA-III	RVEA	KnEA	Two_Arch2	Two_Arch2-RP
DTLZ1	3	8.3791e-1 =	8.3869e-1 =	8.3897e-1 =	7.5661e-1 -	8.2936e-1 -	**8.3773e-1**
	5	9.7422e-1 +	9.7383e-1 +	**9.7457e-1** +	5.9979e-1 -	9.6753e-1 =	9.6812e-1
	8	9.9603e-1 =	9.6258e-1 =	**9.9754e-1** +	2.9033e-1 -	9.9006e-1 -	9.9457e-1
	15	8.0659e-1 -	9.9649e-1 =	9.9805e-1 =	0.0000e+0 -	9.9021e-1 -	**9.9594e-1**
DTLZ2	3	5.5947e-1 =	**5.5949e-1** +	5.5942e-1 =	5.3730e-1 -	5.5753e-1 -	5.5944e-1
	5	**7.9478e-1** +	7.9455e-1 =	7.9477e-1 +	7.7046e-1 -	7.3619e-1 -	7.9439e-1
	8	9.2378e-1 =	9.1680e-1 -	**9.2387e-1** =	8.8641e-1 -	7.5771e-1 -	9.2371e-1
	15	9.4180e-1 -	9.8130e-1 -	**9.9073e-1** +	7.8354e-1 =	7.0490e-1 -	9.9048e-1
DTLZ3	3	2.5968e-1 =	3.6115e-1 +	2.1350e-1 =	**4.2167e-1** +	1.2781e-1 =	1.8697e-1
	5	5.0225e-1 =	3.6639e-1 =	4.6009e-1 =	4.1800e-1 =	**5.0437e-1** =	4.6801e-1
	8	6.7231e-1 =	3.1121e-1 -	**9.0146e-1** =	0.0000e+0 -	6.6900e-1 -	7.8111e-1
	15	2.3817e-1 -	2.9271e-1 -	**9.8615e-1** +	0.0000e+0 -	2.4225e-3 -	8.5409e-1
DTLZ4	3	3.8979e-1 -	5.3177e-1 =	5.5938e-1 =	5.4225e-1 -	5.5725e-1 -	**5.5945e-1**
	5	6.6025e-1 -	7.8712e-1 =	**7.9475e-1** =	7.7595e-1 -	7.1788e-1 -	7.9460e-1
	8	8.0588e-1 -	9.0968e-1 -	9.2407e-1 -	9.0180e-1 -	7.4292e-1 -	**9.2512e-1**
	15	9.2544e-1 -	9.8856e-1 -	9.9122e-1 -	9.9060e-1 -	7.3696e-1 -	**9.9312e-1**
DTLZ5	3	1.8191e-1 =	1.9392e-1 =	1.5092e-1 -	1.9102e-1 -	**1.9990e-1** +	1.9322e-1
	5	**1.2594e-1** +	1.0122e-1 -	1.0592e-1 -	8.2080e-2 -	1.1342e-1 -	1.1827e-1
	8	**1.0429e-1** +	8.8252e-2 -	9.0883e-2 -	7.0891e-2 -	9.3247e-2 -	9.6932e-2
	15	**9.4694e-2** +	6.8796e-2 -	9.0965e-2 =	2.9581e-2 -	9.0842e-2 =	9.1011e-2
DTLZ6	3	1.8187e-1 -	1.9053e-1 -	1.4456e-1 -	1.8732e-1 =	**2.0002e-1** +	1.9199e-1
	5	**1.2593e-1** +	7.6789e-2 -	9.6798e-2 -	6.2438e-2 -	1.0664e-1 -	1.1588e-1
	8	**1.0404e-1** +	2.2720e-2 -	9.2946e-2 -	5.6818e-3 -	9.1120e-2 -	9.6274e-2
	15	**9.4560e-2** +	5.6812e-3 -	9.0870e-2 =	0.0000e+0 -	9.0930e-2 =	7.3531e-2
DTLZ7	3	2.5405e-1 -	2.6687e-1 -	2.6321e-1 -	2.7587e-1 +	**2.7691e-1** +	2.7008e-1
	5	1.1332e-1 -	2.3584e-1 -	2.0345e-1 -	2.5173e-1 -	2.2600e-1 -	**2.5752e-1**
	8	4.2628e-4 -	1.9447e-1 -	1.5065e-1 -	1.3705e-1 -	1.4614e-1 -	**2.0447e-1**
	15	8.9008e-5 -	7.8028e-2 -	2.3709e-2 -	0.0000e+0 -	1.8334e-2 -	**1.5170e-1**
+/-/=		8/13/7	3/17/8	4/12/12	2/23/3	3/20/5	

Table 2 lists the IGD values achieved by Two_Arch2-RP and five other MaOEAs on the DTLZ benchmark. From the results, Two_Arch2-RP outperforms MOEA/D, NSGA-III, RVEA, KnEA, and Two_Arch2 on 13, 17, 12, 20, and 16 test instances, respectively. It means that Two_Arch2 performs better on most test problems. Two_Arch2-RP is superior to other algorithms on DTLZ4 and DTLZ7 with 5, 15, and 20 objectives. On DTLZ5 and DTLZ6, it seems that Two_Arch2-RP cannot achieve promising results. For the remaining test problems, Two_Arch2-RP obtains better results than other algorithms on most problems. From the comparison between Two_Arch2 and Two_Arch2-RP, Two_Arch2-RP is not worse than Two_Arch2 on four problems DTLZ1-DTLZ4, while Two_Arch2 outperforms Two_Arch2-RP on DTLZ7.

Table 2. IGD values of different MaOEAs on the DTLZ benchmark set.

Problem	M	MOEA/D	NSGA-III	RVEA	KnEA	Two_Arch2	Two_Arch2-RP
DTLZ1	3	2.1014e-2 =	2.0898e-2 =	2.0871e-2 =	4.6714e-2 -	2.8584e-2 -	**2.1167e-2**
	5	6.3729e-2 -	6.3931e-2 -	**6.3278e-2** =	2.0264e-1 -	6.3684e-2 =	6.3567e-2
	8	**9.4457e-2** +	1.3658e-1 -	9.7160e-2 =	1.4905e+0 -	1.0196e-1 -	9.9812e-2
	15	1.3953e-1 -	1.5255e-1 -	1.3529e-1 =	4.5222e+0 -	1.4025e-1 -	**1.3585e-1**
DTLZ2	3	**5.4467e-2** +	5.4471e-2 +	5.4502e-2 -	7.1972e-2 -	6.1349e-2 -	5.4495e-2
	5	1.9490e-1 -	1.9491e-1 -	1.9491e-1 -	2.1257e-1 -	1.9840e-1 -	**1.9461e-1**
	8	3.2504e-1 -	3.3090e-1 -	**3.1543e-1** +	3.8433e-1 -	3.7854e-1 -	3.2305e-1
	15	**5.2620e-1** +	5.6021e-1 =	5.2718e-1 +	6.5776e-1 -	5.7239e-1 -	5.3896e-1
DTLZ3	3	6.8596e-1 =	3.0730e-1 +	6.6131e-1 =	**2.1442e-1** =	1.1578e+0 =	1.0210e+0
	5	5.6829e-1 =	9.4118e-1 =	6.5303e-1 =	7.1411e-1 =	**4.3235e-1** =	6.2101e-1
	8	5.3580e-1 =	1.3787e+0 -	**3.3651e-1** +	1.1083e+2 -	5.6466e-1 -	5.1902e-1
	15	1.1003e+0 -	2.2401e+0 -	**5.3870e-1** +	7.3166e+2 -	2.9172e+0 -	6.4374e-1
DTLZ4	3	4.0428e-1 =	1.1539e-1 =	**5.4479e-2** +	6.9425e-2 -	5.9986e-2 -	5.4502e-2
	5	4.4858e-1 -	2.0913e-1 =	**1.9491e-1** -	2.1149e-1 -	2.0007e-1 -	1.9499e-1
	8	5.8590e-1 -	3.5197e-1 -	**3.1669e-1** +	3.7326e-1 -	3.8354e-1 -	3.3115e-1
	15	7.1109e-1 -	5.4729e-1 +	**5.2724e-1** +	5.4983e-1 -	5.5912e-1 -	5.4789e-1
DTLZ5	3	3.3791e-2 =	1.2554e-2 +	7.7188e-2 -	1.2533e-2 +	**6.2824e-3** +	1.4449e-2
	5	**3.2717e-2** +	1.6608e-1 -	1.7278e-1 -	1.6750e-1 -	7.1251e-2 =	7.9479e-2
	8	**2.5854e-2** +	2.5353e-1 -	3.8159e-1 -	2.9163e-1 -	1.4042e-1 =	1.3464e-1
	15	**5.2243e-2** +	4.8460e-1 -	6.6010e-1 -	4.7520e-1 -	3.4839e-1 -	1.0144e-1
DTLZ6	3	3.3880e-2 =	1.8823e-2 =	8.6757e-2 -	1.9318e-2 =	**6.4486e-3** +	1.6874e-2
	5	**3.2277e-2** +	3.1187e-1 -	1.2586e-1 -	3.8620e-1 -	1.1654e-1 -	1.1705e-1
	8	**2.5219e-2** +	8.6808e-1 -	2.8082e-1 -	9.9349e-1 -	3.3145e-1 -	1.3456e-1
	15	**5.2214e-2** +	1.3389e+0 -	2.3700e-1 -	2.0564e+0 -	6.2387e-1 -	1.2950e-1
DTLZ7	3	1.5448e-1 -	9.3874e-2 -	1.0754e-1 -	7.0940e-2 +	**5.9699e-2** +	8.6154e-2
	5	6.3330e-1 -	3.3633e-1 -	5.6856e-1 -	3.0831e-1 =	**2.9330e-1** +	3.0555e-1
	8	1.8129e+0 -	7.8299e-1 +	1.5803e+0 -	6.3476e-1 +	**6.1458e-1** +	9.4596e-1
	15	3.1950e+0 +	4.7336e+0 +	3.0472e+0 +	2.3032e+0 +	**1.7998e+0** +	8.0033e+0
+/-/=		10/13/5	6/17/5	8/12/8	4/20/4	6/16/6	

Figure 2 gives the distribution of solutions obtained by Two_Arch2-RP and five other MaOEAs on DTLZ4 with 8 objectives. As shown in Fig. 2, only NSGA-III, RVEA, KnEA, and Two_Arch2-RP achieve the true PF. However, for the HV metric, Two_Arch2-RP obtains the best results. For the IGD metric, Two_Arch2-RP is worse than RVEA, but Two_Arch2-RP is better than the rest of the four comparison algorithms. The reason is that RVEA can better maintain diversity by adaptive strategies to adjust the reference vectors on the DTLZ4 problem.

From the values of HV and IGD, Two_Arch2-RP achieves better results than five other MaOEAs on most test instances. It means that the proposed modifications on Two_Arch2 is effective in terms of diversity and convergence.

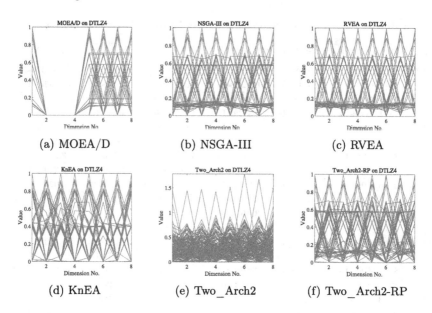

Fig. 2. The final solution set of MOEA/D,NSGA-III, RVEA,KnEA ,Two_Arch2, and Two_Arch2-RP on DTLZ4 with 8objectives.

5   Conclusion

In this paper, we suggest a modified Two_Arch2 (called Two_Arch2-RP) algorithm based on reference points for many-objective optimization. In Two_Arch2-RP, there are two main modifications: 1) a simplified RP-dominance and cosine distance are used to replace the indicator $I_{\epsilon+}$ to update CA; and 2) the niche-preserving operation based reference points is combined with the Pareto dominance to update DA. To verify the effectiveness of Two_Arch2-RP, seven DTLZ benchmark problems with 3, 5, 8, and 15 objectives are tested.

Performance of Two_Arch2-RP is compared with Two_Arch2, MOEA/D, RVEA, KnEA, and NSGA-III. According to the HV and IGD values, Two_Arch2-RP outperforms five other MaOEAs on most test instances. Results demonstrate that the proposed modifications can effectively improve the performance of Two_Arch2. This paper only uses the DTLZ benchmark. more benchmark problems will be tested. Especially for MaOPs with irregular PFs, more experiments will be investigated to determine whether Two_Arch2-RP is still effective.

Acknowledgment. This work was supported by National Natural Science Foundation of China (No. 62166027), and Jiangxi Provincial Natural Science Foundation (Nos. 20212ACB212004, 20212BAB202023, and 20212BAB202022).

References

1. Deb, K.: Multi-objective optimisation using evolutionary algorithms: an introduction. In: Multi-Objective Evolutionary Optimisation for Product Design and Manufacturing. Springer (2011) 3–34. https://doi.org/10.1007/978-0-85729-652-8_1
2. Deb, K., Pratap, A., Agarwal, S., Meyarivan, T.: A fast and elitist multiobjective genetic algorithm: NSGA-II. IEEE Trans. Evol. Comput. **6**(2), 182–197 (2002)
3. Coello, C.A.C., Pulido, G.T., Lechuga, M.S.: Handling multiple objectives with particle swarm optimization. IEEE Trans. Evol. Comput. **8**(3), 256–279 (2004)
4. Fleming, P.J., Purshouse, R.C., Lygoe, R.J.: Many-objective optimization: an engineering design perspective. In: International Conference on Evolutionary Multi-Criterion Optimization, Springer, pp. 14–32 (2005). https://doi.org/10.1007/978-3-540-31880-4_2
5. Yuan, Y., Xu, H.: Multiobjective flexible job shop scheduling using memetic algorithms. IEEE Trans. Autom. Sci. Eng. **12**(1), 336–353 (2013)
6. Ramirez, A., Romero, J.R., Ventura, S.: A survey of many-objective optimisation in search-based software engineering. J. Syst. Softw. **149**, 382–395 (2019)
7. Zitzler, E., Thiele, L.: Multiobjective evolutionary algorithms: a comparative case study and the strength pareto approach. IEEE Trans. Evol. Comput. **3**(4), 257–271 (1999)
8. Wang, S., Wang, H., Wei, Z., Wu, J., Liu, J., Zhang, H.: Many-objective artificial bee colony algorithm based on decomposition and dimension learning. In: International Conference on Neural Computing for Advanced Applications, Springer, pp. 150–161 (2022). https://doi.org/10.1007/978-981-19-6135-9_12
9. Wang, S., Wang, H., Wei, Z., Liao, F., Wang, F.: An enhanced subregion dominance relation for evolutionary many-objective optimization. In: International Conference on Neural Computing for Advanced Applications, Springer, pp. 220–234 (2022). https://doi.org/10.1007/978-981-99-5844-3_16
10. Wang, H., Wang, S., Wei, Z., Zeng, T., Ye, T.: An improved many-objective artificial bee colony algorithm for cascade reservoir operation. Neural Comput. Appl. **35**(18), 13613–13629 (2023)
11. Wang, H., Wei, Z., Yu, G., Wang, S., Wu, J., Liu, J.: A two-stage many-objective evolutionary algorithm with dynamic generalized pareto dominance. Int. J. Intell. Syst. **37**(11), 9833–9862 (2022)
12. Wang, H., Jiao, L., Yao, X.: Two_arch2: an improved two-archive algorithm for many-objective optimization. IEEE Trans. Evol. Comput. **19**(4), 524–541 (2014)
13. Zitzler, E., Künzli, S., et al.: Indicator-based selection in multiobjective search. In: PPSN. Volume 4., Springer, pp. 832–842 (2004). https://doi.org/10.1007/978-3-540-30217-9_84
14. Wang, Z., Yao, X.: An efficient multi-indicator and many-objective optimization algorithm based on two-archive. arXiv preprint arXiv:2201.05435 (2022)
15. Elarbi, M., Bechikh, S., Gupta, A., Said, L.B., Ong, Y.S.: A new decomposition-based NSGA-II for many-objective optimization. IEEE Trans. Syst. Man Cybern.: Syst. **48**(7), 1191–1210 (2017)
16. Praditwong, K., Yao, X.: A new multi-objective evolutionary optimisation algorithm: the two-archive algorithm. In: 2006 International Conference on Computational Intelligence and Security. Volume 1., IEEE, pp. 286–291 (2006)
17. Cai, L., Qu, S., Cheng, G.: Two-archive method for aggregation-based many-objective optimization. Inf. Sci. **422**, 305–317 (2018)

18. Deb, K., Jain, H.: An evolutionary many-objective optimization algorithm using reference-point-based nondominated sorting approach, part I: solving problems with box constraints. IEEE Trans. Evol. Comput. **18**(4), 577–601 (2013)
19. Zhang, Q., Li, H.: MOEA/D: a multiobjective evolutionary algorithm based on decomposition. IEEE Trans. Evol. Comput. **11**(6), 712–731 (2007)
20. Ye, T., et al.: An improved two-archive artificial bee colony algorithm for many-objective optimization. Expert Syst. Appl. **236**, 121281 (2024)
21. Zhang, X., Tian, Y., Jin, Y.: A knee point-driven evolutionary algorithm for many-objective optimization. IEEE Trans. Evol. Comput. **19**(6), 761–776 (2014)
22. Cheng, R., Jin, Y., Olhofer, M., Sendhoff, B.: A reference vector guided evolutionary algorithm for many-objective optimization. IEEE Trans. Evol. Comput. **20**(5), 773–791 (2016)
23. Tian, Y., Cheng, R., Zhang, X., Jin, Y.: Platemo: a Matlab platform for evolutionary multi-objective optimization [educational forum]. IEEE Comput. Intell. Mag. **12**(4), 73–87 (2017)
24. Bader, J., Zitzler, E.: Hype: an algorithm for fast hypervolume-based many-objective optimization. Evol. Comput. **19**(1), 45–76 (2011)

Floorplanning of VLSI by Mixed-Variable Optimization

Jian Sun[1], Huabin Cheng[2], Jian Wu[3], Zhanyang Zhu[3], and Yu Chen[1(✉)]

[1] School of Science, Wuhan University of Technology, Wuhan 430070, China
ychen@whut.edu.cn
[2] Department of Basic Science, Wuchang Shouyi University, Wuhan 430064, China
[3] School of Information Engineering, Wuhan University of Technology, Wuhan 430070, China

Abstract. By formulating the floorplanning of VLSI as a mixed-variable optimization problem, this paper proposes to solve it by a memetic algorithm, where the discrete orientation variables are addressed by the distribution evolutionary algorithm based on a population of probability model (DEA-PPM), and the continuous coordination variables are optimized by the conjugate sub-gradient algorithm (CSA). Accordingly, the fixed-outline floorplanning algorithm based on CSA and DEA-PPM (FFA-CD) and the floorplanning algorithm with golden section strategy (FA-GSS) are proposed for the floorplanning problems with and without fixed-outline constraint. Numerical experiments on GSRC test circuits show that the proposed algorithms are superior to some celebrated B*-tree based floorplanning algorithms, and are expected to be applied to large-scale floorplanning problems due to their low time complexity.

Keywords: VLSI · floorplanning · distribution evolutionary algorithm · conjugate sub-gradient algorithm

1 Introduction

Floorplanning is a critical stage in the physical design of very large-scale integration circuit (VLSI) that determines the performance of VLSI chips to a large extent [1]. It is a complex optimization problem with multiple objectives and constraints, which makes it challenging to develop high-performance algorithms for floorplanning of VLSI [2].

Floorplanning algorithms generally fall into two categories: the floorplanning algorithm based on combinatorial optimization model (FA-COM) and the floorplanning algorithm based on analytic optimization model (FA-AOM). Representing the relative positions of macros by combinatorial coding structures such as the B*-tree, the sequential pair, etc., one can formulate the floorplanning problem as a combinatorial optimization problem, which is then addressed by metaheuristics in the FA-COMs [3–6]. The combinatorial codes representing relative positions of macros can be naturally decoded into the compact floorplans complying with the non-overlapping constraints, however, the combinatorial explosion contributes to poor performances of FA-COM on large-scale cases. Accordingly, the problem size could be reduced by clustering or partitioning strategies,

© The Author(s), under exclusive license to Springer Nature Singapore Pte Ltd. 2024
K. Li and Y. Liu (Eds.): ISICA 2023, CCIS 2146, pp. 137–151, 2024.
https://doi.org/10.1007/978-981-97-4393-3_12

which in turn makes it hard to converge to the global optimal results of the investigated large-scale floorplanning problems [7, 8].

FA-AOMs address analytical floorplanning models by continuous optimization algorithms, which contributes to their lower time complexities on large-scale cases [9, 10]. Since the optimization results of continuous optimization algorithms do not fulfill the non-overlapping constraints for most cases, a FA-AOM usually consists of the global floorplanning stage and the legalization stage, the first optimizing the overall evaluation index, and the second tuning the positions of macros to eliminate constraint violations of results. Li *et al.* [11] proposed an analytic floorplanning algorithm for large-scale floorplanning cases, where the fixed-outline global floorplanning was implemented by optimizing the electrostatic field model of global placement. In the legalization stage, horizontal constraint graphs and vertical constraint graphs were constructed to eliminate overlap of floorplanning results. Huang *et al.* [12] presented an improved electrostatics-based analytical method for fixed-outline floorplanning, which incorporates module rotation and sizing driven by wirelength.

Since some of the evaluation indexes of global floorplanning are not smooth, additional smooth approximation to the optimization objective function could be incorporated to achieve fast convergence of gradient-based optimization algorithms. However, the approximation procedure not only introduces extra time complexity of the FA-AOM, but also leads to its local convergence to an optimal solution significantly different from that of the original non-smooth model. Accordingly, the conjugate subgradient algorithm [13] is employed in this paper to deal with the continuous variables representing coordinates of modules. Meanwhile, we address the orientation of modules by discrete variables, and formulate the floorplanning problem as a mixed-variable optimization problem.

Rest of this paper is organized as follows. Section 2 introduces some preliminaries. Then, the proposed algorithms developed for floorplanning problems with and without fixed-outline constraints are presented in Sects. 3 and 4, respectively. Numerical experiment is performed in Sect. 5 to demonstrate the competitiveness of the proposed algorithms, and Sect. 6 concludes this paper.

2 Preliminaries

2.1 Problem Statement

Given a collection of rectangular modules $V = \{v_1, v_2, \cdots, v_n\}$ and a set of edges (networks) $E = \{e_1, e_2, \cdots, e_m\}$, the VLSI floorplanning problem tries to minimize the total wirelength and the floorplan area by placing modules in approximate positions. Denote the center coordinates of module v_i be (x_i, y_i), and its orientation is represented by r_i. The orientation of modules is confirmed by clockwise rotation, and we set $r_i = j$ if the rotation angle is $\theta_i = j\pi/2$, $j = 0, 1, 2, 3$, $i = 1, \ldots, n$. A floorplan of VLSI is represented by the combination of vectors x, y and r, where $x = (x_1, \ldots, x_n)$, $y = (y_1, \ldots, y_n)$, $r = (r_1, \ldots, r_n)$. Subject to the constraint of placing non-overlapping modules with a fixed outline, the floorplanning problem is formulated as

$$\min\ W(x,y)$$

$$s.t. \begin{cases} D(x,y,r) = 0, \\ B(x,y,r) = 0, \end{cases} \tag{1}$$

Where $W(x,y)$ is the total wirelength, $D(x,y,r)$ is the sum of overlapping area, and $B(x,y,r)$ is the sum of width beyond the fixed outline. By the Lagrange multiplier method, it can be transformed into an unconstrained optimization model

$$\min\ f(x,y,r) = \alpha W(x,y) + \lambda \sqrt{D(x,y,r)} + \mu B(x,y,r), \tag{2}$$

where α, λ, μ are parameters to be confirmed. Here, the square root of $D(x,y,r)$ is adopted to ensure that all indexes to be minimized are of the same dimension.

Total Wirelength $W(x,y)$: The total wirelength is here taken as the total sum of half-perimeter wirelength (HPWL)

$$W(x,y) = \sum_{e \in E} (\max_{v_i \in e} x_i - \min_{v_i \in e} x_i + \max_{v_i \in e} y_i - \min_{v_i \in e} y_i). \tag{3}$$

Sum of Overlapping Area $D(x,y,r)$: The sum of overlapping area is computed by

$$D(x,y,r) = \sum_{i,j} O_{i,j}(x,r) \times O_{i,j}(y,r), \tag{4}$$

where $O_{i,j}(x,r)$ and $O_{i,j}(y,r)$ represent the overlapping lengths of modules i and j in the X-axis and Y-axis directions, respectively. Denoting $\Delta_x(i,j) = |x_i - x_j|$, we know

$$O_{i,j}(x,r) = \begin{cases} \min(\hat{w}_i, \hat{w}_j), & \text{if } 0 \le \Delta_x(i,j) \le \frac{|\hat{w}_i - \hat{w}_j|}{2}, \\ \frac{\hat{w}_i - 2\Delta_x(i,j) + \hat{w}_j}{2}, & \text{if } \frac{|\hat{w}_i - \hat{w}_j|}{2} < \Delta_x(i,j) < \frac{\hat{w}_i + \hat{w}_j}{2}, \\ 0, & \text{if } \Delta_x(i,j) \ge \frac{\hat{w}_i + \hat{w}_j}{2}, \end{cases} \tag{5}$$

where $\hat{w}_i$ is confirmed by

$$\hat{w}_i = \begin{cases} w_i, & \text{if } r_i \in \{0,2\}, \\ h_i, & otherwise, \end{cases} \quad i \in \{1,2,\ldots,n\}. \tag{6}$$

Denoting $\Delta_y(i,j) = |y_i - y_j|$, we have

$$O_{i,j}(y,r) = \begin{cases} \min(\hat{h}_i, \hat{h}_j), & \text{if } 0 \le \Delta_y(i,j) \le \frac{|\hat{h}_i - \hat{h}_j|}{2}, \\ \frac{\hat{h}_i - 2\Delta_y(i,j) + \hat{h}_j}{2}, & \text{if } \frac{|\hat{h}_i - \hat{h}_j|}{2} < \Delta_y(i,j) < \frac{\hat{h}_i + \hat{h}_j}{2}, \\ 0, & \text{if } \Delta_y(i,j) \ge \frac{\hat{h}_i + \hat{h}_j}{2}, \end{cases} \tag{7}$$

where $\hat{h}_i$ is confirmed by

$$\hat{h}_i = \begin{cases} h_i, & \text{if } r_i \in \{0,2\}, \\ w_i, & otherwise, \end{cases} \quad i \in \{1,2,\ldots,n\}. \tag{8}$$

Sum of Width beyond the Fixed Outline $B(x, y, r)$: For floorplanning problems with fixed-outline, the positions of modules must meet the following constraints:

$$\begin{cases} 0 \le x_i - \hat{w}_i/2, x_i + \hat{w}_i/2 \le W^*, \\ 0 \le y_i - \hat{h}_i/2, y_i + \hat{h}_i/2 \le H^*, \end{cases}$$

where W^* and H^* are the width and the height of square outline, respectively. Let

$$b_{1,i}(x, r) = \max(0, \hat{w}_i/2 - x_i), \ b_{2,i}(x, r) = \max(0, x_i + \hat{w}_i/2 - W^*),$$

$$b_{1,i}(y, r) = \max(0, \hat{h}_i/2 - y_i), \ b_{2,i}(y, r) = \max(0, y_i + \hat{h}_i/2 - H^*),$$

$\hat{w}_i$ and $\hat{h}_i$ are confirmed by (6) and (8), respectively. Accordingly, $B(x, y, r)$ can be confirmed by

$$B(x, y, r) = \sum_{i=1}^{n} (b_{1,i}(x, r) + b_{2,i}(x, r) + b_{1,i}(y, r) + b_{2,i}(y, r)). \tag{9}$$

which is smoothed by

$$\tilde{B}(x, y, r) = \sum_{i=1}^{n} (b_{1,i}^2(x, r) + b_{2,i}^2(x, r) + b_{3,i}^2(y, r) + b_{4,i}^2(y, r)). \tag{10}$$

Let $\beta = 0$, we get for legalization of the global floorplanning result the optimization problem

$$\min \tilde{f}(x, y, r) = \lambda_0 D(x, y, r) + \mu_0 \tilde{B}(x, y, r). \tag{11}$$

Algorithm 1: $u^* = CSA(f, u_0, k_{max}, s_0)$

Input: Objective function $f(u)$, Initial solution u_0, Initial step control parameter s_0, Maximum iterations k_{max};
Output: Optimal solution u^*;
1. $g_0 \in \partial f(u_0)$, $d_0 = 0$, $k \leftarrow 1$;
2.   **while** termination-condition 1 is not satisfied **do**
3.   calculated sub-gradient $g_k \in \partial f(u_{k-1})$;
4.   calculated Polak-Ribiere parameters $\eta_k = \dfrac{g_k^T(g_k - g_{k-1})}{\| g_{k-1} \|_2^2}$;
5.   computed conjugate directions $d_k = -g_k + \eta_k d_{k-1}$;
6.   calculating step size $a_k = s_{k-1}/\| d_k \|_2$;
7.   renewal solution $u_k = u_{k-1} + a_k d_k$, update u^*;
8.   update step control parameters $s_k = q s_{k-1}$;
9.   **end**

2.2 The Conjugate Sub-gradient Algorithm for Optimization of the Coordinate

Zhu *et al.* [13] proposed to solving the non-smooth continuous optimization model of the global placement by the conjugate sub-gradient algorithm (CSA). With an initial solution u_0, the pseudo code of CSA is presented in Algorithm 1. Because the CSA is not necessarily gradient-descendant, the step size has a significant influence on its convergence performance. The step size is determined by the norm of the conjugate directions together with the control parameter s_k, which is updated as $s_k = qs_{k-1}$. As an initial study, we set $q = 0.997$ in this paper. The *termination-condition 1* is satisfied if k is greater than a given budget k_{max} or several consecutive iterations fails to get a better solution.

2.3 The Distribution Evolutionary Algorithm for Optimization of the Orientation

Besides the coordinate vectors x and y, the floorplan is also confirmed by the orientation vectors r. Then, optimization of the orientation vectors contributes to a combinatorial optimization problem.

The estimation of distribution algorithm (EDA) is a kind of metaheuristics that can address the combinatorial optimization problem well, but its balance between global exploration and local exploitation is a challenging issue [14]. Xu *et al.* [15] proposed for the graph coloring problem a distribution evolutionary algorithm based on a population of probability model (DEA-PPM), where a novel probability model and the associated orthogonal search are introduced to achieve well convergence performance on large-scale combinatorial problems. The core idea of DEA-PPM for floorplanning is to simulate the probability distribution of orientations by constructing a probability matrix

$$q = (\vec{q}_1, \ldots, \vec{q}_n) = (q_{i,j})_{4 \times n}, \tag{12}$$

where $\vec{q}_j$ representing the probability that module j satisfies

$$||\vec{q}_j||_2^2 = \sum_{i=1}^{k} q_{ij}^2 = 1, \quad \forall j = 1, \ldots, n. \tag{13}$$

Then, the random initialization of q generates a distribution matrix

$$q(0) = (1/2)_{4 \times n}. \tag{14}$$

The implementation of DEA-PPM is based on distributed population $Q(t) = (q^{[1]}, \ldots, q^{[np]})$ and solution population $P(t) = (p^{[1]}, \ldots, p^{[np]})$, which are employed here for the probability distributions and instantiations of orientation, respectively. Global convergence of DEA-PPM is achieved by an orthogonal search on $Q(t)$, and the local exploitation are implemented in both the distribution space and the solution space.

3   The Fixed-Outline Floorplanning Algorithm Based on CSA and DEA-PPM

3.1   Framework

Algorithm 2: FFA-CD

Input: $f(x,y,r)$, $\tilde{f}(x,y,r)$.

Output: Optimal coordinate vector (x^*, y^*) and orientation vector r^*.

1.   initialize $Q(0)$ by (14), and generate $P(0)$ by sampling $Q(0)$;

2.   initialize X, Y by LHS, initialize the step control parameter s;

3.   let
$$(x^*, y^*, p^*) = \arg\ \min f(x,y,r), x \in X, y \in Y, r \in P(0);$$
set q^* as the distribution q corresponding to p^*;

4.   set $t \leftarrow 1$, $\alpha \leftarrow 1$, $\lambda \leftarrow 20$, $\mu \leftarrow 100$, $\lambda_0 \leftarrow 1$, $\mu_0 \leftarrow 1$, $k_{max} \leftarrow 50$;

5.   **while** termination-condition 2 is not satisfied **do**

6. $Q'(t) = OrthExpQ(Q(t-1), P(t-1))$;

7. $P'(t) = SampleP(Q(t), P(t-1))$;

8. $(P(t), X, Y, s) = UpdateXY(P'(t), X, Y, s)$;

9. $Q(t) = \text{Re}\,fineQ(P'(t), P(t), Q'(t))$;

10. $t = t+1$;

11.   **end**

In this paper, the FFA-CD is proposed to solve the problem of fixed-outline floorplanning, where the DEA-PPM is employed to optimize the orientations of the modules and the CSA is used to optimize the corresponding coordinates of the modules.

The framework of FFA-CD is presented in Algorithm 2. It starts with initialization of the distribution and solution populations $Q(0)$ and $P(0)$, where $P(0)$ consists of orientation combinations of modules. Meanwhile, the corresponding population X and Y of module coordinate is initialized by Latin hypercube sampling (LHS) [16]. Combining the orientation and coordinates of modules, we get the best coordinate vectors x^* and y^*, as well as the corresponding orientation vector r^*. Then, the *while loop* of FFA-CD is implemented to update $Q(t)$ and $P(t)$, where the CSA is deployed in *UpdateXY* to get the best module coordinate.

3.2   Evolution of the Distribution Population

In order to better explore the distribution space, DEA-PPM carries out orthogonal exploration for individuals in $Q(t)$. Algorithm 3 gives the flow of orthogonal exploration, which aims to change m worst individuals in Q by orthogonal transformation performed on c columns of a distribution matrix. Here, m is a random integer in $[1, np/2]$ and c is a random integer in $[1, n/10]$.

In Algorithm 4, the intermediate distribution population $Q'(t)$ is further updated to get $Q(t)$. Given $q^{[i]} \in Q'(t)$, it is updated using $v'[i]$ and $v^{[i]}$, two orientation combinations selected from P' and P, respectively. Columns of $q^{[i]}$ are updated using either a exploitation strategy or a disturbance strategy presented as follows.

Algorithm 3: $Q' = OrthExpQ(Q, P)$

Input: Q, P.

Output: Q'.

1. sorting Q by fitness values of corresponding individuals of Q;
2. take Q_w as the collection of m worst individuals of Q;
3. $Q' = Q \setminus Q_w$;
4. **for** $q \in Q_w$ **do**
5. $q' \leftarrow q$;
6.   randomly select c columns $\vec{q}_{jl} (l = 1, \cdots, c)$ from q';
7.   **for** $l = 1, \ldots, c$ **do**
8.   generate a random orthogonal matrix M_l;
9. $\vec{q}_{jl}' = M_l \vec{q}_{jl}'$;
10.   **end**
11. $Q' = Q' \cup q'$;
12. **end**

Algorithm 4: $Q = \mathrm{Re}\, fineQ(P', P, Q')$

Input: P', P, Q';

Output: Q;

1. **for** $i = 1, \ldots, np$ **do**
2. $q^{[i]} \in Q'$, $v'^{[i]} \in P'$, $v^{[i]} \in P$;
3.   **for** $j = 1, \ldots, n$ **do**
4.   set $rnd_j \sim U(0,1)$;
5.   **if** $rnd_j \leq p_0$ **then**
6. $\vec{r}_j^{[i]}$ is generated by the exploitation strategy (Eqs. (15) and (16));
7.   **else**
8. $\vec{r}_j^{[i]}$ is generated by the exploitation strategy (Eq. (17));
9.   **end if**
10.   **end for**
11. $r^{[i]} = (\vec{r}_1^{[i]}, \ldots, \vec{r}_n^{[i]})$;
12. **end for**
13. $Q = \bigcup_{i=1}^{np} r^{[i]}$;

The Exploitation Strategy: To update the jth column of $q^{[i]}$, it is first renewed as

$$r_{i,j}^{[i]} = \begin{cases} \sqrt{\alpha_0 + (1 - \alpha_0)(q_{i,j}^{[i]})^2}, & \text{if } l = v_j^{[i]} \\ \sqrt{(1 - \alpha_0)(q_{i,j}^{[i]})^2}, & \text{if } l \neq v_j^{[i]} \end{cases} \quad l = 1, \ldots, 4. \tag{15}$$

where $v_j^{[i]}$ is the jth component of $v^{[i]}$. Then, an local orthogonal transformation is performed as

$$\begin{bmatrix} r_{l_1,j}^{[i]} \\ r_{l_2,j}^{[i]} \end{bmatrix} = U(\Delta\theta_j) \times \begin{bmatrix} r_{l_1,j}^{[i]} \\ r_{l_2,j}^{[i]} \end{bmatrix}, \tag{16}$$

where $l_1 = v_j'^{[i]}$, $l_2 = v_j^{[i]}$. $U(\Delta\theta_j)$ is an orthogonal matrix given by

$$U(\Delta\theta_j) = \begin{bmatrix} \cos(\Delta\theta_j) & -\sin(\Delta\theta_j) \\ \sin(\Delta\theta_j) & \cos(\Delta\theta_j) \end{bmatrix},$$

The Disturbance Strategy: In order to prevent the distribution population from premature, let $l_0 = v_j^{[i]}$, the disturbance strategy is performed as

$$r_{i,j}^{[i]} = \begin{cases} \dfrac{\lambda(q_{l_0,j}^{[i]})^2}{1-(1-\lambda)(q_{l_0,j}^{[i]})^2} & , \text{if } l = l_0, \\[3mm] \dfrac{(q_{l,j}^{[i]})^2}{1-(1-\lambda)(q_{l_0,j}^{[i]})^2} & , \text{if } l \neq l_0. \end{cases} \tag{17}$$

3.3   Optimization of the Floorplan with a Fixed Outline

Algorithm 5: $P' = SampleP(Q, P)$

Input: P, Q;
Output: P';
1.   **for** $i = 1,\ldots,np$ **do**
2. $q^{[i]} \in Q$, $v^{[i]} \in P$;
3.   **for** $j = 1,\ldots,n$ **do**
4. $rnd_j \sim U(0,1)$;
5.   **if** $rnd_j \leq r$ **then**
6.   sampling $\bar{q}_j^{[i]}$ to get $v_j'^{[i]}$;
7.   **else**
8. $v_j'^{[i]} = v_j^{[i]}$;
9.   **end**
10.   **end**
11. $v'^{[i]} = (v_1'^{[i]},\ldots,v_n'^{[i]})$
12.   **end**
13. $P' = \bigcup_{i=1}^{np} v'^{[i]}$

The floorplan is represented by the orientation vector r and the coordinate vectors x and y. In FFA-CD, the evolution of orientation vectors is implemented by iteration of solution population $P(t)$, and the corresponding coordinate vectors are optimized by the function *UpdateXY*.

Initialization of the Module Orientation. According to the principle of DEA-PPM, the solution population $P'(t)$ is obtained by sampling the distribution population $Q'(t)$. To accelerate the convergence process, the sampling process is performed with inheritance as the process illustrated in Algorithm 5.

Optimization of Module Position. With the orientation of modules confirmed by the solution population, the position of the modules is optimized by Algorithm 6. For a combination of position vector $(x^{[i]}, y^{[i]})$, the global floorplanning is first implemented by optimizing f; Then, the weights of the constraint items is increased to legalize the floorplan approach by lines 4–7, or the legalization process is implemented by lines 9–10. The legalization process based on constraint graphs [11] are implemented *Graph()*, which is presented in Algorithm 7. To prevent X and Y from falling into inferior local solutions, the coordinates are reinitialized if no better solution is obtained for several times.

Algorithm 6: $(P, X, Y, s) = UpdateXY(P', X, Y, s)$

Input: P', X, Y, s;
Output: P, X, Y, s
1. **for** $i = 1, \ldots, np$ **do**
2. $\quad (x^{[i]}, y^{[i]}) = CSA(f_1, x^{[i]}, y^{[i]}, 50, s)$;
3. $\quad$ **if** $d_0 > \delta_1$ **then**
4. $\quad\quad \lambda = \min(1.5\lambda, \lambda + 30)$;
5. $\quad\quad$ **if** $c_0 > \delta_2$ **then**
6. $\quad\quad\quad \mu = \min(1.1\mu, \mu + 10)$;
7. $\quad\quad$ **end**
8. $\quad$ **else**
9. $\quad\quad (x^{[i]}, y^{[i]}) = CSA(\tilde{f}, x^{[i]}, y^{[i]}, 1000, \max(s/2, 50))$;
10. $\quad\quad (x^{[i]}, y^{[i]}) = Graph(x^{[i]}, y^{[i]})$;
11. $\quad$ **end**
12. **end**
13. $s = \max(0.95 \cdot s, s_{min})$;
14. **if** no better solution is obtained for several times **do**
15. $\quad$ reinitialize X, Y;
16. **end**

The legalization process is implemented in Algorithm 7 as follows. Let (x'_i, y'_i) be the lower-left coordinate of block v_i. v_i is *to the left of* v_j if it holds

$$O_{i,j}(y) > 0, O_{i,j}(y) > O_{i,j}(x), x'_i < x'_j;$$

v_i is *to the below of* v_j if

$$O_{i,j}(x) > 0, O_{i,j}(x) > O_{i,j}(y), y'_i < y'_j.$$

Denote I_i and J_i as the left-module set and the lower-module set of module i, respectively. Then, the x- and y-coordinates of module i are updated by

$$x_i' = \begin{cases} \max_{\forall v_j \in I_i} \left(x_j' + w_j \right), & \textit{if } I_i \neq \varnothing, \\ 0, & \textit{otherwise;} \end{cases} \tag{18}$$

$$y_i' = \begin{cases} \max_{\forall v_j \in J_i} \left(y_j' + h_j \right), & \textit{if } J_i \neq \varnothing, \\ 0, & \textit{otherwise.} \end{cases} \tag{19}$$

Algorithm 7: $(x^*, y^*) = Graph(x, y)$

Input: (x, y).

Output: (x^*, y^*).

1. sorting all modules according to the x-coordinates of the bottom-left corner and denote them as $\{v_1, v_2, \cdots, v_n\}$;
2. **for** $i \leftarrow 1$ **to** n **do**
3. update x_i' and x^* according to formula (18);
4. **end**
5. sorting all modules according to the y-coordinates of the bottom-left corner and denote them as $\{v_1, v_2, \cdots, v_n\}$;
6. **for** $i \leftarrow 1$ **to** n **do**
7. update y_i' and y^* according to formula (19);
8. **end**

---

4 The Floorplanning Algorithm Based on the Golden Section Strategy

While the analytical optimization method is applied to the floorplanning problem without fixed-outline, it is a challenging task to minimize the floorplan area. In this paper, we proposed FA-GSS, where minimization of the floorplan area is achieved by consecutively narrowing the contour of fixed outline.

Minimization of the floorplan area $S(x, y, r)$ is equivalent to minimizing the blank ratio

$$\gamma = \frac{S(x, y, r) - A}{A} * 100\%, \tag{20}$$

where A is the sum of areas of all modules. As presented in Algorithm 8, we use the golden section strategy to continuously reduce the area of the fixed contour. Given the initial white rate γ_{max} and γ_{min}, where the fixed-outline floorplanning is feasible for γ_{max} but infeasible for γ_{min}, we set

$$\gamma_m = 0.618 * (\gamma_{max} - \gamma_{min}) + \gamma_{min}.$$

If a legal layout can be obtained for γ_m, then $\gamma_{max} = \gamma_m$; otherwise, we set $\gamma_{min} = \gamma_m$. Repeat the section process until $\gamma_{max} - \gamma_{min} < \epsilon$.

Algorithm 8: FA-GSS

Input: $f(x, y, r)$, $\tilde{f}(x, y, r)$.
Output: Optimal solution (x^*, y^*, r^*).

1. initialize $Q(0)$ according to formula (7), and sample to generate $P(0)$;
2. initialize X, Y, step control parameter s;
3. set the best solution in $P(0)$ is p^*, the corresponding distribution matrix is q^*, and the module coordinates are (x^*, y^*);
4. $t \leftarrow 1$, $\lambda_0 \leftarrow 1$, $\mu_0 \leftarrow 10$, $k_{max} \leftarrow 50$;
5. initialize the maximum whitespace ratio γ_{max} and minimum whitespace ratio γ_{min};
6. **while** $\gamma_{max} - \gamma_{min} < \varepsilon$ **do**
7. $\alpha \leftarrow 1$, $\lambda \leftarrow 20$, $\mu \leftarrow 100$, $\gamma_m = 0.382 \times (\gamma_{max} - \gamma_{min}) + \gamma_{min}$;
8.   calculate the width W^* and height H^* of the fixed profile;
9.   **while** termination-condition 2 is not satisfied **do**
10. $Q'(t) = OrthExpQ(Q(t-1), P(t-1))$;
11. $P'(t) = SampleP(Q(t), P(t-1))$;
12. $(P(t), X, Y, s) = UpdateXY(P'(t), X, Y, s)$;
13. $Q(t) = \mathrm{Re}\,fineQ(P'(t), P(t), Q'(t))$;
14.   set the best solution for the current X, Y, P, and call it x', y', p';
15.   t=t+1, update p^*, q^*, (x^*, y^*), $k_{max} = 35$;
16.   **end**
17.   **if** $\tilde{f}(x', y', p') = 0$ **then**
18. $\gamma_{max} = \gamma_m$;
19.   **else**
20. $\gamma_{min} = \gamma_m$;
21.   **end**
22. **end**

---

5   Experimental Results and Analysis

To verify the performance of the proposed algorithm, we conducted experiments on the well-known test benchmark GSRC. For all test circuits, the I/O pads are fixed at the given coordinates, and the modules of all circuits are hard modules. All experiments are developed in C++ programming language program, and run in Microsoft Windows 10 on a laptop equipped with the AMD Ryzen 7 5800H @ 3.2 GHz and 16 GB system memory.

5.1   Wirelength Optimization with Fixed-Outline Constraints

We first test the performance of FFA-CD on the fixed-outline cases. It is compared with the well-known open source layout planner Parquet-4.5 [17], where the floorplan is represented by the B*-tree and the simulated annealing algorithm to solve the combinatorial optimization model of floorplanning.

According to the given aspect ratio R, the width W^* and height H^* of the fixed contour are calculated as [18]

$$W^* = \sqrt{(1 + \gamma)A/R}, \ H^* = \sqrt{(1 + \gamma)AR}, \tag{21}$$

where A is the summed area of all modules, and γ is the white rate defined in (20). The experiment set the white rate as $\gamma = 15\%$, the aspect ratio R as 1, 1.5, 2, and the population number as 5. For different aspect ratios, each experiment was independently run 10 times, and the results were shown in Table 1.

Table 1. Performance comparison for the fixed-outline cases of GSRC test problems.

GSRC	R	Parquet-4.5			FFA-CD		
		SR	HPWL	CPU	SR	HPWL	CPU
n10	1.0	60	**55603**	**0.04**	100	55774	0.11
	1.5	60	**55824**	**0.04**	100	56696	0.20
	2.0	80	58247	**0.04**	90	**58236**	0.31
n30	1.0	100	172173	**0.28**	100	**160208**	0.41
	1.5	90	173657	0.34	100	**164237**	**0.28**
	2.0	100	174568	**0.32**	100	**166133**	0.54
n50	1.0	100	209343	0.68	100	**185793**	**0.55**
	1.5	100	211591	0.79	100	**189878**	**0.41**
	2.0	100	208311	0.78	100	**195398**	**0.71**
n100	1.0	100	334719	2.10	100	**293578**	**0.89**
	1.5	100	340561	2.26	100	**300079**	**1.05**
	2.0	100	347708	2.26	100	**308811**	**1.02**
n200	1.0	100	620097	9.03	100	**521140**	**2.38**
	1.5	100	625069	9.07	100	**529918**	**2.53**
	2.0	100	649728	9.24	100	**541565**	**2.71**
n300	1.0	100	768747	19.08	100	**588118**	**3.73**
	1.5	100	787527	19.16	100	**606548**	**3.85**
	2.0	100	847588	19.65	100	**626658**	**4.21**

Numerical results demonstrate that FFA-CD outperforms Parquet-4.5 on cases with more than 50 modules, but runs a bit slow for some of the small cases, which is attributed to the compact floorplan of Parquet-4.5. The combinatorial floorplan implemented by Parquet-4.5 could lead to smaller HPWL and shorter runtime, but its performance would degrade significantly while the problem size increases. The iteration mechanism based on CSA ensures that FFA-CD can explore the floorplan space more efficiently. At the same time, DEA-PPM is introduced to explore the rotation strategy, which increases the flexibility of the floorplan and greatly improves the success rate of small-scale problems. Consequently, the success rate of FF-CD was better than or equal to Parquet-4.5 for all cases. Meanwhile, better results on wirelength and tuntime is obtained in several different aspect ratios for the larger-scale cases (n50–n100).

5.2 Minimization of Wirelength and Area Without Fixed-Outline Constraints

For floorplanning problems without fixed contour constraints, FA-GSS is used to optimize the wirelength and area. The proposed FA-GSS is compared with Parquet-4.5 and the Hybrid Simulated Annealing Algorithm (HSA) [19], where the population size is set as 5, and we get $\epsilon = 0.2\%$. Due to the different magnitude of wirelength and area, the cost function to minimized for the floorplanning problem without fixed outline is taken as

$$Cost = 0.5 * \frac{W}{W_{\min}} + 0.5 * \frac{S}{S_{\min}}, \tag{22}$$

where W_{min} and S_{min} are the minimum values of W and A, respectively.

The results in Table 1 show that all examples obtain better wirelength and shorter time when the aspect ratio is 1. So, we take $R = 1$ in FA-GSS for all test cases. For benchmarks in GSRC, the average $Cost$, and runtime (CPU) of ten independent runs are collected in Table 2.

The experimental results show that FA-GSS outperforms both Parquet-4.5 and HAS except for the n30 case. Although FA-GSS runs a bit slower than Parquet-4.5 when they are tested by the n30 case, FA-GSS has the smallest rate of increase in run time as the module size increases. This means that FA-GSS is expected to achieve excellent results on larger circuits.

Table 2. Performance comparison for the GSRC test problems without fixed-outline constraints.

GSRC	Parquet-4.5		HAS		FA-GSS	
	Cost	CPU (s)	Cost	CPU (s)	Cost	CPU (s)
n10	1.0885	**0.03**	1.0799	0.11	**1.0688**	0.17
n30	1.1040	**0.19**	**1.0881**	0.86	1.0959	0.69
n50	1.0871	**0.47**	1.0797	2.15	**1.0750**	1.29
n100	1.1034	**1.61**	1.1040	7.94	**1.0648**	3.53
n200	1.1301	**6.23**	1.1628	37.70	**1.0713**	8.96
n300	1.1765	**12.57**	1.2054	78.21	**1.0715**	15.31

In order to make the algorithm more explanatory, we compare the proposed algorithm with the latest research results in the industry. Liu *et al.* [20] combines reinforcement learning and genetic algorithm to propose GA+RL, which also adopts B*-tree and is compared with traditional hybrid genetic algorithm and simulated annealing (GA+SA) algorithm. Only interconnection relationships between modules are considered here. The experimental data comes from the literature, and the parameter setting of FA-GSS algorithm is consistent with the above (Table 3).

Table 3. GA+SA, GA+RL, FA-GSS area and wirelength minimization results.

GSRC	GA+SA		GA+RL		FA-GSS	
	HPWL (e5)	S (e5)	HPWL (e5)	S (e5)	HPWL (e5)	S (e5)
n10	**0.13**	**2.24**	0.14	2.41	0.15	2.46
n30	0.38	**2.25**	**0.37**	2.27	0.40	2.35
n50	0.91	2.20	0.89	**2.14**	**0.85**	2.23
n100	1.69	1.99	1.44	**1.97**	**1.13**	2.01
n200	3.66	2.07	3.66	2.03	**2.28**	**1.98**
n300	6.98	3.36	6.59	3.29	**3.25**	**3.09**

6 Conclusion

In this paper, we formulate the flooplanning problem of VLSI as a mixed-variable optimization problem, where the discrete variables represent module orientations and the coordinates of modules are incorporated by continuous variables. Then, the DEA-PPM is introduced to get the module orientation, and coordinate variables are optimized by the CSA. Experimental results show that the proposed FFA-CD and FA-GSS, respectively developed for floorplanning problems with and without fixed-outline, can generally outperforms the floorplanning algorithms designed based on the B*-tree and the simulated annealing. Attributed to their low time complexity, the proposed algorithms are expected to address large-scale floorplanning problems effectively.

Acknowledgement. This research was partially supported by the National Key R\& D Program of China (No. 2021ZD0114600) and the Guiding Project of Scientific Research Plan of Hubei Provincial Department of Education (No. B2022394).

References

1. Srinivasan, B., Venkatesan, R., Aljafari, B., et al.: A novel multicriteria optimization technique for VLSI floorplanning based on hybridized firefly and ant colony systems. IEEE Access **11**, 14677–14692 (2023)
2. Weng, Y., Chen, Z., Chen, J., et al.: A modified multi-objective simulated annealing algorithm for fixed-outline floorplanning. In: 2018 IEEE International Conference on Automation, Electronics and Electrical Engineering (AUTEEE), pp. 35–39. IEEE Press, Piscataway (2018)

3. Chen, J,. Zhu, W.: A hybrid genetic algorithm for VLSI floorplanning. In: 2010 IEEE International Conference on Intelligent Computing and Intelligent Systems, pp. 128–132. IEEE Press, Piscataway (2010)
4. Du, S., Xia, Y., Chu, Z., et al.: A stable fixed-outline floorplanning algorithm for soft module. J. Electron. Inf. Technol. **36**(5), 1258–1265 (2014)
5. Zou, D., Wang, G., Pan, G., et al.: A modified simulated annealing algorithm and an excessive area model for floorplanning using fixed-outline constraints. Front. Inf. Technol. Electron. Eng. **17**(11), 1228–1244 (2016)
6. Ye, Y., Yin, X., Chen, Z., et al.: A novel method on discrete particle swarm optimization for fixed-outline floorplanning. In: 2020 IEEE International Conference on Artificial Intelligence and Information Systems (ICAIIS), pp.591–595. IEEE Press, Piscataway (2020)
7. Yan, J.Z., Chu, C.: DeFer: deferred decision making enabled fixed-outline floorplanning algorithm. IEEE Trans. Comput. Aided Des. Integr. Circuits Syst. **29**(3), 367–381 (2010)
8. Ji, P., He, Kun., Wang, Z., et al.: A quasi-newton-based floorplanner for fixed-outline floorplanning. Comput. Oper. Res. **129**(6), 105225 (2010)
9. Lin, J.M., Hung, Z.X.: UFO: unified convex optimization algorithms for fixed-outline floorplanning considering pre-placed modules. IEEE Trans. Comput. Aided Des. Integr. Circuits Syst. **30**(7), 1034–1044 (2011)
10. Huang, Z., Li, X., Zhu, W.: Optimization models and algorithms for placement of very large scale integrated circuits. Oper. Res. Trans. **25**(3), 15–36 (2021)
11. Li, X., Peng, K., Huang, F., et al.: PeF: poisson's equation based large-scale fixed-outline floorplanning. IEEE Trans. Comput. Aided Des. Integr. Circuits Syst. **42**(6), 2002–2015 (2023)
12. Huang, F., Liu, D., Li, X., Yu, B., Zhu, W.: Handling orientation and aspect ratio of modules in electrostatics-based large scale fixed-outline floorplanning. In: 2023 IEEE/ACM International Conference on Computer Aided Design (ICCAD), pp. 1–9. IEEE Press, Piscataway (2023)
13. Zhu, W., Chen, J., Peng, Z., et al.: Nonsmooth optimization method for VLSI global placement. IEEE Trans. Comput. Aided Des. Integr. Circuits Syst. **34**(4), 642–655 (2015)
14. Zhou, S., Sun, Z.: A survey on estimation of distribution algorithms. Acta Automatica Sinica **02**, 113–124 (2007)
15. Xu, Y., Cheng, H., Xu, N., et al.: A distribution evolutionary algorithm for the graph coloring problem. Swarm Evol. Comput. **80**, 101324 (2023)
16. Xu, J., Lu, H., Zhao, J., et al.: Self-adaptive gaussian keyhole imaging butterfly optimization algorithm based on latin hypercube sampling. Appl. Res. Comput. **39**(9), 2701–2708, 2751 (2022)
17. Adya, S.N., Markov, I.L.: Fixed-outline floorplanning: enabling hierarchical design. IEEE Trans. Very Large Scale Integr. Syst. **11**(6), 1120–1135 (2003)
18. Zou, D., Hao, G., Pan, G., et al.: An improved simulated annealing algorithm and area model for the fixed-outline floorplanning with hard modules. In: 2015 3rd International Symposium on Computational and Business Intelligence (ISCBI), pp. 21–25. IEEE Press, Piscataway (2015)
19. Chen, J., Zhu, W., Ali, M.M., et al.: A hybrid simulated annealing algorithm for nonslicing VLSI floorplanning. IEEE Trans. Syst. Man Cybern. Part C (Appl. Rev.) **41**(4), 544–553 (2011)
20. Liu, K., Gu, J., Zhu, Z.: A hybrid reinforcement learning and genetic algorithm for VLSI floorplanning. In: 2023 15th International Conference on Machine Learning and Computing (ICMLC 2023), pp. 412–418. Association for Computing Machinery, New York (2023)

A Multi-population Hierarchical Differential Evolution for Feature Selection

Jian Guan[1,2], Fei Yu[1,2(✉)], and Zhenya Diao[1,2]

[1] School of Physics and Information Engineering, Minnan Normal University,
Zhangzhou 363000, China
yufei@whu.edu.cn

[2] Key Lab of Intelligent Optimization and Information Processing, Minnan Normal University,
Zhangzhou 363000, China

Abstract. In order to address the curse of dimensionality and reduce data processing costs, feature selection (FS) is an essential component. Since the process of selecting features is an optimization problem and prone to getting trapped in local optima, FS methods based on evolutionary computation (EC) can effectively tackle such problems. Therefore, this paper proposes a multi-population hierarchical differential evolution (MPDE) to solve the FS problem. In MPDE, inspired by the chicken swarm optimization (CSO) algorithm, a new multi-population communication mechanism is introduced, where individuals in each sub-population are hierarchically divided, and different levels of individuals adopt corresponding individual enhancement strategies to improve algorithm performance. MPDE is compared with five advanced EC-based FS methods on 18 datasets. The experiments demonstrate that MPDE can achieve higher classification accuracy while selecting fewer features, thus proving its effectiveness in solving the FS problem.

Keywords: Differential evolution · feature selection · multi-population mechanism · evolutionary computation

1 Introduction

Due to the rapid development of information science and the advent of the big data era, data processing and analysis have become exceptionally challenging. Taking classification problems in the real world as an example, there may exist redundant and irrelevant features in massive data, which can decrease the accuracy of classification and increase computational costs. Therefore, research on feature selection (FS) has become crucial. FS, also known as variable or attribute selection, is often used as an essential part of data preprocessing. It aims to select the primary features related to the predictive model's performance while eliminating irrelevant or redundant features. This can enhance the model's classification accuracy, reduce the risk of overfitting, and decrease computational costs. Most importantly, it provides interpretability for the model, which is crucial for many application scenarios.

In previous research, the methods of FS mainly include filters, wrappers, and embedded methods [1]. Filter methods primarily use the correlation between features and the

© The Author(s), under exclusive license to Springer Nature Singapore Pte Ltd. 2024
K. Li and Y. Liu (Eds.): ISICA 2023, CCIS 2146, pp. 152–164, 2024.
https://doi.org/10.1007/978-981-97-4393-3_13

target variable or statistical indicators to evaluate the importance of each feature. Examples of commonly used methods include Pearson correlation coefficient, chi-square test, and analysis of variance (ANOVA). Wrapper methods combine machine learning algorithms with the feature selection process, repeatedly training models to select important features. Common methods used in machine learning include K-nearest neighbor (KNN), support vector machines (SVMs), and artificial neural networks (ANNs) [2]. As for embedded methods, they are a combination of filter and wrapper methods. Firstly, features are evaluated based on certain evaluation criteria, and then wrapper methods are used to select relevant features.

However, traditional FS methods still have issues with low accuracy and high computational costs when dealing with large-scale data. Due to the significant advancements in global optimization, evolutionary computation has been widely applied. Therefore, feature selection methods based on Evolutionary Computation (EC) have attracted widespread attention from researchers. Classical evolutionary algorithms such as particle swarm optimization (PSO) [3], genetic algorithms (GA) [4], ant colony optimization (ACO) [5], genetic programming (GP) [6], and differential evolution (DE) have been extensively applied to FS. We define the FS problem as an optimization issue, with feature subsets acting as individuals within the algorithm. By using the established objective function, we evaluate the fitness of each individual (feature subset), and subsequently select the most promising solutions (optimal feature subsets) based on their fitness values. Ultimately, we employ a classifier to assess the chosen optimal feature subset, resulting in classification accuracy after the FS process.

In EC, DE is a global optimization algorithm proposed by Storn and Price [7], characterized by its simple structure and ease of operation. To our knowledge, the application of DE in the realm of feature selection is relatively limited. Therefore, in order to better integrate DE with FS and improve the performance of FS, in this paper, we propose a multi-population hierarchical differential evolution algorithm for feature selection (MPDE). Within MPDE, multiple sub-populations are included, each with its independent search space and optimal solution. To ensure populations con-verge rapidly to the global optimum, we introduce a novel sub-population communication mechanism to exchange information among them. Additionally, within these sub-populations, three layers of individuals exist, each playing distinct roles to maintain the exploration and exploitation of the population. The proposed algorithm was compared with several current EC-based FS methods on multiple datasets. Experimental results further demonstrate the efficacy of MPDE in addressing the FS problem. The main contributions of this paper are as follows:

1. This paper adopts multiple populations simultaneously and proposes a novel communication mechanism among subpopulations.
2. By ranking the fitness values, each population is divided into three layers. Each layer has different contributions to the population to balance exploration and exploitation.
3. MPDE is compared with several advanced feature selection methods based on evolutionary algorithms through experimental comparisons.
4. The organization of this paper is as follows. Section 2 briefly introduces the background and related work. Section 3 focuses on describing the proposed method in this

paper. Comparative experiments and result analysis are conducted in Sect. 4. Finally, the conclusion and future prospects will be provided in Sect. 5.

2 Background and Related Work

2.1 Differential Evolution

DE is a simple structure and easy to implement evolutionary algorithm, which mainly includes three operations: mutation, crossover and selection. The initial population of DE was randomly generated by $X = rand * (ub - lb) + lb$, where ub and lb are the maximum and minimum boundaries of the population, respectively.

(1) **Mutation:** The mutation operation is a crucial step in DE. The original vector $X_{r1,G}$ undergoes the mutation operator F to obtain a mutated vector $V_{i,G+1}$, defined as follows.

$$V_{i,G+1} = X_{r1,G} + F * \left(X_{r2,G} - X_{r3,G} \right) \tag{1}$$

where r_1, r_2 and r_3 are unique integers randomly selected from the range $[1, NP]$, NP is the population size; $F \in [0, 2]$ is a mutation operator.

(2) **Crossover:** In the crossover operation, the best genes are selected between the mutated vector $V_{i,G+1}$ and the original vector $X_{r1,G}$ to obtain a crossover vector $U_{i,G+1}$, defined as follows.

$$u^j_{i,G+1} = \begin{cases} v^j_{i,G+1}, & \text{if } rand \leq CR \text{ or } j = j_{rand} \\ x^j_{i,G}, & \text{otherwise} \end{cases} \tag{2}$$

where $CR \in [0, 1]$ is a crossover operator; $rand$ is a random numbers in the interval $(0, 1)$; and j and j_{rand} are respectively the j-th dimension and a dimension randomly chosen in the interval $[1, D]$, where D is the dimension of the individual.

(3) **Selection:** The selection operation uses greedy selection to obtain the final trial vector $X_{i,G+1}$ between the original vector $X_{r1,G}$ and the crossover vector $U_{i,G+1}$, defined as follows.

$$X_{i,G+1} = \begin{cases} U_{i,G+1}, & \text{iff } (U_{i,G}) \leq f(X_{i,G}) \\ X_{i,G}, & \text{otherwise} \end{cases} \tag{3}$$

where $f(U_{i,G})$ and $f(X_{i,G})$ are the fitness values of the crossover and original vectors, respectively, and the better ones are retained as trial vectors by comparing the fitness values.

2.2 Related Work

In the past two decades, many evolutionary algorithms (EAs) have been applied to feature selection (FS). According to different encoding methods, Xue et al. [8] categorized EC-based FS methods into four types: vector representation methods represented by GA and PSO, graph representation methods represented by ACO, tree representation methods represented by GP, and matrix representation methods utilizing sparse matrix transformation.

(1) **Vector representation:** Qiu et al. [9] proposed a multi-population particle swarm optimization (MSPSO) combined with FS; Xue et al. [10] treated feature selection as a multi-objective optimization problem and used multi-objective DE to optimize the objective functions; Salima et al. [11] used an enhanced crow search algorithm (ECSA) to select feature subsets. Mohammad et al. [12] combined opposition-based learning (OBL) and an improved salp swarm algorithm for FS.

(2) **Graph representation:** Ke et al. [13] used the ACO with limited pheromone information to detect the edges of different features. Vieira et al. [14] proposed ACO with two populations for collaborative optimization of feature selection. Khushaba et al. [15] combined ACO with DE to search for the optimal feature subset.

(3) **Tree representation:** Hunt et al. [16] proposed a wrapper feature selection method based on genetic programming (GP). Mei et al. [17] found that GP has feature construction capabilities while performing feature selection and improves interpretability.

(4) **Matrix representation:** Li et al. [18] combined sparse learning with feature selection, where each row of the matrix represents a feature.

Besides the representative algorithms mentioned in the aforementioned four solution representations, there are many other EC applications for FS. On one hand, FS is viewed as a single-objective optimization problem primarily focusing on classification accuracy. For instance, Too et al. [19] proposed a pbest best guided binary PSO for addressing electromyograph signal feature issues (PBPSO), offering valuable assistance for clinical and rehabilitation purposes. Since current PSO-based FS methods use a fixed-length representation, Tran et al. [20] introduced a variable-length PSO representation method for tackling the FS problem (VLPSO), which, to some extent, enhanced the performance of PSO while saving significant computational costs. As FS operates as an optimization challenge, multiple optimal feature subsets may exist. To obtain numerous optimal feature subsets, Wang et al. [21] put forward an FS method assisted by feature clustering (FCNDE). It employs the maximum information coefficient (MIC) to categorize features and is embedded within DE to select multiple optimal feature subsets. On the other hand, FS is treated as a multi-objective optimization problem, evaluating both classification accuracy and the number of selected features. For instance, Usman et al. [22] perceived FS as a multi-objective optimization issue, combining the non-dominated sorting GA (NSGA-III) with mutual information (MI) to tackle the FS problem, considering both classification performance and the number of selected features. In [23], Hancer et al. first applied neighborhood component analysis (NCA) to evolutionary feature selection and incorporated NCA into the objective function of the differential algorithm to address FS problems in both single-objective and multi-objective scenarios. Similarly, Agrawal et al. [24] recognized features in the search space probabilistically during initialization and proposed a multi-objective DE based on niching and convergence archiving to select optimal feature subsets.

3 The Proposed Method

3.1 A Novel Population Communication Mechanism

In EAs, the coevolution of multiple populations is a good method to improve algorithm performance. When facing the problem of the algorithm getting stuck in local optima, dividing the population into multiple subpopulations allows each subpopulation to explore the optimal solution in its own region. When one subpopulation gets trapped in a local optimum, the updates of the remaining subpopulations are not affected, thereby reducing the likelihood of the entire algorithm getting stuck in local optima. However, communication between populations in algorithms with multiple populations is crucial, and effective communication between subpopulations can improve the convergence speed of the algorithm.

In previous studies, the common communication mechanism between multiple populations is the external archive mechanism [25]. In this paper, inspired by the chicken swarm algorithm (CSO) [26], we propose a novel communication mechanism among multiple populations. In CSO, the population consists of three types: roosters, hens, and chicks. In order to find food faster, each rooster occupies a specific area and leads several hens and chicks in search of food. Therefore, the number of roosters represents the number of subpopulations, while hens are individuals with fitness second only to roosters. They learn from the roosters of their own subpopulation and compete with the hens from other subpopulations, thus achieving communication between different subpopulations. As for the chicks, they forage alongside their mothers. Specifically, roosters play a leadership role in the population to quickly find the optimal solution, communication between subpopulations primarily relies on hens, and chicks maintain the exploratory ability of the population.

Based on the above content, we divide each subpopulation of MPDE into three tiers: elite, common, and inferior individuals. The number of elite individuals represents the number of subpopulations, communication between subpopulations is facilitated by common individuals, and inferior individuals enhance population diversity. The model of MPDE is illustrated in Fig. 1. Specifically, in Fig. 1, the population is divided into four sub-populations. The most prevalent are the common individuals (represented by triangles). The common individuals in each sub-population can learn from the elite individuals (represented by dark blue pentagons) within their own sub-population and from the common individuals of other sub-populations. That is, when updating the common individuals, they will choose elite individuals and other common individuals as parent individuals, achieving a communication mechanism similar to that of CSO. For the elite individuals and the inferior individuals (represented by dots), they respectively serve the roles of guiding the population towards more promising regions and maintaining the diversity of the population. The details regarding the update of each individual can be found in Sect. 3.2, and the details of the entire algorithm are in Algorithm 1.

3.2 Individual Enhancement Strategy

Different individuals contribute differently to the population, and we modify the updates for the three types of individuals, which correspond to the mutation operation in DE.

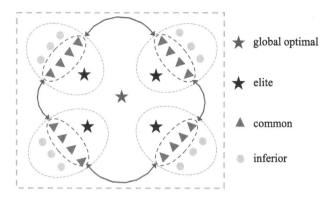

Fig. 1. The model of multiple populations and hierarchical individuals.

Since elite individuals play a crucial role in the population, in Eq. (4), only elite individuals are subjected to an appropriate Gaussian perturbation.

$$X_{i,G+1} = X_{r1,G} * (1 + N(0, \sigma)) \tag{4}$$

$$\sigma = \begin{cases} 1, & if\ f(X_i) \leq f(X_k) \\ e^{\frac{f(X_k)-f(X_i)}{f(X_i)+\varepsilon}}, & otherwise \end{cases} \tag{5}$$

where $f()$ is a fitness value; *epsilonnnnn* is a small value used to prevent the denominator from being zero, and $N(0, \sigma)$ is Gaussian perturbation.

In common individuals, the learning weights are different for learning from elite individuals and common individuals from other subpopulations. We utilize adaptive learning weights and crossover operators, and the individual update is shown in Eq. (6):

$$V_{i,G+1} = X_{i,G} + F'_{i,G} * \left(X_{r1,G} - X_{i,G} \right) + F'_{i,G} * \left(X_{r2,G} - X_{i,G} \right) \tag{6}$$

where $X_{r1,G}$ and $X_{r2,G}$ are elite and common individuals randomly selected from the population, respectively; $F'_{i,G}$ and $F''_{i,G}$ are their corresponding learning factors, and $F'_{i,G} + F''_{i,G} = 1$.

For inferior individuals, in order to maintain population diversity, they focus on exploration within the subpopulation while learning from common individuals. Therefore, the individuals are updated using the learning factor generated by the logical mapping in Eq. (7).

$$\begin{cases} V_{i,G+1} = X_{i,G} + K_G * \left(X_{r3,G} - X_{i,G} \right) \\ K_{G+1} = \quad \mu K_G * \left(X_{r3,G} - X_{i,G} \right) \end{cases} \tag{7}$$

where $X_{r3,G}$ is randomly selected from the common individuals; and K_G is a chaotic sequence in the interval (0,1).

In order to ensure better communication between common individuals and other subpopulations, as well as to improve the overall performance of MPDE, adaptive adjustment strategies are employed for the learning factors in common individuals and the crossover operator in the population [27], as shown below:

$$F'_{i,G+1} = \begin{cases} F'_l + rand * F'_u, & if \ rand \leq \gamma_1 \\ F'_{i,G}, & otherwise \end{cases} \tag{8}$$

$$CR_{i,G+1} = \begin{cases} rand, & if \ rand \leq \gamma_2 \\ CR_{i,G}, & otherwise \end{cases} \tag{9}$$

where $rand \in (0, 1)$ is a random value, and γ_1, γ_2, F'_l, and F'_u are fixed values of 0.1, 0.1, 0.1, and 0.9, respectively.

3.3 Mpde-Fs

In this study, we use a continuous encoding method to represent solutions, where 0 and 1 indicate whether a feature is selected. Each dimension of an individual is a value in the range [0, 1], and a threshold is set ($thres = 0.5$). If $X_i^j > thres$, we consider the feature selected($X_i^j = 1$); otherwise, it is not selected($X_i^j = 0$). We evaluate the final feature subset using the KNN method (see Eq. 10), where $Error$ represents the error rate. Num_{FS} and $Count$ are the selected feature count and total feature count, respectively. α and β are the corresponding weights, and $\alpha + \beta = 1$. The pseudo-code for MPDE-FS is in Algorithm 1.

$$FE = \alpha * Error + \beta * \frac{Num_{FS}}{Count} \tag{10}$$

Algorithm 1 MPDE-FS

Input: Training data, Population size (NP), Dimension (D), Maximum number of iterations (*Max_Iter*), initial mutation operator (F_0), initial crossover operator (CR_0);
Output: optimal solution $f(x^*)$;
1: Randomly generate initial populations and evaluate fitness;
2: **While** $T \leq Max\_Iter$ **do**
3:     Rank the populations according to fitness values;
4:     **for** $i = 1$ to subpopulation **do**
5:         Elite individuals are updated using Eq.(4);
6:         Common and inferior individuals are updated using Eq.(6) and (7);
7:         Individuals other than elites take the crossover operation via Eq.(2);
8:         Subpopulation individuals are updated using Eq.(3);
9:     **end for**
10:    $T = T + 1$;
11:    Eq.(8) and (9) are used to update F and CR, respectively;
13: Getting the selected features;
14: Load the test data and evaluate the classification accuracy;

---

4   Experimental Results and Analysis

4.1   Dataset

We selected 14 UCI datasets and 4 large-scale datasets for testing, and their brief information is shown in Table 1. For each dataset, we evaluated feature subsets using 10-fold cross-validation, randomly selecting 70% of the instances for training and 30% for testing.

Table 1. The brief information of dataset.

Name	Instances	Features	Classes
Glass	214	9	7
Wine	178	13	3
Heart	303	13	5
Zoo	101	16	7
Parkinsons	195	22	2
Dermatology	366	34	6
Ionosphere	351	34	2
LungCancer	32	56	3
Movement-libras	360	91	15
Musk1	476	166	2
Arrhythmia	452	279	16
LSVT	126	309	2
SCADI	70	205	7
Madelon	2600	500	2
Yale	165	1024	15
Colon	62	2000	2
TOX-171	171	5748	4
Leukemia	72	7070	2

4.2 Experimental Settings

To test the performance of MPDE, comparative experiments were conducted with five advanced EC-based feature selection methods, namely "TLPSO [28]", "BASO [29]", "ECSA [11]", "ISSA [12]", and "PLTVACIW-PSO [30]". For each algorithm, we set the population size to 10 and the maximum number of iterations to 100, and ran the experiments independently for 20 times. The other parameters of the algorithms were set according to the original literature. All experimental results were obtained using MATLAB2020a on the Windows system.

4.3 Experimental Results

This section presents the experimental results. Table 2 and Table 3 show the average classification accuracy and the average number of selected features achieved by the algorithm on all datasets, respectively. The convergence curves of each algorithm on some datasets are shown in Fig. 2.

According to the data in Table 2, MPDE method only fails to achieve the highest classification accuracy on a few datasets, while BASO outperforms other methods on the "wine", "Parkinsons", "Ionosphere", "LungCancer", and "SCADI" datasets. However,

Table 2. The comparative results of classification accuracy between MPDE and advanced algorithms on the datasets.

Name	Type	MPDE	TLPSO	BASO	ECSA	ISSA	PLTVACIW-PSO
Glass	Mean	**0.7344**	0.7250	0.7289	0.7141	0.5891	0.7164
	std	0.0336	0.0363	0.0403	0.0430	0.0738	0.0449
Wine	Mean	0.9736	0.9472	**0.9764**	0.9604	0.7226	0.9500
	std	0.0216	0.0249	0.0192	0.0202	0.1030	0.0502
Heart	Mean	**0.6315**	0.6045	0.6264	0.5978	0.4961	0.6045
	std	0.0249	0.0337	0.0136	0.0311	0.0459	0.0238
Zoo	Mean	0.9750	0.9433	0.9667	0.9683	0.8450	0.9500
	std	0.0262	0.0391	0.0265	0.0253	0.0642	0.0382
Parkinsons	Mean	0.9155	0.8991	**0.9164**	0.9086	0.8336	0.8948
	std	0.0185	0.0276	0.0264	0.0149	0.0472	0.0230
Dermatology	Mean	**0.9902**	0.9855	0.9799	0.9879	0.8332	0.9883
	std	0.0064	0.0071	0.0106	0.0081	0.0978	0.0085
Ionosphere	Mean	0.9300	0.9052	**0.9405**	0.8790	0.8352	0.9105
	std	0.0263	0.0246	0.0188	0.0158	0.0238	0.0186
LungCancer	Mean	0.9600	0.9300	**0.9800**	0.9100	0.4600	0.8900
	std	0.0821	0.0979	0.0616	0.1021	0.2349	0.1373
Movement-libras	Mean	**0.8153**	0.7713	0.7833	0.7440	0.7046	0.7880
	std	0.0276	0.0336	0.0351	0.0339	0.0393	0.0333
Musk1	Mean	**0.9370**	0.9204	0.9085	0.8968	0.8349	0.9303
	std	0.0226	0.0199	0.0209	0.0254	0.0295	0.0188
Arrhythmia	Mean	**0.6896**	0.6693	0.6726	0.6415	0.5930	0.6800
	std	0.0273	0.0182	0.0153	0.0214	0.0250	0.0198
LSVT	Mean	**0.8216**	0.6838	0.8081	0.6811	0.6122	0.6473
	std	0.0406	0.0795	0.0410	0.0493	0.0838	0.0647
SCADI	Mean	0.8000	0.7679	**0.8250**	0.7714	0.6750	0.7714
	std	0.0440	0.0650	0.0365	0.0548	0.0750	0.0548
Madelon	Mean	**0.8320**	0.7826	0.7906	0.7590	0.6763	0.8075
	std	0.0301	0.0159	0.0176	0.0133	0.0527	0.0172
Yale	Mean	**0.6959**	0.6602	0.6806	0.6469	0.5612	0.6735
	std	0.0550	0.0571	0.0450	0.0607	0.0598	0.0369

(continued)

Table 2. (*continued*)

Name	Type	MPDE	TLPSO	BASO	ECSA	ISSA	PLTVACIW-PSO
Colon	Mean	**0.9417**	0.8583	0.9333	0.8333	0.7583	0.8500
	std	0.0611	0.0859	0.0580	0.1115	0.1236	0.0960
TOX-171	Mean	**0.8588**	0.7882	0.8186	0.7647	0.7010	0.8235
	std	0.0385	0.0546	0.0353	0.0633	0.0917	0.0402
Leukemia	Mean	**0.9786**	0.9393	0.9714	0.9071	0.8607	0.9464
	std	0.0408	0.0419	0.0427	0.0660	0.0912	0.0799

Table 3. The comparative results of the number of selected features between MPDE and advanced algorithms on the datasets.

Name	MPDE	TLPSO	BASO	ECSA	ISSA	PLTVACIW-PSO
Glass	**3.60**	3.65	4.00	4.75	4.15	4.30
Wine	**3.65**	4.95	**3.65**	7.25	6.60	4.70
Heart	3.60	4.05	**3.00**	6.75	5.90	4.45
Zoo	6.30	6.35	**6.15**	10.35	8.25	6.80
Parkinsons	2.95	4.60	**2.60**	10.85	10.90	4.60
Dermatology	**10.55**	13.40	11.60	21.95	15.80	14.45
Ionosphere	**3.35**	12.05	4.35	16.10	16.75	10.95
LungCancer	**3.85**	18.50	10.45	33.45	27.90	15.90
Movement-libras	**20.30**	37.40	17.15	57.75	43.95	38.60
Musk1	50.15	78.80	**34.20**	111.55	81.20	75.15
Arrhythmia	48.60	125.50	**43.40**	162.80	128.65	115.15
LSVT	**4.65**	117.75	27.60	163.25	137.80	106.95
SCADI	**13.65**	69.90	28.00	117.40	99.10	65.10
Madelon	**17.10**	242.95	105.35	355.85	228.55	239.75
Yale	**155.85**	478.95	187.95	701.05	352.45	452.00
Colon	**23.80**	904.30	207.85	1094.50	373.20	857.60
TOX-171	**964.60**	2800.35	1581.75	4232.05	1453.95	2759.35
Leukemia	**115.45**	3310.25	736.15	3659.70	587.65	3239.35

it is worth noting that although MPDE does not always achieve the optimal classification accuracy on some datasets, the difference with the best-performing method is not significant. Additionally, MPDE achieves considerable classification accuracy on the four large-scale datasets and outperforms other methods by a large margin. This indicates that MPDE has great potential in feature selection for large-scale datasets. Combining the selected feature numbers in Table 3, MPDE selects the fewest features on large-scale datasets, significantly fewer than other methods, such as "Colon" and "Leukemia". For other datasets, MPDE achieves the highest classification accuracy with the minimum number of features, while BASO only selects the fewest features on a few datasets.

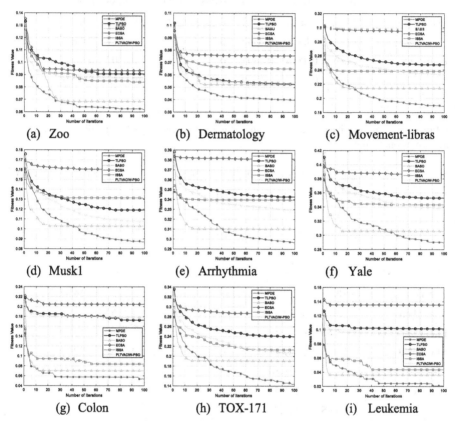

Fig. 2. The convergence curves of 20 independent runs for MPDE and advanced algorithms on datasets.

To understand the convergence process of the algorithm, the convergence curves of the algorithm on some representative datasets are plotted in Fig. 2. It can be clearly seen that MPDE achieves good performance on these datasets. Compared to MPDE, other algorithms converge faster but are prone to getting trapped in local optima. At the same time, we notice that on the "Movement-libras", "Musk1", "Arrhythumia", and "TOX-171" datasets, MPDE did not reach convergence and showed a gradually decreasing trend. Therefore, MPDE has better optimization potential on these datasets. Due to the maintenance of good population diversity in MPDE, this is also the main factor leading to its slower convergence compared to other algorithms.

5 Conclusions

Inspired by CSO, in this paper, we propose a multi-population hierarchical DE for solving the FS problem. Based on the characteristics of CSO populations, we divide the DE population into three types: elite, ordinary, and inferior individuals, with elite individuals leading each sub-population, and common and inferior individuals randomly assigned to each sub-population. In MPDE, communication between sub-populations is mainly carried out through ordinary individuals, while elite and inferior individuals balance the exploitation and exploration of sub-populations. To improve the performance of the algorithm, individual enhancement strategies are used to enhance the optimization ability of each individual. Finally, MPDE is compared with several advanced EC-based FS methods. The experimental results show that MPDE is superior in terms of both classification accuracy and the number of selected features on multiple datasets. Therefore, the introduction of MPDE effectively solves the FS problem.

In future research, we will enhance the information exchange and feedback capability between sub-populations in MPDE to further improve the algorithm's performance. Additionally, MPDE will be applied to solve larger-scale FS problems and handle FS as a multi-objective problem to better address real-world issues.

References

1. Liu, H., Motoda, H., Setiono, R., Zhao, Z.: Feature selection: an ever evolving frontier in data mining. In: Liu, H., Motoda, H., Setiono, R., Zhao, Z., eds.: Proceedings of the Fourth International Workshop on Feature Selection in Data Mining, vol. 10 of Proceedings of Machine Learning Research., Hyderabad, India, PMLR, pp. 4–13 (21 Jun 2010)
2. Liu, H., Zhao, Z.: In: Manipulating data and dimension reduction methods: Feature selection, vol. 9781461418009, pp. 1790–1800. Springer, New York (nov (2012)
3. Yu, F., Tong, L., Xia, X.: Adjustable driving force based particle swarm optimization algorithm. Inf. Sci. **609**, 60–78 (2022)
4. Deng, W., et al.: An enhanced fast non-dominated solution sorting genetic algorithm for multi-objective problems. Inf. Sci. **585**, 441–453 (2022)
5. Chen, W.N., Tan, D.Z., Yang, Q., Gu, T., Zhang, J.: Ant colony optimization for the control of pollutant spreading on social networks. IEEE Trans. Cybern. **50**(9), 4053–4065 (2020)
6. Nag, K., Pal, N.R.: A multiobjective genetic programming-based ensemble for simultaneous feature selection and classification. IEEE Trans. Cybern. **46**(2), 499–510 (2016)
7. Storn, R., Price, K.: Differential evolution - a simple and efficient heuristic for global optimization over continuous spaces. J. Global Optim. **11**, 341–359 (1997)
8. Jiao, R., Nguyen, B.H., Xue, B., Zhang, M.: A survey on evolutionary multiobjective feature selection in classification: approaches, applications, and challenges. IEEE Trans. Evol. Comput. 1 (2023)
9. Qiu, C.: A novel multi-swarm particle swarm optimization for feature selection. Genet. Program Evol. Mach. **20**(4), 503–529 (2019)
10. Xue, B., Fu, W., Zhang, M.: Differential evolution (de) for multiobjective feature selection in classification. In: Proceedings of the Companion Publication of the 2014 Annual Conference on Genetic and Evolutionary Computation, Vancouver, BC, Canada, Association for Computing Machinery, pp. 83–84 (2014)
11. Ouadfel, S., Abd Elaziz, M.: Enhanced crow search algorithm for feature selection. Expert Syst. Appl. **159**, 113572 (2020)

12. Tubishat, M., Idris, N., Shuib, L., Abushariah, M.A.M., Mirjalili, S.: Improved salp swarm algorithm based on opposition based learning and novel local search algorithm for feature selection. Expert Syst. Appl. **145**, 113122 (2020)

13. Ke, L., Feng, Z., Ren, Z.: An efficient ant colony optimization approach to attribute reduction in rough set theory. Pattern Recogn. Lett. **29**(9), 1351–1357 (2008)

14. Vieira, S.M., Sousa, J.M., Runkler, T.A.: Two cooperative ant colonies for feature selection using fuzzy models. Expert Syst. Appl. **37**(4), 2714–2723 (2010)

15. Khushaba, R.N., Al-Ani, A., AlSukker, A., Al-Jumaily, A.: A combined ant colony and differential evolution feature selection algorithm. In: Dorigo, M., Birattari, M., Blum, C., Clerc, M., Stützle, T., Winfield, A.F.T. (eds.) ANTS 2008. LNCS, vol. 5217, pp. 1–12. Springer, Heidelberg (2008). https://doi.org/10.1007/978-3-540-87527-7_1

16. Hunt, R., Neshatian, K., Zhang, M.: A genetic programming approach to hyper-heuristic feature selection. In: Bui, L.T., Ong, Y.S., Hoai, N.X., Ishibuchi, H., Suganthan, P.N. (eds.) Simulated Evolution and Learning, pp. 320–330. Heidelberg, Springer, Berlin Heidelberg, Berlin (2012)

17. Mei, Y., Chen, Q., Lensen, A., Xue, B., Zhang, M.: Explainable artificial intelligence by genetic programming: a survey. IEEE Trans. Evol. Comput. **27**(3), 621–641 (2023)

18. Li, X., Wang, Y., Ruiz, R.: A survey on sparse learning models for feature selection. IEEE Trans. Cybern. **52**(3), 1642–1660 (2022)

19. Too, J., Abdullah, A.R., Saad, N.M., Tee, W.: Emg feature selection and classification using a pbest-guide binary particle swarm optimization. Computation **7**(1), 12 (2019). https://doi.org/10.3390/computation7010012

20. Tran, B., Xue, B., Zhang, M.: Variable-length particle swarm optimization for feature selection on high-dimensional classification. IEEE Trans. Evol. Comput. **23**(3), 473–487 (2019)

21. Wang, P., Xue, B., Liang, J., Zhang, M.: Feature clustering-assisted feature selection with differential evolution. Pattern Recogn. **140**, 109523 (2023)

22. Usman, A.M., Yusof, U.K., Naim, S., Musa, N., Chiroma, H.: Multiobjective filter-based feature selection using nsgaiii with mutual information and entropy. In: 2020 2nd International Conference on Computer and Information Sciences (ICCIS). 1–7 (2020)

23. Hancer, E., Xue, B., Zhang, M.: An evolutionary filter approach to feature selection in classification for both single- and multi-objective scenarios. Knowl.-Based Syst. **280**, 111008 (2023)

24. Agrawal, S., Tiwari, A., Yaduvanshi, B., Rajak, P.: Feature subset selection using multimodal multiobjective differential evolution. Knowl.-Based Syst. **265**, 110361 (2023)

25. Zhang, Q., Li, H.: Moea/d: A multiobjective evolutionary algorithm based on decomposition. IEEE Trans. Evol. Comput. **11**(6), 712–731 (2007)

26. Xianbing Meng, Yu., Liu, X.G., Zhang, H.: A new bio-inspired algorithm: Chicken swarm optimization. In: Tan, Y., Shi, Y., Coello, C.A., Coello, (eds.) Advances in Swarm Intelligence, pp. 86–94. Springer International Publishing, Cham (2014). https://doi.org/10.1007/978-3-319-11857-4_10

27. Brest, J., Bo˘skovi´c, B., Greiner, S., Zumer, V., Mauˇcec, M.S.: Performance comparison of self-adaptive and adaptive differential evolution algorithms. Soft Comput. **11**(7) 617–629 (2006)

28. Azar, A., Khan, Z., Amin, S., Fouad, K.: Hybrid global optimization algorithm for feature selection. Comput. Materials Continua **74**, 2021–2037 (2022)

29. Too, J., Rahim Abdullah, A.: Binary atom search optimisation approaches for feature selection. Connect. Sci. **32**(4), 406–430 (2020)

30. Qiu, C., Liu, N.: A novel three layer particle swarm optimization for feature selection. J. Intell. Fuzzy Syst. **41**, 2469–2483 (2021)

Research on State-Owned Assets Portfolio Investment Strategy Based on Improved Differential Evolution

Dong Ji[1](✉) and Dandan Cui[2]

[1] Nantong Guorong Asset Operation Co., Ltd., Nantong, China
ji_d@sina.cn
[2] Nantong Yuanchuang Technology Investment Company, Nantong, China

Abstract. The state-owned assets portfolio problem is a nonlinear programming problem, and the traditional algorithm can not effectively find the optimal solution, so the effective solution has become a hot issue. Differential evolution algorithm has some disadvantages, such as slow convergence speed and easy to fall into local optimal solution. In this paper, a new differential evolution algorithm based on cluster analysis is proposed. The improved algorithm is used to make portfolio investment of state-owned assets. Firstly, the cluster analysis method is used to cluster the populations of the difference algorithm, extract the representative element individuals, replace the poor individuals in the original population with new individuals, remove the redundant information in the population, and optimize and update the population, so that the whole population can converge to the global optimal solution quickly and accurately. The experimental results show that the differential evolution algorithm with cluster analysis strategy can not only effectively inhibit premature convergence and improve convergence speed, but also has the characteristics of simplicity, efficiency and strong robustness. On the basis of satisfying the investment objectives and constraints, it also reflects the different kinds of income and risk needs of investors, and has good practice.

Keywords: Differential Evolution · Clustering Algorithm · Investment Portfolio

1 Introduction

State-owned assets play a pivotal role in realizing the government's economic functions. To promote high-quality development, such investments should focus on bolstering economic vitality, innovation, and competitiveness. This aids in addressing regional economic disparities and establishes industrial momentum and exemplary effects, propelling the overall growth of the regional economy. Given the finite nature of state-owned capital, innovative investment approaches, like equity participation or shares, can be utilized to harness social capital and initiate the "catfish effect" in regional economic development.

This study delves into the challenge of state-owned asset portfolio investment purely from an economic efficiency perspective. Portfolio theory explores how "rational investors" optimize their portfolios. These investors either aim to maximize returns for

© The Author(s), under exclusive license to Springer Nature Singapore Pte Ltd. 2024
K. Li and Y. Liu (Eds.): ISICA 2023, CCIS 2146, pp. 165–177, 2024.
https://doi.org/10.1007/978-981-97-4393-3_14

a set level of risk or minimize risk for a specified return level. The objective is to furnish an optimal investment strategy that maximizes investor gains.

The Differential Evolution (DE) algorithm is a potent tool for global optimization problems due to its simplicity, efficiency, and adaptability for multidimensional challenges. However, its traditional form sometimes gets trapped in local optima because of its evolutionary methods. Balancing global and local search is critical for the algorithm's success. Thus, refining this optimization algorithm holds significant research importance, especially in economic applications.

Given the complex nature of the portfolio problem, traditional algorithms often fall short in finding optimal solutions. This paper employs an enhanced differential evolution algorithm to tackle this issue, analyzing the efficacy of the new algorithm and how various parameters influence the outcome.

2   Research Background

Portfolio optimization is fundamental in modern finance, focusing on rational investment decisions under uncertain risks and returns to maximize returns and minimize risks. Harry Markowitz's Mean-Variance model in 1952 pioneered securities portfolio theory. However, the sheer data volume in real securities markets makes portfolio decisions complex, rendering them NP-hard. Traditional algorithms struggle with these problems, prompting increased reliance on swarm intelligence algorithms for better optimization. This paper leverages an enhanced swarm differential evolution algorithm to optimize state-owned asset portfolios.

The Differential Evolution (DE) algorithm, a type of swarm intelligence algorithm, is recognized for its simplicity and efficiency across various scientific and engineering domains. Yet, it isn't flawless; issues include slow convergence and tendencies towards local optima. Also, control parameters selection, like the crossover factor, scaling factor, and population size, profoundly influence the algorithm's final optimization. Many scholars have proposed enhancements, such as parameter modifications [1], varied variation directions [2], and hybrid DE algorithms [3], with hybrid variants emerging as research hotspots.

In genetic algorithms, the crossover operator broadens the solution search by generating diverse offspring populations, emphasizing its essential role. However, a challenge arises with real-number coded crossover strategies: offspring from two parent crossovers often stay confined within boundaries set by the parents, limiting the search breadth. Over-relying on mutation to escape these confines risks pushing the genetic algorithm into mere random search territory [6]. Certain literature introduced novel crossover operators addressing this limitation [7], providing offspring the chance to venture beyond parental boundaries and enhancing search areas. Regrettably, these findings lack quantitative theoretical analysis and empirical support.

Recently, the DE algorithm's merits, including robustness and computational simplicity, have amplified its application in deriving optimal or near-optimal solutions to varied problems. It operates using three main operators: selection, crossover, and mutation. While it showcases strong global convergence and aptness for multidimensional challenges, pitfalls like easy divergence into local optima remain. Balancing global and local searches is paramount. This research focuses on refining the classical DE algorithm, targeting portfolio optimization. Experimental simulations juxtaposed with the classic algorithm substantiate the proposed algorithm's effectiveness. Still, issues like slow convergence persist, prompting continuous improvement endeavors by researchers.

3 Differential Evolution

The difference algorithm studied in this paper is to minimize the objective function f $(X), X = (x^1, x^2, ..., x^D) \in R^D$, the feasible solution space for $\Omega = \prod_{j=1}^{D} [L_j, U_j]$, including U_j and L_j respectively, of upper and lower bounds of the first j a variable. As a swarm intelligence algorithm based on random search, differential evolution algorithm first randomly generates and gradually evolves a population of candidate solutions of size NP according to the pre-set population size and problem size, where each solution is a D -dimensional vector, which can be expressed as:

$$X_{i,G} = \left(x_{i,G}^1, x_{i,G}^2, ..., x_{i,G}^D \right) \tag{1}$$

where $= 1, 2, ..., NP$, where NP represents the population size and G represents the number of current iterations.

In the differential evolution algorithm, the initial population is a set composed of all individuals randomly generated in the feasible region $[L_j, U_j]$of each dimension of the solution vector. The j-dimensional component of the i-th individual is generated in the way shown in equation (2).

$$x_i^j = Lj + rand(0, 1)*(U_j - L_j) \tag{2}$$

Rand(0,1) represents a random number on the interval [0,1] that follows a uniform distribution. After the population initialization, the differential evolution algorithm began to undergo a series of cyclic evolution operations, such as Mutation, Crossover and Selection, as follows.

3.1 Mutation

The differential evolution algorithm generates mutation vector Xi,G, also called target vector, through mutation operation. The mutation mode of differential evolution algorithm is denoted as "DE/x/y/z", where: x represents whether the current mutated vector is "best" or "random". y represents the number of difference vectors used; z represents the way in which the operation is crossed. "DE/rand/1" is the most widely used mutation operation, "DE/rand/2", "DE/ current-to-rand/1", "DE/best/1","DE/current-to-best/1", "de/current-to-best/1", "de/current-to-best/1". "DE/rand-to-best/1" is another variant operation. To better understand the mutation operation, consider the following general form:

$$v_{i,G} = \beta_{i,G} + F * \delta_{i,G} \tag{3}$$

Where F is the variation scaling factor;β is the basis vector and δ is the difference vector.

3.2 Cross Operation

The variation vector $V_{i,G}$, next to the variation vector $V_{i,G}$ and initial solution $X_{i,G}$ performs crossover operation generates a test vector $u_{i,G}$, There are two types of cross operation, binomial hybridization and exponential hybridization, of which binomial hybridization is more widely used,

$$u_{i,G}^j = \begin{cases} V_{i,G}^j \text{ if } r \text{ and}(0, 1) \leq Crorj = j_{rand} \\ x_{i,G}^j \text{ otherwise} \end{cases} \tag{4}$$

n this formula, $Cr \in [0, 1]$ is the crossover probability, and $rand(0,1)$ is the random number that follows a uniform distribution between $[0,1]$.

3.3 Selecting Operation

Differential evolution algorithm is using a one-on-one competition mechanism, namely the cross between test vector and the original solution vectors of crossover operation, from each pair of $u^j_{i,G}$ and $x^j_{i,G}$ vector choose to adapt to the value of higher into the next generation of group, choose the way is as follows:

$$x_{i,G}^j = \begin{cases} u_{i,G}^j \text{ if } f\left(u_{i,G}^j\right) < f\left(x_{i,G}^j\right) \\ x_{i,G}^j \text{ otherwise} \end{cases} \tag{5}$$

3.4 Improved Differential Evolution

The performance of the differential evolution algorithm hinges on its ability to balance global exploration and local exploitation, which in turn depends on the appropriate selection of control parameters like population size, scaling factor, and crossover probability. Remarkably, compared to other evolutionary algorithms, differential evolution requires fewer parameter adjustments.

Numerous recent studies have sought to enhance the differential algorithm. Literature [8] introduced a combination of differential evolution and particle swarm optimization (DEPSO), leveraging statistical learning. This approach selects evolutionary strategies based on the relative success rates of both algorithms in prior iterations. Literature [9] incorporated simulated annealing into the classical differential algorithm, allowing the acceptance of sub-optimal solutions probabilistically. Literature [10] integrated a gravity search strategy to enhance the experimental vector quality during selection, while [11] focused on adaptively adjusting algorithm parameters and scaling factors.

Despite these advancements, premature convergence remains a challenge, and some enhancements have unintentionally increased both algorithm evaluation times and overall processing times. Cluster analysis, as highlighted in [12], categorizes datasets based on varied attributes, enabling redundant information removal, dimension reduction, and focused algorithmic operations using representative elements. Building on this, this paper proposes an improvement to the differential algorithm using cluster analysis, yielding promising results.

3.5 Clustering Analysis Algorithm

Cluster analysis refers to the grouping of data objects in the absence of classification labels based on information found in the data describing the characteristics of the objects and their interrelationships. The goal of clustering is that objects within a group are as similar to each other as possible and objects in different groups are as unrelated as possible. The greater the similarity within the group and the greater the difference between the groups, the better the clustering effect.

3.5.1 Idea of Algorithom

Cluster analysis in the context of the differential evolution algorithm involves categorizing the population into distinct groups and selecting the most informative representative element from each group for operation. Using these representative elements, the aim is to reconstruct a population that is superior to the initial one, guiding the population towards the global optimum. Implementing clustering early on can be problematic. Initially, the distribution of solutions is wide-ranging and holds substantial information. Clustering too soon may lead to the loss of this vital information, causing the population to converge prematurely or fall into local optima. A significant risk is that the optimal solution, even if currently poorly adapted, might be excluded, resulting in failure to achieve the problem's final solution. Thus, the timing of when to employ cluster analysis is crucial, and it's important to note that clustering also requires computational resources. Based on experience, it's not necessary to cluster frequently. Cluster analysis serves as an acceleration technique in two ways: it can hasten the convergence rate post-clustering, and it can mitigate the risk of information loss. After clustering, the next steps involve rapidly converging the representative elements using classical methods and reintegrating them into the population. To do this, the fitness values of each cluster center are calculated and ranked. If a cluster center's fitness value outperforms the worst solution in the original population, the former replaces the latter.

3.5.2 Steps of an Algorithm

The basic steps of the differential evolution algorithm based on cluster analysis are given below:

1. **Initialization:** Initialization of population and algorithm parameters.
2. **Classical Operations:** Carry out the same mutation operation, crossover operation and selection operation as the classical differential evolution algorithm.
3. **Clustering:** After a certain number of iterations, a specific clustering analysis algorithm is used to cluster the whole population, and the whole population is grouped into class K.
4. **Element Extraction and Replacement:**
   (a) Extract the representative elements of theseK categories respectively, calculate the function value of these K representative elements.
   (b) IfK function values < the original population K individual function value with the worst:

– Use the representative element to replace the poor individual in the original population.

End if

5. Iteration Check:

6. If the number of iterations < Maximum number of iterations:

– Go back to Step 1.

7. Else: output the optimal solution and the optimal fit.
The algorithm flow is shown in Fig. 1.

(b) Else: output the optimal solution and the optimal fit.

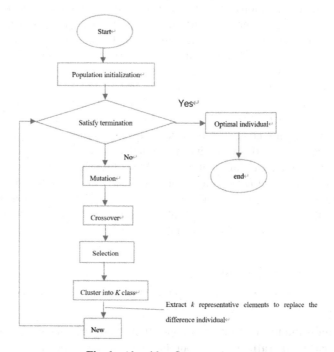

Fig. 1. Algorithm flow

As can be seen from Fig. 2, the differential evolution algorithm based on K-means cluster analysis has poor performance and is not very stable compared with the other two algorithms. The cluster analysis method based on K-medoids shows a good effect in the collaborative research with differential evolution algorithm, but the differential evolution algorithm based on hierarchical cluster analysis is the best among the three mixed algorithms. It can not only improve the convergence speed of the differential algorithm, but also will not fall into the local optimal. The above simulation experiments and result analysis show that the optimization ability of hierarchical cluster analysis method is usually superior to the other two clustering algorithms for both single-modal and multi-modal function optimization problems. Therefore, the subsequent hybrid algorithms in this paper adopt the integration of hierarchical cluster analysis method and differential evolution algorithm.

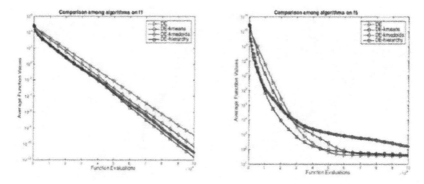

Fig. 2. Different algorithms run comparison

In order to better analyze the performance of the differential Clustering algorithm, such as convergence speed, convergence speed and convergence accuracy, this paper combines the hierarchical clustering analysis method with the standard differential evolution (DE) algorithm to form the hybrid evolutionary algorithm DE-CLU. In this paper, the widely used standard differential algorithm DE/rand/1 and a classical improved DE deformation (JDE) differential evolution algorithm are used as comparison objects, and the performance of the hybrid differential evolution algorithm based on hierarchical clustering is studied and compared with the two classical DE algorithms.

It can be seen from the evolution curve in Fig 3 that the differential evolution algorithm with hierarchical clustering in the f1, f2, f6 and f10 functions has improved its convergence speed and convergence accuracy compared with the previous algorithms. The evolutionary algorithm based on hierarchical clustering can very effectively make the standard DE converge quickly and obtain the global optimal solution for the function f8. The results show that the convergence speed and accuracy of the algorithm can be improved by integrating the hierarchical cluster analysis method, both for the widely used benchmark difference algorithm and the classical improved version of differential evolution.

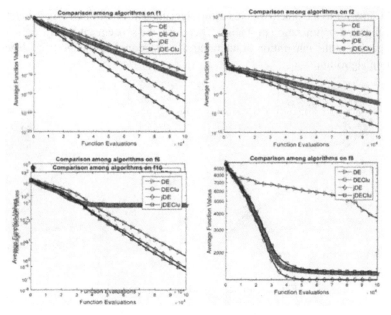

Fig. 3. Various algorithms run comparison

4 State-Owned Assets Portfolio Investment Strategy Model Based on Improved Differential Evolution

When constructing a portfolio, it is necessary to consider the expected returns and risks of the portfolio, and the differences between different strategies and methods are reflected in the measurement of returns and risks. Many scholars have put forward their own unique methods for the measurement of returns and risks, and some scholars estimate the future returns of portfolio investments through fuzzy random variables. A portfolio selection model considering both random factors and fuzzy information is established. Some scholars also creatively put forward a new portfolio risk measurement, that is, the uncertainty of portfolio fuzzy return to measure the risk of stocks and portfolios. Because the vector form of fuzzy number can also consider the impact of risk when measuring return, its dual risk measurement can perfectly help us to more accurately measure the return and risk of the investment portfolio we need to build. Therefore, this paper chooses to introduce the fuzzy concept to study the portfolio strategy.

The cluster differential evolution algorithm designed in the previous section is applied to the portfolio investment of state-owned assets. Binary coding is adopted for differential evolution, and each investment item is divided into two situations: vote or no vote, and the value of vote is 1, and the value of no vote is 0. Adaptive value function is introduced into the measure of fuzzy factors, here is the calculation method of portfolio returns and variance.

Calculate the return on investment i-fuzzy random variable x_i

$$\overline{X_i} = \left(\frac{1}{T} \sum_{t=1}^{T} m_{X_{it}}, \frac{1}{T} \sum_{t=1}^{T} l_{X_{it}}, \frac{1}{T} \sum_{t=1}^{T} \gamma_{X_{it}} \right) \tag{6}$$

The return of portfolio P is:

$$r_p = \sum_{i=1}^{n} w_i \overline{X_i} = \left(\sum_{i=1}^{n} w_i \frac{1}{T} \sum_{t=1}^{T} m_{X_{it}}, \sum_{i=1}^{n} w_i \frac{1}{T} \sum_{t=1}^{T} l_{X_{it}}, \sum_{i=1}^{n} w_i \frac{1}{T} \sum_{t=1}^{T} \gamma_{X_{it}} \right) \tag{7}$$

The variance of the portfolio is:

$$Q_p^2 = \sum_{i=1}^{n} \sum_{j=1}^{n} w_i w_j \sigma_{ij} \tag{8}$$

The risks of the portfolio are:

$$\sigma_p = \sqrt{Q_p^2} = \sqrt{\sum_{i=1}^{n} \sum_{j=1}^{n} w_i w_j \sigma_{ij}} \tag{9}$$

Where w_i and w_j are the weights of investment i and investment j in the portfolio, respectively, and σ_{ij} is the covariance of investment i and fuzzy returns of item j. The adaptation function of differential evolution is to maximize the return of portfolio P and minimize portfolio risk.

$$F = \alpha r_p - \beta \sigma_p \tag{10}$$

5   Experiment and Result

The experimental data came from Wind database. The scale of the experiment was considered, At the same time, considering the differences in industry, we selected the daily maximum price, daily minimum price and daily closing price of 10 constituent stocks of Shanghai Stock Exchange 300 Index (Guosen Securities, Lep Medical, BYD, Tomson BiHealth, Poly Real Estate, Aier Eye, AVIC Electronics, Industrial Bank, Chongqing Water, China Railway Construction) from January 1, 2017 to January 1, 2020. The stock returns are converted into fuzzy returns. The fuzzy returns of 10 stocks are shown in Table 1: The portfolio strategy obtained by plugging the data into the cluster difference algorithm is shown in Fig. 4:

Table 1. Experimental stock fuzzy returns

stock	fuzzy income
Lepu Medical	(0.8598*10–3,0.0184,0.0179)
BYD	(−0.0463*10–3,0.0138,0.0144)
Tomson BiHealth	(0.5096*10–3,0.0145,0.0137)
Poly Real Estate	(0.9132*10–3,0.0167,0.0157)
Aier Eye	(1.8731*10–3,0.0169,0.0146)
AVIC Electronics	(−0.3958*10–3,0.0155,0.0161)
Industrial Bank	(0.4322*10–3,0.0084,0.0090)
Chongqing Water	(−0.2183*10–3,0.0079,0.0072)
China Railway Construction	(-0.1564*10–3,0.0120,0.0121)
Guosen Securities	(−0.2554*10–3,0.0126,0.0141)

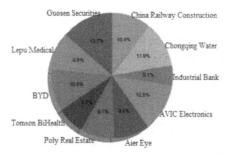

Fig. 4. Stock rights weight

The portfolio strategy obtained by using the difference algorithm to substitute data is shown in Fig. 5:

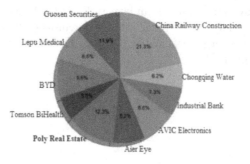

Fig. 5. Stock rights weight

When evaluating the experimental results, the annualized interval returns of the above 10 stocks from January 1, 2021 to January 1, 2022 can be obtained through data query, as shown in Table 2:

Table 2. Annualized range return rate

Stock	Range return (annualized)
Lepu Medical	−17.7556
BYD	322.2361
Tomson BiHealth	52.9131
Poly Real Estate	3.3213
Aier Eye	152.3972
AVIC Electronics	39.5070
Industrial Bank	10.4971
Chongqing Water	−3.9952
China Railway Construction Guosen Securities	−20.6759
	11.0381

According to the data in the table above, using respectively introduced clustering difference model with the introduction of the difference of clustering model portfolio strategy can work out the two strategies in January 1, 2021 to January 1, 2022 yields as follows:

R1 is the introduction of the difference of clustering model portfolio strategy. R1 = 11.0381% * 13.7% − 17.7556% * 9.8% + 322.2361% * 10.8% + 52.9131% * 7.2% + 3.3213% * 9.1% + 152.3972% * 9.4% + 39.5070% * 12.5% + 10.4971% * 5.1% − 3.9952% * 11.9% − 20.6759% * 10.4% = 55.86%

R2 is the introduction of the difference of clustering model portfolio strategy. R2 = 11.0381% * 11.9% − 17.7556% * 8.5% + 322.2361% * 9.6% + 52.9131% * 5.8% + 3.3213% * 12.3% + 152.3972% * 8.2% + 39.5070% * 8.8% + 10.4971% * 7.3% − 3.9952% * 6.2% − 20.6759% * 21.3% = 46.30%

It can be clearly and intuitively seen that there are differences in returns between the investment strategy constructed by the difference model with the introduction of clustering and the portfolio strategy constructed by the difference model without the introduction of clustering. The return of the portfolio strategy constructed by the difference model with the introduction of clustering is obviously higher than that of the investment portfolio strategy without the introduction of cluster.

6   Conclusion

In this paper, the differential evolution algorithm is studied, the advantages and disadvantages of the differential evolution algorithm are analyzed, and an improved scheme is proposed, that is, in view of the shortcomings of the differential evolution algorithm, such as slow convergence speed and easy to fall into the local optimal solution, a differential evolution algorithm incorporating cluster analysis is proposed. The cluster analysis method is used to cluster the population of the difference algorithm, remove the redundant information in the population, optimize and update the population, so that the whole population can converge to the global optimal solution quickly and accurately.

In this paper, portfolio optimization problem is studied. Portfolio problem is a highly complex nonlinear optimization problem, and it is difficult for traditional optimization algorithms to effectively find the optimal solution. In this paper, the proposed clustering difference algorithm is used to solve the problem, and a good result is obtained.

References

1. Zorarpacı, E., Özel, S.A.: A hybrid approach of differential evolution and artificial bee colony for feature selection. Expert Syst. Appl. **62**, 91–103 (2016). https://doi.org/10.1016/j.eswa.2016.06.004
2. Punyakum, V., Sethanan, K., Nitisiri, K., Pitakaso, R., Gen, M.: Hybrid differential evolution and particle swarm optimization for Multi-visit and Multi-period workforce scheduling and routing problems. Comput. Electron. Agric. **197**, 106929 (2022). https://doi.org/10.1016/j.compag.2022.106929
3. Dixit, A., Mani, A., Bansal, R.: DEPSOSVM: variant of differential evolution based on PSO for image and text data classification. Int. J. Intell. Comput. Cybern. **13**(2), 223–238 (2020). https://doi.org/10.1108/IJICC-01-2020-0004
4. Zhou, X.-G., Zhang, G.-J.: Differential evolution with underestimation-based multimutation strategy. IEEE Trans. Cybern. **49**(4), 1353–1364 (2019). https://doi.org/10.1109/TCYB.2018.2801287
5. Sun, Z., Cao, D., Ling, Y., Xiang, F., Sun, Z., Wu, F.: Proton exchange membrane fuel cell model parameter identification based on dynamic differential evolution with collective guidance factor algorithm. Energy **216**, 119056 (2021). https://doi.org/10.1016/j.energy.2020.119056
6. "Research on IRP of Perishable Products Based on Mobile Data Sharing Environment: Computer Science & IT Journal Article I IGI Global." https://www.igi-global.com/article/researchon-irp-of-perishable-products-based-on-mobile-data-sharing-environment/268845. Accessed 13 Sep 2023
7. Cheng, S., Wang, Z.: Solve the IRP problem with an improved discrete differential evolution algorithm. Int. J. Intell. Inform. Database Syst. **12**(1–2), 20–31 (2019). https://doi.org/10.1504/IJIIDS.2019.102324.
8. Biazi, L.E., et al.: A differential evolution approach to estimate parameters in a temperature-dependent kinetic model for second generation ethanol production under high cell density with Spathaspora passalidarum. Biochem. Eng. J. **161**, 107586 (2020). https://doi.org/10.1016/j.bej.2020.107586
9. Wang, S.L., Morsidi, F., Ng, T.F., Budiman, H., Neoh, S.C.: Insights into the effects of control parameters and mutation strategy on self-adaptive ensemble-based differential evolution. Inf. Sci. **514**, 203–233 (2020). https://doi.org/10.1016/j.ins.2019.11.046
10. Yang, X., Li, J., Peng, X.: An improved differential evolution algorithm for learning high-fidelity quantum controls. Sci. Bull. **64**(19), 1402–1408 (2019). https://doi.org/10.1016/j.scib.2019.07.013

11. Caraffini, F., Kononova, A.V., Corne, D.: Infeasibility and structural bias in differential evolution. Inf. Sci. **496**, 161–179 (2019). https://doi.org/10.1016/j.ins.2019.05.019
12. Liu, J., Li, Q., Chen, W., Cao, T.: A discrete hidden Markov model fault diagnosis strategy based on K-means clustering dedicated to PEM fuel cell systems of tramways. Int. J. Hydrogen Energy **43**(27), 12428–12441 (2018). https://doi.org/10.1016/j.ijhydene.2018.04.163

A Particle Swarm Optimization Algorithm with Dynamic Population Synergy

Qianqian Dong, Wei Li[(✉)], and Fufa He

School of Information Engineering, Jiangxi University of Science and Technology,
Ganzhou 341000, China
liwei@jxust.edu.cn

Abstract. Particle swarm optimization (PSO) algorithms have been successfully applied to all kinds of optimization problems. However, the standard PSO algorithm can easily fall into local optimal regions when solving complex problems. It is difficult for the particles to get out of these regions, resulting in the inability to obtain the global optimal solution. This paper proposes a dynamic population synergy particle swarm optimization algorithm (DPSPSO) to solve the above problem. The DPSPSO algorithm dynamically divides the population into three sub-populations based on the function-adapted value ordering strategy. Each sub-population has a different task depending on its potential. Moreover, the algorithm employs an integrated learning strategy. This strategy aims to fully utilize the practical information provided by the particles to prevent premature convergence of the particles. Finally, this paper compares the proposed algorithm with five state-of-the-art PSO variants on the CEC2022 test function to investigate the algorithm's effectiveness. The experimental results show that DPSPSO can achieve competitive performance compared to the other PSO variants.

Keywords: Optimization problems · Particle swarm optimization · Dynamically dividing populations · Collaborative learning

1 Introduction

As society advances, numerous optimization problems in social production become increasingly intricate. This results in exponential growth in computational expenses and more incredible difficulty in solving them with traditional computational methods. Intelligent optimization algorithms draw on the intelligent behavior of groups. Its fewer parameters, efficient implementation, and high performance have sparked considerable scientific interest in tackling complex problems. Drawing inspiration from the foraging habits of birds and fish, particle swarm optimization algorithms have been extensively studied and applied for straightforward implementation and effectiveness [1].

However, the convergence accuracy and convergence efficiency of the PSO are not sufficient when solving complex problems with multidimensional structures. The particle only considers its own historical best position *(pbest_i)* and the whole best position of the entire population *(gBest)* when updating its position. The learning strategy is relatively

© The Author(s), under exclusive license to Springer Nature Singapore Pte Ltd. 2024
K. Li and Y. Liu (Eds.): ISICA 2023, CCIS 2146, pp. 178–191, 2024.
https://doi.org/10.1007/978-981-97-4393-3_15

simple to implement and largely ignores the effective information provided by the other particles in the population. However, this strategy can lead to a rapid loss of diversity in the population, causing the algorithm to converge prematurely once the global optimal position *(gBest)* is constrained to a particular local optimal solution. This will inevitably lead to the entire population being attracted to this local optimal solution and unable to jump out of the region, ultimately preventing the entire algorithm from arriving at a proper global optimal solution.

To solve the defect of the PSO algorithm in solving complex problems with premature convergence, this paper propose a particle swarm optimization algorithm with dynamic population synergy. The main contribution of DPSPSO focuses on the two strategies used in this paper. The dynamic population strategy divides the whole population into three subpopulations: Learning paradigm subpopulation, exploitation subpopulation, and search subpopulation. An integrated learning strategy is used for the particles in each subpopulation to reduce the frequent acquisition of information from the global optimal position.

The main work of the remaining four sections is as follows: Sect. 2 presents the basic PSO and related works. Section 3 describes the algorithm proposed in this paper. Section 4 compares five state-of-the-art PSO variants in detail. It further discusses the rationality and effectiveness of DPSPSO. Finally, Sect. 5 shows the conclusion of this paper.

2 Background

2.1 Basic Particle Swarm Optimization

In the basic PSO algorithm, each particle has three D-dimensional vectors during evolution. In each iteration, the position vector of the partice is $x_i = (x_1, x_2, ..., x_D)$, the corresponding velocity vector is $v_i = (v_1, v_2, ..., v_D)$ and $pBest_i = (pbest_i^1, pbest_i^2, ..., pbest_i^D)$ is the particle's own historical best position. The speed and direction of a particle are influenced by its velocity ad the quality of the solution it represents in the search space can be inferred frm its position. This indicates the importance of position in relation to the effectiveness of the solution produced. During the iterative procedure, a particle records its best function adaptation value achieved at a specific location in its best vector history. If a better location is discovered, the best vector is continuously updated throughout the iteration. In addition to this, $gBest = (gbest^1, gbest^2, ..., gbest^D)$ denotes the global optimal vector for the entire population. It will induce the particles to converge toward the global optimal region. Therefore, the core idea of PSO is to continuously update the particles through each evolution until the global optimal solution or acceptable solution of the objective function is found.

2.2 Related Work

Many researchers have conducted extensive research on PSO to achieve a balance between global and local search in particle swarm algorithms. The primary avenues for advancement include parameter adaptive control, devising distinct neighborhood topologies, refining particle learning strategies, and incorporating additional algorithms. Shi et al. [2] proposed a fuzzy adaptive particle swarm optimization algorithm by defining ω using fuzzy inference and achieved excellent results. Zhan et al. [3] proposed an OLPSO algorithm based on orthogonal learning using an orthogonal algorithmic formula to construct a better particle as a leader from the population's historical best experience to guide the population towards a better particle from the population's historical best experience as a leader to guide the population to evolve in a better direction of evolution. The research results show that this OLPSO algorithm based on an orthogonal experimental strategy has faster global convergence and better optimization performance. Lynn et al. [4] proposed a heterogeneous comprehensive learning algorithm using two subpopulations called HCLPSO. In HCLPSO, one subpopulation focuses on exploration while the other subpopulation focuses on development. In 2019, Lin et al. [5] introduced the strategy of toroidal topology and linear tuning of parameters into the genetic learning algorithm GLPSO, and the results showed that replacing the global topology with a toroidal topology for particle generation improves the diversity and employability of GLPSO, and there is a large improvement in the convergence ability. In 2022, Li et al. [6] proposed a two-stage hybrid learning particle swarm optimization, where the learning strategies used in each stage emphasize exploration and exploitation, respectively. The first phase proposes a learning strategy based on Manhattan remote learning to increase population diversity. In the second phase, an excellent example of a learning strategy performs local optimization operations on the population. Compared to other cutting-edge PSO variants, DHLPSO performs very competitively in handling global optimization problems.

3   Proposed DPSPSO

Aiming at the defectives of the PSO algorithm in obtaining less effective information from other particles, this paper proposes a dynamic population synergy particle swarm optimization based on dynamic multiple popular mechanisms and collaborative learning ideas to obtain better optimization results. The algorithm dynamically divides the entire population into three subpopulations: The learning paradigm subpopulation, the development subpopulation, and the search subpopulation, each with its own learning strategies and tasks. To better utilize the information provided by the entire population, the integrated learning strategies are integrated to increase the utilization of practical information and reduce the frequency of obtaining information from the global optimal position. The algorithm finally achieves a higher convergence accuracy.

3.1   Dynamic Population Strategy

The particles are initially randomized in the solution space in the PSO algorithm. All particles belong to the same population and utilize the same learning strategy that depends solely on themselves and the global optimal particle. During the early stages, the particles prioritize global searching, extensively exploring the optimal solution area. In the later stages of identifying a suitable region, the particle concentrates on local evolution and converges toward the optimal position within a suitable region. The DPSPSO algorithm divides the entire population into three subpopulations based on ranking the function's fitness values. These are the learning paradigm subpopulation, evolution subpopulation, and search subpopulation.

The particles in learning paradigm subpopulation possess significant potential to find globally optimal solutions. Therefore, they can serve as learning objects for other particles. At the beginning of the algorithm, there are fewer particles near the optimal solution. Therefore, the top 50% of the particles are part of the learning paradigm subpopulation. With an increase in iterative evolutions, the number of particles near the optimal solution gradually grows. As a result, the top 80% of the particles belong to the learning paradigm subpopulation.

The task of these particles in evolution subpopulation is to further focus on the local search and converge to the optimal solution in their region as much as possible. Initially, the top 70% of particles belong to the development subpopulation, and as the number of iterations increases, more and more particles will enter the subpopulation for a later detailed search. Eventually, the top 90% of particles will join the subpopulation.

The task of these particles search subpopulation is to help the whole population escape the local optimal region as much as possible and thus find the global optimal solution. The particles of this subpopulation usually update their velocity and then perform an intensive search to get their position directly toward the learning paradigm object. Initially, particles ranked after 70% belong to the search subpopulation. As the number of iterative evolutions increases, the number of particles gradually decreases, and eventually, only particles ranked after 90% belong to this subpopulation.

In DPSPSO, the subpopulation size changes dynamically, and each particle is not permanently assigned to a particular subpopulation. At each iteration, particles are reranked based on their current fitness value. Learning from a random particle in the learning paradigm subpopulation efficiently utilizes the information provided by other particles. By enabling the searching subpopulation to conduct an in-depth exploration, the diversity of the population can be augmented, and the local optimal area can be eradicated to avoid premature convergence. The equation for the alteration in the quantity of particles for every subpopulation is presented below:

$$y_1 = N * 0.5 + \left(\frac{N * 0.3}{maxFES} \right) * FES \tag{1}$$

$$y_2 = N * 0.7 + \left(\frac{N * 0.2}{maxFES} \right) * FES \tag{2}$$

where y_1 is the number of particles in the learning exemplar population, y_2 is the count of particles in the development population, N is the count of particles in the entire population, $maxFES$ is the maximum number of function evaluations and FES is the number of evolutions in the current algorithm.

Standard PSO has a critical control parameter ω, usually a fixed value of 0.9. Its primary function is to define the effect of the previous velocity of the particles on the evolutionary direction of the particles. It is shown that a more significant velocity is favorable for global search. At the same time, a lower speed is favorable for local and intensive development, so this paper defines the inertia weight ω is defined as a dynamic variable, which decreases with increasing iteration number. The specific formula is as follows:

$$maxGEN = \frac{maxFES}{N} \tag{3}$$

$$\omega = \omega_{ini} - \frac{\omega_{sub}}{maxGen} * GEN \tag{4}$$

where ω_{ini} and ω_{sub} is the initial weight value and the reduced weight value. GEN is the current number of evolutions of the particle, and $maxGen$ is the maximum number of evolutions of the particle. Initially, ω can be set to a larger value (e.g., 0.9), in which case the particle is able to explore globally in the solution space, and then set ω gradually decreases to a smaller value (e.g., 0.2), in which case the particle will be more exploitative, favoring the fine-grained search of local regions.

3.2 Integrated Learning Strategies

Although there are many variants of the PSO, premature convergence in solving multimodal problems is still the main defect of the PSO. This is primarily because each particle only learns from its *pBest* and the *gBest* of the entire population during updates. This learning method enables the standard PSO algorithm to converge rapidly. When the *gBest* of the entire population and the *pBest* of the particle are located on either side of the current particle position may cause the particle's motion to oscillate. If *gBest* is associated with the particle's *pBest$_i$* on the same side, the particle will undoubtedly move in that direction move. Once *gBest* and *pBest$_i$* are in the same of the local optimum region. The particle may be unable to jump out of that local optimal region.

Therefore, Liang et al. [9] proposed a CLPSO algorithm based on an integrated learning strategy in 2006. In CLPSO algorithm, each particle learns not from its own historically optimal or globally optimal particles, but through the combination of different particles. The velocity update formula of the algorithm is as follows:

$$V_i^d(t+1) = \omega * V_i^d(t) + c * rand_i^d * (pBest_{fi(d)}^d - X_i^d) \tag{5}$$

where $f_i = [f_i(1), f_i(2), \ldots, f_i(D)]$ is the value of the corresponding dimension of the learning object.

Inspired by the integrated learning strategy, this paper proposes a new integrated learning strategy by combining the integrated learning strategy in CLPSO with the dynamic population strategy. This strategy can enhances the diversity of the particle population and broadens the scope of information accessible to the particles during updating. It steers the particles toward improving in a beneficial direction by utilizing the adequate information supplied by the entire population. The DPSPSO velocity update formula is outlined below:

$$V_i^d(t+1) = \omega * V_i^d(t) + c * rand_i^d * (pBest_{fi(d)}^d - X_i^d) \tag{6}$$

where ω is updated using Eq. (4), $c = 1.49445$, $rand_i^d$ is a random matrix of size 1 $\times$ D. The values of the matrix are all random numbers within [0,1). $pBest_{fi(d)}^d$ is an exemplary object of learning selected using an integrated learning strategy that incorporates dynamic populations. Unlike the strategy in the CLPSO algorithm, the DPSPSO algorithm implements a dynamic division into subpopulations, with a specific learning exemplar subpopulation serving as a learning object for other particles. During the updating of the learning object for each dimension of each particle, only those particles classified as belonging to the learning exemplar subpopulation are chosen for learning. These particles have a relatively more significant potential for exploitation and can be learning objects for other particles.

In the DPSPSO algorithm, when a particle belongs to the search subpopulation, it undergoes an enhanced search. This aims to use the valid information previously provided by the whole population better and make the particle directly close to the position of the object of the learning paradigm. The following formula is used to update the particle's position:

$$X_i^d(t+1) = X_i^d(t) + 10 * c * rand_i^d * (pBest_{fi(d)}^d - X_i^d) \tag{7}$$

when particle i does not belong to the search subpopulation. The position is updated using the following formula:

$$X_i^d(t+1) = X_i^d(t) + V_i^d(t+1) \tag{8}$$

3.3 Framework of DPSPSO

The DPSPSO algorithm uses a rank-constrained integrated learning strategy to select the learning example object, regardless of the subpopulation to which the particle belongs. It uses Eq. (6) to update its velocity to use more of the effective messages provided by the population. Equation (7) is then used to update the particle's position if it belongs to the search subpopulation. Equation (8) updates the particle's position if it belongs to the exploitation subpopulation. The pseudo-code of DPSPSO is shown in Algorithm 1.

Algorithm 1: The pseudo-code of the proposed DPSPSO

Input: $D, N, c, \omega_{ini}, \omega_{sub}, X_{max}, X_{min}$

Output: $gBest$

1 Initialize the $X_0 = (X_1^0, X_2^0, ..., X_N^0)$ and $V_0 = (V_1^0, V_2^0, ..., V_N^0)$ of the population;

2 Calculate function fitness and calculate $pBest$ and $gBest$ from the fitness values;

3 Ascending order based on fitness, recording the rank of each particle;

4 Generate a learning probability for each particle;

5 Define the formula for calculating the number of each subpopulation;

6 Select a learning example object for each dimension of each particle;

7 While $FEs < maxFEs$

8   Update particle velocity using Eq. (6);

9   if the particle belongs to the search subpopulation

10   Update the particle positions according to Eq. (7);

11   else

12   Update the particle positions according to Eq. (8);

13   End if

14   Calculate the fitness of the new position and find $pBest$ and $gBest$;

15   Update the ranking of each particle and the size of the subpopulations;

16   for 1 to N

17   Select a new learning paradigm for each particle;

18   for 1 to D

19   Record the values that will be learned for each dimension;

20   End for

21   End for

22 End While

From Lines 1–6, focus on the initialization of the population, initializing the position and velocity of the particle population $X_0 = (X_1^0, X_2^0, ..., X_N^0)$ and calculating the function fitness to compute $pBest$ and $gBest$. The particles are sorted in ascending order according to the function fitness. Each particle's ranking is recorded, focusing on the learning probability of each particle. The formula for calculating the number of each subpopulation and the learning paradigm for each particle dimension is also considered.

From Lines 8–13, the particle velocities are updated using Eq. (6), the positions of the particles in the search subpopulation are updated using Eq. (7), and the positions of the other particles are updated using Eq. (8).

From Lines 14–22, calculate the fitness value of the particle's new position, find $pBest$ and $gBest$, and update the ranking of each particle. Choose a new learning paradigm for each particle and record the values that the particle will learn for each dimension.

4 Experimental Study

4.1 Experimental Setting

In this section, we evaluate the performance of the DPSPSO algorithm using the CEC2022 test function [11]. CEC2022 includes four categories with twelve test functions, of which F1 is an unimodal function, F2-F5 are basic functions, F6-F8 are hybrid functions, and F9-F12 are composition functions.

In this study, we used DPSPSO to run 30 times on each function and the termination criteria to reach 200000 for 10-D and 1000000 for 20-D (*maxFEs*) evaluation of the objective function for each run. The population size of the DPSPSO algorithm is set to 40. The ω is updated using Eq. (4), $c = 1.49445$. The search range is $[-100, 100]^D$. This paper uses a computer with a Win_64 bit, Intel (R) Core (TM) i9-12900H CPU@ 2.50 GHz and 16 GB RAM. The population size of the algorithm is set based on most of the literature and historical experience.

4.2 Optimization Performance

The performance of DPSPSO on the test set through error (the difference between the global optimal value determined by the algorithm and the objective function value) is analyzed in this section. The error values obtained for 10-D and 20-D are tabulated in Tables 1 and 2. The error was evaluated as 0 when the difference between the ideal and the best solutions found by the method was 1E-8 or less.

In Table 1, the F1 function of DPSPSO on 10-D has good robustness and excellent performance. In basic functions, DPSPSO can be solved to the optimal value on the F3 and F5 functions and has good convergence accuracy and stability performance. The F2 and F4 functions perform slightly worse than the other basic functions. In hybrid functions, the F7 function gives better function values and averages close to the optimal value of the function. Although the F6 and F8 functions cannot obtain the optimal value, they can improve the convergence accuracy. In composition functions, although the function situation is complex and there are multiple local optimums, all of which fail to obtain the optimal solution, the algorithm is more stable and can obtain a relatively stable solution.

Table 1. The DPSPSO algorithm performance on 10-D

Function	Best	Worst	Mean	Std
F1	0.00E + 00	0.00E+00	0.00E+00	0.00E+00
F2	5.51E−02	8.92E+00	1.26E+00	2.05E+00
F3	0.00E+00	0.00E+00	0.00E+00	0.00E+00
F4	2.98E+00	1.02E+01	6.72E+00	1.79E+00
F5	0.00E+00	0.00E+00	0.00E+00	0.00E+00
F6	5.23E+01	4.43E+02	1.78E+02	8.51E+01
F7	6.35E−03	1.45E+00	1.67E−01	2.80E−01
F8	2.80E+00	2.21E+01	1.11E+01	7.57E+00
F9	2.29E+02	2.29E+02	2.29E+02	0.00E+00
F10	1.00E+02	1.00E+02	1.00E+02	2.72E−02
F11	2.19E+01	3.00E+02	1.38E+02	1.01E+02
F12	1.59E+02	1.64E+02	1.62E+02	1.41E+00

In Table 2, the F1 function of DPSPSO on 20-*D* still has good stability. In basic functions, the F3 function performs as well as it does on 10-*D* to obtain the optimal value and the F5 function can also obtain a solution that approaches the optimal value. The performance of the F2 function in 20-*D* could be better than that in 10-*D*, which is mainly due to the characteristics of solving the problem. The F2 function is relatively smooth near the optimal value, and the solution difference is too small. The F4 functions perform slightly worse than the other basic functions. In hybrid functions, all functions perform slightly lower on 20-*D* than on 10-*D* and do not yield optimal values. This also shows that the increase in the number of dimensions significantly affects the algorithm's optimization ability. In composition functions, it is worth noting that the F9 function has better convergence accuracy on 20-*D* than on 10-*D* and that F11 stabilizes into a local optimum.

Table 2. The DPSPSO algorithm performance on 20-D

Function	Best	Worst	Mean	Std
F1	0.00E+00	0.00E+00	0.00E+00	0.00E+00
F2	1.14E+01	4.91E+01	4.41E+01	8.75E+00
F3	0.00E+00	0.00E+00	0.00E+00	0.00E+00
F4	1.59E+01	3.55E+01	2.49E+01	5.14E+00
F5	0.00E+00	4.03E−05	4.19E−06	8.80E−06
F6	2.75E+02	2.25E+03	1.01E+03	5.56E+02
F7	6.54E+00	3.06E+01	2.49E+01	4.25E+00
F8	2.16E+01	2.32E+01	2.25E+01	3.04E−01
F9	1.81E+02	1.81E+02	1.81E+02	0.00E+00
F10	9.86E−01	3.57E+02	1.47E+02	8.58E+01
F11	3.00E+02	3.00E+02	3.00E+02	0.00E+00
F12	2.31E+02	2.37E+02	2.34E+02	1.56E+00

4.3 Comparison with Improved PSO Algorithms

To validate the performance of the DPSPSO algorithm on the CEC2022 test function, DPSPSO is compared with five state-of-the-art algorithms. These algorithms are state-of-the-art algorithms that have been proposed in the past and have attracted the attention of many researchers: CLPSO [7], HCLPSO [4], DSPSO [8], XPSO [9] and TLPSO [10]. To ensure fairness, each algorithm is run 30 times independently and records the mean and standard deviation.

Table 3 shows the comparison results of the five algorithms in 10-*D*, and Table 4 shows the results in 20-*D*. We calculate the average ranking for different algorithms on each function and determine the algorithm's performance ranking based on the average

Table 3. Comparisons of DPSPSO with PSO variants on on 10-D

Function		CLPSO	HCLPSO	DSPSO	XPSO	TLPSO	DPSPSO
F1	Best	8.08E+01	3.60E−14	**0.00E+00**	3.79E−15	**0.00E+00**	**0.00E+00**
	Mean	4.02E+01	3.16E−14	**0.00E+00**	1.44E−14	**0.00E+00**	**0.00E+00**
	Rank	6	5	**1**	4	**1**	**1**
F2	Best	2.18E+00	**8.09E−01**	1.90E+00	5.34E+00	5.93E+00	1.26E+00
	Mean	1.89E+00	**1.53E+00**	3.41E+00	4.29E+00	1.76E+00	2.05E+00
	Rank	4	**1**	3	5	6	2
F3	Best	5.26E−08	1.25E−13	**0.00E+00**	1.52E−14	**0.00E+00**	**0.00E+00**
	Mean	3.48E−08	8.63E−14	**0.00E+00**	3.93E−14	**0.00E+00**	**0.00E+00**
	Rank	6	5	**1**	4	**1**	**1**
F4	Best	1.16E+01	8.60E+00	**3.58E+00**	7.59E+00	8.29E+00	6.72E+00
	Mean	2.43E+00	3.84E+00	**1.95E+00**	3.12E+00	3.97E+00	1.79E+00
	Rank	6	5	**1**	3	4	2
F5	Best	2.12E−01	2.12E−13	**0.00E+00**	7.96E−14	**0.00E+00**	3.94E−13
	Mean	8.43E−02	1.26E−13	**0.00E+00**	6.08E−14	**0.00E+00**	8.10E−13
	Rank	6	4	**1**	3	**1**	5
F6	Best	**4.55E+01**	1.57E+02	1.82E+02	1.15E+03	5.46E+03	1.78E+02
	Mean	**2.87E+01**	1.14E+02	3.92E+02	9.63E+02	1.29E+03	8.51E+01
	Rank	**1**	2	4	5	6	3
F7	Best	**4.78E−02**	1.84E+00	8.76E+00	8.36E+00	5.82E+00	1.67E−01
	Mean	**4.17E−02**	4.99E+00	9.74E+00	9.57E+00	8.79E+00	2.80E−01
	Rank	**1**	3	6	5	4	2
F8	Best	9.69E+00	1.16E+01	1.30E+01	1.55E+01	**6.72E+00**	1.11E+01
	Mean	4.77E+00	9.15E+00	9.95E+00	8.98E+00	**9.39E+00**	7.57E+00
	Rank	2	4	5	6	**1**	3
F9	Best	**2.26E+02**	2.29E+02	2.29E+02	2.35E+02	2.29E+02	2.29E+02
	Mean	**2.04E+01**	2.97E−13	1.84E−01	2.66E+01	0.00E+00	0.00E+00
	Rank	**1**	2	5	6	2	2
F10	Best	**1.20E+00**	9.45E+01	1.64E+02	1.73E+02	1.00E+02	1.00E+02
	Mean	**1.44E+00**	1.79E+01	5.34E+01	5.26E+01	5.26E−01	2.72E−02
	Rank	**1**	2	5	6	4	3
F11	Best	6.24E+01	**5.31E−13**	4.33E+01	5.00E+01	5.55E+00	1.38E+02
	Mean	3.08E+01	**1.72E−13**	1.14E+02	1.14E+02	3.04E+01	1.01E+02

(continued)

Table 3. (*continued*)

Function		CLPSO	HCLPSO	DSPSO	XPSO	TLPSO	DPSPSO
	Rank	5	**1**	3	4	2	6
F12	Best	1.64E+02	1.65E+02	1.65E+02	1.65E+02	1.62E+02	**1.62E+02**
	Mean	7.69E−01	6.49E−01	7.93E−01	5.10E−01	4.67E−01	**1.41E+00**
	Rank	3	4	6	5	2	**1**
Average rankings		3.23	2.92	3.15	4.31	2.62	2.38
Algorithm Rankings		5	3	4	6	2	1

ranking's size. The 10-D optimal value shows that the DPSPSO search effect ranks first overall. It is shown that the DPSPSO algorithm has better convergence and helps improve the solution accuracy.

Table 4. Comparisons of DPSPSO with PSO variants on 20-D

Function		CLPSO	HCLPSO	DSPSO	XPSO	TLPSO	DPSPSO
F1	Best	3.13E+01	6.44E−14	2.65E−14	4.74E−14	5.68E−15	**3.79E−15**
	Mean	1.10E+01	2.47E−14	2.88E−14	2.15E−14	1.73E−14	**1.44E−14**
	Rank	6	5	3	4	2	**1**
F2	Best	**2.11E+01**	2.86E+01	4.90E+01	4.88E+01	4.84E+01	4.57E+01
	Mean	**1.09E+01**	2.09E+01	7.65E−01	1.53E+01	1.54E+00	5.75E+00
	Rank	**1**	2	6	5	4	3
F3	Best	8.34E−14	1.14E−13	8.72E−14	1.05E−02	**7.20E−14**	1.14E−13
	Mean	5.11E−14	0.00E+00	4.89E−14	5.74E−02	**5.57E−14**	0.00E+00
	Rank	2	4	3	6	**1**	4
F4	Best	3.27E+01	2.19E+01	**1.01E+01**	1.90E+01	1.99E+01	2.52E+01
	Mean	7.01E+00	5.72E+00	**3.81E+00**	7.09E+00	7.72E+00	6.14E+00
	Rank	6	4	**1**	2	3	5
F5	Best	3.78E+00	4.44E−01	3.01E−02	6.42E−01	**0.00E+00**	1.39E−05
	Mean	1.74E+00	6.29E−01	8.70E−02	7.40E−01	**0.00E+00**	4.18E−05
	Rank	6	4	3	5	**1**	2
F6	Best	**9.11E+01**	2.87E+02	9.30E+02	3.62E+03	6.80E+06	8.24E+02
	Mean	**3.56E+01**	3.29E+02	1.20E+03	3.75E+03	1.25E+07	5.31E+02
	Rank	**1**	2	4	5	6	3

(*continued*)

Table 4. (*continued*)

Function		CLPSO	HCLPSO	DSPSO	XPSO	TLPSO	DPSPSO
F7	Best	2.20E+01	**2.02E+01**	2.26E+01	2.47E+01	2.28E+01	2.42E+01
	Mean	2.05E+00	**5.76E+00**	4.99E+00	8.23E+00	3.90E+00	6.01E+00
	Rank	2	**1**	3	6	4	5
F8	Best	2.14E+01	**2.09E+01**	2.86E+01	3.33E+01	2.12E+01	2.25E+01
	Mean	2.20E−01	**3.48E−01**	3.04E+01	3.62E+01	5.96E−01	4.34E−01
	Rank	3	**1**	5	6	2	4
F9	Best	1.81E+02	1.81E+02	1.81E+02	1.82E+02	1.81E+02	**1.81E+02**
	Mean	1.49E−06	4.38E−13	2.37E−01	1.51E−01	8.67E−14	**8.67E−14**
	Rank	4	3	5	6	2	**1**
F10	Best	**5.80E−02**	1.34E+01	1.88E+02	1.93E+02	1.09E+02	1.61E+02
	Mean	**3.06E−01**	2.98E+01	1.20E+02	1.42E+02	3.45E+01	8.16E+01
	Rank	**1**	2	5	6	3	4
F11	Best	**1.05E+02**	1.93E+02	3.40E+02	3.43E+02	3.43E+02	3.00E+02
	Mean	**8.43E+01**	1.31E+02	4.98E+01	5.04E+01	1.02E+02	2.80E−13
	Rank	**1**	2	4	6	5	3
F12	Best	2.38E+02	2.39E+02	2.57E+02	2.65E+02	2.44E+02	**2.34E+02**
	Mean	2.27E+00	4.00E+00	1.01E+01	1.57E+01	3.68E+00	**2.06E+00**
	Rank	2	3	5	6	4	**1**
Average rankings		2.69	2.54	3.62	4.85	2.85	2.77
Algorithm Rankings		2	1	5	6	4	3

The combination of Tables 3 and 4 shows the DPSPSO algorithm is slightly inferior to CLPSO and HCLPSO on 20-*D*, its overall performance is still competitively optimized in 10-*D* and 20-*D*, which strongly proves that the algorithm has a significant improvement effect.

4.4 Discussion of Convergence

To examine the difference between the DPSPSO algorithm and the five comparison algorithms regarding the convergence speed. In this paper, these six algorithms have been tested for convergence by selecting four representative functions F1, F3, F6 and F12 in the CEC2022 test function.

Among the four representative test functions, XPSO is the algorithm that converges the fastest and can quickly reach the better region. For the F1 function, the DSPSO algorithm exhibits a convergence speed similar to that of XPSO, followed by HCLPSO, and then the DPSPSO outperforms TLPSO and CLPSO. As for the F3 problem, the XPSO algorithm demonstrates a much faster convergence speed than the other algorithms, followed by CLPSO. The DPSPSO has a slightly slower convergence speed than CLPSO, but it still outperforms the other three algorithms. For the F6 problem, the convergence speeds of XPSO and TLPSO are comparable to the algorithm proposed in this paper. The same holds for the F12 function, which is similar to the F6 function. Overall, the results demonstrate that the DPSPSO algorithm proposed in this paper has a top-three convergence speed among the four test functions. Although the convergence speed of DPSPSO is not dominant for another algorithm, it can be seen from Tables 3 and 4 that the overall performance of CDLPSO is better than another algorithm. In summary, the DPSPSO algorithm is not superior in convergence speed, but DPSPSO has better solution quality for complex problems. The convergence speed of CDLPSO is faster than that of some algorithms. CDLPSO is still an excellent algorithm through a comprehensive evaluation of the algorithm's solving quality and convergence speed for complex problems.

In short, the DPSPSO algorithm has a better optimization effect on most of the test functions for the following reasons: Firstly, by integrating the optimization strategies in two different directions, the method achieves two goals. It ensures the diversity of the population and increases the channels for the particles to obtain information. Additionally, it makes more effective use of the information provided by the whole population. Then, by requiring the particles to learn only from the particles in the learning exemplar subpopulation, the algorithm constructs the historical optimal dimensional information of multiple better particles into exemplar particles. This allows the optimal information to be shared among different individuals. As a result, the population can still obtain a higher accuracy solution even under complex environmental conditions. Therefore, the DPSPSO algorithm proposed in this paper can achieve optimization results that meet the objective requirements for solving most problems.

5   Conclusion

In this paper, we initially examine the strengths and limitations of the conventional PSO algorithm. Subsequently, we present an algorithm for particle swarm optimization termed DPSPSO. The algorithm dynamically divides the constant total population into three subpopulations based on the ranking mechanism of the function adaptation value. The particles in each subpopulation are assigned different tasks by their potentials. An integrated learning strategy with a ranking restriction is introduced to improve the information exchange between particles and balance the algorithm's global development and local search capabilities. Finally, the algorithm's performance is assessed through simulation experiments on the CEC2022 test function, demonstrating exceptional convergence accuracy and robustness of the DPSPSO algorithm.

References

1. Kennedy, J., Eberhart, R.C.: Particle swarm optimization. In: Proceedings of ICNN'95-International Conference on Neural Networks, vol. 4, pp. 1942–1948. IEEE (1995)
2. Shi, Y., Eberhart, R.C.: Fuzzy adaptive particle swarm optimization. In: Proceedings of the 2001 Congress on Evolutionary Computation (IEEE Cat. No.01TH8546) (2001)
3. Zhan, Z.H., Zhang, J., Li, Y., et al.: Orthogonal learning particle swarm optimization. IEEE Trans. Evol. Comput. **15**(6), 832–847 (2010)
4. Lynn, N., Suganthan, P.N.: Heterogeneous comprehensive learning particle swarm optimization with enhanced exploration and exploitation. Swarm Evol. Comput. **24**, 11–24 (2015)
5. Li, A., Wei, S., Hongshan, Y., et al.: Global genetic learning particle swarm optimization with diversity enhancement by ring topology. Swarm Evol. Comput. **44**, 571–583 (2019)
6. Li, W., Chen, Y., Cai, Q., et al.: Dual-Stage hybrid learning particle swarm optimization algorithm for global optimization problems. Complex Syst. Model. Simul. **2**(4), 288–306 (2022)
7. Liang, J.J., Qin, A.K., Suganthan, P.N., et al.: Comprehensive learning particle swarm optimizer for global optimization of multimodal functions. IEEE Trans. Evol. Comput. **10**(3), 281–295 (2006)
8. Zhang, X., Wang, X., Kang, Q., et al.: Differential mutation and novel social learning particle swarm optimization algorithm. Inform. Sci.: An Int. J. **480**, 109–129 (2019)
9. Xia, X., Gui, L., He, G., et al.: An expanded particle swarm optimization based on multiexemplar and forgetting ability. Inf. Sci. **508**, 105–120 (2020)
10. Zhang, X., Lin, Q.: Three-learning strategy particle swarm algorithm for global optimization problems. Inf. Sci. **593**, 289–313 (2022)
11. Ahrari, A., Elsayed, S., Sarker, R., et al. Problem definition and evaluation criteria for the CEC'2022 competition on dynamic multimodal optimization. In: Proceedings of the IEEE World Congress on Computational Intelligence (IEEE WCCI 2022), pp. 18–23. Padua, Italy (2022)

Preference-Based Multi-objective Optimization Algorithms Under the Union Mechanisms

Yi Zhong[1](✉) and Lanlan Kang[2]

[1] School of Information Engineering, Jiangxi University of Science and Technology,
Ganzhou 341000, Jiangxi, China
[2] School of Information Engineering, Jiangxi University of Science and Technology,
Ganan Institute of Science and Technology, Ganzhou 341000, Jiangxi , China
victorykll@163.com

Abstract. Preference multi-objective optimization (PMOP) is hot problem in the field of current optimization. The searching objection of the PMOP which is local target pareto region according to preference information different from general multi-objective optimization. To improve the performance of the PMOP, a new Preference-based multi-objective optimization algorithms under the union mechanisms (UM-NSGAII) is proposed. Two strategies are proposed in UM-NSGAII. Firstly, initial population generated from limiting it to a certain range, which can reap a population dominated by preference information.Mutation individual is also come into being a certain range, which can reap progeny of populations closer to the pareto frontier corresponding to preference information. Secondly, Angle preference strategy is to identify the angle formed by a certain point with an arbitrary point and the origin as an angle preference region, where individuals in this region are selected in preference to individuals outside the region, this allows individuals evolving towards the target pareto region to be retained, facilitating rapid optimization searching.

Keywords: angular preference · preference vectors · multi-objective optimization · angular preference radius

1 Introduction

With the development of society and the progress of science and technology. Multi-objective optimization wider use in economy [1], scientific research [2], transportation [3] and other fields.

In real applications, the decision maker is interested in a part of the pareto region rather than the entire pareto boundary. When the decision maker state in advance a reference point, the algorithm runs to obtain the region of interest (ROI) of the decision maker, which behaves as a part of the whole Pareto region, and hence there is no need to search and find the whole Pareto best frontier [4]. The idea of combining individual preferences to guide the optimization of an algorithm, narrowing the search region, and converging quickly is defined as preference multi-objective optimization. Many

© The Author(s), under exclusive license to Springer Nature Singapore Pte Ltd. 2024
K. Li and Y. Liu (Eds.): ISICA 2023, CCIS 2146, pp. 192–206, 2024.
https://doi.org/10.1007/978-981-97-4393-3_16

preferences multi-objective optimization algorithms have been proposed, which can be divided into three categories. The first category is the preference point guidance approach that uses preference points (the expectation of each function in the decision maker's region of interest) to divide the preference region, such as g-NSGAII algorithms can alleviate the pressure of candidate solution domination selection, but the position of the preference points in the objective space affects the optimization performance of the algorithm [5, 6]. The second type combines preference point guidance and preference region guidance to divide the priority regions, with preference points guiding individual optimization in the early stage and preference regions to guide evolution in the later stage [7, 8]. The third category is preference optimization in combination with the Euclidean distance to the preference point, which further refines the dominance rank for the same pareto non-dominance ranked individual according to the distance from the Euclidean distance to the preference point, based on the NSGAII algorithm, both use binary crossover and mutation, however, individuals with close Euclidean distances are preferentially selected leading to over-convergence of the population aggregation [9–11].

Nevertheless, these algorithms either use a single preference information with poor applicability or rely too much on the merits of preprocessing parameters. Therefore, based on the NSGAII algorithm, this paper proposes a Preference-based multi-objective optimization algorithms under the union mechanism, which firstly improves the initialization strategy to generate individuals with clear directionality, then controls the variant individuals in a certain range to avoid slowly approaching the pareto frontier, and achieves the optimization effect of strong directionality and fast convergence by using the angular information of the preference points and the preference vectors, which is preliminarily Angular preference strategy is used, which is the movement of the population towards the preference point, but does not dominate with g as it does, and the individuals by preference point dominated(g non-dominated and g dominated) are taken as the preferred selection object, and the hybrid preference bootstrap strategy is used in the later stage for the optimization of the search iteration. The experimental results show that the algorithm proposed in this paper can make the population converge to the preference region quickly and obtain the proper size of the preference solution set.

This paper is divided into five sections, the first section is an introductory section that describes the research on PMOP. The second section has known related concepts and definitions. The third section details the new strategies, innovations proposed in this paper. The fourth section is the experimental comparative analysis. The fifth section is the summary.

2 Related Concepts

2.1 Multi-objective Optimization Problem

The multi-objective optimization problem can be transformed into the following minimal problem.

$$minF(x) = min\{f_1(x), f_2(x), \ldots, f_m(x)\}, x \in R$$
$$st. \begin{cases} h_i(x) \leq 0, i = 1, \ldots, p \\ g_j(x) \leq 0, j = 1, \ldots, q \end{cases} \tag{1}$$

In formula (1): $x = (x_1, x_2, x_3 \ldots x_n)$ is an N-dimensional decision variable. R is the decision space. m is number of objective functions. $h_i(x)$ and $g_i(x)$ are linear inequality constraints.

The pareto definition of mop:

Pareto Dominate. x_a, x_b are individuals of population, if and only if the following 2 conditions are satisfied, which is defined x_a pareto dominates x_b.

$$\begin{cases} f_i(x_a) \leq f_i(x_b), \forall i = 1, \ldots m. \\ \exists j \in \{1, 2, \ldots, m\}, f_j(x_a) < f_j(x_b). \end{cases} \quad (2)$$

Pareto Best Solution. In the viable space, the x_a denotes a solution of mop. When Eq. (2) is satisfied, x_a is called the pareto best solution of this optimization problem.

$$\neg \exists x \in R, x \prec x_a \quad (3)$$

Pareto Best Solution Set. Decentralized individuals do not dominate each other.

$$P = \{x \in R | \neg \exists x_a \in R, x_a \prec x\}. \quad (4)$$

2.2 G-dominance

G-dominance [5] was proposed by Molina et al.. It is defined a new individual dominance relationship by the decision maker's preference points, which realizes a further subdivision of the group dominance relationship, where the target space is divided into preference and non-preference regions.

p is the decision maker's preference point and y is targeting vector in the target space, then there exists the following definition

$$\begin{cases} \text{mark}_p(y) = 1, \text{ if } y_i \leq p_i, \forall i = 1, \ldots m \\ \text{mark}_p(y) = 1, \text{ if } y_i \geq p_i, \forall i = 1, \ldots m \\ \text{mark}_p(y) = 0, \text{ others} \end{cases} \quad (5)$$

By the definition of $\text{mark}_p(y)$, for 2 target vectors y_a, $y_b \in R$, it is defined that y_a g-dominates y_b, if one of the following conditions holds.

$$\begin{cases} \text{mark}_p(y_a) > \text{mark}_p(y_b). \\ \text{if } \text{mark}_p(y_a) = \text{mark}_p(y_b), \text{ then } y_{a_i} \leq y_{b_i}, (\forall i = 1, \ldots, m), \\ \text{there exists at least one } i \text{ such that } y_{a_i} < y_{b_i}. \end{cases} \quad (6)$$

2.3 Relevant Definitions

Definition 1 (preference point)
The set of objective values consisting of the decision maker's best expected value for each objective serves as a preference point in the objective space and is usually denoted as $p(f_{1p}, f_{2p}, \ldots, f_{mp})$.

Definition 2 (preference vector)
In the target space, a vector with the origin pointing to a preference point is defined as a preference vector. The schematic diagram of the division of the preference region in the target space, where Op (the line connecting the origin O and p) is the preference vector in Fig. 1.

Definition 3 (individual preference distance)
The perpendicular distance between y_i corresponding to any individual x_i in the population and the preference vector is defined as the preference deviation of individual x_i, denoted as $Dis(x_i)$, and computed as:

$$Dis(x_i|\,p) = |y_i| * \left(1 - \left(\frac{y_i * op}{|y_i| * |op|}\right)^2\right)^{\frac{1}{2}} \quad j = 1, 2, \ldots, pop \tag{7}$$

According to Eq. (7), $Dis(x_a|p)$ in Fig. 1 is the preference distance of population individual x_i.

Definition 4 (preference radius)
To obtain a set of preference solutions [12] that satisfy the proper range of preferences of the decision maker, an upper bound on the preference deviation is set, which is defined as the preference radius, in Fig. 1.

And the preference region combines with NSGAII [13, 14] to form a new dominance. First, the population is sorted by non-domination sort, then the individuals of the same dominance rank is sorted by preference region, and finally the individuals of the same region and the same rank are sorted according to the crowding degree, which is the preference bootstrapping strategy[8].

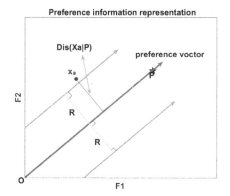

Fig. 1. Preference information.

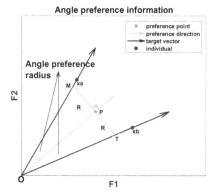

Fig. 2. Angle preference information.

3   UM-NSGAII

First, this section states the new definition proposed in this paper. Then it states the initialization strategy and the angle preference strategy.

3.1   Angle Preference Information Display

Definition 5 (Individual preference angle)
Taking the vector from the coordinate origin O to the preference point p as the axis, let there exist an individual x_a. Then the angle between the vector Ox_a and the vector Op is the preference angle of x_a, denoted as angle_x_a.

$$angle\_x_a = acos\left\{ \frac{dot(Ox_a, Op)}{norm(Ox_a)*norm(Op)} \right\} \tag{8}$$

dot(*) is the vector inner product function. Norm(*) is the function to find the number of paradigms.

Defeintion 6 (preference angle radius)
OP is perpendicular to the line segment MT of length 2R.The angle of MOP is the preference angle radius, marking angle_R, in Fig. 2.

$$angle\_R = asin\left\{ \frac{R}{\text{sqrt}(p(1)^2 + p(2)^2)} \right\} \tag{9}$$

asin(*) is the sin(*) inverse function. p(1) is the first objective function value and p(2) is the second objective function value.

3.2   Target Initialization Strategy

According to the chart to analyze, the traditional NSGAII is usually initialized randomly, however, such an initialization will produce many random solutions with random distribution of uncertainty, purposelessness and instability, and the merit of the initial population seriously affects the performance of the algorithm. It is known that iterative optimization is to select the current best population individuals. In this paper, the initialization is limited the area at the preference point, so that the initialized population will all fall within the pareto-dominated region under this operation, and this strategy is applied to the ZDT series of test functions.

$$\text{chromo}(i, 1) = p(1) + (x\_max(1) - p(1)) * \text{rand}(1) \tag{10}$$

chromo(i,1) denotes the decision variable, p(1) is the first value of the preference point and x_max(*) is the upper limit of the decision variable.

The idea of decision variable decomposition is used in NSGAII to achieve the purpose of gradually approaching the pareto frontier by gradually narrowing the decision variables, and analyzing the process of generating decision variables, it is not difficult to find that the decision variables are generated randomly [15]. Then to achieve the purpose of shortening the optimization process without losing the diversity of the group, we can control the decision variables in a smaller range. In this paper, we adopt the strategy of narrowing the range of values of the decision variables to produce a better solution closer to the pareto frontier, which is suitable to be applied to the ZDT series of test functions, a is a random number.

$$\begin{cases} a = a, & a \leq 0.5 \\ a = a - 0.5, & a > 0.5 \end{cases} \tag{11}$$

$$chromo(i, j) = x_{\min(j)} + \left(x_{\max(j)} - x_{\min(j)}\right) * a \tag{12}$$

chromo(i,j) denotes the decision variable of individual.

3.3 Angle Preference Dominance Strategy

Decision makers always want to get solution set that satisfies the preference range for their decision-making choices. However, the usual preference point cannot control the solution set preference range, which makes the optimization result get too many nondominated solutions and puts decision selection pressure on the final decision. To control the size of the preference solution set, this paper combines the preference vector and the preference radius provided by the decision maker, calculates the preference angle, divides the target space into new preference regions, and then uses pareto non-dominated sorting to prioritize the non-dominated solutions in the angle preference region to achieve the effect of increasing convergence of the preference individuals. According to the relationship between the individual preference angle and the size of the preference angle radius, the target space is divided into corresponding angle preference regions, denoted as

$$\begin{cases} angle\_x_a \leq angle\_R, then x_a\, in\, angle\, preference\, area \\ angle\_x_a > angle\_R, then x_a\, out\, of\, angle\, preference\, area \end{cases} \tag{13}$$

Algorithm 2: Angle preference region segmentation algorithm

Input: mixed population input_chromo, p.

1 calculate angle_R   //calculate angle preference radius .

2 for each $x_j \in$ input_chromo, do

3   calculate angle_x_j.

4   if angle_$x_j <=$ angle_R

5   chromo1 $\cup$ { chromo(x_j)}

6   else

7   chromo2 $\cup$ { chromo(x_j)}

8   end if

9 end for

Output: chromo1, chromo2.

The target space is divided into angle preference region and non-angle preference region, which are two parts. In this way, the dominance relationship of the population will be changed, assuming that any individuals x_a and x_b in the population, when any one of the following conditions is satisfied, define x_a Angle preference dominance (Ad-dominance) x_b.

1) x_a belongs to the angle preference region and x_b is in the non-angle preference region.
2) x_a and x_b are in the same angle preference region, and x_a pareto dominates x_b .
3) x_a and x_b are in the same angle preference region, and x_a, x_b are in the same pareto dominance hierarchy, x_a is in the preference region, x_b is not in the preference region

.

According to the angle preference domination strategy, the angle preference domination algorithm pseudo-code is shown in algorithm 3.

Algorithm 3. angle preference dominance algorithm

Input: population size N, preference points p, chromo1, chromo2.

1 if num(chromo1)<=N then

2   num1=N-num(chromo1)

3   non-dominated sort chromo2,the first num1 best individuals are selected and put into temp1.

4   chromo=temp1 U chromo1

5 else

6   non-dominated sort chromo1, the top N best individuals are selected and put into chromo.

7 end if

Output: chromo.

3.4   Algorithmic Structure of This Paper

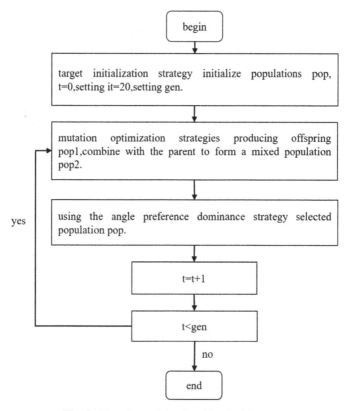

Fig. 3. Flowchart of the algorithm in this paper

4 Experimental Design and Comparative Analysis

4.1 Experimental Setting

Parameter Setting

The algorithm proposed in this paper(UM-NSGAII) is compared experimentally with g-NSGAII and r-NSGAII.

According to the literature [16], the r-NSGAII algorithm sets the weight of each objective as the average weight, and the non-domination threshold δ is set to 0.1. For the comparative analysis, the R of the preference vector guidance in this paper is also set to 0.1. The number of preceding period iterations of UM-NSGAII is set to 20, and the preference radius is set to 0.1. In the settings of the genetic operation operators of the three algorithms, simulated binary crossover and polynomial mutation are selected, and the same crossover probability and probability of mutation are set, which are 0.99 and 0.1, respectively.

Test Function

The test functions selected in this paper are ZDT and DTLZ test function[17, 18]. For the positional relationship of the reference point, this paper categorizes the positional relationship of the reference point into three cases, i.e., infeasible domain, pareto surface and feasible domain, and the specific settings are shown in Table 1.

Table 1. Preference points reference location for this article

functions	infeasible	pareto surface	feasible	pop	gen	dimensions
ZDT1	(0.10, 0.20)	(0.50, 0.30)	(0.80, 0.80)	100	200	30
ZDT2	(0.20, 0.40)	(0.60, 0.64)	(0.90, 0.90)	100	200	30
ZDT3	(0.20, 0.20)	(0.24, 0.28)	(0.50, 0.60)	100	200	30
DTLZ2	(0.20, 0.30, 0.40)	(0.50, 0.70, 0.50)	(0.70, 0.80, 0.80)	200	300	10

4.2 Comparative Analysis of Experiments

Evaluation Function
In this paper, all the running parameters of the algorithm are unified and run independently for 30 times. To evaluate the convergence performance of the algorithm, IGD is used in this paper [19], which is defined as follows.

$$IGD(P^*, P) = \frac{\sum_{v \in P^*} \min_{w \in P}(d(v, w))}{|P^*|} \tag{14}$$

where P^* is the real pareto set and $|P^*|$ is the number of individuals in P^*. P is the best pareto best solution set obtained by the algorithm. And $\min(d(v,w))$ is the minimum Euclidean distance from individual v in P^* to P. IGD is used to evaluate the comprehensive performance of the algorithm by calculating the average value of the minimum distance from the set of points on the real Pareto surface to the obtained population. The smaller the value of IGD, the closer the obtained solution set is to the Pareto best surface, which means the algorithm has a better convergence performance.

Comparative Experiments with Reference Points at Distinct Locations
As can be seen from Table 2, when the reference point is on the feasible region, the IGD values obtained by this paper's algorithm on functions are smaller than those obtained by the other algorithms, which indicates that UM-NSGAII has better distribution and convergence.

From Table 3, when the reference point is located on the real pareto surface, the IGD values obtained which indicates that UM-NSGAII has the effect of obtaining a more best solution set, it shows better distribution and convergence.

Table 2. The IGDs obtained when the reference preference point is on the feasible region.

Test Functions	g-NSGAII	r-NSGAII	UM-NSGAII
ZDT1	2.8636e−2	2.5317e−1	**1.808e−3**
ZDT2	6.9261e−2	3.9198e−1	**1.426e−3**
ZDT3	1.8545e−1	4.7070e−1	**1.688e−3**
DTLZ2	2.4199e−1	4.8728e−1	**3.600e−3**

Table 3. The IGDs obtained when the reference preference point is at the pareto front.

Test Functions	g-NSGAII	r-NSGAII	UM-NSGAII
ZDT1	3.144E−01	3.121E−01	**4.580E−03**
ZDT2	3.732E−01	3.686E−01	**3.010E−03**
ZDT3	4.805E−01	4.470E−01	**1.150E−03**
DTLZ2	5.530E−01	5.398E−01	**5.740E−03**

Table 4. The IGD obtained when the reference preference point is in the infeasible domain.

Test Functions	g-NSGAII	r-NSGAII	UM-NSGAII
ZDT1	8.605E−02	3.693E−01	**1.956E−03**
ZDT2	7.005E−02	3.507E−01	**1.985E−03**
ZDT3	3.004E−01	4.281E−01	**1.200E−03**
DTLZ2	1.884E−01	5.012E−01	**4.804E−03**

As can be seen from Table 4, UM-NSGAII has the least IGD values, which also shows that the solution set obtained by the algorithm in this paper is the closest to the pareto.

As can be seen in Fig. 4, the algorithm in this paper can converge to the smallest distance from pareto in 20 generations and obtain a moderate set of preferred solutions.

The set of preference solutions obtained from the three algorithms run on the ZDT test function. When the preference points are in the infeasible domain, the convergence of the preferred solution sets obtained by this algorithm and the r-NSGAII, g-NSGAII are comparable. However, the distributions of solution sets generated by the three algorithms were found to be dissimilar. UM-NSGAII can produce a solution set that is suitable in size and fulfills the decision-making requirements. The solution set produced by g-NSGAII is broadly distributed, whereas the solution set of r-NSGAII is mostly clustered at a single point, and both sets fail to satisfy preference requirements. Thus, the algorithm presented in this paper is more efficient in the infeasible domain.

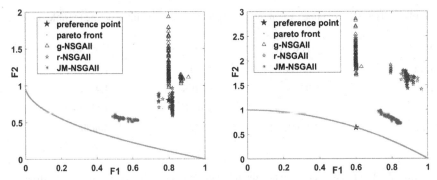

Fig. 4. Each algorithm iterates over ZDT1 (0.8, 0.8), ZDT2 (0.6, 0.64) for 20 generations to obtain the preference solution set

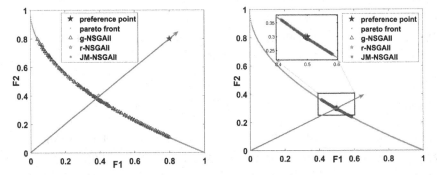

Fig. 5. The set of preference solutions obtained on the ZDT1((0.8,0.8), left), ((0.5,0.3), right) test functions

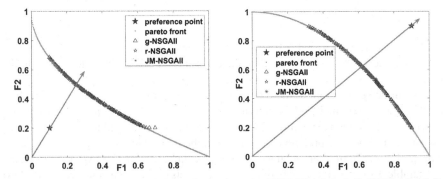

Fig. 6. The set of preference solutions obtained on the ZDT1((0.1,0.2), left),ZDT2 ((0.9,0.9), right) test functions

When the preference point is in the true frontier, the preference solution sets obtained by running the three comparative algorithms are found that the g-NSGAII algorithm and the r-NSGA algorithm are greatly affected by the location of this preference point,

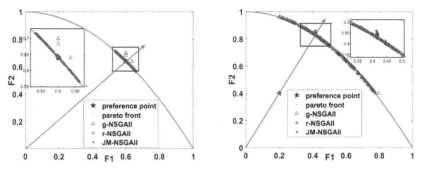

Fig. 7. The set of preference solutions obtained on the ZDT2((0.6,0.64), left),ZDT2 ((0.2,0.4), right) test functions

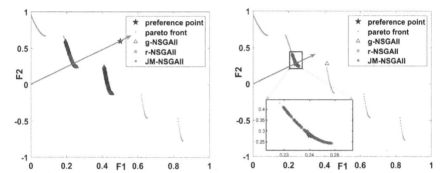

Fig. 8. The set of preference solutions obtained on the ZDT3((0.5,0.6), left),ZDT3 ((0.24,0.28), right) test functions

and the solution sets are over-converged, and the preference solution set that satisfies the decision maker's range of preferences cannot be obtained. In contrast, the algorithm JM-NASGAII in this paper can obtain the preference solution set with better convergence and satisfy the preference range of the decision maker.

When the preference points are in the feasible domain. The solution set is appropriately sized and converges well. However, the solution set based on the g-NSGAII covers the whole real boundary, which is not favorable to the decision maker's final decision choice, while the algorithm in this paper controls the size of the preferred solution set well, obtains a part of the non-dominated solutions, and meets the requirements of the decision maker. r-NSGAII is over-converged in the preferred solution set obtained from ZDT1, ZDT2.

The preference solution sets obtained by running the three algorithms on the three-objective DTLZ2 benchmark function are shown in Fig. 9. The preference points are located on their true frontiers, and it can be seen that as the number of target dimensions increases, the convergence performances of r-NSGAII, g-NSGAII and the algorithm in this paper are comparable, but the distribution of the solution sets is too centralized, and there is a lack of diversity of preferences, which is not convenient for the decision maker to make a final decision. While the algorithm in this paper obtains the preferred solution

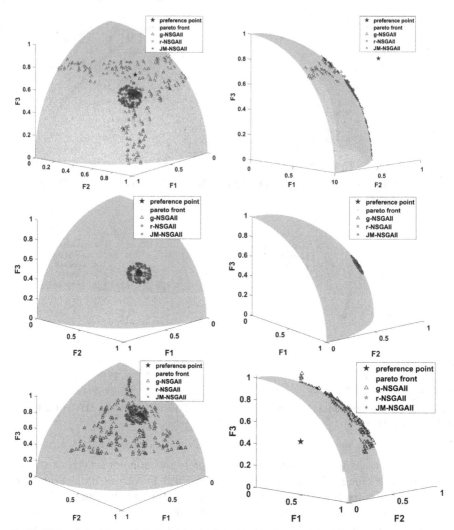

Fig. 9. The set of preferred solutions obtained by running the three algorithms on the DTLZ2 test function, with the three left panels in front view and the three right panels side view

sets with a certain range. When the preference points are in the feasible domain, the solution set obtained by the g-NSGAII algorithm not only has poor convergence, but also the distribution of the preference solution set is too dispersed. The results obtained by the r-NSGAII converge near the frontier, but the solution set is clustered together. The preference solution set obtained by the algorithm of the present paper converges on the true frontier and is distributed in the specified preference region. When the preference points are in the infeasible domain, the solution set obtained by the g-NSGAII algorithm converges poorly, and the distribution of the preference solution set is too dispersed. The r-NSGAII and this paper's algorithm have comparable convergence performances, but the aggregation of the solution set of the r-NSGAII is more centralized, but this paper's algorithm obtains the preference solution set with a certain range.

5 Summary

This paper proposes UM-NSGAII, which adopts the evolutionary strategy based on the predominance of angular preference in the early stage, evolves to the specified number of generations, obtains the candidate populations with better convergence of preference, and reduces the cost of searching the unnecessary objective space, and adopts the preference vector steering mechanism in the late stage, which combines with the radius of preference given by decision makers to control the size of the preference solution set.

Through the experimental comparison of the three algorithms, the IGD evaluation function is adopted, and the horizontal and vertical comparison of the data shows that the algorithm proposed in this paper can obtain a good preference solution set. And the experimental results also show that the preference solution set obtained by this paper's algorithm is not affected by the specific location of the preference points, and at the same time, the size of the preference solution set can be controlled by changing the radius of preference, to obtain the preference solution set with good convergence, uniform distribution, and approximation to the real pareto frontier. As society develops, the applications of preferred multi-targeting are becoming richer and the role demonstrated will grow.

References

1. Tikadar, D., Gujrati, A.M., Guria, C.: Safety, economic, environmental and energy-based criteria for multi-objective optimization of natural gas desulfurization process: an industrial case study. J. Natural Gas Sci. Eng. **95**, 104207 (2021)
2. Liu, Z.J., Fan, G.Y., Zhang, S.C., et al.: Multi-objective optimization of distributed energy systems incorporating multiple energy storage – an example of a near-zero energy community. Building Sci. **38**(8), 44–53 (2022)
3. Gupta, P., Mehlawat, M.K., Aggarwal, U., et al.: An integrated AHP-DEA multi-objective optimization model for sustainable transportation in mining industry. Resour. Policy **74**, 101180 (2021)
4. Liping, W., Meiling, F., Qicang, Q., Minglei, Z., Feiyue, Q.: A review of research on preference multi-objective evolutionary algorithms. J. Comput. **42**, 1289–1315 (2019)
5. Molina, J., Santana, L.V., Hernández-Díaz, A.G., et al.: G-dominance: reference point-based dominance for multi-objective metaheuristics. Eur. J. Oper. Res. **197**(2), 685–692 (2009)
6. Dai, Y.B.: Research on a nonlinear predictive control algorithm for multi-objective optimization. Electr. Drives **45**(11), 62–67 (2015)
7. Dai, Y.B., Chen, H.T.: A hybrid bootstrap preference multi-objective particle swarm optimization algorithm. Control. Eng. **26**(3), 549–554 (2019)
8. Liang, H., Wang, Y., Lin, W., et al.: A two-stage hybrid bootstrap preference multi-objective optimization algorithm. J. Chongqing Univ. Posts Telecommun. (Nat. Sci. Edn.) **34**(05), 836–848 (2022)
9. Said, L.B., Bechikh, S., Ghédira, K.: The r-dominance: a new dominance relation for interactive evolutionary multicriteria decision making. IEEE Trans. Evol. Comput. **14**(5), 801–818 (2010)

10. Qiu, F.-Y., Wu, Y.-S., Qiu, Q.-C., Wang, L.-P.: Many-objective evolutionary algorithm based on bipolar preferences dominance: many-objective evolutionary algorithm based on bipolar preferences dominance. J. Softw. **24**(3), 476–489 (2014). https://doi.org/10.3724/SP.J.1001.2013.04273

11. Zheng, J., Xie, Z.: A study on how to introduce decision makers' preferences with angular information. J. Electron. **42**(11), 2239–2246 (2014)

12. Wang, S.-F.: Research on preference multi-objective evolutionary algorithm for dividing regions by adaptive preference radius. Xiangtan University (2018)

13. Zhao, Z., Fan, B., Huo, H., Sun, L.: Production scheduling optimization based on hybrid NSGA2 algorithm. Combined Mach. Tools Autom. Mach. Technol. **11**, 159–163 (2022)

14. Majharulislam, B., Line, P., Ulla, K., et al.: Application of non-dominated sorting genetic algorithm (NSGA-II) to increase the efficiency of bakery production: a case study. Processes **10**(8), 1623 (2022)

15. Wenqi, C., Hua, G., Chengwang, X., Wei, W., Jiamin, P., Guanglin, L.: An improved NSGA II algorithm for enhancing diversity. Guangxi Sci. **28**(4), 353–362 (2021)

16. Tan, W., Qiu, Q., Yu, W., et al.: A decomposed multi-objective evolutionary algorithm based on neighborhood improvement. Small Microcomput. Syst. **41**(12), 2543–2549 (2020)

17. Filatovas, E.: Synchronous R-NSGA-II: an extended preference-based evolutionary algorithm for multi-objective optimization. Informatica **26**(1), 33–50 (2015)

18. Xu, Z., Zhu, S.: Multi-strategy fusion of improved multi-objective particle swarm algorithms. J. Measur. Sci. Instrumen. **13**(3), 284–299 (2022)

19. Gu, Q.H., Luo, J.L., Li, X.H.: A multi-objective evolutionary algorithm based on small habitats. Comput. Eng. Appl. **59**(1), 126–139 (2023)

Auto-Enhanced Population Diversity with Two Options

Yangcong Ou[1,2(✉)], Ming Yang[1,2], and Jing Guan[3]

[1] School of Computer Science, China University of Geosciences,
Wuhan 430078, China
1051135835@qq.com
[2] Hubei Key Laboratory of Intelligent Geo-Information Processing,
China University of Geosciences, Wuhan 430078, China
[3] China Ship Development and Design Center, Wuhan 430064, China

Abstract. This paper proposes an auto-enhanced population diversity with two options (TO-AEPD) for addressing the premature convergence of CMA-ES. TO-AEPD is founded on a modification of AEPD. By quantifying the population distribution in each dimension, TO-AEPD allows the identification of population convergence or stagnation instances. When convergence or stagnation is identified, the population must be diversified to acceptable levels. Two population diversification enhancement methods are used to balance exploration and exploitation. TO-AEPD selects the population diversification strategy with exploration when the current round of optimization does not produce a new optimal solution or when convergence accuracy is achieved. Otherwise, population diversification with an exploitation strategy is implemented. TO-AEPD-CMAES describes the CMAES that incorporates TO-AEPD. Experimentation demonstrated that TO-AEPD-CMAES outperforms several competing algorithms on benchmarks, significantly enhancing the algorithm's performance and validating the model's viability and efficacy.

Keywords: Evolution strategy · Covariance matrix adaptation · Population diversity auto-enhancement · Population adaptation

1 Introduction

In many optimization problems that cannot be solved using traditional gradient-based optimization algorithms due to non-triviality or the black-box nature of the optimization problem, heuristic algorithms can be instrumental in such problems. In the field of heuristic algorithms, Evolutionary algorithms (EAs) [1] is a large class of algorithms that do not require domain knowledge of the problem but only need to be able to compute the fitness of the optimization objective to be applied. Evolution strategy (ES) [2–4] is a trendy class of evolutionary algorithms that typically sample using a probability distribution. The algorithm is an evolution strategy algorithm proposed by Hansen. The covariance matrix

© The Author(s), under exclusive license to Springer Nature Singapore Pte Ltd. 2024
K. Li and Y. Liu (Eds.): ISICA 2023, CCIS 2146, pp. 207–219, 2024.
https://doi.org/10.1007/978-981-97-4393-3_17

adaptation evolution strategy (CMAES) [5–7] samples the solution space of the optimization problem using a Gaussian distribution and updates the Gaussian distribution according to some sample selection mechanism. The sampling and updating process continues iteratively until the optimization process is stopped by searching for a satisfactory solution, reaching a stopping condition such as the maximum number of samples.

However, several properties of black-box optimization problems may lead to premature convergence of CMAES. Various approaches to increase the probability of finding global optima have been proposed. Many of them belong to niching approaches and restart strategies [8]. A representative approach of the first category is the CMAES with fitness sharing [9,10], where the niche radius is adapted during the search to keep several running individual CMAES instances at a certain distance from each other, thus maintaining some diversity. The second category of restart strategies is similar to the first one since restarts can also be viewed as a parallelized search, but instead in time than space [11–13].

Therefore, the key to solving optimization problems using CMAES lies in escaping convergence and acquiring more excellent exploration capabilities.

To address the abovementioned issue, we propose an auto-enhanced population diversity approach with two options (TO-AEPD). The foundation of TO-AEPD is the auto-enhanced population diversity (AEPD) [14] technique, which detects moments when a population is converging or stagnating by analyzing its distribution in each dimension. When convergence or stagnation is observed in a dimension, the population is appropriately diversified to overcome the stagnation issue. Although AEPD effectively exploits the population, its exploration aspect remains insufficiently investigated. To achieve a better balance between exploration and exploitation, we introduce TO-AEPD. TO-AEPD offers two options representing exploration and exploitation, enhancing the algorithm's ability to discover optimal solutions. Exploration involves investigating unexplored regions in the search space to uncover potential solutions, while exploitation focuses on utilizing the knowledge of promising solutions to improve their quality further. A high exploitation rate often leads to faster convergence. To achieve a better trade-off between exploration and exploitation, we employ conditional judgments that allow the algorithm to adaptively select between exploration and exploitation strategies. This adaptive approach enables satisfactory performance on global optimization problems. Finally, we integrate TO-AEPD with the CMAES algorithm, resulting in the TO-AEPD-CMAES algorithm.

The rest of the paper is organized as below. Section 2 introduces the mechanism of the auto-enhanced population diversity with two options. The related TO-AEPD-CMAES algorithm is presented in Sect. 3. Section 4 presents experimental results. Finally, The conclusion is presented in Sect. 5.

2   Auto-Enhanced Population Diversity with Two Options

TO-AEPD suggests an automatic strategy for enhancing population diversity, with two options representing exploration and exploitation. TO-AEPD can select the population diversity enhancement strategy adaptively during operation.

Using the symbol h, which takes the value 0 or 1, the algorithm specifies which population diversity enhancement strategy to implement.

2.1 Identification of Population Convergence and Stagnation

Before enhancing the population's diversity automatically, it is necessary to determine if the population requires automatic diversity enhancement. Below are methods for determining population stagnation and convergence.

Identification of Population Convergence. Convergence occurs when a population reaches a local optimum as a result of a loss in diversity. To measure population diversity in each dimension, the mean and standard deviation of individual variables in the jth dimension at the Gth generation is calculated by

$$m_{j,G} = \frac{1}{NP} \sum_{i=1}^{NP} x_{i,j,G} \tag{1}$$

$$std_{j,G} = \sqrt{\frac{1}{NP} \sum_{i=1}^{NP} (x_{i,j,G} - m_{j,G})^2} \tag{2}$$

where $m_{j,G}$ and $std_{j,G}$ represent the population's mean and standard deviation in the jth dimension at generation G. NP denotes the population size, which refers to the number of individuals in the population. The standard deviation quantifies population distribution, so $std_{j,G}$ is used to quantify population diversity in the jth dimension. When $std_{j,G}$ decreases, the population in the jth dimension becomes less diverse. When $std_{j,G}$ is sufficiently small, we can ascertain that the population has converged in the jth dimension. $\hat{r}_{j,G}$ is a marker that indicates if the population has converged in the jth dimension at the Gth generation.

$$\bar{r}_{j,G} = \begin{cases} 1 & \text{if } std_{j,G} \leq \omega_{j,G} \\ 0 & \text{otherwise} \end{cases} \tag{3}$$

If $std_{j,G}$ does not exceed $\omega_{j,G}$, $\bar{r}_{j,G}$ is set to 1 to indicate that the population has reached convergence in the jth dimension. The following formula calculates $\omega_{j,G}$:

$$\omega_{j,G} = \max(T_{\text{pd}}, \theta_{j,G}). \tag{4}$$

where

$$\theta_{j,G} = \begin{cases} |m_{j,G} - MR_j| \cdot T_{\text{init}} & \text{if } std_{j,G} \leq T_{\text{init}} \text{ and } h = 0 \\ T_{\text{init}} & \text{otherwise} \end{cases}, \tag{5}$$

$T_{\text{pd}} = 10^{-10}$ and $T_{\text{init}} = 10^{-3}$. The value of $m_{j,G}$ immediately before to the last diversity enhancement procedure for the jth dimension is denoted by MR_j. The initial value of MR_j is set to $m_{j,G}$ in the first generation.

By comparing $\omega_{j,G}$ and $std_{j,G}$, we can ascertain the convergence of the population in the jth dimension. The value of $\omega_{j,G}$ is dynamically determined. The exploitation strategy is employed when h equals 0, resulting in a relatively confined population distribution. Under the exploitation strategy, the population converges toward the same position as in the previous iteration. In such cases, $\theta_{j,G}$ is set to $|m_{j,G} - MR_j| \cdot T_{\text{init}}$, reducing the value of $\omega_{j,G}$ and enabling the population to be fully exploited before diversification occurs. In other scenarios, $\theta_{j,G}$ is set to T_{init}, ensuring that the population diversifies within a short period after the onset of convergence while facilitating early identification of convergence.

Identification of Population Stagnation. Stagnation refers to a state where the population distribution remains unchanged within all dimensions during optimization. This paper employs the mean and standard deviation indicators to detect population stagnation. Specifically, suppose the mean and standard deviation of the population do not exhibit any significant changes across consecutive generations. In that case, it can be inferred that the population has reached a stagnant state.

$\lambda_{j,G}$ denotes the number of consecutive generations in which the values of $m_{j,G}$ and $std_{j,G}$ remain unchanged. The value of $\lambda_{j,G}$ for the jth dimension at the Gth generation is calculated by

$$\lambda_{j,G} = \begin{cases} \lambda_{j,G-1} + 1 & \text{if } m_{j,G} = m_{j,G-1} \text{ and} \\ & \quad std_{j,G} = std_{j,G-1} \\ 0 & \text{otherwise} \end{cases} \tag{6}$$

and $\lambda_{j,0} = 0$. We introduce an integer constant, denoted as UN to determine population stagnation. Population stagnation is identified when there is no change in the population distribution for UN consecutive generations. The symbol $\hat{r}_{j,G}$ signifies that the population has stagnated in the jth dimension at the Gth generation. The value of $\hat{r}_{j,G}$ for the jth dimension at the Gth generation is calculated by

$$\hat{r}_{j,G} = \begin{cases} 1 & \text{if } \lambda_{j,G} \geq UN \\ 0 & \text{otherwise} \end{cases} \tag{7}$$

where, UN is equated with $\sqrt{NP}$.

2.2 Enhancement of Population Diversity

When the population converges ($\overline{r}_{j,G} = 1$) or stagnates ($\hat{r}_{j,G} = 1$) in the jth dimension, it becomes necessary to introduce diversity in that particular dimension. $r_{j,G}$ determines whether the population should be diversified in the jth dimension during the Gth generation.

$$r_{j,G} = \begin{cases} 1 & \text{if } \hat{r}_{j,G} = 1 \text{ or } \overline{r}_{j,G} = 1 \\ 0 & \text{otherwise} \end{cases} \tag{8}$$

The population requires diversification only when $r_{j,G}$ equals 1 in all dimensions. The variable z_G indicates whether the population needs to be diversified in generation G.

$$z_G = \begin{cases} 1 & \text{if } \sum_{j=1}^{D} r_{j,G} = D \\ 0 & \text{otherwise} \end{cases} \tag{9}$$

When z_G equals 1, it indicates the need for a population diversification operation. First, it is necessary to determine which population diversification strategy to implement, and the value of h indicates which population diversification strategy to implement

$$h = \begin{cases} 1 & \text{if } T_{\text{pd}} = \theta_{j,G} \quad \forall j \in \{1, 2, \ldots, D\} \text{ or} \\ & y_{\text{best}} = y_{\text{last}} \\ 0 & \text{otherwise} \end{cases} \tag{10}$$

y_{best} represents the global optimal solution, while y_{last} represents the global optimal solution before the most recent population diversification operation. When T_{pd} equals $\theta_{j,G}$ across all dimensions, the population converges to a narrow range. When y_{best} is equal to y_{last}, it indicates that from the beginning of this round of optimization until convergence or stagnation, there has yet to be a superior solution. Exploration is required when either of the two preceding conditions is met. The mean M of the search distribution is set to x_{best}, and the step size σ is set to $up - low$ for CMAES. up and low are predefined minimum and maximum values. Setting the step size σ to a larger value represents a broader range for the next generation of the population, which is more conducive to exploration.

When neither of the conditions mentioned above is met, exploitation must be carried out. We set the search distribution's mean to x_{best} and its step size to $(up - low) \cdot 0.1$. $(up - low) \cdot 0.1$ is a smaller value, and setting step size σ to a smaller value causes the next generation of the population to be smaller and more amenable to exploitation. In addition, during the next round of optimization, the worst individual $x_{\text{worst},G}$ in the population generated by the subsequent generation will be substituted by the globally optimal individual x_{best} each time. This is performed to expedite convergence during exploitation and enhance overall optimization.

Algorithm 1 illustrates the framework of the auto-enhanced population diversity with two options (TO-AEPD).

3 TO-AEPD-CMAES

TO-AEPD-CMAES, the proposed algorithm, incorporates the preceding TO-AEPD mechanism. The Algorithm 2 describes the fundamental structure of TO-AEPD-CMAES.

Algorithm 1. Auto-Enhanced Population Diversity with Two Options (TO-AEPD)

Input: P: population generated in generation G;
Output: M: the mean of the search distribution; σ: step size;
1: Use Eq. (1) and Eq. (2) compute $m_{j,G}$ and $std_{j,G}$ for each dimension $j = 1, 2,...,$ D, respectively;
2: **for** each dimension $j = 1, 2,..., D$ **do**
3:    Compute $r_{j,G}$ using Eq. (8);
4: **end for**
5: Compute z_G using Eq. (9);
6: **if** z_G equals 1 **then**
7:    Compute h using Eq. (10);
8:    $M = x_{\text{best}}$;
9:    **if** h equals 1 **then**
10:       $\sigma = up - low$;
11:    **else**
12:       $\sigma = (up - low) \cdot 0.1$;
13:    **end if**
14: **end if**

---

4   Experimental Studies

In order to verify the validity of the proposed algorithm TO-AEPD-CMAES, comparative experiments on a set of benchmarks are conducted in this paper.

4.1   Benchmark Test Functions

IEEE CEC 2022 test problems are used to evaluate the functionality of TO-AEPD-CMAES. CEC 2022 is the most recent test set for single-objective constrained constraint optimization functions. CEC 2022 benchmark functions include twelve functions of four types: unimodal functions (f_1), basic functions (f_2–f_5), hybrid functions (f_6–f_8), and composition functions (f_9–f_{12}).

The functions f_1–f_5 are derived by applying a shift and partial or complete rotation to a base function. For the functions f_6–f_8, the variables are arbitrarily divided into subcomponents, and fundamental functions are applied to each subcomponent individually. Each hybrid function comprises a unique combination of fundamental functions. In the case of the functions f_9–f_{10}, they effectively integrate the characteristics of the subfunctions while preserving the consistency of the global or local optima. Moreover, the composition functions exclusively employ shifted and rotated functions as fundamental components. This complexity adds a significant performance challenge for the algorithm.

In this paper, two sets of experiments were devised, one set comparing TO-AEPD-CMAES with CMAES and the other set comparing the proposed TO-AEPD-CMAES with two variants of CMAES: ZOCMAES [15,16] and NIPOP-CMAES [17]. ZOCMAES is the cutting-edge CMAES-based algorithm in the

---

Algorithm 2. TO-AEPD-CMAES

Input: D: dimension of the problem; up: predefined maximum; low: predefined minimum;

Output: y_{best}: the globally optimal solution;
1: Initialize internal parameters and constants;
2: $G \leftarrow 0$;
3: **while** termination criteria not met **do**
4:     **for** $i = 1, \ldots, N$ **do**
5:         Sample $t_i \sim \mathcal{N}(0, C_G)$;
6:         Compute candidate solution $x_i = M_G + \sigma_G t_i$;
7:         Boundary handling on x_i;
8:         Evaluate the objective function $f(x_i)$;
9:     **end for**
10:    **if** h equals 0 **then**
11:       Replace the individual $x_{\text{worst},G}$ with the globally optimal individual x_{best};
12:    **end if**
13:    Sort the candidate solutions by their objective function values;
14:    Select the top μ solutions
15:    Compute the weighted mean $M_{G+1} = \sum_{i=1}^{\mu} w_i x_{i:G}$;
16:    Update the evolution path $p_{c,G+1}, p_{s,G+1}$;
17:    Adapt the covariance matrix: C_{G+1};
18:    Adapt step size: σ_{G+1};
19:    Implement the auto-enhancement of population diversity using Algorithm 1;
20:    $G \leftarrow G + 1$;
21: **end while**

---

most recent CEC 2022, while NIPOP-CMAES is a variant of the traditional CMAES.

To determine the statistical significance of algorithmic differences, the Wilcoxon rank [18] sum test is applied with a significance level of 0.05 between TO-AEPD-CMAES and each compared peer algorithm.

4.2 Parameter Settings

All test functions are conducted in the experiments using either 10 or 20 variables, with a search space ranging from $[-100, 100]^D$. Following the competition rules, each function is independently executed 30 times, and the termination criteria are set at 200,000 evaluations of the objective function for the 10-D case and 1,000,000 evaluations for the 20-D case in each run.

The ZOCMAES experimental results were obtained directly from CEC 2022. Source code for NIPOP-CMAES from [19].

4.3 Comparisons with CMAES

Table 1 summarizes the average objective function error value and standard deviation derived from 30 independent trials of the two compared algorithms. The significantly superior outcomes are highlighted in bold. b, n, and l represent the number of functions in which TO-AEPD-CMAES performs significantly better, statistically equivalent to, or substantially worse than its competitors, respectively.

Table 1. Average fitness values $\pm$ standard deviations on the CEC 2022 functions over 30 independent runs.

D	Fun	TO-AEPD-CMAES	CMAES
10	f_1	1.33e-14 $\pm$ 2.45e-14	**0.00e+00 $\pm$ 0.00e+00**↓
	f_2	**1.52e-13 $\pm$ 4.60e-13**	2.85e+00 $\pm$ 3.53e+00↑
	f_3	**3.55e-04 $\pm$ 3.06e-04**	4.09e-01 $\pm$ 6.66e-01↑
	f_4	**3.38e+00 $\pm$ 8.10e-01**	1.27e+01 $\pm$ 4.62e+00↑
	f_5	1.52e-14 $\pm$ 3.93e-14	1.87e-01 $\pm$ 4.46e-01
	f_6	**1.29e+00 $\pm$ 1.00e+00**	1.40e+01 $\pm$ 1.53e+01↑
	f_7	**8.60e+00 $\pm$ 6.90e+00**	4.43e+01 $\pm$ 2.50e+01↑
	f_8	**6.69e+00 $\pm$ 7.88e+00**	3.38e+01 $\pm$ 3.31e+01↑
	f_9	2.29e+02 $\pm$ 9.67e-08	**2.29e+02 $\pm$ 1.45e-13**↓
	f_{10}	**9.73e+01 $\pm$ 1.64e+01**	1.69e+02 $\pm$ 5.67e+01↑
	f_{11}	2.11e-09 $\pm$ 5.08e-10	1.77e+02 $\pm$ 1.49e+02
	f_{12}	**1.63e+02 $\pm$ 6.98e-01**	1.67e+02 $\pm$ 3.39e+00↑
	$b/n/l$	–	8/2/2
20	f_1	6.63e-14 $\pm$ 2.15e-14	**2.27e-14 $\pm$ 2.83e-14**↓
	f_2	**4.93e-13 $\pm$ 1.56e-12**	3.63e+01 $\pm$ 2.22e+01↑
	f_3	**9.45e-04 $\pm$ 4.38e-04**	5.85e-01 $\pm$ 9.51e-01↑
	f_4	**1.06e+01 $\pm$ 1.80e+00**	3.27e+01 $\pm$ 9.06e+00↑
	f_5	9.47e-14 $\pm$ 6.03e-14	5.54e-01 $\pm$ 8.87e-01↑
	f_6	**1.15e+01 $\pm$ 6.01e+00**	4.89e+01 $\pm$ 2.80e+01↑
	f_7	**2.64e+01 $\pm$ 3.50e+00**	9.23e+01 $\pm$ 4.91e+01↑
	f_8	**2.04e+01 $\pm$ 3.16e+00**	6.44e+01 $\pm$ 6.69e+01↑
	f_9	1.81e+02 $\pm$ 7.78e-08	**1.81e+02 $\pm$ 5.78e-14**↓
	f_{10}	**1.00e+02 $\pm$ 5.01e-02**	1.05e+03 $\pm$ 7.37e+02↑
	f_{11}	**6.00e+01 $\pm$ 1.22e+02**	2.93e+02 $\pm$ 8.68e+01↑
	f_{12}	**2.40e+02 $\pm$ 3.15e+00**	2.76e+02 $\pm$ 1.90e+01↑
	$b/n/l$	–	10/0/2

On the test instances with 10-D, TO-AEPD-CMAES achieves the best performance on 8 test instances, while its peer competitors CMAES achieve two best results respectively; on the test instances with 20-D, TO-AEPD-CMAES achieves the best performance on 10 test instances, while its peer competitors

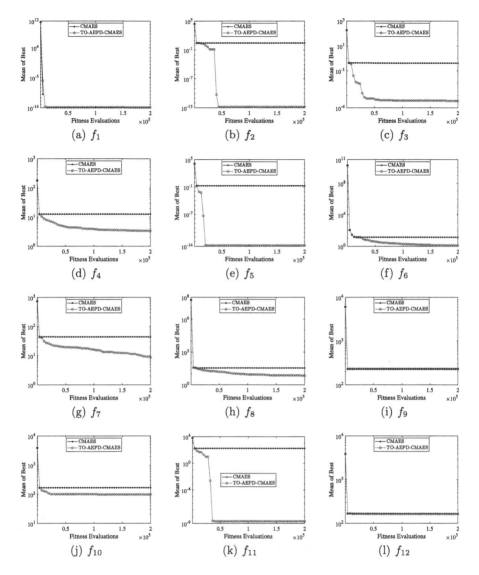

Fig. 1. CEC 2022 average convergence of 30 independent runs of each function in 10-*D*

CMAES achieve two best results respectively. Even in marginally worse situations, competitive performance outperforms the optimal method.

TO-AEPD-CMAES outperforms CMAES on all functions except f_1 and f_{11} for 10-*D*. TO-AEPD-CMAES has clear advantages over f_2, f_3, f_4, f_6, f_7, and f_8. TO-AEPD-CMAES outperforms CMAES for all functions except f_1 and f_{11} for 20-*D*. TO-AEPD-CMAES has distinct advantages over f_2, f_3, f_5, and f_{11}.

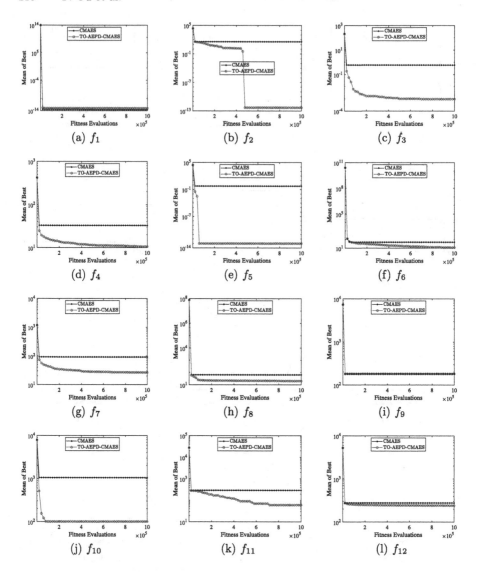

Fig. 2. CEC 2022 average convergence of 30 independent runs of each function in 20-D

Figures 1 and 2 depict the convergence trajectories of TO-AEPD-CMAES and CMAES in 10-D and 20-D, respectively.

On f_2, f_3, f_4, f_5, f_6, f_7, f_8, f_{10}, and f_{11}, CMAES lapses into local optimality, whereas the TO-AEPD-CMAES algorithm applying the TO-AEPD mechanism successfully escapes local optimality and finds a superior solution. Although TO-AEPD-CMAES discovers a superior solution on f_{12} than CMAES, the difference is negligible.

4.4   Comparisons with State-of-the-Art Algorithms

We also conduct an experimental comparison of TO-AEPD-CMAES with ZOC-MAES and NIPOP-CMAES in CEC 2022. NIPOP-CMAES is the classic CMAES variant with restart strategies, while ZOCMAES is a CMAES variant recently proposed in CEC 2022.

Table 2. Average fitness values ± standard deviations on the CEC 2022 functions over 30 independent runs.

D	Fun	TO-AEPD-CMAES	NIPOP-CMAES	ZOCMAES
10	f_1	**1.33e-14 ± 2.45e-14**	3.65e-10 ± 7.91e-11↑	1.20e-06 ± 1.92e-06↑
	f_2	**1.52e-13 ± 4.60e-13**	3.74e-10 ± 9.21e-11↑	1.84e+00 ± 3.15e+00↑
	f_3	3.55e-04 ± 3.06e-04	**6.46e-10 ± 1.00e-10**↓	7.19e-01 ± 1.27e+00↑
	f_4	3.38e+00 ± 8.10e-01	**3.32e-02 ± 1.82e-01**↓	1.33e+00 ± 1.23e+00↓
	f_5	**1.52e-14 ± 3.93e-14**	3.37e-10 ± 8.67e-11↑	1.60e-01 ± 2.51e-01↑
	f_6	1.29e+00 ± 1.00e+00	**4.27e-01 ± 1.13e-01**↓	2.18e+00 ± 4.21e+00
	f_7	**8.60e+00 ± 6.90e+00**	1.18e+01 ± 9.88e+00↑	1.23e+01 ± 1.13e+01↑
	f_8	**6.69e+00 ± 7.88e+00**	1.01e+01 ± 9.33e+00↑	2.09e+01 ± 9.06e+00↑
	f_9	2.29e+02 ± 9.67e-08	**1.86e+02 ± 1.16e-13**↓	2.29e+02 ± 4.25e-02
	f_{10}	9.73e+01 ± 1.64e+01	8.78e+01 ± 5.56e+01	1.34e+02 ± 5.66e+01↑
	f_{11}	**2.11e-09 ± 5.08e-10**	2.22e-09 ± 2.78e-09↑	1.27e-06 ± 2.61e-06↑
	f_{12}	**1.63e+02 ± 6.98e-01**	1.77e+02 ± 1.54e+01↑	1.65e+02 ± 1.45e+00↑
	$b/n/l$	–	7/1/4	10/1/1
20	f_1	**6.63e-14 ± 2.15e-14**	4.98e-10 ± 6.34e-11↑	4.67e-07 ± 1.14e-06↑
	f_2	**4.93e-13 ± 1.56e-12**	5.21e-10 ± 7.50e-11↑	3.13e+01 ± 2.37e+01↑
	f_3	9.45e-04 ± 4.38e-04	6.27e-10 ± 1.70e-10↓	5.01e-02 ± 1.39e-01↑
	f_4	1.06e+01 ± 1.80e+00	**6.89e-10 ± 1.11e-10**↓	2.06e+01 ± 1.31e+01↑
	f_5	**9.47e-14 ± 6.03e-14**	4.64e-10 ± 6.13e-11↑	2.98e-03 ± 1.63e-02↑
	f_6	1.15e+01 ± 6.01e+00	**4.88e-01 ± 3.67e-02**↓	4.16e+01 ± 1.57e+01↑
	f_7	2.64e+01 ± 3.50e+00	**2.08e+01 ± 1.12e+00**↓	9.53e+01 ± 3.91e+01↑
	f_8	2.04e+01 ± 3.16e+00	**1.85e+01 ± 6.00e+00**↓	5.49e+01 ± 6.71e+01↑
	f_9	1.81e+02 ± 7.78e-08	**1.65e+02 ± 2.89e-14**↓	1.81e+02 ± 1.64e-04↓
	f_{10}	**1.00e+02 ± 5.01e-02**	2.50e+02 ± 1.36e+02↑	7.34e+02 ± 1.04e+03↑
	f_{11}	**6.00e+01 ± 1.22e+02**	2.70e+02 ± 9.15e+01↑	2.97e+02 ± 6.15e+01↑
	f_{12}	2.40e+02 ± 3.15e+00	**2.00e+02 ± 5.45e-04**↓	2.48e+02 ± 1.00e+01↑
	$b/n/l$	–	5/0/7	11/0/1

The results of the experiment are presented in Table 2. For the 10-D case, the TO-AEPD-CMAES algorithm outperformed the other two for the f_1, f_2, f_5, f_7, f_8, f_{11}, and f_{12} functions. In the 20-D case, the TO-AEPD-CMAES algorithm also demonstrated superior performance for the f_1, f_2, f_5, f_{10}, and f_{11} functions among the three algorithms.

The preceding experiments demonstrate that TO-AEPD-CMAES is competitive with other outstanding CMAES variants.

5   Conclusion

An auto-enhanced population diversity with two options addresses the problem of premature convergence of CMAES to the problem. The two techniques for enhancing population diversity can be utilized for both exploration and exploitation. When no new optimal solution is generated or convergence accuracy is reached in this round of optimization, the exploration population diversification strategy is chosen. Otherwise, the exploitation strategy of population diversification is implemented. TO-AEPD-CMAES is compared with CMAES, ZOCMAES, and NIPOP-CMAES in the CEC 2022 test suite. The experimental results demonstrate that TO-AEPD-CMAES outperforms the other benchmark test algorithms and that TO-AEPD-CMAES can more effectively address the global optimization problem.

Although TO-AEPD-CMAES solves some of the abovementioned issues, it still confronts considerable obstacles. When the current exploration strategy is employed, repetitive exploration of known areas is a waste of resources. Future work will be required to determine how to conduct exploration more efficiently.

Acknowledgement. This work is supported by the Open Research Project of the Hubei Key Laboratory of Intelligent Geo-Information Processing (Grant No. KLIGIP-2021B04).

References

1. Sarker, R., Mohammadian, M., Yao, X.: Evolutionary Optimization, vol. 48. Springer, Cham (2002)
2. Bäck, T., Fogel, D.B., Michalewicz, Z.: Handbook of evolutionary computation. Release **97**(1), B1 (1997)
3. Beyer, H.G., Schwefel, H.P.: Evolution strategies-a comprehensive introduction. Nat. Comput. **1**, 3–52 (2002)
4. Hansen, N.: The CMA evolution strategy: a comparing review. In: Towards a New Evolutionary Computation: Advances in the Estimation of Distribution Algorithms, pp. 75–102 (2006)
5. Hansen, N., Ostermeier, A.: Adapting arbitrary normal mutation distributions in evolution strategies: the covariance matrix adaptation. In: Proceedings of IEEE International Conference on Evolutionary Computation. IEEE, pp. 312–317 (1996)
6. Hansen, N., Ostermeier, A.: Completely derandomized self-adaptation in evolution strategies. Evol. Comput. **9**(2), 159–195 (2001)
7. Hansen, N.: The CMA evolution strategy: a tutorial. arXiv preprint arXiv:1604.00772 (2016)

8. Varelas, K., et al.: A comparative study of large-scale variants of CMA-ES. In: Auger, A., Fonseca, C., Lourenço, N., Machado, P., Paquete, L., Whitley, D. (eds.) Parallel Problem Solving from Nature–PPSN XV: 15th International Conference, Coimbra, Portugal, 8–12 September 2018, Proceedings, Part I 15, pp. 3–15. Springer, Cham (2018). https://doi.org/10.1007/978-3-319-99253-2_1

9. Shir, O.M., Emmerich, M., Bäck, T.: Adaptive niche radii and niche shapes approaches for niching with the CMA-ES. Evol. Comput. **18**(1), 97–126 (2010)

10. Preuss, M.: Niching the CMA-ES via nearest-better clustering. In: Proceedings of the 12th Annual Conference Companion on Genetic and Evolutionary Computation, pp. 1711–1718 (2010)

11. Auger, A., Hansen, N.: A restart CMA evolution strategy with increasing population size. In: 2005 IEEE Congress on Evolutionary Computation, vol. 2, pp. 1769–1776. IEEE (2005)

12. Hansen, N.: Benchmarking a bi-population CMA-ES on the BBOB-2009 function testbed. In: Proceedings of the 11th Annual Conference Companion on Genetic and Evolutionary Computation Conference: Late Breaking Papers, pp. 2389–2396 (2009)

13. Hansen, N., Ros, R.: Benchmarking a weighted negative covariance matrix update on the BBOB-2010 noiseless testbed. In: Proceedings of the 12th Annual Conference Companion on Genetic and Evolutionary Computation, pp. 1673–1680 (2010)

14. Yang, M., Li, C., Cai, Z., Guan, J.: Differential evolution with auto-enhanced population diversity. IEEE Trans. Cybern. **45**(2), 302–315 (2014)

15. Ning, Y., Jian, D., Wu, H., Zhou, J.: Zeroth-order covariance matrix adaptation evolution strategy for single objective bound constrained numerical optimization competition. In: Proceedings of the Genetic and Evolutionary Computation Conference Companion, pp. 9–10 (2022)

16. Liu, S., Chen, P.Y., Kailkhura, B., Zhang, G., Hero III, A.O., Varshney, P.K.: A primer on zeroth-order optimization in signal processing and machine learning: principals, recent advances, and applications. IEEE Signal Process. Mag. **37**(5), 43–54 (2020)

17. Loshchilov, I., Schoenauer, M., Sebag, M.: Alternative restart strategies for CMA-ES. In: Coello, C.A.C., Cutello, V., Deb, K., Forrest, S., Nicosia, G., Pavone, M. (eds.) PPSN 2012. LNCS, vol. 7491, pp. 296–305. Springer, Heidelberg (2012). https://doi.org/10.1007/978-3-642-32937-1_30

18. Cuzick, J.: A Wilcoxon-type test for trend. Stat. Med. **4**(1), 87–90 (1985)

19. Loshchilov, I.: CMA-ES with restarts for solving CEC 2013 benchmark problems. In: 2013 IEEE Congress on Evolutionary Computation, pp. 369–376. IEEE (2013)

Exploration of Computer Vision

Robot Global Relocation Algorithm Based on Deep Neural Network and 3D Point Cloud

Yan Chen$^{(\boxtimes)}$, Zhengying Li, and Wenbin Qiu

College of Mathematics and Informatics, College of Software Engineering, ·
South China Agricultural University, GuangZhou, China
13380225519@163.com

Abstract. When a mobile robot is restarted or "hijacked" due to an emergency situation, it cannot determine its position and orientation in a known map using its own sensors in places without a GPS signal, such as indoors or outdoors with tall buildings, trees, and other dense areas. In both of these cases, the robot will be unable to find its position in the known map, resulting in a relocation failure. To address the relocation problem of mobile robots, this paper proposes a mobile robot global relocation method that combines deep neural networks and 3D point clouds, based on SLAM (Simultaneous Localization and Mapping) technology. The main idea is to use spherical projection to process the point cloud data, then feed it into a neural network to extract descriptors for similarity measurement, and perform pose calculation to accomplish relocalization. The application of SLAM technology enables the robot to achieve simultaneous localization and map building based on sensor data, accurately determining its own position even in the absence of GPS signals. Extensive experiments have been conducted to validate the proposed method, demonstrating its effectiveness in reducing relocation time and memory costs, as well as achieving relatively high relocation accuracy. Therefore, this method not only enhances the robustness of the system but also improves the feasibility of the overall system.

Keywords: Relocation · Robot · Deep neural networks · Point cloud · SLAM

1 Introduction

In recent years, with the booming development of intelligent manufacturing, the application of mobile robots has become increasingly extensive, and related technologies have gradually become the focus of scientific research. During the process of autonomous navigation, if bit initialization or tracking fails, the robot needs to find its own location on the map using the sensor information it carries. This process is called repositioning.

There are two situations where a mobile robot may need to be repositioned. The first is when the robot is performing a task and its battery is depleted or it is forced to restart. The second is when the robot is "kidnapped", meaning that external factors such as collisions or human interference interrupt the robot's original posture during autonomous task execution. Although the robot saves its posture information before

© The Author(s), under exclusive license to Springer Nature Singapore Pte Ltd. 2024
K. Li and Y. Liu (Eds.): ISICA 2023, CCIS 2146, pp. 223–230, 2024.
https://doi.org/10.1007/978-981-97-4393-3_18

the sudden event, its posture will change significantly after the interruption, leaving the robot unable to determine its location after restarting.

There are various approaches for the mobile robot global localization problem. Fox et al. proposed the Markov Chain Monte Carlo Localization (MCL) algorithm [1], which is the mainstream algorithm in the field of localization and has proven to be effective. However, MCL cannot solve the "kidnapping" problem. Thrun et al. proposed the Adaptive Monte Carlo Localization (AMCL) algorithm based on MCL with random sampling to address the "kidnapping" problem [2]. However, AMCL may take longer to perform global localization without prior information and is prone to relocalization errors in environments with similar scene features, such as densely wooded areas.Engel et al. improved the ORB-SLAM-based robot outdoor relocation algorithm by extracting more scene features and improving the robustness of the algorithm [3]. However, this algorithm is computationally intensive and is also affected by lighting conditions. Li Zhongfa et al. proposed a database-based localization method that records the localization value in a database and uses it for particle initialization when the matching fails [4], improving the speed of the AMCL algorithm. However, the problem of robot "kidnapping" still cannot be solved with this method. To address the "kidnapping" problem, Hess W et al. created a pure localization interface in the open-source Cartographer SLAM (Simultaneous Localization and Mapping) algorithm [5]. However, this approach takes too long and may fail to relocalize in a single environmental scene. For mobile robots navigating autonomously in large scenes without GPS signal, Manhui Sun et al. proposed a spatial database method that allows the robot to store and use laser-built map information for subsequent repositioning, which accurately describes the spatial information and improves repositioning accuracy [6].

In localization based on 3D point clouds and deep neural networks, Uy et al. proposed the PointNetVLAD model [7], which combines PointNet and VLAD (Vector of Locally Aggregated Descriptors). This model can be used for large-scale scene recognition and relocation and has good performance in point cloud retrieval and relocation tasks. Komalasari et al. proposed a method that combines deep neural networks with visual-inertial odometry for precise repositioning in dynamic environments [8]. This method improves the performance of traditional visual odometers by learning residual networks. Sunderhauf et al. proposed a semantic SLAM method based on a deep neural network [9]. Through the combination of semantic information and 3D point cloud information, indoor relocation and map construction are realized. This method has high positioning accuracy in indoor environments.

To address the above problems where a mobile robot may need to be repositioned due to unexpected conditions during autonomous navigation tasks, this paper proposes a global repositioning method that combines deep neural networks and 3D point clouds. This method can quickly and accurately reposition the robot and enhance the overall robustness of the system.

2   Method

The most common method for repositioning robots is using the Global Positioning System (GPS). However, when a robot is indoors or in dense urban areas with tall buildings or dense trees, there may be no GPS signal or interference, rendering GPS-based repositioning impossible. In this paper, we propose a robot global repositioning method that combines deep neural networks and 3D point clouds to address these issues. The overall process of the proposed method is illustrated in Fig. 1:

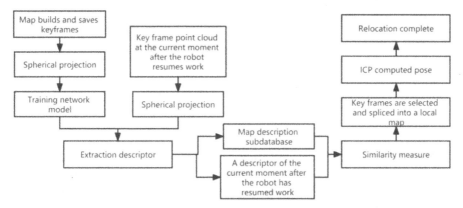

Fig. 1. Architecture of the global repositioning method for mobile robots combining deep neural networks and 3D point clouds.

2.1   Spherical Projection

Spherical projection is a technique used to represent 3D point cloud data as 2D image data. In the field of autonomous driving, a large amount of point cloud information is received every second, and processing this data computationally is a highly demanding task. By representing the point cloud as an image, storage space and computational requirements can be significantly reduced, making it a more efficient approach.

2.2   Network Model

The network proposed in this paper uses the idea of a self-coding network [10], as shown in Fig. 2, which consists of two parts, Encoder and Decoder, with the aim of having the output of Decoder x' replicate the input x of Encoder as much as possible.

In this paper, the design neural network is noted as VOC, and the specific structure is shown in Fig. 3:

In the specific task, the depth of this network model can be adjusted according to the specific requirements, and the descriptor size of the VOC network extracted in this paper is designed $1 \times 8 \times 8$. The reason for taking low-dimensional feature vectors is that in the subsequent similarity metric calculation, it not only consumes less time, but also the low-dimensional data storage space is small, thus to improve the overall system performance.

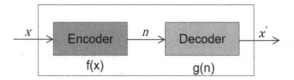

Fig. 2. Self-coding network model structure.

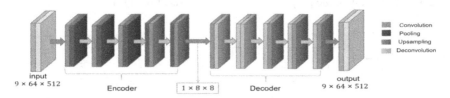

Fig. 3. VOC network structure model.

2.3 Similarity Metric

The similarity degree algorithm used in this paper is cosine similarity, which uses the cosine value of two vectors in vector space as a measure of the magnitude of individual differences, and when the cosine value is closer to 1, it indicates that the closer the angle between the two vectors is to 0 degrees, then the more similar the two vectors are. In the triangle of vector representation, the descriptor of the scene captured after the robot failure resumes working is assumed to be $q = (q_1, ..., q_8)_{1\times 8}$ and the descriptor database of the map is $P_{n\times 8} = \begin{pmatrix} p_{11} & \cdots & p_{18} \\ \vdots & \ddots & \vdots \\ p_{n1} & \cdots & p_{n8} \end{pmatrix}_{n\times 8}$, take one of the vectors $p_i = (p_{1i}, ..., p_{i8})_{1\times 8}$, then the vector q and the vector p_i are calculated in the following way:

$$\cos\theta = \frac{\sum_{i=1}^{n} q_i \Delta p_i}{\sqrt{\sum_{i=1}^{n} q_i^2} \times \sqrt{\sum_{i=1}^{n} p_i^2}} \tag{1}$$

2.4 Displacement Calculation

The ICP (Iterative Closest Point) method is a point cloud registration algorithmthat estimates the overall rigid body transformation matrix by finding the transformation relationship between the closest point pairs in two point clouds, thereby aligning the two point clouds [11]. It minimizes the distance difference between two point clouds through iteration to achieve the registration goal between point clouds, as detailed below.

ICP Algorithm Ideas

Iterative closest point (ICP) is a classical point cloud matching algorithm, the idea is to construct an objective function and iterate through the least squares method until the error is less than a certain threshold or the number of iterations is reached, to get the rotation matrix R and translation vector t. The core of ICP algorithm is to optimize an objective function to get the optimal rotation (R) and translation (t), the optimized objective function is as follows:

$$E(R, t) = \frac{1}{n} \sum_{i=1}^{n} ||x_i - (Ry_i + t)||^2 \tag{2}$$

First, we calculate the center of mass of the two point clouds, which is the average value of the point clouds, and then calculate the coordinates of the center of mass as the origin of the point in the two point clouds, calculate H and decompose it by SVD to obtain the rotation R and translation t, which is the position of the mobile robot in the map, and complete the repositioning.

3 Experimental Results and Analysis

In this paper, we use LIDAR data (data_object_velodyne) from the Kitti [12] public dataset for training and testing, and three subseries of raw data, Road, City, Person, are used for validation experiments.

3.1 A Discussion of Subsequence Relocation Time Overhead and Memory Overhead

For the relocation experiments, we used the sequences of Road, City, and Person from the Kitti dataset as target maps. Table 1 provides an overview of the system memory usage and the time overhead required for relocation when the experiments were conducted on each sub-sequence.

Table 1. Time overhead and memory overhead of subsequence relocation.

Data set type	Number of data sets	Raw point cloud occupies system memory/GB	Point cloud descriptors occupy system memory/Kb	Time taken for repositioning/s
Road	688	1.5	172	5.19
House	978	2.14	245	10.67
City	1323	2.9	331	19.93

It can be seen that when the time consumption is small and the robot has discontinuous poses, the system searches and matches the current frame with the map point cloud database and performs the pose-solving to get the most accurate pose, at least in terms of real-time.

3.2 Discussion and Analysis of the Effect of Candidate Frame Length

In the similarity matching, the length of the candidate frame is set, and this subsection analyzes the influence of the candidate frame length on the repositioning accuracy, and a series of experiments are conducted. House_dataset", "city_dataset". The lengths of the candidate frames are taken as 8, 9, 10, 11, and 12 for the comparison test, and the experimental results are shown in Fig. 4(a), (b), and (c).

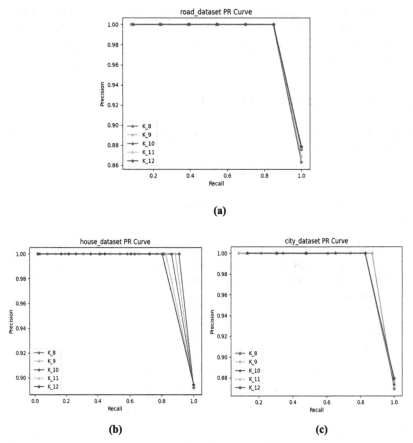

Fig. 4. (a). PR curve of road_dataset. **(b).** PR curve for house_dataset. **(c).** PR curve for city_dataset.

As can be seen from the figure, the experimental sample data of this dataset, road_dataset, is relatively small. The maximum recall of the other two datasets is relatively stable in general, and the experiments illustrate that it is not the case that the longer the extraction length of the key frame interval is, the higher the matching accuracy of relocation is.

3.3 Discussion and Analysis of Key Frame Continuity and Non-continuity

For a particular scene, the continuous data captured by the sensors are all descriptions of the scene, and if the matching results in a continuous sequence, then it means that the matching solution over is reliable. This section addresses the question of whether the key frames are continuous or not to investigate the effect of the continuity of the matching sequence on the repositioning accuracy. The experimental results are evaluated using PR curves, which are shown in Fig. 5(a), (b), (c).

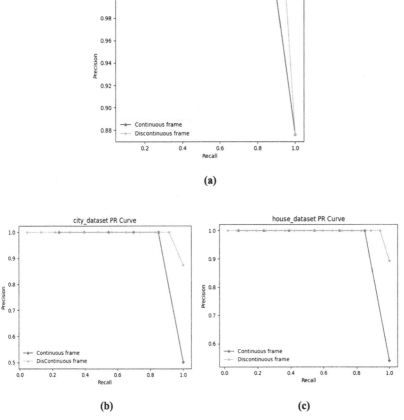

Fig. 5. (a) PR curve of road_dataset. (b) PR curve for City_dataset. (c) PR curve for House_dataset.

It is envisioned that the experimental results should be better with continuous frames and better accuracy of relocation, but on the contrary, it is obvious that the method of this experiment has a higher recall in the case of discontinuous frames. These situations occur because after adding the continuity constraint, the discriminant condition of the system becomes harsh, thus discarding some less rigorous matching results, and although the recall rate decreases, the localization matching results will have a higher confidence level.

4   Conclusion and Work

In this paper, we introduce a novel repositioning method for mobile robots that combines deep neural networks with 3D point clouds. To evaluate its effectiveness, we conducted extensive experiments on three datasets: "road_dataset", "house_dataset", and "city_dataset". Our results demonstrate that the proposed repositioning method offers improved positioning accuracy, reduced time overhead, and efficient memory storage. Moreover, our method provides a new approach to address the challenge of mobile robot repositioning when GPS signals are not available.

References

1. Fox, D., Burgrand, W., Thrun, S.: MarKov localization foe mobile robots in dynamic environments. J. Artif. Intell. Res. **11**, 391–427 (1999)
2. Thrun, S., Fox, D., Burgard, W.: Robust Monte Carlo localization for mobile robots. Artif. Intell. **128**(1), 99–141 (2001)
3. Mur-Artal, R., Montiel, J.M.M., Tardos, Juan D.: ORB-SLAM: a versatile and accurate monocular SLAM system. IEEE Trans. Robot. **31**(5), 1147–1163 (2015)
4. Li, Z.F., Yang, G., Ma, L., et al.: Research on repositioning of substation inspection robots. Comput. Sci. **47**(S1), 599–602 (2020)
5. Hess, W.: Real-time loop closure in 2D LIDAR SLAM. In: Kohler, D., Rapp, H., et al. (eds.) 2016 IEEE International Conference on Robot and Automation (ICRA) (2016)
6. Sun, M., Yang, S., Yi, X., et al.: GIS and SLAM-based autonomous navigation for robots in large range environments. J. Instrum. **38**(3), 586–592 (2017)
7. Uy, M.G., et al.: PointNetVLAD: deep point cloud based retrieval for large-scale place recognition. In: Proceedings of the IEEE Conference on Computer Vision and Pattern Recognition (CVPR) (2018)
8. Komalasari, K., et al.: Deep virtual inertial odometry: leveraging deep residual networks for visual-inertial odometry in dynamic environments. In: Proceedings of the IEEE International Conference on Robotics and Automation (ICRA) (2020)
9. Wu, X., et al.: Semantic 3D Localization and Mapping with Deep Neural Networks. In: Proceedings of the IEEE Conference on Computer Vision and Pattern Recognition (CVPR) (2018)
10. Rumelhart, D.E., Hinton, G.E., Williams, R.J.: Learning representations by back-propagating errors. Nature **323**(6088), 533–536 (1986)
11. Smith, J.: A method for registration of 3D shapes. Int. J. Comput. Vision **25**(3), 100–120 (2008)
12. Geiger, A., Lenz, P., Stiller, C., Urtasun, R.: Vision meets robotics: the KITTI dataset. Int. J. Robot. Res. **32**(11), 1231–1237 (2012)

Safety Zone and Its Utilization in Collision Avoidance Control of Industrial Robot

Yongcai Zhang[1,2], Yih Bing Chu[1(✉)], and Tian Jiang[3]

[1] Department of Electrical and Electronic Engineering, FETBE, UCSI University,
Kuala Lampur, Malaysia
chuyb@ucsiuniversity.edu.my
[2] College of Artificial Intelligence, Dongguan City University, Guangdong, China
[3] College of Mathematics and Information, South China Agricultural University, Guangdong,
China

Abstract. With the development of the digital economy, small-batch customized production has been increasing, leading to a growing application of industrial robots. In these scenarios, new situations arise due to increased debugging frequency and insufficient expertise of debugging personnel. This often results in human errors during actual operation and debugging, leading to control parameters exceeding the safe range. As a consequence, collisions occur between industrial robots and workpieces, causing severe damage to both the equipment and the samples.

Through the study of safety zones and robot collision avoidance systems, we propose an effective method – the concept of a virtual safety zone, without the need for additional external equipment or increased investments. This approach ensures a dual layer of safety between the robot and the samples, reducing the risk of collisions, easing the operator's tasks, and enhancing the likelihood of safe robot operations and successful debugging.

Keywords: Safety Zone · Industrial Robot · Collision Avoidance · Motion Control System · Nachi Simulation System

1 Introduction

Currently, with the rapid development of the digital economy, industrial robots are being widely adopted in the manufacturing industry. They have demonstrated significant advantages in enhancing production efficiency, quality control, and cost savings, making them a crucial pathway to achieving intelligent production. In high-end manufacturing sectors like automotive, robots have become common assembly tools, effectively optimizing the entire assembly process and significantly improving the efficiency of assembly lines. With the development of small-batch and customized production models, the continuous operation time of industrial robots has been substantially reduced, while the frequency of adjustments and debugging has greatly increased. Concurrently, the cost of specialized personnel is rising, leading to an increased reliance on on-site operators for debugging and operating robot systems, resulting in a higher probability of human errors.

© The Author(s), under exclusive license to Springer Nature Singapore Pte Ltd. 2024
K. Li and Y. Liu (Eds.): ISICA 2023, CCIS 2146, pp. 231–246, 2024.
https://doi.org/10.1007/978-981-97-4393-3_19

A notable incident at the 2016 China High-Tech Fair exemplifies the potential consequences of human errors during robot operation [1]. Due to an operator mistakenly pressing the "forward" button instead of the "backward" button, the robot collided with the glass display stand, causing the glass to shatter and injuring a bystander.

Currently, collision avoidance technology for industrial robots primarily relies on sensors and computer vision techniques. These technologies detect the status of the robot and its surrounding environment, enabling industrial robot path planning and collision prevention. With advancements in edge computing and artificial intelligence, these technologies can be combined with collision avoidance techniques to unlock even greater potential. For instance, utilizing deep learning and pattern recognition for object detection and target tracking, or employing machine learning and reinforcement learning to enhance algorithm accuracy and real-time capabilities. However, these approaches often entail substantial hardware and software investments, leading to high maintenance costs.

This paper explores an alternative method: providing an effective approach to achieve dual safety between the robot and the workpieces without the need for additional external devices or increased investments. The objective is to reduce collision risks, alleviate the difficulties faced by operators, and enhance the safety and efficiency of robot operations and debugging. Taking into consideration the specific project requirements, we have selected the MZ07 robot from Nachi Fujikoshi Corporation as the subject of our research. Nachi Fujikoshi robots have found widespread application in intelligent manufacturing industries globally, particularly with numerous successful cases in human-robot collaboration projects [2].

Nachi Fujikoshi Corporation, originating from its establishment in machine tool automation and hydraulic control technology, is one of the earliest industrial robot manufacturers in Japan, beginning its operations in 1968. Presently, as an indispensable partner in automotive production lines, it has earned high praise and trust from manufacturing sites worldwide. The Nachi Fujikoshi robots are capable of accomplishing a wide range of tasks, including high-speed precision work, heavy lifting, welding, and various assembly requirements.

In recent years, Nachi Fujikoshi has been promoting the "world's fastest lightweight compact robot" – MZ07, which possesses several key features [2]:

(1) Improved productivity through the world's fastest motion performance. With a lightweight and highly rigid body and unique control technology, the robot achieves 37% higher motion speed compared to the previous models of the same category, making it the fastest in the world and contributing to enhanced productivity across various production processes.

(2) Lightweight, compact design, and wide range of motion. Despite having a compact installation area of less than the size of an A4 paper, it can reach a maximum arm length capable of carrying 7 kg payloads. Weighing around 30 kg, its lightweight enables manual handling, saving space while providing a wide range of motion. This allows for easy integration of the robot into new equipment or replacing existing robotic systems.

(3) Simplified cable wiring. The hollow wrist design ensures tidy cable routing, especially for the cables connected to the wrist's top part. By routing all cables through the hollow section, the risk of interference with surrounding devices is significantly reduced.

(4) Standard high-level dust and drip protection. The body and wrist have improved dust and drip protection performance (protection grade IP67), making it suitable for harsh environments, such as deburring processes with flying chips or mechanical processing processes with coolant splashes.

(5) Rich selection of specifications and flexible installation conditions. Users can choose from seven types of standard hand units and further select application-specific options like visual and force sensors. The installation options include floor-mounted, wall-mounted, overhead, and tilted configurations, catering to all requirements of replacing manual operations.

(6) Compact, high-performance controller "CFD". Supporting the MZ07, the small-sized controller "CFD" can be installed vertically, contributing to space-saving. The controller includes system software such as PLC and offline simulation tools, providing both high performance and user-friendliness.

Figure 1 depict the main body and handheld controller of Nachi MZ07, respectively.

Fig. 1. Main Body and Handheld Controller of Nachi MZ07.

The logical structure of the robot, as shown in Fig. 2, integrates various I/O boards within the controller.

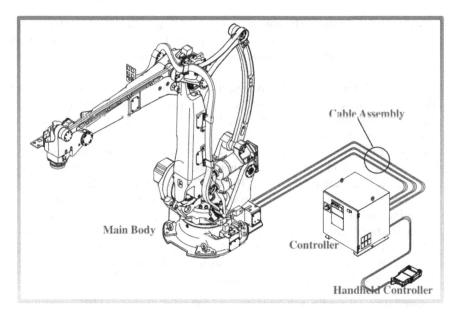

Fig. 2. Logical Structure of Nachi MZ07.

2 Current Sources of Collision Risk in Practical Industrial Robot Applications

2.1 Substituting Intelligent I/O for Hardware I/O to Solve the Issue of Insufficient Host I/O Bus Quantity in Industrial Robots

Communication between industrial robots and equipment relies primarily on I/O ports, where each I/O point corresponds to a state, and each I/O point requires a dedicated wire. In the controller, this is referred to as I/O area mapping, the I/O logic mapping area is shown in Fig. 3. The typical number of I/O ports in industrial robot host I/O buses ranges from 32 to 128. However, when multiple devices need to collaborate in modern intelligent production lines, requiring real-time communication of their working status and positions, the demand for I/O ports increases exponentially, often reaching around 1,000 ports. In such cases, the conventional I/O communication method becomes impractical. Additionally, controlling through I/O ports poses challenges in later maintenance and quick reconfiguration for small-batch customized intelligent manufacturing.

In the practical application of Nachi Fujikoshi robots, typical welding scenarios require the use of more than 200 I/O signals. Using hard-wired connections in this situation results in bulky and unwieldy I/O cable bundles, making it difficult to move and complicating the task of locating and repairing damaged I/O wires, significantly increasing the time spent on this process. Moreover, when the equipment needs to be relocated from the workstation or workshop, it becomes challenging to achieve rapid removal.

As industrial robot applications have advanced, new communication methods have been developed, involving various devices and the industrial robot controller connected

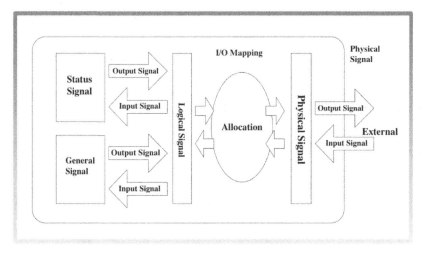

Fig. 3. I/O Logic Mapping Area.

through TCP/IP networks. In Nachi Fujikoshi robot's motion control system, robot programs are divided into task programs (motion programs) and user task programs (communication programs). By employing Socket communication (robot socket communication) in the Ethernet communication mode, data exchange for hundreds of I/O signals can be easily accomplished using a single Ethernet cable, with an upper limit of 65,536 signals and no limitations on the spatial distance, only requiring a stable network connection. When relocating the equipment, the controller can be placed elsewhere while retaining the original mechanical structure, simply requiring the plugging and unplugging of the Ethernet cable, providing convenience and efficiency. The primary functions of Socket communication (robot socket communication) include exchanging data between the robot and a buffer. Dedicated application commands are used for operations involving data exchange between the robot and the buffer. Key commands include Sockcreate (creating socket number), Sockbind (specifying the waiting port number for the socket number), Sockwait (waiting for external connections with the specified port number), Sockconnect (establishing communication with an upper-level computer with a specified IP address), Sockrecv (receiving information sent by an upper-level computer), Socksendstr (returning processed data to the upper-level computer after internal data processing), and Sockclose (closing the socket communication after completing one communication to release memory resources).

Through this method, communication between the robot and production equipment avoids the traditional hard-wired I/O approach and adopts Ethernet communication, enabling software-based control of every robot action, significantly enhancing communication flexibility between the robot and equipment while reducing costs. However, at the same time, it introduces a new challenge: during debugging, even a small programming error can lead to erroneous robot execution and pose a collision risk.

2.2 Implementing Safety Light Curtains to Ensure Robot Safety in the Surrounding Environment

In practical application scenarios, safety light curtains are used to establish a safety zone within the robot's work area, the safety light curtain is shown in Fig. 4. If anyone enters this zone during robot operation, it triggers an alarm and immediately stops robot motion, ensuring safe robot operation and preventing collision risks with personnel [3].

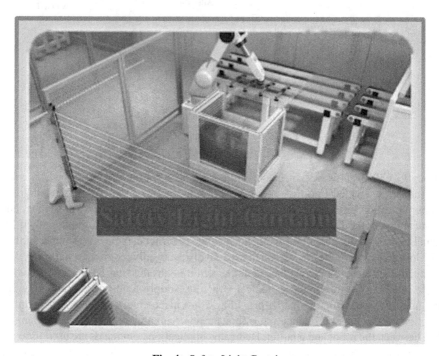

Fig. 4. Safety Light Curtain.

However, this method alone does not reduce the collision risk of industrial robots with workpieces during the debugging process. Once a collision occurs between the robot and the workpiece, it not only affects work efficiency but also damages both the workpiece and the robot, and in severe cases, may disrupt the entire production line and even endanger the operator's safety [4].

2.3   Reducing Collision Risks Between Robots and Workpieces Through Mechanical Collision Prevention Devices

Currently, commonly used robot collision prevention [5] devices is shown in Fig. 5, the devices have the following issues:

(1) They can only prevent collisions in one or two directions.
(2) These collision prevention mechanisms do not significantly unload or weaken collision forces. In the event of a strong collision force, the prevention mechanism may be damaged, failing to fulfill its intended purpose.
(3) During sudden acceleration of the robotic arm, due to non-balanced forces, instability may occur, affecting motion precision [6].

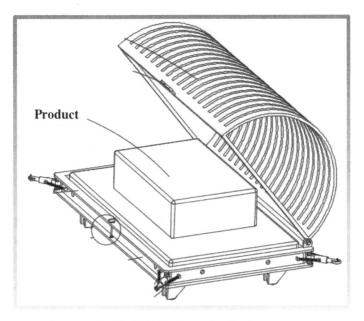

Fig. 5. Collision Prevention Devices.

2.4   Real-Time Detection of Robot

Real-Time Detection of Robot Status and Surrounding Environment through Sensors and Computer Vision, Combined with Edge Computing and Artificial Intelligence, for Real-Time Industrial Robot Path Planning and Collision Detection to Reduce Collision Risks during Robot Debugging [8]. However, the application of these technologies requires increased investment, and maintenance costs remain substantial [9] (Fig. 6).

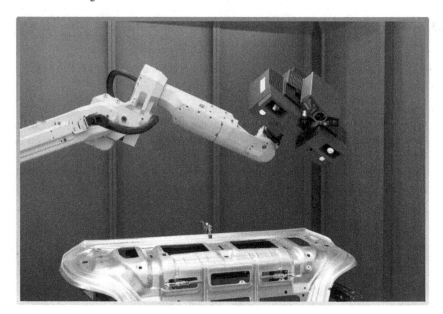

Fig. 6. Collision Prevention Device Based on Vision and AI.

3 Reducing Robot Collision Risk with Workpieces During Debugging Through Safety Zones

3.1 Robot Safety Zones

The concept of robot safety zones has always been highly regarded, regardless of the application scenario. Industrial robots need to maintain a relative safe distance from objects to avoid hardware damage and ensure personnel safety. This safety zone encompasses not only the distance between robots but also involves safe trajectory planning between robots and equipment, as well as ensuring the safety of operators. Introducing safety zones can address potential threats.

In the context of industrial robot debugging scenarios already installed on the production line, the implementation of safety zones, at both the hardware and algorithm levels, ensures that the industrial robot cannot enter the designated safety area during production line debugging. Additionally, various workpieces are positioned within the safety zone. By doing so, collision risks between the robot and workpieces during debugging can be avoided, even in the case of human input errors.

Robot safety zones can be further divided into interference zones and work safety zones. The interference zone occurs when the operating range of one robot overlaps with that of other devices due to the robots' close proximity. In such cases, collisions may occur when they simultaneously move within the overlapping area. On the other hand, the safe area is a region where the robot and other devices will not overlap at any time, ensuring the robot's safe operation.

3.2   Ensuring Robot Safety During Debugging Through Safety Zones

During production line debugging, equipment maintenance personnel may need to send relevant signals to various devices. In this process, there is a possibility of sending erroneous signals to the robot. Under the influence of these incorrect signals, the robot might collide with workpieces on the production line, potentially causing damage to the robot or the workpieces, and even leading to personnel injuries [9]. To address this issue, setting up safety zones in advance or on-site can help ensure safety during robot debugging. When the robot needs to be debugged, the safety zone can be activated manually or automatically. In such a case, as the robot approaches the safety zone, it will automatically stop its motion and trigger an alarm in the system. The robot will remain immobilized until the engineer resolves the alarm signal. After completing the debugging process, the safety zone can be quickly deactivated manually or automatically without affecting the robot's original motions, ensuring that the robot does not collide with the workpieces during production line debugging [10].

Of course, this method may not be effective in situations where the robot's control system or motion hardware malfunctions.

The implementation of safety zones allows for a higher level of debugging safety without the need for additional hardware and software investments or increased maintenance costs [11]. In cases where industrial robots are extensively used, but skilled professionals are relatively scarce, this approach offers a viable solution to ensure production line and equipment debugging safety [12].

A comparative summary of various methods for robot collision avoidance, with reference to Table 1.

Table 1. Comparison of Robot Collision Avoidance Methods.

Collision Avoidance Method	Investment Scale	Effect	Maintenance Cost
Mechanical Collision Avoidance	Medium	Moderate	Low
Light Curtain Collision Avoidance	Medium	Moderate	Low
Sensor Collision Avoidance	Medium	Good	Low
Vision System Collision Avoidance	Large	Good	High
Safety Zone Collision Avoidance	Small	Moderate	Low

4   Simulation and Practical Application

4.1   Simulation and Actual Robot Verification

Through the simulation software provided by Nachi-Fujikoshi Corp, the main parameters of simulating robots are shown in Fig. 7. Simulations of production line debugging with workpieces are conducted, and after successful simulation, validation is performed using real robots. Two scenarios were designed based on actual production conditions:

lateral collision and top pressure of a cuboid workpiece, as well as lateral collision and top pressure of a spherical workpiece [13]. Through multiple rounds of simulation and physical robot testing, the predetermined objectives were achieved both in the simulation system and in real-world testing, yielding the desired outcomes. These validations ensure that during production line debugging, there is no risk of collision between the robot and the workpiece [14]. The simulation and real-world testing of lateral collision and top pressure for the cuboid workpiece are shown in Fig. 8, while the simulation and real-world testing of lateral collision and top pressure for the spherical workpiece are depicted in Fig. 9.

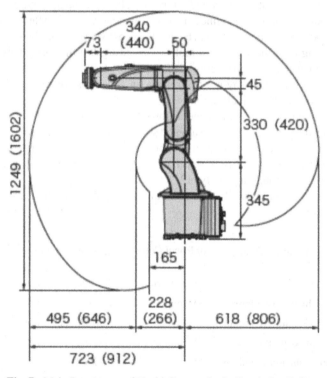

Fig. 7. Main Parameters of Nachi Corporation's Simulating Robots.

4.2 Practical Application

The results from simulation and actual testing demonstrate that the safety zone is effective in preventing collisions between the robot and the workpiece during production line debugging. Based on these findings, practical application and validation are required on the actual production line. Taking into account the production reality, the practical application and validation were carried out on the existing printer's main axis production line. The printer's main axis is a critical component in the printer, serving as a key part in the paper feeding module and ensuring print quality with high precision requirements.

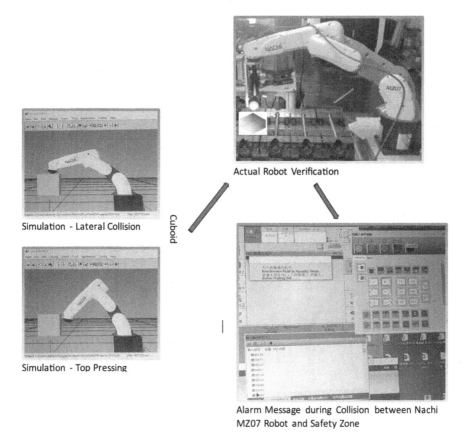

Fig. 8. Simulation and Actual Testing of Lateral Collision and Top Pressure for Cuboid Workpiece.

This automated production line is operated by robots and has previously encountered frequent collisions between the robot and the workpiece due to human errors during line debugging, resulting in robot damage, tool damage, workpiece damage, or production stand damage [15].

For this validation plan, the method of collecting data for one week before and after the upgrade was adopted. The collected parameters include: the number of adjustments made, time spent on each adjustment, the quantity of damaged products during adjustments, and the number of personnel involved in the adjustments [16]. Data collection started from November 1, 2022, and continued for seven consecutive days, followed by the system upgrade to ensure the normal operation of the production line. Subsequently, data was collected continuously for another seven days starting from December 1, 2022. The specific data is presented in Table 2.

The data was analyzed using line and bubble charts, as shown in Figs. 10 and 11. From Fig. 10, it can be observed that after the upgrade, with minimal changes in the

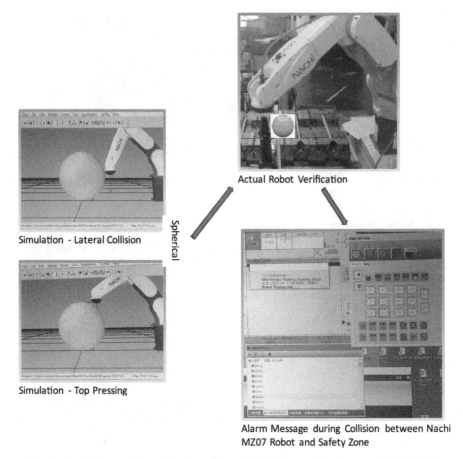

Fig. 9. Simulation and Actual Testing of Lateral Collision and Top Pressure for Spherical.

frequency of adjustments, there is a noticeable decrease in the number of damaged products, the duration of adjustments, and the required number of personnel for adjustments. Particularly significant is the remarkable reduction in the number of damaged products.

From Fig. 11, it is evident that the bubble area representing the number of damaged products significantly decreases after the upgrade.

Further average statistical analysis was conducted on the data in Table 2., and the specific results are presented in Table 3.

The data was analyzed using bar and line charts, as illustrated in Fig. 12. It can be observed from Fig. 12 that, with the frequency of adjustments remaining unchanged, there is a noticeable reduction in the duration of adjustments, the number of damaged products, and the required personnel for adjustments after the upgrade. Particularly significant is the 94% decrease in the number of damaged products, effectively mitigating collision risks between the robot and the workpiece during robot debugging.

Table 2. Production Line Debugging Data for One Week Before and After the Upgrade.

Prod Date	Debugging Number	Debugging Duration	Product Damage Number	Debugging Worker Number
2022-11-01	4	4	10	3
2022-11-02	2	2	6	3
2022-11-03	3	3	6	3
2022-11-04	2	2	8	3
2022-11-05	4	3	8	3
2022-11-06	3	4	6	3
2022-11-07	3	4	6	3
2022-12-01	2	1	1	2
2022-12-02	3	2	1	2
2022-12-03	3	2	0	2
2022-12-04	4	2	0	2
2022-12-05	4	2	0	2
2022-12-06	2	1	1	2
2022-12-07	3	1	0	2

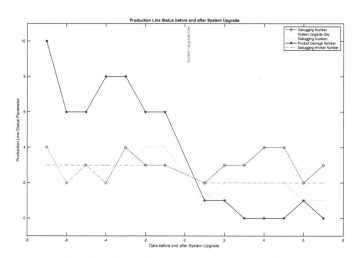

Fig. 10. Line Chart of Data Before and After the Upgrade.

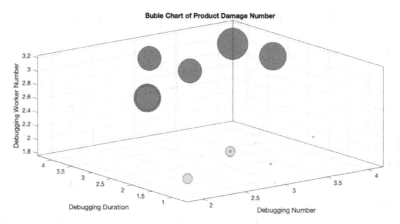

Fig. 11. Bubble Chart of Data Before and After the Upgrade (Red denotes before the upgrade, and Green denotes after the upgrade). (Color figure online)

Table 3. Average Statistical Table of Production Line Parameters Before and After the Upgrade.

Average of Parameter	Without Proposed System	With Proposed System	Percentage Changes
Debugging Number	3	3	0%
Debugging Duration	3.14	1.57	50%
Product Damage Number	7.14	0.43	94%
Debugging Worker Number	3	2	33%

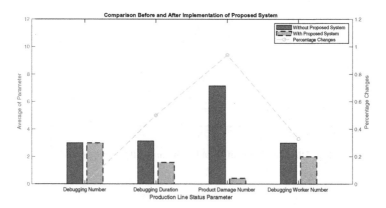

Fig. 12. Statistical Analysis Chart of Production Line Debugging Parameters.

5   Existing Issues and Prospects

The method of employing safety zones effectively avoids collision risks during robot production line debugging. However, two main problems persist:

(1) In case of a malfunction in the motion control system, the safety zone method may become ineffective.
(2) When two robots collaborate on tasks, the safety zone method can lead to increased waiting time, thereby reducing production efficiency.

With the development of Industry 4.0 and smart manufacturing, industrial robots will play an increasingly crucial role in future manufacturing industries. Nevertheless, not all robots are equipped with collision avoidance systems, and only a few possess sensory capabilities like vision or hearing. In the future, the research on robot collision avoidance technology can be further enhanced by optimizing the functionality of safety zones. Exploring improved models and implementation methods for collision avoidance technology, while integrating artificial intelligence and edge computing, can utilize deep learning and other techniques to enhance algorithm accuracy and real-time performance. This will lead to the advancement of robot collision avoidance technology and its applications, ensuring the safety of production line debugging and promoting the widespread adoption of robots. Furthermore, it is essential to consider the cost burden on the system to prevent excessive financial pressure on enterprises, thus hindering large-scale implementation.

6   Conclusion

Through various theoretical analyses and simulation tests on safety zones, it is believed that virtual safety zones can effectively prevent collisions in Nachi robots. Through practical validation on the production line, the Nachi robot system, incorporating virtual safety zones, can effectively prevent collisions between robots and workpieces during line debugging. This simplifies the work difficulty for maintenance personnel, achieves intelligent and rapid product model changes, enhances flexibility in operations, and ensures production safety while realizing intelligent production.

Acknowledgments. This work is supported by Key Field Special Project of Guangdong Provincial Department of Education with No. 2021ZDZX3019.

References

1. Mohammed, A., Schmidt, B., Wang, L.: Active collision avoidance for human–robot collaboration driven by vision sensors. Int. J. Comput. Integr. Manuf. **30**(9), 970–980 (2017)
2. Nachi Product Info: Website of NACHI-FUJIKOSHI COPR. https://www.nachi-fujikoshi.co.jp/eng /rob/index.html
3. Nikolakis, N., Maratos, V., Makris, S.: A cyber physical system (CPS) approach for safe human-robot collaboration in a shared workplace. Robot. Comp.-Integr. Manuf. **56**, 233–243 (2019)

4. Wang, L.: Collaborative robot monitoring and control for enhanced sustainability. Int. J. Adv. Manuf. Technol. **81**, 1433–1445 (2015)
5. Lo, S.Y., Cheng, C.A., Huang, H.P.: Virtual impedance control for safe human-robot interaction. J. Intell. Rob. Syst. **82**, 3–19 (2016)
6. Michalos, G., Makris, S., Tsarouchi, P., et al.: Design considerations for safe human-robot collaborative workplaces. Proc. CIrP **37**, 248–253 (2015)
7. Schmidt, B., Wang, L.: Depth camera based collision avoidance via active robot control. J. Manuf. Syst. **33**(4), 711–718 (2014)
8. Pérez, L., Diez, E., Usamentiaga, R., et al.: Industrial robot control and operator training using virtual reality interfaces. Comput. Ind. **109**, 114–120 (2019)
9. Bosscher, P., Hedman, D.: Real-time collision avoidance algorithm for robotic manipulators. Ind. Robot Int. J. **38**(2), 186–197 (2011)
10. Dröder, K., Bobka, P., Germann, T., et al.: A machine learning-enhanced digital twin approach for human-robot-collaboration. Proc. CIrP **76**, 187–192 (2018)
11. Avanzini, G.B., Ceriani, N.M., Zanchettin, A.M., et al.: Safety control of industrial robots based on a distributed distance sensor. IEEE Trans. Control Syst. Technol. **22**(6), 2127–2140 (2014)
12. Graham, J.H., Meagher, J.F., Derby, S.J.: A safety and collision avoidance system for industrial robots. IEEE Trans. Ind. Appl. **1**, 195–203 (1986)
13. Parsons, H.M.I.: Human factors in industrial robot safety. J. Occup. Accid. **8**(1–2), 25–47 (1986)
14. Zanchettin, A.M., Ceriani, N.M., Rocco, P., et al.: Safety in human-robot collaborative manufacturing environments: metrics and control. IEEE Trans. Autom. Sci. Eng. **13**(2), 882–893 (2015)
15. Magrini, E., Ferraguti, F., Ronga, A.J., et al.: Human-robot coexistence and interaction in open industrial cells. Robot. Comput.-Integr. Manuf. **61**, 101846 (2020)
16. Zhang, Y.C.: Analysis of robot control system based on dual-machine coordination. Integr. Circuit Appl. **39**(07), 138–139 (2022)

Entropy of Interval Type-2 Fuzzy Sets and Its Application in Image Segmentation

Jianqiao Shen[1], Haijun Qian[2(✉)], and Huabei Nie[1]

[1] DongGuan City College, DongGuan 523106, China
[2] ZhuHai City Polytechnic, Zhuhai 519000, China
331096487@qq.com

Abstract. To overcome the deficiencies of entropy of interval type-2 fuzzy sets, a new type of entropy of interval type-2 fuzzy sets is established. Influence of membership degree's fuzziness and mean value on entropy of interval type-2 fuzzy sets is considered, and that the proposed entropy of interval type-2 fuzzy sets satisfies the four axioms is proved. The proposed entropy of interval type-2 fuzzy sets measures uncertainty of interval type-2 fuzzy sets comprehensively, which is used in image segmentation. Numerical examples show the rationality and practicality of the proposed entropy of interval type-2 fuzzy sets.

Keywords: interval type-2 fuzzy sets · entropy of interval type-2 fuzzy sets · entropy · image segmentation

1 Introduction

Interval type-2 fuzzy sets (IT2FS) is a extension of traditional type-1 fuzzy sets (T1FS), and the character of it is that the membership degree of IT2FS is fuzzy sets, which can effectively process the uncertainty of language and data. This character of IT2FS made it has more obvious performance than the corresponding performance of T1FS in the high uncertainty situation, which cause extensive concern of scholars [1–3]. A current hot research area is how to measure the uncertainty features of IT2FS. Intuitionistic fuzzy set theory (IFS) is equivalent the IT2FS. Burillo is the first scholar who put forward the concept of fuzzy entropy of interval type-2, which only taking into account the membership itself fuzziness of entropic effects, failure to effectively distinguish membership bounds of difference equal to the entropy of IT2FS size, especially when the degradation of IT2FS is T1FS, entropy is 0, which do not accord with people's intuition [4, 5]. Szmidt proposed four axioms which interval type-2 entropy should meet, and constructed the interval type-2 fuzzy entropy by computing the most distance and recent distance of IFS [6]; Zeng proposed a new interval type-2 fuzzy entropy, which satisfies the four axioms, but only consider the influence of the entropy from the average value of membership degree, failure to effectively distinguish entropic value of IT2FS which have the same average degree of membership [7]; Wu defined the interval type-2 fuzzy entropy as a IT2FS, which less rarely used in practice [8]; Deng Yanquan put forward a new interval type-2 fuzzy entropy, but the proving process is error, and does not satisfy the axiom [9].

© The Author(s), under exclusive license to Springer Nature Singapore Pte Ltd. 2024
K. Li and Y. Liu (Eds.): ISICA 2023, CCIS 2146, pp. 247–253, 2024.
https://doi.org/10.1007/978-981-97-4393-3_20

According to the shortcomings of above method in calculation the interval type-2 fuzzy entropy, this paper proposes a new method to compute the interval type-2 fuzzy entropy, not only satisfies the four axioms, but also takes into account the influence of entropy from membership degree and average value of fuzzy entropy, and measures the IT2FS uncertainty information in a more comprehensive and objective measurement. The new method to compute the interval type-2 fuzzy entropy is applied to image segmentation, which achieved good results.

2 Interval Type-2 Fuzzy Sets

Definition 1. Let X be a nonempty set of ordinary, called mapping $A : X \rightarrow [I], x \rightarrow [A^-(x), A^+(x)]$ is the IT2FS on the X, in which, $x \in X, A^-(x) \leq A^+(x), A(x)$ is the membership function of IT2FS, $A^-(x)$ and $A^+(x)$ are the lower bound and upper bound of IT2FS $[A^-(x), A^+(x)]$.

If use integral symbol to express the IT2FS, then A can expressed: $A = \int_{x \in X} [A^-(x), A^+(x)]/x$, the integral represents the corresponding fuzzy set of arbitrary value x in domain X.

Theorem 1. Suppose A and B are IT2FS, they have following properties:

(1) The union set of A and B is:

$$A \cup B = \int_{x \in X} [A^-(x) \vee B^-(x), A^+(x) \vee B^+(x)]/x$$

(2) The intersection of A and B is:

$$A \cap B = \int_{x \in X} [A^-(x) \wedge B^-(x), A^+(x) \wedge B^+(x)]/x$$

(3) The complement of A is:

$$A^c = \int_{x \in X} [1 - A^+(x), 1 - A^-(x)]/x$$

3 Interval Type-2 Fuzzy Entropy

Definition 2. Real function $E : IT2FS(X) \rightarrow R^+$, is called interval type-2 fuzzy entropy, and meet the following four axiom:
   (T1) If A is a classic collection (non fuzzy set), then $E(A) = 0$;
   (T2) $E(A) = E(A^C), \forall A \in IT2FS$.
   (T3) $E(A) \leq E(B)$, if A is less fuzzy then B:
   ① When $B^+(x_i) + B^-(x_i) \leq 1$,

$$A^-(x_i) \leq B^-(x_i); A^+(x_i) \leq B^+(x_i)$$

② When $B^+(x_i) + B^-(x_i) \geq 1$,

$$A^-(x_i) \geq B^-(x_i); A^+(x_i) \geq B^+(x_i)$$

(T4) If $A^-(x_i) + A^+(x_i) = 1$, then $E(A) = 1$.

Burillo proposed a new concept of interval type-2 fuzzy entropy, the basic form is:

$$E(A) = \frac{1}{N} \sum_{i=1}^{N} \left(A^+(x_i) - A^-(x_i)\right) \tag{1}$$

Formula (1) has fully considered the fuzziness of membership degree. If the larger the gap value between the upper bound and the lower bound, the bigger of the entropy; But Formula (1) did not consider the effects on the entropy from the membership mean (half of the upper and lower bounds), especially when IT2FS was degraded into T1FS. At that moment, the entropy is 0, which is not accord with the concept of T1FS, and don't meet the axiom T3 and T4 in Definition 2.

Zeng proposed a computing method of interval type-2 fuzzy entropy based on the condition that interval type-2 fuzzy entropy meet the four axioms:

$$E(A) = 1 - \frac{1}{N} \sum_{i=1}^{N} \left|A^-(x_i) - A^+(x_i) - 1\right| \tag{2}$$

Formula (2) has fully considered the effects on the entropy from the membership mean. When the value of membership degree closing to 0.5, the entropy is becoming big, which accord with the four axioms in Definition 2. Formula (2) also did not consider the effects from the membership fuzzy to the interval type-2 fuzzy entropy.

Theorem 2.   Definiton 2 can conduct the following formula.

$$E(A) = \frac{N - \sum_{i=1}^{N} \left|A^-(x_i) + A^+(x_i) - 1\right| \cdot \left(1 - \left(A^+(x_i) - A^-(x_i)\right)\right)}{N + \sum_{i=1}^{N} \left|A^-(x_i) + A^+(x_i) - 1\right| \cdot \left(1 - \left(A^+(x_i) - A^-(x_i)\right)\right)} \tag{3}$$

Formula (3) is a fuzzy entropy of A (A is an interval type-2 fuzzy entropy).

Proof. (1) If A is a classic set, then the following expressions are established.
$A^-(x_i) = A^+(x_i) = 0$ and $A^-(x_i) = A^+(x_i) = 1$,
Substitute the expressions to the Formula (3), then, $E(A) = 0$.
(2) Because expression $(A^c)^-(x_i) = 1 - A^+(x_i), (A^c)^+(x_i) = 1 - A^-(x_i)$ established, then,

$$E(A^c) = \frac{N - \sum_{i=1}^{N} \left|1 - A^-(x_i) - A^+(x_i)\right| \cdot \left(1 - \left(A^+(x_i) - A^-(x_i)\right)\right)}{N + \sum_{i=1}^{N} \left|1 - A^-(x_i) - A^+(x_i)\right| \cdot \left(1 - \left(A^+(x_i) - A^-(x_i)\right)\right)}$$

$$= E(A)$$

(3) Both A and B are IT2FS, and B is more fuzzy then A, the following 2 situations are discussed:

① When $B^+(x_i) + B^-(x_i) \leq 1$, $A^-(x_i) \leq B^-(x_i)$, $A^+(x_i) \leq B^+(x_i)$, below expressions established: $A^-(x_i) + A^+(x_i) \leq B^-(x_i) + B^+(x_i) \leq 1$, $E(A) = \dfrac{2N}{N+\sum_{i=1}^N \left((1-A^+(x_i))^2 - (A^-(x_i))^2\right)} - 1$ $E(B) = \dfrac{2N}{N+\sum_{i=1}^N \left((1-B^+(x_i))^2 - (B^-(x_i))^2\right)} - 1$, because of $1 - A^+(x_i) \geq 1 - B^+(x_i)$, it can be conducted the expression: $E(A) \leq E(B)$.

②When, $B^+(x_i) + B^-(x_i) \geq 1$, $A^-(x_i) \geq B^-(x_i)$, $A^+(x_i) \geq B^+(x_i)$, the following expressions conducted: $A^-(x_i) + A^+(x_i) > B^-(x_i) + B^+(x_i) > 1$ $E(A) = \dfrac{2N}{N+\sum_{i=1}^N \left((1-A^-(x_i))^2 - (A^+(x_i))^2\right)} - 1$, $E(B) = \dfrac{2N}{N+\sum_{i=1}^N \left((1-B^-(x_i))^2 - (B^+(x_i))^2\right)} - 1$, because of $1 - A^-(x_i) \leq 1 - B^-(x_i)$, it can be conducted the expression: $E(A) \leq E(B)$.

(4) When $A^-(x_i) + A^+(x_i) = 1$, $E(A) = 1$.

The interval type-2 fuzzy entropy conducted by Formula (3) not only satisfies the four axioms, and combines the advantages of Burillo and Zeng method, but also overcomes the insufficient of Burillo and Zeng method, and conducts a more comprehensive measure on uncertainty information of IT2FS.

4   Application Examples

Example 1. Calculate the entropy of A and B (Both A and B are interval type-2 fuzzy entropy)

$$A = \frac{[0.3, 0.6]}{x_1}, B = \frac{[0.1, 0.4]}{x_1}$$

The entropies of A and B are calculated by Formula (1) ~ (3), and the results show in Table 1.

Table 1. Comparison of three methods.

Calculate method	A	B
Burillo	0.300	0.300
Zeng	0.900	0.500
New method	0.869	0.481

Table 1 shows that the Burillo method can not effectively distinguish the entropy size of A and B when the difference between the upper and lower bound of membership degree of A is equal to that of B, and the entropy of A is very small. Theoretically, if the average of membership close to 0.5, the entropy value will close to 1. Both the Zeng method and the new method proposed in this paper can compare the entropy size of A and B, and can achieve the biggest value of entropy while the average membership closing to 1.

Example 2. Calculate the entropy value of A and B (both A and B are interval type-2 fuzzy entropy).

$$A = \frac{[0.3, 0.6]}{x_1}, B = \frac{[0.2, 0.7]}{x_1}$$

Use formula (1) ~ (3) to calculate the entropy value of A and B, and the comparison result shows in Table 2.

Table 2. Comparison of three methods.

Calculate method	A	B
Burillo method	0.300	0.500
Zeng method	0.900	0.900
New method	0.869	0.905

Table 2 shows that the Zeng method can not compare the entropy size of A and B when they have the same membership degree. The main reason is that the Zeng method has not considering the membership degree. Although the Burillo method can compare the entropy size, the entropy value is obviously small, which is contrary to the fact that entropy value should near 1 when the average of membership degree closing to 0.5. The new method proposed in this paper can compare the entropy size correctly, and the entropy value is close to 1.

5 Image Segmentation Based on Interval Type-2 Fuzzy Entropy

According to the fuzzy set theory, a $M \times N$ dimension and with L gray level image can be seen as a $M \times N$ dimensional fuzzy matrix. Use x_{ij} to express the gray level of (i, j)-th pixel, and x_{max} express the max gray level, then, a image (X) can be expressed:

$$X = \left[\left(\mu_{ij}^-(x_{ij}), \mu_{ij}^+(x_{ij}) \right) \right]_{M \times N}, i = 1, \cdots, M, i = 1, \cdots, N \tag{4}$$

$\mu_{ij}^-(x_{ij})$ and $\mu_{ij}^+(x_{ij})$ are the membership function of gray interval type-2 fuzzy sets of the (i, j)-th pixel in matrix.

$$\mu_{ij}^-\left(x_{ij}, a^-, b^-, c^-\right) = \begin{cases} 0, x_{ij} \leq a^- \\ 2\left(\frac{x_{ij}-a^-}{c^--a^-}\right)^2, a^- < x_{ij} \leq b^- \\ 1 - 2\left(\frac{x_{ij}-a^-}{c^--a^-}\right)^2, b^- < x_{ij} \leq c^- \\ 1, x_{ij} < a^- \end{cases} \tag{5}$$

$$\mu_{ij}^+\left(x_{ij},a^+,b^+,c^+\right) = \begin{cases} 0, x_{ij} \leq a^+ \\ 2\left(\frac{x_{ij}-a^+}{c^+-a^+}\right)^2, a^+ < x_{ij} \leq b^+ \\ 1-2\left(\frac{x_{ij}-a^+}{c^+-a^+}\right)^2, b^+ < x_{ij} \leq c^+ \\ 1, x_{ij} < a^+ \end{cases} \tag{6}$$

In above expression, the parameters values are:

$b^+ = \left(a^+ + c^+\right)/2, b^- = \left(a^- + c^-\right)/2, a^+ \leq a^-, c^+ \leq c^-$.

Make $a^- = 1.1a^+, c^- = 1.1c^+$, then the algorithm of image segmentation of interval type-2 fuzzy entropy as following:

Step 1: Initialization parameter a+:$a^+ = L_{\min}$;

Step 2: Calculate parameter c+ and b+: $c^+ = a^+ + \Delta b, b^+ = a^+ + \Delta b/2$;

Step 3: Calculate the interval type-2 fuzzy entropy by formula (5), formula (6) and (3);

Step 4: If $a^+ \leq L_{\max} - \Delta b$, then, $a^+ ++$, go to Step 2;

Step 5: Calculate the corresponding a+ and b+ of max interval type-2 fuzzy entropy, and determine the threshold of segment;

Step 6: Divide image.

In the above algorithm, value of Δb is predetermined. In theoretically, its value range is: $0 < \Delta b \leq L_{\max} - L_{\min}$, and in practical application, the value range can be obtained by experience.

(a) original Image of rice   (b)segmentation result by Burillo

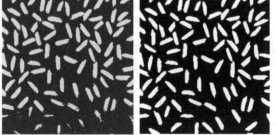

(c) segmentation result by Zeng   (d) result by new method

Fig. 1. Segmentation result in rice image.

Example 3.   Do the global threshold segmentation in classical rice image by interval type-2 fuzzy entropy. The experiment as following: CPU, AMD Sepron Processor 2600 +; Memory, DDR 768M; Operating system: Windows XP; Simulation, Matlab7.0. The comparison result is show in Fig. 1.

Figure 1 shows some fact: In the Burillo method, some of the information losing, and was mixed by much noise point; In the Zeng method, noise decreased, but part of information of the lower half part of image losing; The segment result by new method proposed in this paper are obviously better than other two methods.

6   Conclusion

This paper proposed a new algorithm of interval type-2 fuzzy entropy, which meets the four axioms, and considering the influence on the entropy from the fuzzy and mean value. Does objective measure on uncertainty information of IT2FS, and applied in image segmentation. Numerical examples and simulation experience show the rationality and practicality of the new method.

Acknowledgment. This work is supported by:

1. The Key Field Special Project of Guangdong Provincial Department of Education with No. 2021ZDZX1029.

2. The 2021 Key scientific research platforms and projects in ordinary colleges and universities in Guangdong Province, ID: 2021ZDZX3019.

References

1. Liang, Q., Mendel, J.M.: Interval type-2 fuzzy logic systems theory and design. IEEE Trans. Fuzzy Syst. **8**(5), 535–550 (2000)
2. Mendel, J.M., John, R.I., Liu, F.: Interval Type-2 Fuzzy Logic Systems Made Simple. IEEE Trans. Fuzzy Syst. **14**(6), 808–821 (2006)
3. Pan, Y., Huang, D., Sun, Z.: II Type fuzzy control overview. Control Theory Appl. **28**(1), 13–23 (2011)
4. Yang, W., Yin, M.: Several L–fuzzy relationship. Journal of Liaoning Normal University (NATURAL SCIENCE EDITION) **28**(2), 143–145 (2005)
5. Burillo, P., Bustince, H.: Entropy on intuitionistic fuzzy sets and on interval-valued fuzzy sets. Fuzzy Sets Syst. **78**(3), 305–316 (1996)
6. Szmidt, E., Kacprzyk, J.: Entropy for intuitionistic fuzzy sets. Fuzzy Sets Syst. **118**, 467–477 (2001)
7. Zeng, W.Y., Li, H.X.: Relationship between similarity measure and entropy of interval valued fuzzy sets. Fuzzy Sets Syst. **157**(11), 1447–1484 (2006)
8. Wu, D.R., Mendel, J.M.: Uncertainty measures for interval type-2 fuzzy sets. Inf. Sci. **177**(23), 5378–5393 (2007)
9. Deng, T., Wang, Z., Wang, P.: Research on Control and Decision. **27**(3), 408-412 (2012)
10. Zeng, J., Guo, D.: Masquerade intrusion detection algorithm based on interval type-2 fuzzy entropy set. Chin. J. Electron. **36**(4), 777–787 (2008)

An Improved Algorithm for Facial Image Restoration Based on GAN

Jibo Zhang[1], Jia Yuan[2], Dongbo Zhang[2(✉)], and Lu Xiang[3]

[1] School of Software Engineering, Software Engineering Institute of Guangzhou,
Guangzhou 510990, China
[2] School of Computer and Information Engineering, Guangdong Polytechnic of Industry and
Commerce, Guangzhou 510510, China
db_zhang2013@163.com
[3] College of Automation, Zhongkai University of Agriculture and Engineering,
Guangzhou 510070, Guangdong, China

Abstract. The paper proposed a two-stage facial image restoration method based on structure and texture network, which includes sketch restore network and texture generation network. The sketch restore network strives to restore the sketch structure in the missing area of the image, and the texture generation network generates the texture information of the missing area based on the sketch structure of the sketch restore network and the pixels of the known area. The method is not only able to successfully generate semantically reasonable and visually realistic content for missing regions, but also allows users to manipulate the structural properties of synthetic content in missing regions. Experimental results show that the proposed method not only has good performance, but also is more flexible.

Keywords: image restoration · face inpainting · structure network

1 Introduction

In recent years, with the continuous development of deep learning, especially the emergence of [1] (Generative Adversarial Networks, GAN), face image synthesis technology has made breakthrough progress. For example, StyleGAN [2] model proposed by Nvidia can unconditionally synthesize realistic high-resolution face images; and FaceShifter [3] confrontation face change model proposed by Microsoft Institute can realize high-quality video face replacement. In general, these face image synthesis technologies based on generative adversarial networks can be divided into two categories: unconditional face image synthesis and conditional face image synthesis. Unconditional face image synthesis refers to the synthesis of face images by fitting the potential distribution of training data from the random noise vector. Because the generation process is uncontrollable, the value is limited. Conditional face image synthesis refers to the generation of corresponding face images according to given conditions. According to different conditions, it can be divided into face editing [4], face restore [5], face domain conversion [6] and face exchange [7] and so on. These condition-based face synthesis tasks are essentially

© The Author(s), under exclusive license to Springer Nature Singapore Pte Ltd. 2024
K. Li and Y. Liu (Eds.): ISICA 2023, CCIS 2146, pp. 254–265, 2024.
https://doi.org/10.1007/978-981-97-4393-3_21

very relevant, for example, face editing and face restore are only local areas of synthetic face, and face restore itself is also a face editing method [8]; face exchange is a special face editing operation, used to edit the identity information of face image. The goal of face image restore is to restore and rebuild the damaged or missing areas in the image. The content of the reconstruction does not need to be completely consistent with the original image, but only needs to be consistent with the texture and structure of the known areas around, so as to make it look real visually. Therefore, the face image restore task is multi model in nature, and there are many feasible solutions that are different from the original image content. However, the current existing face restore method [9] aims to improve the quality of image restore, but lacks attention to the controllability of the restore process and the diversity of restore results. On the other hand, for face images, an important application of image restoration is face editing. So more need to ensure the restore results visual authenticity on the basis of the controllability of restore process and the diversity of restore results, for example, the user wants to remove the face image black glasses, when image restore way the editing, task, need to remove by editing mask, then according to the image, the area of information synthesis of new content, but because of the current image restore method can only generate a single result and restore process is uncontrollable, so the face editing ability is very limited, which seriously limits the actual value. In addition, since the various fields of conditional face image synthesis are very relevant in nature, the research on the controllability and result diversity of face image restore process can also provide new ideas and new methods for other related fields, such as: face domain conversion [6], face aging [10], etc. Therefore, the study on how to realize the controllability and diversity of face image restore not only has a very important scientific research value, but also has a very high application value.

Recently, many image restore methods based on generative adversarial networks have emerged and achieved promising results. These methods [11, 12] regard the image restore task as a conditional image generation problem, adopt end-to-end encoding-decoding network structure, and jointly train the restore model with reconstruction loss and resistance loss. The reconstruction loss model can make reasonable inferences about the missing area according to the existing information, and the resistance loss ensures that the content of the synthesis is consistent with the known area around. Depending on whether the restore process is controllable, these image restore methods based on generative adversarial networks can be divided into two categories: uncontrollable image restore and controllable image restore. Uncontrollable image restore method [13, 14]. To ensure the visual authenticity of the restore results as the main goal, committed to improve the quality of image restore, this kind of method is not only suitable for face images, but also for the restore of natural scene images. Uncontrollable image restore method lays the foundation for the research of controllable image restore. However, this method can only generate a single restore result, and the restore process is uncontrollable and users cannot interact. Therefore, for face images, uncontrollable methods can only restore damaged areas, or remove some unwanted semantics such as glasses, its face editing ability is very limited.

2 Related Theory

Generative adversarial network (GAN) was proposed [1] by Ian Goodfellow in 2014, which aims to optimize model parameters to fit the potential distribution of real data through adversarial training. The proposal of generative adversarial network model greatly promotes the development of generative model and makes image synthesis and other related fields enter a new stage of development. The original generative adversarial network model GAN consists of two parts: generator G and discriminant D, generator G tries to generate samples as realistic samples as possible so that the discriminant cannot distinguish the true from the false. The discriminator D tries to distinguish whether the input sample comes from real data or from model generation. This can use an interesting metaphor to clarify the process of confrontation between the generator and the discriminator. Generators are similar to counterfeiters, trying to make counterfeits and use them undetected, while discriminators are similar to police officers, trying to detect them. The two compete (fight) against each other and constantly improve their technical capabilities to gain an advantage in the competition, which will make the counterfeit money more and more like the real money until the police cannot detect it.

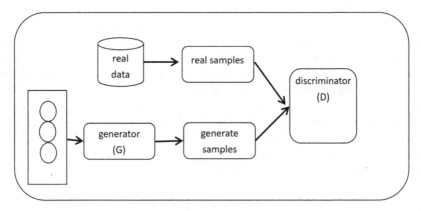

Fig. 1. The basic framework of GAN

The basic framework of generating adversarial network is shown in Fig. 1. The input of generator G is a set of random noise vectors that meet the specified prior distribution, and the input of discriminant D has two categories: generated sample and real sample. The generator G optimizes the model parameters to fit the potential distribution of the real data by adversarial training, thus using random noise vectors to synthesize samples that fit the potential distribution of the real data. Then the generated sample is sent to the discriminator D together with the real data, and the discriminator D is the difference between the true and the input as far as possible. In the original generative adversarial network model, the discriminator needs to make a binary discrimination, with the generated data judged 0 (false) and the real data judged 1 (true). The generator G strives to generate as real as possible to make the discriminator D true. In this antagonistic process, both the generative model and the judgment model are optimized, and their performance is improved. The optimization ends when the final judgment model cannot distinguish between its data sources. At this point, the generator G can fit the distribution of the real data, and generate samples that are difficult to distinguish from the real data. The adversarial process between generator G and discriminator D can be regarded as a minimal game (two-player minimax game), whose mathematical expression is defined as follows:

$$\min_{G} \max_{D} V(D, G) = E_{x \sim P_{data}}[\log D(x)] + E_{z \sim P_z}[\log(1 - D(G(z)))] \tag{1}$$

P_{data} where z and x represent the random noise vector and the real sample, respectively, x can be generated by random sampling from the real data distribution, and z can be generated by the specified prior distribution Pz sampling. Generating adversarial networks adopt an alternating iterative update approach to optimize the parameters of generator G and discriminant D. According to Eq. (1), the maximum and minimum optimization process can be separated, first fixing the generator G and maximizing the solution V (D,G) solution D, then fixing the discriminator D and minimizing V (D, G) solution G.

3   Proposed Algorithms

Recently, with the continuous development of convolutional neural networks and generative adversarial network [6], deep learning-based image restore research has made substantial progress. These methods [5, 13, 15] image restore task as conditional image generation problem, using the end-to-end encoding-decoding (Encoder-Decoder) network structure, using reconstruction loss and against loss joint training restore model, reconstruction loss makes model according to the known letter, the missing area content make reasonable inference, against loss to ensure the content of the synthesis is consistent with the surrounding known area. Although these methods can generate new content in highly structured image areas (such as eyes or nose), they tend to produce distorted structures and blurred textures. This is mainly because the structure and texture are two indispensable parts of the image, and the image information represented by the two is very different. In particular, the synthesis of semantic information directly affects the generation of the corresponding texture information, and the distorted semantic structure will inevitably lead to the visual unreal of the synthesized texture information. However, the existing methods use the same objective function to model the two parts of information, which leads to neither good structure information nor texture information, making the resulting restore structure distortion and texture blur. In addition, because these methods adopt end-to-end coding-decoding network structure, with missing image as the input restore model, directly in the output of the restore model, the whole restore process completely controlled by the model, it is difficult to develop flexible user interaction, which seriously limits the potential application of these methods.

In order to solve the above problems, this paper proposes a two-stage face image restore method based on the structure and texture network, respectively, modeling the structure and texture information of the image in the missing area, which can not only generate the continuous structure and clear texture restore results, but also provides the potential for manipulating the image restore process. Because the image restore task is divided into two stages: structure restore and texture generation, breaking the limitation of the current end-to-end model to fully independently control the restore process, users can fine-tune the restored structural information to control the structural attributes of the generated content in the missing area, so as to achieve diversified restore results. The overall process of the method proposed in this paper is shown in Fig. 2. The method consists of three modules: sketch structure restore module, structural attribute migration module and texture generation module.

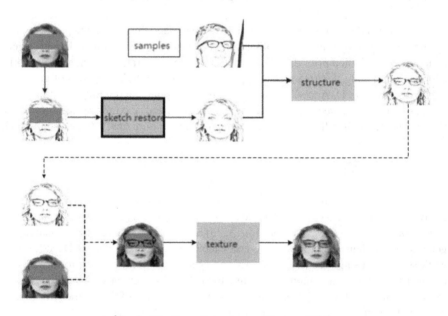

Fig. 2. The flow of structure and texture GAN

3.1   Sketch Structure Restore Network

On the pixel value is consistent with the original image, the constraint is too strict and ignores the reconstruction of the high level of semantic structure information, and the sketch mainly contains the structure information in the original image, more need to reconstruct the high level of semantic structure information, so we introduce the perceptual loss training model makes it able to make reasonable inferences of the missing area semantic structure according to the known information. Furthermore, we only used a global discriminator for adversarial training to ensure that the synthesized content is consistent with the surrounding known regions, unlike the GL method, using the local discriminator in the sketch structure restore network sGAN severely affects the stability of adversarial training and leads to poor results. Network structure: For sGAN, generator G1 uses a network structure similar to the GL [14] method, but contains fewer network layers. This is because the sketch restore process in sGAN is a simpler task and only needs to generate the sketch information of RGB images. More specifically, the generator G1 consists of three parts: two subsampled convolutional layers, four expanded convolution layers (Dilated Convolution), and two upsampled convolution layers. Expansion convolution was used to increase the receptive fields of neurons to provide a broader view to capture the overall structure of the sketch.

The global discriminant D1 consists of four subsampled convolutional layers and a fully connected layer, and then processes the features of the network output using the Sigmoid activation function. Different from the GL method, the global discriminator and local discriminator are used for adversarial training, only the global discriminator D1 is used in sGAN, because the sketch mainly contains the structural information of the face image, which is easier to be distinguished by the discriminator, so that the adversarial training cannot be well executed. Furthermore, the local discriminator takes as input a sketch block centered on the missing region, which makes the local discriminator focus too much on the missing region and ignoring the agreement between known and missing regions. Instead, the global discriminator takes the entire image as the input. It ensures that the synthesis content is consistent with the surrounding known areas, and the adversarial training can also be conducted more stably and effectively. The improved network structure is shown in Fig. 3.

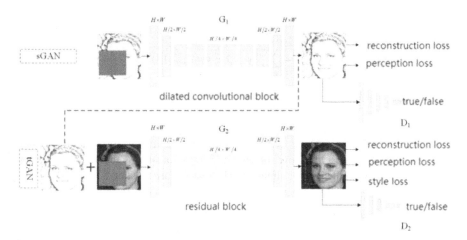

Fig. 3. The network structure of sGAN and tGAN

$s \otimes (1 - M)$ $\hat{s} \otimes$ Loss function: the definition x is a real image, and s represents its corresponding sketch structure. M is a binary restore mask with a value of 1 indicating the broken area and a value of 0 indicating the known area. Generator G1 uses a pending fix sketch as network input to generate a restoreed sketch, here representing point-by-point multiplication. The target loss function used to train the generating network G1 mainly consists of three parts: reconstructed loss function, perceptual loss function and global adversarial loss function. We will introduce these three losses in detail respectively.

(1) Reconstruction loss

The reconstruction loss provides pixel-level constraints, and we first define the reconstruction loss with the Euclidean distance between s and: $\hat{s}\xi_r$

$$\xi r = \|s - \hat{s}\|_2 \tag{2}$$

$\hat{s} = G_1(s \otimes (1 - M))$ Among them, the reconstruction loss makes the generator G1 learn to make reasonable inferences about the content of the missing region based on the existing information, but the results generated by the reconstruction loss constraint are consistent with the original image in pixel value, which is too strict and ignores the reconstruction of high-level semantic structure information.

(2) Perception loss

$\hat{s}$ Different from the reconstruction loss by directly calculating the Euclidean distance between the pixels in the original sketch s and the restore result. Perception loss [16] first uses the pre-trained VGG-16 network to map these images to the high-dimensional feature space to extract high-level semantic structural information, and then calculates Euclidean distances in the high-dimensional feature space. The perceived loss enables the generative network G1 to attempt to model high levels of semantic structural information, which is particularly suitable for the generation of sketch structures. Perceived loss is defined as follows: ξ_{perc}

$$\xi_{perc} = \sum_{i=1}^{N} \|\varphi_i(s) - \varphi_i(s_{comp})\|_2 \tag{3}$$

$s_{comp} = s \otimes (1 - M) + \hat{s} \otimes M$ φ_i Where, representing the i-th layer of the pre-trained VGG-16 network, N is the total number of layers used to construct the perceptual loss. In the experiments in this paper, using the "conv 1 of the VGG-16 network-1", "conv2-1" "and" conv 3-1" Three layers to extract high levels of semantic structure information. Perceived loss provides a high level of semantic-level constraints that make the restore image consistent with the original image in abstract feature semantics.

(3) Fight against losses

The restore results generated by using only reconstruction loss and perceptual loss are often ambiguous. Therefore, this paper also introduces adversarial losses to ensure that the restore results are visually real and natural by conducting adversarial training between the generative networks and the discriminative networks. As mentioned above, only the global discriminator D1. The adversarial loss used to train the discriminant network D1 is defined as follows:

$$\xi_{adv_{d1}} = \log(D_1(s) + \log(1 - D_1(s_{comp})) \tag{4}$$

The adversarial loss used to train the generative network G1 is defined as follows

$$\xi_{adv_{d1}} = \log(D_1(s_{comp})) \tag{5}$$

In conclusion, the final objective function used to train the generative network G1 is as follows:

$$\min_{G_1} \xi_{G1} = \xi_r + \lambda_{perc}\xi_{perc} - \lambda_{adv}\xi_{adv_{g1}} \tag{6}$$

λ_{perc} λ_{adv} Where, and is the hyperparameters of the model used to balance the effects of different losses. In this experiment, let up, $\lambda_{perc} = 0.05 \lambda_{adv} = 0.004$.

The final objective function used to train the discriminant network D1 is as follows

$$\min_{G1} \xi_{D1} = \xi_{adv_{d1}} \tag{7}$$

3.2 Texture Generation Network

Network structure: as shown in Fig. 3, Although tGAN has a similar network structure to sGAN, But there are several differences: For the tGAN generator G2, The residual block was first used instead of the expanded convolutional layer in the sGAN, This allows for a more efficient use of the sketch information of sGAN synthesized for the missing region, And accelerate the model training process; Second, use the instance normalization layer [17] (Instance Normalization, IN) instead of the batch normalization layer used in the sGAN, The IN normalization layer was initially used for style migration of images, Can better model the color and texture of the image. For discriminant, tGAN, and only use a local discriminator D2 against training, because tGAN can easily produce consistent texture information, make the global discriminator is difficult to distinguish true and false, in addition, the global discriminator set the missing area pixel pixel value to the real image, and then to the whole image as input, which further increases the difficulty of the global discriminant to distinguish true and false, leading to the global discriminant cannot distinguish the true and false image. On the contrary, the local discriminator takes the image block centered on the missing area as input, which makes the local discriminator focus on the restore area, enhances the recognition ability and ensures the stability of training.

Loss function: Since tGAN is trained on two branches, reconstruction and generation, its loss function is divided into two parts. One part is used to optimize the reconstruction branch, mainly including reconstruction loss and perception loss; the other part is used to optimize the generation branch, mainly including adversarial loss and style loss.

(1) Reconstruction of the branch loss

In the reconstruction branch, the texture generator G2 takes the RGB image and the original image as input to output the reconstruction results. The whole process can be described in the following equation: $x \otimes (1 - M)x'$

$$x' = G_2(x \otimes (1 - M) + s \otimes M) \tag{8}$$

ξ_r x' Then the reconstruction loss constraint is consistent with the original image x at the pixel level, and the reconstruction loss is defined as follows:

$$\xi_r ||x - x'||_2 \tag{9}$$

Just as sGAN, the perceptual loss is also used for the reconstruction branch of tGAN to constrain the reconstruction results to be consistent with the original image x at the level of semantic structure, and the perceptual loss is defined as follows: x'

$$\xi_{\text{perc}} = \sum_{i=1}^{N} ||\varphi_i(x) - \varphi_i(x')||_2 \tag{10}$$

(2) Generate the branch loss

$x \otimes (1 - M)$ In the generation branch, the texture generator G2 takes the RGB image to be restoreed and the draft structure generated by sGAN as input to output the restore result, and the whole process can be expressed in the following equation

$$\hat{x} = G_2(x \otimes (1 - M) + \hat{s} \otimes M) \tag{11}$$

Texture information in missing regions synthesized by tGAN via style loss [118]. The definition of style loss is similar to perceptual loss, requiring the pre-trained VGG-16 [119] network to map the image to the high dimensional feature space to extract high level of semantic structure feature information. The difference is that the style loss first models the correlation in the channel dimension, and then calculates the Euclidean distance on the correlation matrix to model the texture style of the image. Style loss is defined as follows: ξ_{style}

$$\xi_{style} = \sum_{i=1}^{N} ||\varphi_1(x)^T \varphi_i(x) - \varphi_i(\hat{x})^T \varphi_i(\hat{x})||_2 \tag{12}$$

φ_i Where, representing the i-layer of the pre-trained VGG-16 network, N is the total number of layers used to construct the style loss. In the experiments, we still used the "conv 1 of the VGG-16 network-1", "conv2-1" and "conv 3-1" Three layers to construct the image of the style texture information.

Finally, tGAN uses the local discriminant D2 to ensure the visual reality of the generated restore results through the antagonistic loss. The adversarial loss used to train the discriminant network D2 is defined as follows:

$$\xi_{adv_{d2}} = \log(D_2(x) + \log(1 - D_2(x_{comp})) \tag{13}$$

In summary, and finally used to train the generator G2. The loss function is as follows:

$$\xi_{adv_{g2}} = \log(D_2(x_{comp})) \tag{14}$$

4 Experimental Analysis

For experiments and analysis in the face image restore public dataset CelebA [122], the CelebA [122] dataset consists of 200k face images with 40 attribute annotations, which can be used to realize the migration of structural properties. During the training process, the face area of size 160160 was first cropped in the center, and then the face image was scaled to 128128 resolution size for the training of the model. For mask M, a mask of size between 4848 and 9696 was randomly generated to guarantee at least one critical facial semantic deletion. During testing, the mask M can be generated randomly or with the location and size specified by the user. For sketch structures, in this paper the classical Gaussian blur method [123] is used to generate the sketch structure corresponding to RGB images. The sketch structure generated by this method is also often used for the image translation task [124–126]. In the training setting, the method proposed in this paper is optimized using Adam [127] optimizer with a learning rate of 0.0001, implemented using the TensorFlow deep learning framework and trained on a single NVIDIA 1080Ti GPU (12GB) with the batch size (Batch Size) set to 32. For sGAN, the generator and the discriminator were iteratively trained 45,000 times, with a learning rate of 0.0001. For tGAN, the generator $\beta_1 = 0.5$, $\beta_2 = 0.99$. And the discriminator was trained 22000 times with a learning rate of 0.0001.

Since the model proposed in this paper divides the face image restore process into sketch structure restore and texture information synthesis.Two phases, thus first a qualitative analysis of the performance of the sketch structure restore network sGAN, followed by knots.Together with sGAN and tGAN for the qualitative evaluation of the entire restore model. The restore results generated by the sketch structure restore network sGAN are shown in Fig. 4. The first column in the figure is the real sketch, the second column is the sketch to be restoreed with mask M, and the third column is the restore results generated by sGAN. It can be seen from the results that sGAN can make full use of the sketch structure information of the known area to synthesize the semantically reasonable and visually real missing content, and maintain the good consistency of the edge of the missing area with the structure of the known area without obvious restore trace; there is no structural distortion of the internal synthesized content in the missing area. This fully demonstrates that sGAN can synthesize semantic structure information of missing regions well and can in modeling semantic structure.

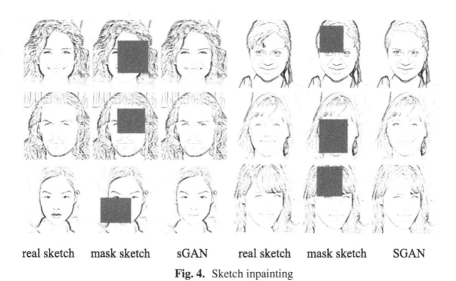

real sketch   mask sketch   sGAN   real sketch   mask sketch   SGAN

Fig. 4. Sketch inpainting

Three different quantitative measures were also quantitatively compared using peak SNR PSNR [18], structural similarity S-SIM [19] and mean L1 error, where higher values of PSNR and SSIM indicate better restore results, and lower values of L1 indicate better restore results. The quantitative comparison results are shown in Table 1, which shows that the proposed methods have outperformed the others in all indicators.

Table 1. Quantitative comparison of facial image inpainting.

algorithm	PSNR	SSIM	Mean loss L_1
GL	19.21	0.806	8.56%
EG	22.26	0.922	6.49%
CA	20.56	0.863	5.97%
Proposed	24.22	0.935	4.65%

5 Conclusion

The paper proposes a new facial image restoration algorithm based on structural and texture networks to solve the problems of structural distortion and texture blur in the process of facial image restoration. The algorithm divides the face restore process into two parts: sketch structure restore and texture information synthesis. Due to the significant difference in image information contained in these two parts, the algorithm proposed in this paper models these two parts of information through sketch restore network (sGAN) and texture generation network (tGAN), respectively. SGAN is committed to repairing the sketch structure of missing areas in images, and tGAN outputs based on sGAN generate texture information for missing areas based on the pixel structure and context of the sketch.

Acknowledgment. This work is supported by the Natural Science Foundation of Guangdong Province under Grant 2021A1515012529 and 2022A1515110633. The key scientific research platform and project of Guangdong Province Department of Education for ordinary colleges and universities under Grant 2022KTSCX159, 2020ZDZX3098, 2022DZZX1052. The project of Guangdong polytechnic of industry and commerce under Grant 2022-TDJ-04 and 2022-gc-05.

References

1. Goodfellow, I., Pouget-Abadie, J., Mirza, M., et al.: Generative adversarial nets. In Advances in Neural Information Processing Systems, pp. 2672–2680 (2014)
2. Karras, T., Laine, S., Aila, T.: A style-based generator architecture for generative adversarial networks. In IEEE Conference on Computer Vision and Pattern Recognition, pp. 4401–4410 (2019)
3. Li, L., Bao, J., Yang, H., et al.: Advancing high fidelity identity swapping for forgery detection. In IEEE Conference on Computer Vision and Pattern Recognition, pp. 5074–5083 (2020)
4. Shu, Z., Yumer, E., Hadap, S., et al.: Neural face editing with intrinsic image disentangling. In IEEE Conference on Computer Vision and Pattern Recognition, pp. 5444–5453 (2017)
5. Pathak, D., Krahenbuhl, P., Donahue, J., et al.: Context encoders: Feature learning by inpainting. In IEEE Conference on Computer Vision and Pattern Recognition, pp. 2536–2544 (2016)
6. Isola, P., Zhu, J.-Y., Zhou, T., et al.: Image-to-image translation with conditional adversarial net- works. In IEEE Conference on Computer Vision and Pattern Recognition, pp. 1125–1134 (2017)

7. Bao, J., Chen, D., Wen, F., et al.: Towards open-set identity preserving face synthesis. In IEEE Conference on Computer Vision and Pattern Recognition pp. 6713–6722 (2018)
8. Ronneberger, O., Fischer, P., Brox, T.: U-Net: Convolutional networks for biomedical image segmentation. In Medical Image Computing and Computer-Assisted Intervention, pp. 234–241 (2015)
9. Liu, H., Jiang, B., Xiao, Y., et al.: Coherent semantic attention for image inpainting. In IEEE International Conference on Computer Vision, pp. 4169–4178 (2019)
10. Wang, Z., Tang, X., Luo, W., et al.: Face aging with identity-preserved conditional generative adversarial networks. In IEEE Conference on Computer Vision and Pattern Recognition, pp. 7939–7947 (2018)
11. Xie, C., Liu, S., Li, C., et al.: Image inpainting with learnable bidirectional attention maps. In International Conference on Computer Vision, pp. 8857–8866 (2019)
12. Nazeri, K., Ng, E., Joseph, T., et al.: EdgeConnect: Structure guided image inpainting using edge prediction. In IEEE International Conference on Computer Vision Workshops (2019)
13. Li, Y., Liu, S., Yang, J., et al.: Generative face completion. In IEEE Conference on Computer Vision and Pattern Recognition, pp. 5892–5900 (2017)
14. Yan, Z., Li, X., Li, M., et al.: Shift-Net: Image inpainting via deep feature rearrangement. In European Conference on Computer Vision, pp. 3–19 (2018)
15. Gatys, L.A., Ecker, A.S., Bethge, M.A.: Neural algorithm of artistic style [J/OL]. CoRR (2015) arXiv preprint arXiv:1508.06576. http://arxiv.org/abs/1508.06576
16. Simonyan, K., Zisserman, A.: Very deep convolutional networks for large-scale image recognition. In International Conference on Learning Representations (2015)
17. Ulyanov, D., Vedaldi, A., Lempitsky, V.S.: Instance normalization: The missing ingredient for fast stylization [J/OL]. (2016) CoRR, arXiv preprint arXiv:1607.08022. http://arxiv.org/abs/1607. 08022
18. Huynh-Thu, Q., Ghanbari, M.: Scope of validity of psnr in image/video quality assessment. Electron. Lett. 44(13), 800–801 (2008)
19. Wang, Z., Bovik, A.C., Sheikh, H.R., et al.: Image quality assessment: from error visibility to structural similarity. IEEE Trans. Image Process. 13(4), 600–612 (2004)

A Study of PyTorch-Based Algorithms for Handwritten Digit Recognition

Kangshun Li[1,2(✉)], Mingchen Xie[1], and Xuhang Chen[1]

[1] College of Mathematics and Informatics, South China Agricultural University, Guangzhou 510642, China
likangshun@dgcu.edu.cn
[2] School of Artificial Intelligence, Dongguan City University, Dongguan 523808, China

Abstract. In this paper, three handwritten digit recognition algorithms based on K-nearest neighbor (KNN), support vector machine (SVM) and convolutional neural network (CNN) are implemented. And compared the performance of the above algorithms in the recognition of handwritten digits using the National Institute of Standards and Technology Modified Database (MNIST) dataset and the USPS dataset. The three algorithms were evaluated in terms of area under the curve (AUC), accuracy (ACC), F1 score (F1), accuracy (PREC) and recall rate (REC), and relative performance index (RPI). We first search for the optimal combination of learning parameters in each algorithm, then evaluate the performance of different algorithms across data sets to obtain the best digit recognition model, and develop a handwritten digit recognition system based on the above algorithm. The system has the functions of numeral input recognition of writing pad, multi-numeral input recognition of picture, and data set making based on input.

Keywords: Handwritten digits · Convolutional neural networks · Pattern recognition

1 Introduction

Handwritten digit recognition is an important part of pattern recognition, especially the off-line handwritten digit recognition used in real-time scenarios. Handwritten digit recognition can be regarded as a special and valuable way of human-computer interaction. There are many application scenarios for handwritten digit recognition in daily life, such as identifying envelopes with postal codes, processing financial statements on a large scale, and processing formal input from banks. Despite having a sample of ten fixed categories, handwritten numerals vary widely among different people or regions. Therefore, in some cases, it is difficult to maintain a relatively high recognition rate. Although many algorithms have been developed for handwritten digit recognition, in handwritten digit recognition systems, poor contrast, blurring of image text, stroke interruption, stroke distortion, and similarities between and within classes all lead to misclassification of models. The early support vector machine calculation gives good results, but it also

© The Author(s), under exclusive license to Springer Nature Singapore Pte Ltd. 2024
K. Li and Y. Liu (Eds.): ISICA 2023, CCIS 2146, pp. 266–276, 2024.
https://doi.org/10.1007/978-981-97-4393-3_22

brings the calculation cost and a lot of time consumption. With the advancement of technology, algorithms in machine learning such as artificial neural network, convolutional neural network, K-nearest neighbor algorithm have shown excellent performance. A number recognition framework utilizing machine learning has been created as a model with practically ideal results.

Some authors [1–4] applied SVM to the classification and recognition of MNIST data sets, and achieved 99.40% [1], 97.70% [2], 98.91% [3], and 98.30% [4]accuracy. Other authors implemented KNN machine learning algorithms on MNIST datasets [5, 6], and achieved 93.86%, 96.8%, 96.67%, and 96.8%, respectively. The recognition accuracy of SVM and KNN is as high as 97% ~ 98%, but for human input, the recognition efficiency of numbers outside the standard database is not high, so it is necessary to improve the recognition accuracy and generalization ability. To overcome this and achieve greater accuracy, many researchers use deep learning networks in their papers. The basic algorithm of deep learning network is convolutional neural network (CNN), which is trained on MNIST data by several authors using different forms of convolutional neural network models, and good results have been obtained. The first convolutional neural network to predict handwritten numbers was proposed by Yann LeCun [7]. The network became known as LeNet-5. Yann LeCun et al. used this unpreprocessed convolutional neural network to achieve an error rate of 0.95%. Author Arora S achieved 98.72% accuracy using Keras-based convolutional neural networks [8]. Many researchers have improved the model on the basis of the common convolutional neural network, so that the new algorithm has improved the accuracy and efficiency. For example, Yang J used an improved convolutional neural network based on dropout and stochastic gradient descent optimizer to achieve a high recognition rate of 99.97% on the MNIST dataset [9], and the time cost of the algorithm is only 21.95% of that of MLP-CNN and 10.02% of SVM. Ali S [10], a convolutional neural network based on deep learning Deeplearning4j (DL4J) framework was used to obtain recognition with high accuracy (99.21%) in a very short computational time. Author Vamvakas G used a two-stage classification technique and proposed a feature extraction technique that relies on recursive subdivision of character images, achieving 99.03% accuracy in MNIST dataset [11]. Recognition of degenerate handwritten numerals using dynamic Bayesian networks has been achieved by Likforman-Sulem on MNIST data and achieves 97.2% accuracy [12]. The gate-guided dynamic deep learning method used achieved 99.66% accuracy on the MNIST dataset.

Some authors have also used convolutional neural networks to test on different types of data sets, such as the CMATERDB dataset [13], and have implemented a deep learning ConvNet model with 97.40% accuracy while implementing a deep neural network with 98.70% accuracy. The Parsian data set (Arabic handwritten numeral set) is applied to the SVM which combines chain code histogram and black and white transition, and the accuracy is 98.55%. The CNN technique, which uses a dataset of handwritten Arabic numerals, achieved 95.7 percent accuracy. Author Ahlawat S proposed a hybrid method of SVM and CNN, which achieved higher accuracy compared with other methods [14]. The author Trivedi A combined genetic algorithm and neural network on the Devanagari handwritten digit set, and the accuracy rate was 96.09% when using genetic algorithm and 96.25% when not using genetic algorithm [15].

Therefore, this paper combines three common algorithms: logistic regression, multi-perceptrons, and support vector machines for simultaneous handwriting problems, compares their advantages and disadvantages, and then looks at whether these models can achieve higher accuracy and why these models present corresponding results.

2 Method

2.1 Model Training Based on K-nearest Neighbor Algorithm

As a lazy and passive algorithm, KNN algorithm does not have a real training process, but stores all the training data and calculates the distance between the test sample and the training set only when it is really necessary to make a decision. Therefore, we only show the testing process of KNN model as shown in Fig. 1.

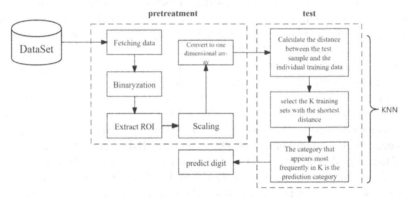

Fig. 1. K-nearest neighbor algorithm model training diagram.

The core of KNN model training lies in the selection of parameter K, and we will find the optimal KNN model between 3–5.

2.2 Design and Training of Convolutional Neural Network Model

We use a seven-layer network with three convolution layers of 3*3 dimensions and two maximum pooling layers of 2*2 as the model of this experiment. ReLu function is selected as the hidden layer excitation function, and Softmax function is used as the final output. The RMSprop algorithm is used as the model optimizer and categorical_crossentropy is used as the multi-classification loss function. The model is shown in Fig. 2.

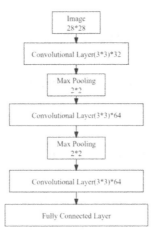

Fig. 2. Convolutional neural network in this paper.

The training process is shown in Fig. 3.

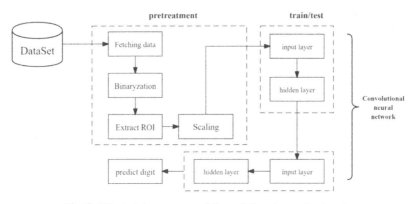

Fig. 3. The training process of Convolutional neural network.

2.3 Design and Training Based on Support Vector Machine Model

For nonlinear data classification. SVM uses the nonlinear mapping of kernel function to extend the classification of nonlinear data to linearly separable data in high dimensional space, and then finds the optimal hyperplane in the high dimensional space. Common kernel functions are polynomial kernel function, Gaussian kernel function, sigmoid kernel function.

In addition, the penalty coefficient C can enhance the generalization ability of SVM. Practical experience shows that the larger the penalty parameter C is, the more stringent the model is for classification, resulting in a larger separation interval and better generalization ability.

Since penalty parameter C and kernel function directly affect the spatial characteristics of the mapping, the core of our prediction model training lies in the selection of kernel function and penalty coefficient C. We will select polynomial kernel function, Gaussian kernel function and sigmoid kernel function as parameters of support vector machine and match them with multiple penalty parameters C. In the support vector machine with many different parameters to find the best performance of the one as our model.

2.4 Design of Handwritten Digit Recognition System Based on Multi-model Voting

Finally, we selected the best performing models from support vector machine, convolutional neural network and K-nearest neighbor algorithm respectively, integrated them into a system, and deployed the system to the server to provide services to users through web pages. The system provides users with real-time digital recognition service by means of writing pad and image file uploading. For the uploaded images, we will also carry out paper edge recognition and multi-digit segmentation to reduce the interference of irrelevant objects. Finally, we collect the handwritten digits provided by the user, the predicted value of the system and the true value of the user's feedback, and make them into a new handwritten digit data set for the iterative update of the model.

1) Pretreatment. Since there are many differences between the input image and the image uploaded by the user in the preprocessing stage, we will use different methods to preprocess the image with different input methods.

Figure 4 shows the process of image preprocessing for users using a handwriting pad. Since the user writes on the writing pad provided by us, the writing scope, number, writing strokes and pictures are all limited to some extent, and the picture background is free of stains (such as uneven and grainy inferior paper), and the style is similar to MNIST. Therefore, compared with the graph provided by the user, its pre-processing process is relatively simple, and it is identified by the target area and adjusted to the acceptable specifications of the model.

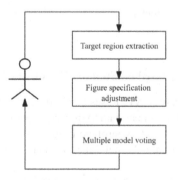

Fig. 4. The image preprocessing process of handwriting tablet.

Figure 5 shows the pre-processing process of the image uploaded by the user.

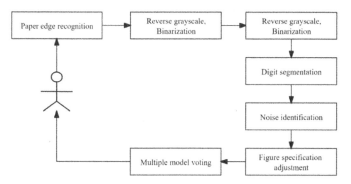

Fig. 5. Upload image preprocessing process.

When the user uses a pencil to write on paper, the strength used for each stroke is not uniform, the font will inevitably appear deep and shallow, and even incomplete, and because the paper quality is different, the picture often has a granular texture, so the preprocessing of such data is often more complicated.

First of all, consider that the user may input a whole sheet of paper as an image, so it is necessary to extract the outline of the paper, which can eliminate irrelevant information, reduce the calculation amount of the system and reduce the interference of irrelevant information.

Then, considering that daily handwritten pictures are often written in black and white, the effect presented by our training data set is often written with a white pen on a black background. In order to make the input pictures close to the style of MNIST and achieve better prediction effect, we will reverse gray scale and reverse binarization of the pictures. And the uneven friction between the pen and the paper often causes the digital lines to have gritty little black spots, making the strokes uneven and smooth. For this reason, we carry out 3*3 ones matrix convolution operation on the image to achieve the expansion effect.

And considering that the user will write multiple numbers, it is also crucial to divide the individual numbers. We use a series of methods based on the contour detection function in the OpenCV. Finally, considering that some large noise points cannot be filtered out in the binarization stage (such as wood chips on environmentally friendly paper, or other stains). Therefore, we perform noise removal based on fill percentage and size for the numbers recognized by the contour algorithm.

2) Multiple model voting. We give the pre-processed images to different models for prediction, and finally vote for the final predicted value through different models. Through the weighted fusion of different models, the system has good robustness, and the working flow is shown in Fig. 6.

According to repeated debugging, the weights of the voting model are SVM = 0.26, KNN 0.26, and CNN 0.48. In other words, the conclusion of CNN can be overturned only if the SVM and KNN algorithm predict the same.

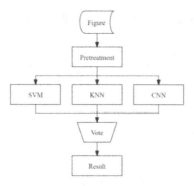

Fig. 6. Voting model.

3) System workflow. Our system will collect the images provided by users and the truth values returned by users to make a dataset for the iterative update of the model, and use Redis database for data persistence. The system workflow is shown in Fig. 7.

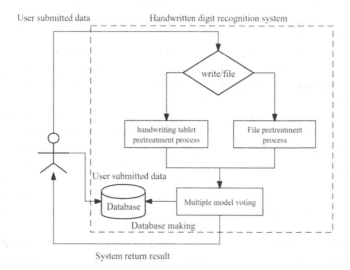

Fig. 7. Workflow.

3 Experimental Results and Analysis

3.1 System Environment and Tool Library

The experimental environment of this paper is Windows10(64bits) operating system, Intel® Xeon® CPU E5 v4@3.70 GHz processor, running memory is 16 GB. The programming language uses Python3.9, uses PyTorch and TensorFlow to implement convolutional neural network, and uses Sklearn function library to implement KNN algorithm and SVM algorithm. Each model was trained using the well-known handwritten numerals data sets MNIST and USPS. MNIST stands for National Institute of Standards and Technology Modified Database dataset, and USPS is a digital dataset automatically scanned from envelopes by the United States Postal Service. We used OpenCV image processing library and Pandas data processing analysis library in data set loading, preprocessing and sharpening. In terms of system construction, we use JavaScript as a front-end language for user interaction and visualization of result presentation, Flask as a web page framework, and Redis as a persistent database for data collection and database production.

3.2 Convolutional Neural Network Model Tuning and Performance Analysis

We combined the parameters from the batch volume and the learning rate of the optimization function, trained it with Epoch 5 and evaluated its accuracy, as shown in Table 1. According to the analysis, it can be concluded that the model has the best recognition rate when the batch processing capacity of convolutional neural network is 128 and the learning rate of optimization algorithm is 0.0001.

Table 1. Parameter combination junction.

learning rate	Batch = 32	Batch = 64	Batch = 128
0.001	0.9901	0.9922	0.9929
0.0001	0.9926	0.9927	0.9930
0.002	0.9899	0.9918	0.9855
0.0005	0.9927	0.9933	0.9921
0.002	0.9873	0.9833	0.9912

3.3 Support Vector Machine Model Tuning and Performance Analysis

According to the above analysis, parameter tuning of support vector machine is mainly reflected in the choice of kernel function and penalty parameter C. Firstly, we start with penalty parameter C, find the optimal solution between 2^{-5} and 2^{10}, and get the following relationship between penalty parameter and accuracy Fig. 8 under the training of 30000 MNIST data.

Secondly, kernel functions and multiple classification strategies (one-versus-one and one-versus-rest) are selected for tuning, and the following relationship is obtained in Table 2.

Based on the above analysis, we conclude that when the penalty parameter C is 500, the RBF kernel function vector machine model has a relatively good performance in terms of accuracy and training efficiency.

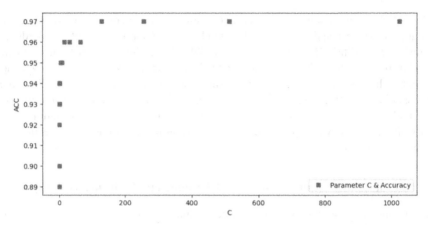

Fig. 8. The relationship between penalty parameter C and accuracy.

Table 2. Accuracy results of SVM under each parameter combination.

Kernel function	OVO		OVR	
	Accuracy	F1-score	Accuracy	F1-score
Linear kernel function	0.931	0.931	0.931	0.931
RBF kernel function	0.97	0.97	0.97	0.961
Sigmoid kernel function	0.927	0.925	0.927	0.925

3.4 K-Nearest Neighbor Algorithm Model Tuning and Performance Analysis

Since KNN algorithm has no actual training stage, the essence of its training is to store all data sets, and the real calculation is concentrated in the prediction stage.

Firstly, we select parameter K. According to experience, we will perform tuning in the range of 2, 3, 4, and 5. The model performance of MNIST under different K parameters is shown in the following Table 3.

Table 3. Performance of the model under different K parameters.

K	Precision	Recall	F1-score
2	0.97	0.97	0.97
3	0.97	0.97	0.97
4	0.97	0.97	0.97
5	0.96	0.95	0.95

There is another parameter distance weight in the KNN model that deserves our attention. Starting this parameter means that the model selects K training sets. The closer the training set is to the test sample, the greater the weight of the final vote. See Fig. 8 for details (Fig. 9).

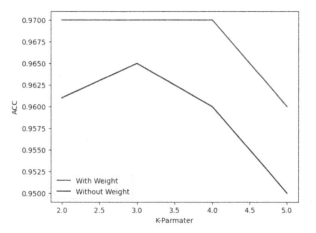

Fig. 9. The relationship between distance weight and accuracy.

In summary, the K-nearest neighbor algorithm performs well on data sets when parameter K is 3 and distance weights are used.

3.5 Generalization Capability Analysis Based on Cross-USPS Datasets

In order to verify the generalization ability of the model, we analyze the model under the optimal parameter combination of each algorithm across the data set. Using the MNIST dataset as a whole as a training set, the models were tested with a new USPS dataset to evaluate the generalization ability of each model.

The accuracy of the convolutional neural network model is 96.4% on the USPS data set processed by image edge zero filling. However, the performance of support vector machine and K-nearest neighbor algorithm is not satisfactory, even in the data set based on image edge zero-fill processing, the accuracy rate only reaches 82% and 81%, and the generalization ability is poor, so we also reduce their weights in the voting mechanism of the system.

4 Conclusion

In this paper, the development of handwritten digit recognition is briefly analyzed. Based on the working principle of convolutional neural network, support vector machine and K-nearest neighbor algorithm, a handwritten digit recognition system with handwriting input, paper edge extraction, multi-digit segmentation and data set making functions is realized. The convolutional neural network proposed in this paper achieves an error recognition rate of 0.8% and 1.6% on the data set and test set, respectively. Secondly, through testing across data sets, the generalization ability of the model based on neural network in this paper is also verified.

Acknowledgement. This work is supported by the Key Field Special Project of Guangdong Provincial Department of Education with No. 2021ZDZX1029.

References

1. Zhang, Y., Wang, S., Ma, Y.: Application comparison of machine and Deep Learning algorithms in handwritten digit recognition. Information Research **47**(05), 60–64 (2021)
2. Zhang, L.: Handwritten digit recognition based on SVM. Electr. Technol. Softw. Eng. **23**, 166–167 (2021)
3. Huang, Y., Lei, B.: Digital recognition system design based on SVM. Info. Technol. Informatiz. **12**, 52–57 (2022)
4. Yang, Y., Xu, X., Chen, X.: Realize the classification of handwritten numerals recognition research based on the SVM technology. Comp. Knowle. Technol. **6**, 195–1962020. https://doi.org/10.14004/j.carolcarrollnkiCKT.2020.0701
5. Hu, J., Fu, X.: Research on handwritten digit recognition based on improved KNN algorithm. J. Wuhan Univ. Technol. (Information and Management Engineering) **41**(01), 22–26 (2019). (in Chinese)
6. Tian, S., Chen, J.: Handwriting digit recognition based on KNN. Agricult. Equipm. Vehicle Eng. **55**(10), 96–97+100 (2017)
7. LeCun, Y., Bottou, L., Bengio, Y., et al.: Gradient-based learning applied to document recognition. Proc. IEEE **86**(11), 2278–2324 (1998)
8. Arora, S., Bhatia, M.P.S.: Handwriting recognition using deep learning in keras. In: 2018 International Conference on Advances in Computing, Communication Control and Networking (ICACCCN), pp. 142–145. IEEE (2018)
9. Yang, J., Yang, G.: Modified convolutional neural network based on dropout and the stochastic gradient descent optimizer. Algorithms **11**(3), 28 (2018)
10. Ali, S., Shaukat, Z., Azeem, M., et al.: An efficient and improved scheme for handwritten digit recognition based on convolutional neural network. SN Applied Sciences **1**, 1–9 (2019)
11. Vamvakas, G., Gatos, B., Perantonis, S.J.: Handwritten character recognition through two-stage foreground sub-sampling. Pattern Recogn. **43**(8), 2807–2816 (2010)
12. Likforman-Sulem, L., Sigelle, M.: Recognition of degraded handwritten digits using dynamic Bayesian networks. Document Recognition and Retrieval XIV SPIE **6500**, 140–147 (2007)
13. Shopon, M., Mohammed, N., Abedin, M.A.: Bangla handwritten digit recognition using autoencoder and deep convolutional neural network. In: 2016 International Workshop on Computational Intelligence (IWCI), pp. 64–68. IEEE (2016)
14. Trivedi, A., Srivastava, S., Mishra, A., et al.: Hybrid evolutionary approach for devanagari handwritten numeral recognition using convolutional neural network. Procedia Computer Science **125**, 525–532 (2018)
15. Ahlawat, S., Choudhary, A.: Hybrid CNN-SVM classifier for handwritten digit recognition. Procedia Comp. Sci. **167**, 2554–2560 (2020)

A High-Quality Video Reconstruction Optimization System Based on Compressed Sensing

Yanjun Zhang, Yongqiang He, Jingbo Zhang, Zhihua Cui[✉], and Xingjuan Cai

School of Computer Science and Technology, Taiyuan University of Science and Technology, Taiyuan, China
cuizhihua@tyust.edu.cn

Abstract. The video compression sensing method based on multi hypothesis has attracted extensive attention in the research of video codec with limited resources. However, the formation of high-quality prediction blocks in the multi hypothesis prediction stage is a big challenge. To solve this problem, this paper constructs a novel compressed sensing-based high-quality video reconstruction optimization system and a range of optimization of prediction blocks models. We combine the high-quality optimization reconstruction of foreground block with the residual reconstruction of background block to improve the overall reconstruction effect of video sequence. The new system mainly includes two parts: the selection of blocks and the optimization of prediction blocks. First, in order to accurately find blocks with large changes and improve the efficiency of the algorithm, we divide blocks into foreground blocks and background blocks by using the difference between the current block and adjacent frame blocks. It can quickly determine the blocks to be optimized. Second, to improve the accuracy of the prediction block, this paper designs multi objective OPBS-NSGA-II model and single objective OPBS-PSO model realizes optimization of prediction blocks. It effectively suppresses the influence of the fluctuation of prediction block on reconstruction and improves the reconstruction performance. In addition, we select the objective function that has a greater impact on the final reconstruction performance through experiments. Experimental results show that the proposed compressed sensing-based high-quality video reconstruction optimization system significantly improves the reconstruction performance in both objective and supervisor quality.

Keywords: Compressed Sensing · Optimization of Prediction Blocks · Selection of Blocks · Many-objective Evolutionary Algorithm

1 Introduction

In 1995, China's first internet company was born, and it has been more than twenty years since then. In these two decades, the Internet has grown exponentially in China. From the beginning of dial-up internet access to now digital routing, the internet has also been enriched with content from text and pictures to videos. Furthermore, online

© The Author(s), under exclusive license to Springer Nature Singapore Pte Ltd. 2024
K. Li and Y. Liu (Eds.): ISICA 2023, CCIS 2146, pp. 277–291, 2024.
https://doi.org/10.1007/978-981-97-4393-3_23

video is taking on an increasingly dominant role in the new era of media. The Internet is changing people's behavioral habits. However, the user demand for high-definition video image is always increasing. Hence, the improvement of reconstruction quality is a steady-state topic in the field of video codecs.

The initial approach of video reconstruction utilized the idea of background subtraction to design the codec [1, 2]. Specifically, the key frame is reconstructed first, then the difference image is reconstructed. Finally, the video sequence can be obtained by sequential accumulation of the key frame and the difference image. However, this method is suitable for compressed sensing (CS) of relatively small spatial video sequences and is not applicable to relatively large spatial video sequences. In addition, the method does not make better use of the redundant information between adjacent frames.

To make better use of picture information in the spatial-temporal domain, distributed compressed video sensing (DCVS) is proposed [3]. It divides frames into key frames and non-key frames, and uses block-based compressed sensing (BCS) to sample each frame. Based on the DCVS, plenty of scholars have conducted a variety of studies. Das et al. [4] studied the coding end. It performs operations such as rotation and weighting on the perceptual matrix corresponding to each block. Considering the impact of reference frames, an improved DCVS system is designed by Zheng et al. [5] to enhance the overall performance of the algorithm. Due to traditional reconstruction methods mainly focus on sparse representation and weight distribution, and ignoring the differences of image itself. In addition, the effect produced by the sparse representation method does not achieve the desired accuracy. Hence, Zheng et al. [6] achieved an image CS method based on multi-level residual reconstruction.

Reconstruction methods based on multiple hypothesis prediction have attracted the attention of scholars. However, how to assign the weights precisely has been a headache for academics. In response to the problem, a new reconstruction method is presented by Zheng et al. [7] to get a better prediction block. It mainly consists of two parts: obtaining the hypothesis set and predicting the weights. However, it is still to be further investigated to optimize the hypothesis set.

The overall procedure of video compression sensing consists of an encoder and a decoder. The operating principle of the end-to-end video codec is shown in Fig. 1. First, the video is compressed and encoded by the content provider and transmitted over the wireless network to the server; Second, encoded video is stored and transmitted by the server and sent to the users. Finally, the consumer decodes it and then it is ready to play.

Content provider   Server side   Content consumer

Fig. 1. The architecture of end-to-end video codecs.

Generally speaking, innovations on the coding side are investigating sampling schemes. Research on how to reduce the amount of data transmitted is the focus of research on the coding side. However, for the decoding side, the best possible reconstruction quality is a stable topic on the decoding side. Hence, to enhance the reconstruction performance of video frames, this paper proposes CS-based high-quality video reconstruction (CS-HQVR) optimization system and a range of optimization of prediction blocks (OPBS) models. The following are the contributions of the paper:

1) Based on the amount of difference between the current block and the blocks of adjacent frames, a block type determination method is designed to decide whether or not to optimize its predicted blocks.
2) A novel multi objective OPBS-NSGA-II model based on compressed sensing is constructed to improve the reconstruction effect.
3) To determine which target has a greater impact on the reconstructed effect of the video, single objective OPBS-PSO model is designed to realize optimization of prediction blocks.
4) Extensive experimental results prove that our designed model improves reconstruction performance in both objective and subjective quality.

The organizational framework for the remaining sections of this article is set out below. The research works of relevant scholars are presented in Sect. 2; Sect. 3 describes the proposed CS-HQVR optimization system and a range of OPBS models in detail; Extensive experiments are conducted in Sect. 4, which compare the paper's approach with three methods to prove the validity of the proposed strategy; Sect. 5 gives a conclusion.

2 Related Work

Compressed sensing is applied in many aspects because of energy efficiency, high transmission efficiency and safety. Such as image encryption, camera sensors, deep learning [8], wireless multimedia sensor networks (WMSN) [9], data compression [10], image processing [11], etc.

With the advancement of multimedia editing technology, copyright protection is receiving more and more attention. Chen et al. [12] proposed a video reversible data hiding scheme and provided accurate clustering results using CS extracted features. In addition, digital images are widely used in people's daily life because of their easy-to-understand and vivid descriptions. Consequently, a series of security issues of digital images have arisen. To increase the security of digital images, an efficient and secure image compression and encryption scheme is designed by Zhu et al. [13]. It takes advantage of the inherent properties of CS to provide strong plaintext sensitivity for compressed encryption schemes with low additional computational cost. However, its reconfiguration robustness could be further improved. To achieve high reconstruction robustness and high security for image encryption and compression, Zhang et al. [14] used chaotic systems and two-dimensional fractional Fourier transform (2D-FRT) for encryption. It avoids the loss of reconfiguration robustness in diffusion encryption simultaneously.

Advances in low-power and low-cost camera sensors and mobile devices have significantly facilitated wireless video streaming applications such as video surveillance, WMSN, the Internet of Things (IoT). However, camera sensors are limited in terms of memory, processing power and energy supply. Therefore, it is necessary to develop low-complexity, energy-efficient coding schemes for camera sensors with extremely tight resource budgets. Amit et al. [15] proposed a new framework for serving wireless video streams to reduce transmission costs. CS inspired various compressed imaging systems. Qiao et al. [16] built a video snapshot compression imaging (SCI) system by using video CS.

Video Compression Sensing (VCS) aims to perceive and recover scene video in a space-time aware manner. It is difficult to implement due to the complexity of the design and optimization. To make better use of temporal and spatial correlations, a hybrid three-dimensional residual block consisting of pseudo- and true three-dimensional sub-blocks is proposed by Zhao et al. [17]. Similarly, in order to explore interframe correlation, Shi et al. [18] proposed a multi-level feature compensation method using convolutional neural networks (CNN). It allows better exploration of intra- frame and inter-frame correlations.

There are many procedures in life that require real-time monitoring, such as real-time monitoring of a driver's mental state, real-time monitoring of a patient's health, etc. Hence, it is necessary to equip the camera with sensors. However, it can easily generate large amounts of data. To reduce the amount of data during transmission, the cs-based Joint Decoding (JD) framework is constructed by Zhao et al. [19]. Simultaneously, in order to enhance the robustness in different video images, Heng et al. [20] proposed a holistic WMSN architecture for image transfer and used a fuzzy logic system (FLS) to resize each corresponding block.

For everyday video encoders such as mobile phones and computer-based encoders, their high computational complexity makes them unsuitable for in-vehicle video compression. To address the problem, an extended compression chain is proposed by Yubal et al. [21]. Similarly for context-specific compression, to enable lower power consumption and faster deployment in coal mine disaster rescue, Xu et al. [22] constructed an adaptive compression method based on Normalized Bhattacharya Coefficients (NBCAC-MHRR) to solve the problem of High Efficiency Video Coding (HEVC) in underground coal mines.

Cloud-based [23, 24] video image uploads are more common in our daily lives. The upload speed is related to the quality of the communication channel, where the amount of data uploaded can be higher if the quality of the communication channel is high and lower if not. Zheng et al. [25] developed a terminal-to-cloud video upload system. It determines the amount of data to be uploaded based on the quality of the channel. Furthermore, a multi-reference frame strategy is proposed to improve the reconstruction. Zhao et al. [26] proposed a new algorithm for efficient reconstruction of video from CS measurements based on the idea of predictive-residual reconstruction. However, it did not consider the effect of the predicted blocks on the reconstruction. The blocks that are more similar to the current block are simply treated as predicted blocks. We conducted a series of studies on the optimization of prediction blocks in our paper. The experimental results demonstrate the influence of the prediction block on the final reconstruction.

3   The Proposed CS-HQVR System

The proposed method of CS-HQVR is illustrated in Fig. 2. It consists of the encoding process and the decoding process.

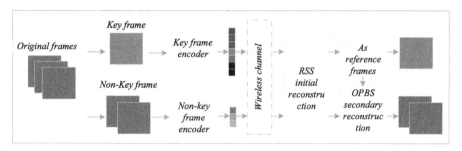

Fig. 2. The overall framework of the CS-HQVR method.

3.1   Encoding Process

During the encoding stage, original frames are divided into several group of pictures (GOPs). In general, there is a key frame in the GOP, and the others are non-key frames. The structure of the GOP is shown in Fig. 3.

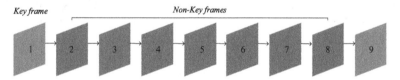

Fig. 3. The structure of the GOP.

As can be seen from Fig. 3, the first frame (orange noted) is the key frame, and the second to eighth frames (blue noted) are non-key frames. Next, all frames are compressed. Key frames and non-key frames use different compression rates. Generally, the compression rate of key frames is higher due to the fact that the key frame is used as a reference frame for other frames during the reconstruction. In addition, the block compression sensing (BCS) is used in encoding stage. The calculation method of the compression process is shown in Eq. (1).

$$y_{h,j} = \Phi t_{h,j} \tag{1}$$

where, $\Phi \in R^{P \times Q}$ ($P << Q$) represents the observation matrix, the random Gaussian observation matrix is applied in this study. P represents the number of sampling values in y_h. t_h represents the h-th frame. $t_{h,j}$ represents the j-th block of the h-th frame. Accordingly, $y_{h,j}$ represents the projection of the j-th block of the h-th original frame. Through Eq. (1), each frame completes its encoding process.

3.2   Decoding Process

3.2.1   The Relationship Between the Predicted Block and the Current Block

The encoded block comes to the decoder via the wireless channel. At the decoding end, this paper uses reweighted residual sparsity (RRS) to reconstruct blocks. However, the performance of reconstruction is not enough to satisfy the need of high quality in society. Hence, this paper proposes optimization of prediction blocks (OPBS) in the second stage of RSS. The relationship between the prediction block and the current block is shown in Fig. 4.

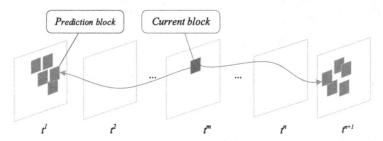

Fig. 4. The description of the relationship between the predicted block and the current block.

The current block is marked with a blue square. It represents a block of the non-key frame. Orange squares are the hypothesis blocks. It consists of the blocks that is most similar to the current block $t_{h,j}$. The prediction block is obtained by assigning different weights to the set of hypothesis blocks. The calculation method of the prediction block is shown in Eq. (2).

$$\tilde{t}_{h,j} = \sum_{i=1}^{m} \alpha_i t_i^{hyp} \qquad (2)$$

where, $\tilde{t}_{h,j}$ represents the prediction block. t_i^{hyp} represents the hypothesis block. α_i represents the corresponding weight of the hypothesis block. Zhao et al. [26] proposed a two-stage video residual sparsity model which mainly includes multiple hypothesis (MH) prediction, residual modeling and optimization. However, it does not take into account the effect of the prediction block on the reconstruction. Thus, OPBS is designed in our paper.

3.2.2   Preliminary Preparation for OPBS

The two questions of what blocks are worth optimizing and how to select them are prerequisites for OPBS. Zheng et al. [25] used residual value between current block and reference block to judgement the type of block. However, the method increased complexity on the encoding side. In addition, due to the sampling rate is different of key frames and non-key frames, it is undesirable to adopt the residual between them. The difference between the current block and block of adjacent frame is more suitable because of the same compression rate in this paper, and the difference can be expressed as in Eq. (3).

$$v = \left\| t_{h,j} - t_{adj,j} \right\|_2 \tag{3}$$

where, $t_{adj,j}$ represents the corresponding block of the current block $t_{h,j}$ in the adjacent frame. v represents the residual between them. In addition, The larger the value of the residual, the more variation in the blocks. Finding the blocks with large residual values is the first thing we need to do. The classification of blocks is shown in Eq. (4).

$$t_{h,j} = \begin{cases} \text{Prospect block } v > \text{THR}_v \\ \text{Background block else} \end{cases} \tag{4}$$

where, THR_v represents the threshold of residual v. When $v > \text{THR}_v$, the current block is selected as a prospect block. Otherwise, $t_{h,j}$ is choose as a background block. In addition, the prospect block will perform the optimization operations of the prediction block. The distribution of v values is shown in Table 1.

Table 1. The distribution of v values

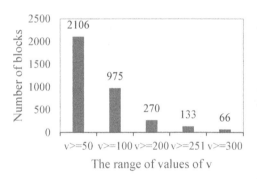

There are 6177 subblocks in each block. It is not advisable to optimize all blocks due to its high time complexity. Hence, this paper only optimizes some of the blocks. As can be seen from Table 1, when $v >= 50$, there are 2106 subblocks. It accounts for about three quarters of t_h. Thus, THR_v is set to 50.

3.2.3 The core components of OPBS

I. Multi objective OPBS-NSGA-II model realizes optimization of prediction blocks.

To enhance the reconstruction performance, the multi-objective OPBS model based on NSGAII is proposed. The distance and similarity between the predicted block and the current block satisfy the condition that they conflict with each other. Hence, they are used in this paper. The calculation method of two objectives is shown in Eq. (5–7).

$$fit1 = \left\| t_{h,j} - \tilde{t}_{h,j} \right\|_2 \tag{5}$$

$$fit2 = SSIM\,(t_{h,j}, \tilde{t}_{h,j}) \tag{6}$$

$$F = [fit1, fit2] \tag{7}$$

where, $fit1$ represents the distance between the prediction block $\tilde{t}_{h,j}$ and current block $t_{h,j}$. $Fit2$ represents the similarity between them. F represents the multi-objective model. Generally, the smaller the distance between the predicted block and the current block, the greater their similarity. In addition, we also use the peak signal-to-noise ratio (PSNR) to measure the reconstruction performance between the original frame and the final reconstruction frame, as shown in Eq. (8).

$$PSNR = 10 \times \log_{10}\left(\frac{(2^n - 1)^2}{MSE}\right) \tag{8}$$

where, n represents the number of bits per sampled value. MSE represents mean squared deviation.

II. OPBS-NSGA-II evolutionary algorithm

The objective of OPBS is to enhance the reconstruction performance. Moreover, the prediction block consists of multiple hypothesis blocks. However, the number of hypothesis blocks is fixed in the most paper. For example, the number of hypothetical blocks set by Zhao et al. [26] is five. They ignore the effect of the number of hypothetical blocks on the final reconstruction. Therefore, the number of hypothetical blocks is used as a decision variable for the OPBS-NSGA-II evolutionary algorithm in our paper. The algorithm flow of OPBS-NSGA-II is shown in Table 2.

III. Single objective OPBS-PSO model realizes optimization of prediction blocks

To verify which objective function has a greater impact on the final reconstruction result. Particle Swarm Optimization (PSO) is used for the optimization of prediction blocks. Equation (5) is selected as an objective in this part, the main steps of OPBS-PSO model are as follows:

Step 1: Calculate the difference between the current block and the adjacent block. Then deter-mine the type of blocks based on the difference.
Step 2: Perform the following loop when the block is prospect block.
Step 3: Initialize the population. The initialization is carried out according to the decision range of the number of hypothetical blocks in the reference frame, and the number of hypothesis blocks is formed for each individual.
Step 4: Perform velocity and position update operations to produce offspring.
Step 5: Solve the objective value according to the designed objective function and the predicti-on block is obtained.
Step 6: Updating personal best values and global best values.
Step 7: Determine whether the maximum number of iterations is satisfied. If so, terminate the operation. Otherwise, return to the Step 4 and continue iterating until the termination condition is met.
Step 8: When the optimization is completed, a relatively better prediction block is obtained.

Equation (6) as an objective is similar to Eq. (5) in its optimization method. Thus, it will not be repeated too much here.

Table 2. The pseudocode of OPBS-NSGA-II algorithm

Algorithm 1: OPBS-NSGA-II pseudocode

Input: Video sequence dataset T_{org}, Search window S, Sparsity D, The block size

　　　　Pat, Number of hypothetical blocks P, Number of hypothetical blocks in

　　　　the reference frame N, Number of iterations $Gmax$

Output: Reconstructed video image T_{fin}

1　　The video image T_{org} is chunked to obtain $t_{h,j}$

2　　Find the comparison block $t_{adj,j}$ of the current block $t_{h,j}$ and calculate

　　　　$v = \left\| t_{h,j} - t_{adj,j} \right\|_2$

3　　**If** $v > THR_v$

4　　　Initializing the population P

5　　　**While** number of iterations $< Gmax$

6　　　　Calculate fitness value F = [fit1, fit2]

7　　　　Non-dominated sorting operation and crowding calculation

8　　　　Genetic operations (Cross (P_t) + Mutation (P_t)) to generate new population

9　　　**End**

10　　　N1 = P. 1; N2 = P. 2; N3 = P. 3; N4 = P. 4;

11　　**Else**

12　　　N1 = 5; N2 = 5; N3 = 5; N4 = 5;

13　　**End**

14　　Reconstruct the prediction block $\tilde{t}_{h,j}$ to get the final video frame T_{fin}

---

4   Simulation Experiment and Analysis

To verify the effectiveness of the proposed CS-HQVR model. Extensive experiments are proposed in this section. Furthermore, all experiments are done with MATLAB R2018b in Windows 10 environment. It should be noted that our experiment is based on previous work.

4.1   Parameter Settings

There are many parameters in the experiment. It includes the parameters related to the encoding process, decoding process, multi-objective and single-objective optimization algorithm. The experimental settings of encoding stage and decoding stage are shown in Table 3. Furthermore, the parameter settings of multi-objective and single-objective models are shown in Table 4.

Table 3. Parameters Setting of encoding and decoding stage

Parameter	Meaning	Value
S _subblock	Size of sub-block	8
Φ	Measurement matrix	Gaussian Random
K _rate	Sampling rate of key frame	0.7
NK _rate	Sampling rate of non-key frame	0.2

Table 4. Parameters Setting of multi-objective and single-objective models

Parameter	Meaning	Value
Pop	Population	50
D	Decision variables	4
I	Iterations	30
F	Objective function	2
Var _min	Lower Bound of Decision Variables	1
Var _max	Upper Bound of Decision Variables	11

In this experiment, each frame is divided into blocks with 8 by 8 pixel. Gaussian Random is used as measurement matrix. In addition, the sampling rate of key frame is 0.7 and the sampling rate of non-key frame is 0.2. In the parameters setting of multi-objective and single-objective models, the population is set to 50, D represents the dimensions of decision variables, I represents the number of iterations, F represents the number of objectives. Moreover, the value range of each decision variable is set as [1, 11].

4.2 Comparison with Classical Algorithms

To verify the validity of the presented method, this paper conducts experiments on the datasets of Forman and Highway, and evaluates the performance with three methods. It includes MC/ME [27], VCSNet [18], and RRS [26]. The comparative results are shown in the figures below.

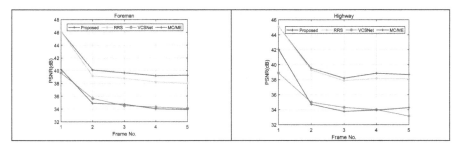

Fig. 5. The CS reconstruction performance of presented algorithm(orange noted) and three comparison methods.

As can be seen in Fig. 5, the proposed method (orange noted) has obvious advantages in enhancing reconstruction performance in the video sequences of foreman and highway. Specifically, for Foreman video sequences, the overall average enhancement of the proposed method over RRS (blue noted) in this paper is 0.80 dB. For Highway video sequences, the improvement effect of the presented method is obviously inferior to that of Foreman video sequences, but it is still an enhancement in a comprehensive view (Table 5).

Table 5. The reconstruction performance comparison of OPBS-NSGA-II and OPBS-PSO on the highway dataset

Frame/Algorithm	OPBS-NSGA-II	OPBS-PSO-fit1	OPBS-PSO-fit2
1	45.1129	45.1129	45.1129
2	39.4741	39.5207	39.5134
3	38.1029	38.189	38.1624
4	38.8664	38.8445	38.8106
5	38.6519	38.6715	38.6432
Average (dB)	40.04171	**40.06772**	40.0485

In this paper, we compare the proposed OPBS-NSGA-II, OPBS-PSO-fit1 and OPBS-PSO-fit2 evolutionary algorithms on the Highway dataset. By comparison, we can conclude that single objective OPBS-PSO is better than multi objective OPBS-NSGA-II evolutionary algorithm if only the final PSNR value is considered. In addition, fit1 has a greater impact on the final reconstruction performance.

To verify the difference between optimized partial blocks and all blocks, experiment is conducted in this paper. As can been seen in Table 6. The orange dots represent the optimization of all blocks. The blue dots represent that only the foreground blocks are optimized. Through the validation on the Forman dataset, it can be seen that the results of OPBS-All blocks are better. However, only foreground blocks are selected for optimization of prediction blocks in our paper. Due to the high time complexity of optimizing all blocks, we choose to optimize only part of the blocks.

Table 6. The performance of optimizing all blocks versus partial blocks on the Foreman dataset

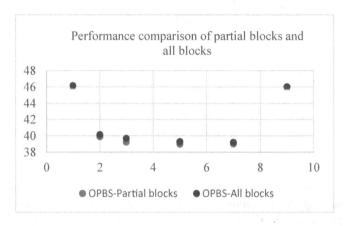

Table 7. The details of the iteration of the third frame of the Paris dataset

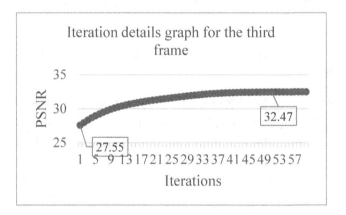

As can be seen from Table 7, we take out the third frame of the Paris video sequence to view a detailed graph of each of its iterations. The PSNR value rises as the number of iterations increases. Furthermore, the maximum PSNR value is reached by the 51st iteration. Hence, the corresponding iterative termination strategies will be constructed later in the work to reduce the time complexity.

This paper takes out the fifth frame of the Foreman video sequence for visual comparison. As we all know, MC/ME is a representative method, and the proposed method achieves a 5.35 dB improvement over MC/ME as can be seen in Fig. 6; VCSNet uses Convolutional Neural Networks (CNN) for video reconstruction, and this paper's method enhances 5.19 dB at a sampling rate of 0.2. In addition, RRS is based on prediction-residual reconstruction, and the presented method improves 1.25 dB over RRS.

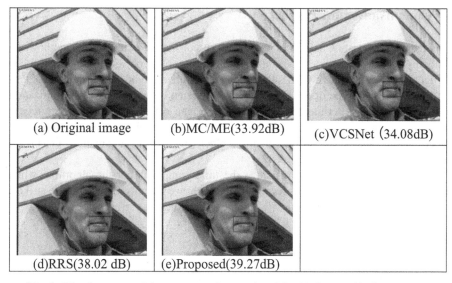

Fig. 6. Visual contrast of the reconstruction results of the 5th frame with diverse ways.

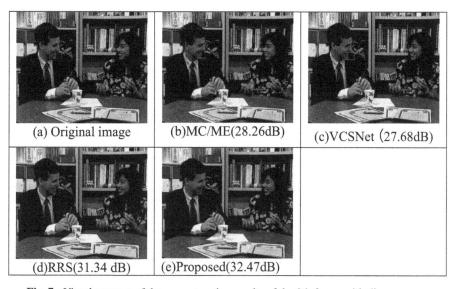

Fig. 7. Visual contrast of the reconstruction results of the 3th frame with diverse ways.

To evaluate the robustness of our algorithm, this paper takes out the third frame of the Paris video sequence for visual comparison. It is can be seen from Fig. 7 that the proposed method achieves a 4.21 dB improvement over MC/ME; This paper's method enhances 4.69 dB at a sampling rate of 0.2 over VCSNet. In addition, the presented method improves 1.03 dB over RRS. The experimental results fully demonstrate that OPBS have strong superiority and robustness in VCS.

5 Conclusion and Prospect

This paper proposes a CS-HQVR framework. It consists of two stages: encoding and decoding. On the decoding side, this paper designs multi objective OPBS-NSGA-II model and single objective OPBS-PSO realizes optimization of prediction blocks. Particularly, the block type of the current block is judged before a series of optimization operations. In the experimental part, we use Forman data set to compare the reconstruction performance of partial block optimization and total block optimization. The experimental results show that optimizing all blocks has obvious advantages, but it brings high time complexity simultaneously.

The next step of work is to reduce the time complexity of the algorithm while improving the video quality. Moreover, the problem we need to solve is how to strike a balance between the time complexity and the reconstruction performance.

References

1. Zheng, J., Jacobs, E.L.: Video compressive sensing using spatial domain sparsity. Opt. Eng. **48**(8), 10 (2009)
2. Zheng, J., Jacobs, E.L.: The application of compressive sensing technique on a stationary surveillance camera system. In: Proceedings of Spie the International Society for Optical Engineering **6941**(69410H), 2008
3. Kang, L.W., Lu, C.S.: Distributed compressive video sensing. In: IEEE International Conference on Acoustics, pp. 1169–1172 (2009)
4. Das, S., Mandal, J.K.: An enhanced block-based compressed sensing technique using orthogonal matching pursuit. Signal Image Video Process **15**(3), 563–570 (2021)
5. Zheng, S., Chen, J., Kuo, Y.H.: An improved distributed compressed video sensing scheme in reconstruction algorithm. Multimedia Tools and Applications **77**(7), 8711–8728 (2018)
6. Zheng, S., Chen, J., Kuo, Y.H.: A multi-level residual reconstruction-based image compressed sensing recovery scheme. Multimedia Tools and Applications **78**(17), 25101–25119 (2019)
7. Zheng, S., Chen, J., Zhang, X.P., Kuo, Y.H.: A new multihypothesis-based compressed video sensing reconstruction system. IEEE Trans. Multimedia **23**, 3577–3589 (2021)
8. Park, W., Kim, M.: Deep predictive video compression using mode-selective uni- and bidirectional predictions based on multi-frame hypothesis. IEEE Access **9**, 72–85 (2021)
9. Banerjee, R., Bit, S.D.: Low-overhead video compression combining partial discrete cosine transform and compressed sensing in WMSNs. Wireless Netw. **25**(8), 5113–5135 (2019)
10. Zhang, R.F., Wu, S.H., Wang, Y., Jiao, J.: High-performance distributed compressive video sensing: jointly exploiting the HEVC motion estimation and the l(1) reconstruction. IEEE Access **8**, 31306–31316 (2020)
11. Cai, X.J., Cao, Y.H., Ren, Y.Q., Cui, Z.H., Zhang, W.S.: Multi-objective evolutionary 3D face reconstruction based on improved encoder-decoder network. Inf. Sci. **581**, 233–248 (2021)
12. Chen, Y.L., Zhou, L.M.N., Zhou, Y.H., Chen, Y., Hu, S.B., Dong, Z.C.: Multiple histograms shifting-based video data hiding using compression sensing. IEEE Access **10**, 699–707 (2022)
13. Zhu, Z.L., Song, Y.J., Zhang, W., Yu, H., Zhao, Y.L.: A novel compressive sensing-based framework for image compression-encryption with S-box. Multime. Tools and Appli. **79**(35–36), 25497–25533 (2020)
14. Zhang, M., et al.: Image compression and encryption scheme based on compressive sensing and fourier transform. IEEE Access **8**, 40838–40849 (2020)

15. Unde, A.S., Pattathil, D.P.: Adaptive compressive video coding for embedded camera sensors: compressed domain motion and measurements estimation. IEEE Trans. Mob. Comput. **19**(10), 2250–2263 (2020)
16. Qiao, M., Meng, Z.Y., Ma, J.W., Yuan, X.: Deep learning for video compressive sensing. APL Phontonics **5**(3), 14 (2020)
17. Zhao, Z.F., Xie, X.M., Liu, W., Pan, Q.Z.: A hybrid-3D convolutional network for video compressive sensing. IEEE Access **8**, 20503–20513 (2020)
18. Shi, W.Z., Liu, S.H., Jiang, F., Zhao, D.B.: Video compressed sensing using a convolutional neural network. IEEE Trans. Circuits Syst. Video Technol. **31**(2), 425–438 (2021)
19. Ebrahim, M., Adil, S.H., Raza, K., Ali, S.S.A.: Block compressive sensing single-view video reconstruction using joint decoding framework for low power real time applications. Applied Sciences-Basel **10**(22), 23 (2020)
20. Heng, S., Aimtongkham, P., Vo, V.N., Nguyen, T.G., So-In, C.: Fuzzy adaptive-sampling block compressed sensing for wireless multimedia sensor networks. Sensors **20**(21), 29 (2020)
21. Barrios, Y., Guerra, R., Lopez, S., Sarmiento, R.: Adaptation of the CCSDS 123.0-B-2 Standard for RGB Video Compression. IEEE J. Select. Topi. App. Earth Observat. Remote Sens. **15**, 1656–1669 (2022)
22. Xu, Y.G., Xue, Y.Z., Hua, G., Cheng, J.W.: An adaptive distributed compressed video sensing algorithm based on normalized bhattacharyya coefficient for coal mine monitoring video. IEEE Access **8**, 158369–158379 (2020)
23. Zhang, Z.X., Zhao, M.K., Wang, H., Cui, Z.H., Zhang, W.S.: An efficient interval many-objective evolutionary algorithm for cloud task scheduling problem under uncertainty. Inf. Sci. **583**, 56–72 (2022)
24. Cai, X.J., Geng, S.J., Wu, D., Cai, J.H., Chen, J.J.: A multicloud-model-based many-objective intelligent algorithm for efficient task scheduling in internet of things. IEEE Internet Things J. **8**(12), 9645–9653 (2021)
25. Zheng, S., Zhang, X.P., Chen, J., Kuo, Y.H.: A high-efficiency compressed sensing-based terminal-to-cloud video transmission system. IEEE Trans. Multim. **21**(8), 1905–1920 (2019)
26. Zhao, C., Ma, S.W., Zhang, J., Xiong, R.Q., Gao, W.: Video Compressive sensing reconstruction via reweighted residual sparsity. IEEE Trans. Circuits Syst. Video Technol. **27**(6), 1182–1195 (2017)
27. Mun, S., Fowler, J.E.: Residual reconstruction for block-based compressed sensing of video. IEEE Data Compression Conference, pp. 183–192. Snowbird (2011)

Conv and Efficient Multi-Scale Attention Module for YOLOv5

Xuan Guo and Weidong Huang[✉]

School of Information Engineering, JiangXi University of Science and Technology,
GanZhou 34100, JiangXi, China
18279766975@139.com

Abstract. Object detection is a popular and tough task, the multiplicity of species and the multi-scales of the same kind contribute to these difficulties, as a result, the performance improvement of various algorithms is minimal. The proposed attention mechanism can improve the performance of the object detection algorithm. However, embedding the attention mechanism may bring negative benefits to the algorithm model, because of the depth of the algorithm model, the attention mechanism can be embedded in many different locations, and different locations will bring different results. To improve the object detection performance of YOLOv5, this paper proposes Conv and Efficient Multi-Scale Attention (CEMA), a new novel module used in YOLOv5, it fusion C3 module and EMA attention. The performance at different locations is compared and analyzed. Experimental results used the VOC2007 + 2012 dataset to verify its performance. It has a 1.76% improvement compared with YOLOv5s, and it also has different degrees of improvement compared with SE, CA, CBAM, and SimAM attention mechanisms.

Keywords: Object detection · YOLOv5 · Attention mechanism

1 Introduction

Following the evolution of deep Convolutional Neural Networks (CNNs), more novel network methods and topologies are employed in the fields of object detection and image classification. When extending the layers of the network, it exhibits remarkable ability. However, stacking more deep convolutional counterparts requires much consumption of memory and computational resources, which is the primary shortcoming for constructing deep CNNs [1, 2]. Another question is when convolutional layers reach a certain number, adding more convolutional layers leads to a higher training error, as reported in [3, 4]. However, ResNet [5] makes it easier to train and optimize deeper networks.

The attention mechanism method has flexible structure characteristics, which allows it can be easily plugged into the backbone architecture of the CNNs, and it also can make CNNs learn more about feature representation. Therefore, the attention mechanism has attracted the attention of many scholars of object detection algorithms. Attention mechanisms are generally divided into three types, the channel attention, spatial attention, and both of them. Squeeze-and-excitation (SE) [6] proposed an SE block, that consists

© The Author(s), under exclusive license to Springer Nature Singapore Pte Ltd. 2024
K. Li and Y. Liu (Eds.): ISICA 2023, CCIS 2146, pp. 292–301, 2024.
https://doi.org/10.1007/978-981-97-4393-3_24

of two structures, Squeeze and Excitation. The squeeze part is responsible for computing the feature maps, and the Excitation according to cross-dimension learns the channel weight. Convolutional block attention module (CBAM) [7] according to fusion the cross-channel and cross-spatial information in the feature maps shows great potential in cross-dimensional attention weights into the input features.

The two-stage object detection algorithm in object detection, such as R-CNN [8], retains the traditional object detection method, according to a sliding window to select the area to be detected, and then extracts features through CNN when the region is determined, and finally uses the trained SVM to classify the features. This sliding window method decreases the computational overhead, most of them stay in the background rather than on the target being detected, that leads it needs 47 s to detect one image. The one-stage object detection solves the problem of slow detection speed, YOLO [9] has increased the detection speed to 45 FPS, and the fast version even reaches 150 FPS. The updated versions also have the extremely fast detection speed. YOLOv5 is one of the most widely used versions, because it can satisfy the requirements of real-time, and it has five different model versions corresponding to different parameter quantities and model sizes, user can choose different modules according to different application scenarios, or make a trade-off between accuracy and speed. Ultralytics also spent the next two years updating the YOLOv5 code and communicating with researchers, making it become the most widely used algorithm for object detection.

This paper proposed Conv and Efficient Multi-Scale Attention (CEMA) module, it is an attention module, by fusion of Efficient Multi-Scale Attention (EMA [10]) with the C3 module, using in YOLOv5 model to improve the performance of YOLOv5 through the EMA attention mechanism, it reshapes the channel dimensions into sub-dimensions, to avoid dimensionality reduction via convolution, and depart the input features into two parallel subnetworks, fuse the output feature maps of the two parallel subnetworks by a cross-spatial learning method. In the end, according to experiment results to compare the performance of putting modules into different locations in the YOLOv5, and also compared with CBAM, CA [11], SE, and SimAM [12] attention mechanisms, and replaced the loss function with Wise-IoU [13]. This paper uses VOC2007 + 2012 datasets to train and demonstrate its effectiveness.

The remainder of this study's chapters are organized as follows. The related work of YOLOv5 and some attention mechanisms are studied in Sect. 2. The proposed module CEMA is described in Sect. 3. In Sect. 4, the experiment results are shown and analyzed. The further research is summarized in Sect. 5.

2 Related Work

2.1 YOLOv5

YOLOv5 is a One-stage object detection model with five versions: YOLOv5s, YOLOv5m, YOLOv5l, YOLOv5x and YOLOv5n, exception of YOLOv5n, the number of parameters in other versions has increased sequentially, with higher accuracy and model complexity, YOLOv5n is provided for edge devices.

YOLOv5 network framework consists of three parts: the backbone network (Backbone), the bottleneck layer network (Neck), and the detection layer (Head). The Backbone consists of a series of modules, convolution modules, C3 modules, and a spatial pyramid pooling fast module (SPPF). Since YOLOv5 algorithm according to multilayer feature maps to prediction, has good results in terms of accuracy and detection speed.

2.2 Attention Mechanism

In a convolutional neural network, the output of each neuron depends only on the output of all the neurons of the previous layer, but in the attention mechanism, the output of each neuron does not only depend on the output of all the neurons of the previous layer but can also be weighted according to different parts of the input data, that is, different weights are given to different parts. This allows the model to pay more attention to the key information in the input sequence, which improves the accuracy and efficiency of the model.

Squeeze-and-excitation attention considers that each of the learned filters operates with a local receptive field, so it is unable to exploit contextual information outside of its region. It according to squeezes global spatial information into a channel descriptor, this is achieved by using global average pooling to generate channel-wise statistics. Its shrinking spatial dimensions $U_c \in H \times W$ as Z_c, calculated by Eq. (1), the next excitation operates is done through two fully connected layers, calculated by Eq. (2), where W_1 and W_2 are learning weight from fully connected layers, δ is ReLU [14] function, σ is Sigmoid function. Finally, the feature map size is restored to the input size by Eq. (3).

$$Z_c = F_{sq}(U_c) = \frac{1}{H \times W} \sum_{i=1}^{H} \sum_{j=1}^{W} U_c(i,j) \tag{1}$$

$$S = F_{ex}(Z, W) = \sigma(g(Z, W)) = \sigma(W_2 \delta(W_1 Z)) \tag{2}$$

$$\tilde{X}_C = F_{scale}(U_c, S_c) = S_C U_c \tag{3}$$

Convolutional Block Attention Module includes a Channel Attention Module (CAM) and a Spatial Attention Module (SAM) two sub-modules, and separate calculations on the channel attention and the spatial attention, this method can save parameters and computing power, therefore, it is a lightweight network module. Input feature maps first pass two parallel MaxPool and AvgPool layers, then the feature map is changed from $C \times H \times W$ to $C \times 1 \times 1$ size, and then passed through the Share MLP module, in which the number of channels is compressed to $1/r$ (Reduction), and then expanded to the original number of channels, and the results of two activations are obtained through the ReLU function. These two output results will pass add operates, and then use a sigmoid activation function to get the output result of Channel Attention. The SAM's input feature maps are CAM's output concatenated with original feature maps, according to MaxPool and AvgPool changed feature maps from $C \times H \times W$ to $C \times 1 \times 1$ size, and concatenate the output, then, use the 7×7 convolution to change the feature map of 1 channel (the experiment proves that the effect of 7×7 is better than that of $3 \times$

3, and then the feature map of Spatial Attention is obtained by Sigmoid function, and finally the output result is multiplied by the original image and changed back to the size of $C \times H \times W$. CAM makes the channel dimension remain unchanged, and the spatial dimension is compressed. This module focuses on meaningful information in the input maps. SAM makes the spatial dimension remain unchanged, and the channel dimension is compressed. This module focuses on the location information of the input maps. The CAM calculated by Eq. (4), and the SAM calculated by Eq. (5), f is convolution operation.

$$M_c(F) = \sigma(MLP(AvgPool(F)) + MLP(MaxPool(F))) \tag{4}$$

$$M_S(F) = \sigma(f^{7*7}([AvgPool(F), MaxPool(F)])) \tag{5}$$

Since SE attention only focuses on the interdependence between construction channels, ignoring the spatial features, and CBAM introduces large-scale convolutional kernels to extract features, but ignores the long-range dependence problem. The Coordinate Attention (CA) was proposed. It not only considers the channel information, but also direction related location information is taken into account. To avoid all spatial information being compressed into the channel, AvgPool is used here. To be able to capture remote spatial interactions with precise location information, the AvgPool is decomposed as follows Eqs. (6,7), the input feature maps with dimensions $C \times H \times W$ were pooled in the X direction AvgPool and Y direction AvgPool, and the feature maps with dimensions $C \times H \times 1$ and $C \times 1 \times W$ were generated. These two parallel 1D feature encoding vectors change into the other vector shape before concatenating two parallel 1D feature encoding vectors across a convolutional layer. These two parallel 1D feature vectors will share a 1×1 convolutional convolution with dimensionality reduction. The 1×1 kernel helps the model capture local cross-channel interaction. Then, CA further uses a 1×1 kernel to factory the output into two parallel 1D feature vectors and stack the Sigmoid function in two parallel ways. Learnt weights of the two parallel ways will be aggregated as the final output. CA according to encoding inter-channel operations, not only retain the precise positional information but also exploit the long-range dependence.

$$Z_c^h(h) = \frac{1}{W} \sum_{0 \leq i < W} x_c(h, i) \tag{6}$$

$$Z_c^w(w) = \frac{1}{H} \sum_{0 \leq j < H} x_c(w, j) \tag{7}$$

CA ignoring the 1×1 kernel has a limited in receptive field, this makes it difficult to model local cross-channel and understand contextual information. Efficient Multi-Scale Attention was proposed to save it. It extracts the shared components of the 1×1 convolution in the CA and names it as the 1×1 branch, and at the same time, to aggregate multi-scale spatial structure information, the 3×3 kernel is placed in parallel with the 1×1 branch to achieve fast response, and named it as the 3×3 branch. For every input feature maps, EMA will divide into G sub-features across the channel dimension direction, which makes the model learn different semantics.

3 Conv and Efficient Multi-Scale Attention

Conv and Efficient Multi-Scale Attention module is proposed in this paper, it works in YOLOv5. YOLOv5 is one of the most widely used algorithms for object detection, it has been embedded by many researchers in a variety of different attention mechanisms. These attention mechanisms were inserted at the end of the C3 module or the Conv module, and in a lot of locations. The CEMA module uses another embedding method, which is fusion with the C3 module, its structure is shown in Fig. 1, and C3 module structure is shown in Fig. 2. Different from embedding directly after the C3 module, CEMA can make full use of the feature maps extracted by the ResNet part in the Bottleneck of C3 module for Concatenate.

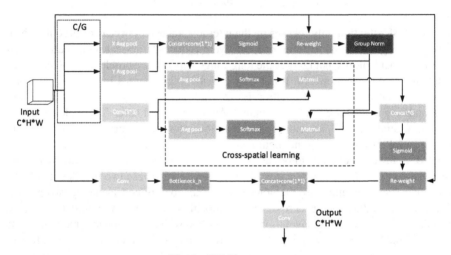

Fig. 1. CEMA structure

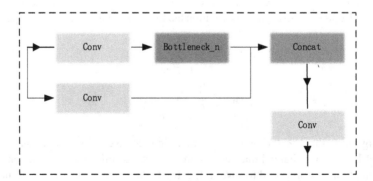

Fig. 2. C3 module structure

In contrast to the YOLOv5 C3 module, the CEMA needs to aggregate multi-scale spatial structure information. CEMA using a larger 3 × 3 convolution kernel to obtain a larger receptive field, and more spatial information will be lost after convolution operations. To retain more spatial information for further cross-spatial operations, the first convolution operations are removed, using the EMA attention mechanism instead of it. The input feature maps are divided into four parallel branches, three of these branches first divide the input features into G sub-features in the channel direction, $G << C$, and the remaining branch passes through a Conv layer and further through Bottleneck modules with a number N, N depends on the network depth of CEMA, concatenate the output features with the output of the other three branches, and then restores the dimension and size of the input feature maps to $C \times H \times W$ through a 1×1 convolution kernel.

Two average pool branches calculated by Eqs. (6,7), to capture precise positional information, use the same method, which was been used in CA, but the channel dimension turns into C/G, then the subsequent cross-spatial learning can learn more contextual information in sub-feature maps.

The further is cross-spatial, cross-spatial learning is the learning of the feature maps in two parallel branches. A branch of this branch concatenates with the main road of the average pool branch, and the main road concatenates with the main road of the average branch after average pooling and Softmax function activate. Since CA ignoring the 1×1 kernel has a limited in receptive field, the 3×3 kernel extends the receptive, aggregate multi-scale spatial structure information. With G sub-features concatenated at the end, the output information will be more than the original only through one convolutional module.

The previous cross-spatial learning output is calculated as follows Eq. (8):

$$F_{B1\times1} = GN(\sigma([XAvgPoolF_{C/G}, YAvgPoolF_{C/G}])) \tag{8}$$

GN [15] is Group Normalization, a deep learning normalization method, when the trained batch size is smaller than 16, GN has a lower error rate than Batch Normalization. The 3×3 branch output calculated as follows Eq. (9):

$$F_{B3\times3} = F_{B1\times1} \cdot soft\max(AvgPool(f^{3*3}, F_{C/G})) \tag{9}$$

YOLOv5 uses GIoU as the loss function, which adds a term after IoU to calculate the minimum bounding rectangle of two boxes, which is used to characterize the distance between two boxes, so as to solve the problem of zero gradient when two targets do not intersect. The equation is (10):

$$GIoU = IoU - \frac{C - (A \cup B)}{C} \tag{10}$$

where C is the area of the minimum bounding rectangle of the two boxes. When IoU = 0, the distance between two boxes can still be well represented. It not only focuses on the overlapping area but also focuses on other non-overlapping areas, which can better reflect the overlap degree of the two boxes.

But, when two boxes A and B belong to the inclusion relation, GIoU will degenerate into IoU and can not distinguish their relative position relation, and since GIoU still depends heavily on IoU, it is difficult to converge due to the large error in both vertical directions. When the two boxes are at the same distance, the area of this part is the smallest in the horizontal and vertical directions, and its contribution to the loss is also smaller, resulting in a poor regression effect in the vertical and horizontal directions. To solve this question, the model uses Wise-IoU as the loss function. WIoU constructs an attention-based bounding box loss, and on this basis, it adds a focusing mechanism by constructing a calculation method of gradient gain. Zanjia Tong [13] got a good performance in YOLOv7, therefore chose WIoU to replace GIoU.

4 Experimental Results and Analysis

In this section, the details of experiments will be shown to verify the proposed CEMA module. The network model used is the Pytorch YOLOv5s repository by Ultralytics. The experiment used the PASCAL VOC2007 + 2012 dataset to train validation. VOC2007 + 2012 has a total of 21505 images and 20 classes, including 17204 for the training set and 4301 for the test set. The results use five measures to evaluate the performance, Parameters to measure the model using parameters number, Floating Point Operations (FLPOs) to measure model complexity, Frames Per Second (FPS) to measure model detection speed, Mean Average Precision (mAP) 0.5 and 0.5:0.95 to measure model precision. The number of groups G is set as 32, all trained epochs set to 200, batch size set to 8. Experiments run on a PC equipped with RTX3070 GPU and AMD Ryzen5 5700x@3.4Ghz with 32 GB memory.

4.1 Contrast of the Same Attention in Different Positions

Since YOLOv5s has 4 C3 modules in the Backbone and 4 C3 modules in the Neck, this ablation experiment first tested the performance of different types of attention in different positions.

This subsection experiment compared the performance of various forms of attention in different positions. The CEMA module has been tested in four positions in the Backbone and four positions in the Neck, but the performance of the second and third positions in the Backbone and Neck are much weaker than that of the first and fourth positions, so they are not listed in the table. The convolution operation will lead to the loss of spatial information, and the shallow feature maps can contain more spatial information than the deeper feature maps.

CEMA and CA, which are more sensitive to spatial information, show better performance as they move closer to the shallow layer, while network depth has no significant effect on SimAM and CBAM, which are less sensitive to spatial information. The shallow features contain more spatial information, which makes CEMA have 0.437% better performance in Backbone than in the Neck. CEMA shows the best performance in the Backbone, with 0.802% better than CBAM, 1.516% better than CA and 0.75% better than SimAM. Other specific data is shown in Table 1.

Table 1. Compared with different position

Attention	mAP0.5 (%)	mAP0.5:0.95 (%)
B-CEMA1	82.306	58.755
B-CEMA4	81.902	58.567
N-CEMA1	81.869	58.557
N-CEMA4	81.210	57.787
B-CBAM1	80.854	57.007
N-CBAM1	81.504	57.856
B-CA1	80.790	56.665
N-CA1	80.645	56.807
B-SimAM1	80.817	56.808
N-SimAM1	81.556	57.743

B is the position in Backbone, N is the position in Neck, number indicates the place in the order.

4.2 Use Multiple Attention Contrasts

Consider whether it is possible to use multiple attention at the same time in different locations, or the same attention in different locations at the same time. This experiment chose to use CEMA in different locations at the same time and use CEMA with CBAM, SE, and SimAM in different locations for experiments to verify our ideas. However, the CA and CEMA front-end are similar, so CA is not selected for this experiment.

Table 2. Compare with multiple attention

Attention	mAP0.5 (%)	mAP0.5:0.95 (%)
B-CEMA14	81.909	58.074
N-CMEA14	81.714	58.525
B-CEMA1-N-SimAM1	80.754	56.793
B-CEMA1-N-CBAM1	81.702	58.363
B-CEMA1-N-SE4	81.889	58.563

B is the position in Backbone, N is the position in Neck, number indicates the place in the order.

From Table 2, it can be seen that using multiple CEMA modules concurrently does not aid in performance enhancement, even if employing various attentions. This is due to the fact that subsequent features already include feature fragments that have been processed by CEMA; adding more instances of the module leads to overfitting and degradation in performance, whereas utilising different attention methods may result in model underfitting and a reduction in performance.

4.3 Compared to Other Attention

This subsection experiment compared the model embedded with the CEMA module with the model embedded with other attention mechanisms. It was also compared with the state-of-the-art YOLOv8s and YOLOv8n [16]. The experimental result used the best result obtained by each attention in the whole experiment, excluding its position relation. The training time of YOLOv8s is the longest, YOLOv8n is the shortest, and there are no other significant differences in the training times. The results of the experimental data are presented in Table 3.

Table 3. Different attention compares results

Method	Param	FLOPs	mAP0.5(%)	mAP0.5:0.95 (%)	Speed (FPS)
YOLOv5s + CEMA + WIoU	7.07M	16.0G	82.323	58.733	140
YOLOv5s + CEMA	7.06M	16.0G	82.306	58.755	128
YOLOv5s + CA	6.74M	14.9G	81.217	56.980	116
YOLOv5s + CBAM	6.9M	15.4G	81.504	57.856	126
YOLOv5s + SE	7.09M	15.9G	81.730	57.876	134
YOLOv5s + SimAM	7.06M	15.9G	81.556	57.743	130
YOLOv5s	7.06M	15.9G	80.564	56.829	137
YOLOv8s	11.13M	28.5G	85.081	66.734	116
YOLOv8n	3.09	8.1G	81.586	61.980	120

It can be seen that the result of embedding the CEMA module has a 1.76% improvement compared with the original YOLOv5s, and it also has different degrees of improvement compared with other attention mechanisms. Compared with CA, it improved by 1.106% in accuracy; with CBAM, that's an increase of 0.819% in accuracy; with SE and SimAM, there was an increase of 1.2% and 1.3% in accuracy, respectively. And it has the best detection speed.

Compared with the state-of-the-art YOLOv8, YOLOv8n has a much smaller model complexity and fewer parameters, despite having an advantage in detection accuracy. In contrast, the situation is entirely opposite when compared to YOLOv8s. This is because of the lighter volume of the C2f module used in the entire YOLOv8 network, which is the reason why YOLOv8 can achieve higher performance with a lower number of parameters and model complexity.

The characteristics of GIoU make the model have large errors during training, whilst WIoU is more stable. Even if WIoU is used instead of GIoU, it is not much better than the best GIoU result, but it has a certain improvement in detection speed.

Finally, the relationship between FLOPs and detection speed is analysed. In many cases, it is thought that the smaller computation will make the detection faster, but in some of these results, this is not the case. For example, YOLOv8n has only 8.1G FLOPs but its detection speed is not the fastest. ShuffleNet V2 [17] recognises the impact of

GPU memory access bandwidth on model inference time, not only model complexity, i.e. FLOPs and the number of parameters on inference time.

5 Conclusion

In this study, the CEMA module is proposed to improve the overall performance of YOLOv5 and has a certain degree of advantage over other embedded attention mechanisms. However, compared with the state-of-the-art YOLOv8 algorithm, there is still a certain gap. However, YOLOv5 and YOLOv8 are from the same company Ultralytics, and their overall framework is roughly the same. Therefore, similar improvements can be adopted in the C2f module of YOLOv8 according to the structure of CEMA to improve the performance of YOLOv8. The further study is trying to integrate EMA into the C2f module to get better performance.

References

1. Chen, L., et al.: SCA-CNN: Spatial and channel-wise attention in convolutional networks for image captioning. In: CVPR (2017)
2. Sandler, M., Howard, A., Zhu, M., Zhmoginov, A., Chen, L.-C.: Mobilenetv2: Inverted residuals and linear bottlenecks. In: CVPR (2018)
3. He, K., Sun, J.: Convolutional neural networks at constrained time cost. In: CVPR (2015)
4. Hinton, G.E., Srivastava, N., Krizhevsky, A., Sutskever, I., Salakhutdinov, R.R.: Improving neural networks by preventing coadaptation of feature detectors. arXiv:1207.0580 (2012)
5. He, K., Zhang, X., Ren, S., Sun, J.: Deep residual learning for image recognition. IEEE (2016)
6. Hu, J., Shen, L., Sun, G.: Squeeze-and-excitation networks. In: CVPR (2018)
7. Woo, S., Park, J., Lee, J.Y., et al.: Cbam: Convolutional block attention module. In: Proceedings of the European conference on computer vision (ECCV), pp. 3–19 (2018)
8. Girshick, R., Donahue, J., Darrell, T., et al.: Rich feature hierarchies for accurate object detection and semantic segmentation. In: Proceedings of the IEEE conference on computer vision and pattern recognition, pp. 580–587 (2014)
9. Redmon, J., Divvala, S., Girshick, R., et al.: You only look once: Unified, real-time object detection. In: Proceedings of the IEEE conference on computer vision and pattern recognition, pp. 779–788 (2016)
10. Ouyang, D., et al.: Efficient multi-scale attention module with cross-spatial learning. In: ICASSP 2023 - 2023 IEEE International Conference on Acoustics, Speech and Signal Processing (ICASSP), pp. 1–5. Rhodes Island, Greece (2023)
11. Hou, Q., et al.: Coordinate attention for efficient mobile network design. In: 2021 IEEE/CVF Conference on Computer Vision and Pattern Recognition. In CVPR (2021)
12. Yang, L., et al.: SimAM: A Simple, Parameter-Free Attention Module for Convolutional Neural Networks. In: International Conference on Machine Learning (2021)
13. Tong, Z., Chen, Y., Xu, Z., Yu, R.: Wise-IoU: Bounding Box Regression Loss with Dynamic Focusing Mechanism. arXiv, abs/2301.10051 (2023)
14. Nair, V., Hinton, G.E.: Rectified linear units improve restricted boltzmann machines. In: ICML (2010)
15. Wu, Y., He, K.: Group Normalization. Int. J. Comput. Vision **128**, 742–755 (2018)
16. Reis, D., Kupec, J., Hong, J., Daoudi, A.: Real-Time Flying Object Detection with YOLOv8. ArXiv, abs/2305.09972 (2023)
17. Zhang, X., Zhou, X., Lin, M., Sun, J.: ShuffleNet: an extremely efficient convolutional neural network for mobile devices. In: 2018 IEEE/CVF Conference on Computer Vision and Pattern Recognition, pp. 6848–6856. CVPR, Salt Lake City, UT, USA (2018)

BRA-YOLO: Object Detection Algorithm with Bi-Level Routing Attention for YOLOv5

Xing Huang and Weidong Huang[✉]

School of Information Engineering, JiangXi University of Science and Technology, GanZhou 34100, JiangXi, China
18279766975@139.com

Abstract. At present, YOLOv5 is the most popular algorithm in single-stage target detection, has covered all areas of society. However, because the neck layer can not effectively integrate the context information content, it is still difficult to identify the small target features incorrectly and omitted. In addition, YOLOv5 also faces the problem of low detection accuracy. In response to the above issues, in this paper, BiFormer attention mechanism is introduced into the Neck C3 module of YOLOv5s model. Based on the above issue, this paper introduces the C3 module with the BiFormer attention mechanism into the Neck of the YOLOv5s model, proposing the BRA-YOLO object detection algorithm. This algorithm aims to optimize the detection ability of the model for small target features and improve the detection accuracy of the model. Experimental results show that compared with YOLOv5s, the BRA-YOLO target detection algorithm has a significant improvement of 1.303% on mAP@0.5%, 1.389% on mAP@0.5:0.95%, and 2.32% on small target detection in general. BRA-YOLO effectively improves the overall accuracy of the model and the accuracy of small target detection.

Keyword: YOLOV5 · BiFormer · Small target detection · Object detection

1 Introduction

In recent years, with the continuous enrichment of deep learning theories and the significant improvement of computer processing speed, the combination of deep learning and artificial intelligence has been popularized in various fields of society. For the industrial field (Industrial digital operation, Automatic driving, Intelligent grasping), the object detection algorithm based on deep learning [1] has been further applied.

The deep learning-based object detection algorithm uses a deep neural network to automatically recognize object features, and has excellent recognition and detection performance. Nowadays, popular deep learning-based object detection algorithms are divided into one-stage object detection algorithm and two-stage object detection algorithm, and the two-stage object detection algorithm is represented by R-CNN [2] and Fast R-CNN [3]. The two-stage object detection algorithm usually divides the object detection task into two parts. At first, the candidate region frames are generated on the object image, and then the target classification and accurate positioning are performed

© The Author(s), under exclusive license to Springer Nature Singapore Pte Ltd. 2024
K. Li and Y. Liu (Eds.): ISICA 2023, CCIS 2146, pp. 302–311, 2024.
https://doi.org/10.1007/978-981-97-4393-3_25

on these candidate regions. These two-stage object detection algorithms usually perform well in accuracy and are appropriate for application scenarios that demand high detection accuracy. However, two-stage object detection algorithms have large computer memory requirements and are unsuitable for real-time detection. The YOLO series algorithm, as a well-known single-stage object detection algorithm, has the advantages of strong real-time processing, a simple reasoning process and fast detection speed compared with the two-stage object detection algorithm. With these advantages enable it can be applied to real-time detection in various fields. In 2016, Redmon et al. [4] published the YOLOv1 network model and proposed a new single-stage object detection algorithm. The core idea of this algorithm is to transform the object detection task into a regression problem and directly regress the coordinates and class probability of the object in the image. YOLOv1 demonstrates the capability of single-stage object detection algorithm, which opens a new starting point of single-stage object detection algorithms.

In recent years, many scholars at home and abroad have made relevant contributions to improve the performance of YOLOv5 [5] by combining attention. Jun Wu et al. [6] aim to achieve lightweight while ensuring that the detection accuracy does not decrease. They introduce CA(Coordinate Attention) et al. [7] to cooperate with the representative Squeeze And Excitation, SE [8]. As a result, a lightweight target detection network is created and the accuracy is improved. Chen et al. [9] propose to introduce CBAM (Convolutional Block Attention Module) [10] into the transition region at the bottom of the network to enhance the ability of the model to extract image features and improve the segmentation accuracy of the network. The deep integration of CBAM and YOLOv5, a dual attention mechanism based on channel and space, by Fan et al. [11] effectively improves the feature assignment problem of a single attention mechanism and improves the model accuracy, but the model is still relatively computationally intensive. Peng Zou et al. [12] proposed to introduce the SE attention mechanism into the Ghost lightweight model to improve the accuracy of model detection and improve the accuracy of irregular driving behavior detection.

The above model algorithm can improve the detection accuracy of the target, but there are still some defects for small targets. Therefore, this paper takes YOLOv5s as the basic model, proposes BRA-YOLO algorithm, and its main work is as follows: To solve the problem that the features of small targets are difficult to detect, BiFormer attention mechanism is integrated into the Neck layer [13]. BiFormer can optimize the detection performance of small targets, which has a good improvement effect on small targets, and can further improve the overall performance of YOLOv5s model.

The structure and organization of this paper are as follows. The YOLOv5 model, BRA attention mechanism, and BiFormer structure involved in BRA-YOLO proposed in this topic are described in detail in Sect. 2. In Sect. 3, the experiment of BRA-YOLO proposed in this paper is conducted and compared with the current popular attention. In the last section, the research content is summarized and analyzed.

2 Related Model YOLOv5

YOLOv5 algorithm model structure is consisted of five parts: Input, Backbone, Head, Neck and Output. YOLOv5s adopted Mosaic data enhancement on the Input side, referring to the CutMix data enhancement method. Mosaic data enhancement was improved from the original two images to four images for Mosaic, and the images were randomly scaled, randomly cropped, and randomly arranged. The use of data enhancement can improve the data set, small, medium, and large target data imbalance. The backbone includes Conv, C3, and SPPF. The Conv layer mainly has the function of feature extraction. The C3 module is mainly responsible for simplifying the network structure and reducing the amount of model calculation. The SPPF module mainly integrates the feature maps of different receptive fields to enrich the expressiveness of the feature maps and further improve the running speed. The Neck module adopts the multi-scale feature fusion structure of the feature pyramid network (FPN)+ path aggregation network (PAN), and the PAN transfers the high-level features by subsampling. The FPN fuses the high-level feature information with low-level feature information by an upsampling operation. The Head part is mainly used for target detection. The Head expands the channel count of the feature maps of different scales obtained in the Neck by 1×1 convolution.

3   Proposed BRA-YOLO

3.1   BRA Attention Mechanism

The core idea of BRA's attention mechanism is that attention can be achieved through dynamic and adaptive sparsity. Based on the target detection model of Transformer architecture, BiFormer model introduces the Bi-level routing attention (BRA) to improve the target detection performance. Transformer [14] was proposed by the Google Brain team in 2017. It uses a self-attention structure to replace the RNN network structure, which is commonly used in NLP tasks. Compared with RNN, Transformer has the advantage that it can perform parallel computation, but for the input of complex information, it will require a lot of computation and occupy a lot of memory space. On this basis, a dynamic sparse attention mechanism, called Bi-level routing attention. It is proposed

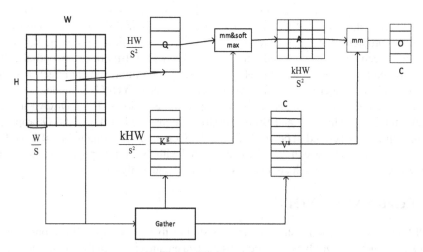

Fig. 1. BRA structure

to achieve flexible computation allocation and content attention, so that it has dynamic query attention sparsity. Its structure is shown in Fig. 1:

The BRA attention mechanism works in three parts:

Splitting the image with the linear map: For an input feature graph X with height H, width W, and dimension C. First, the feature graph X is divided into S × S different regions, each containing $\frac{HW}{S^2}$ feature vectors. At this time, the height of the feature graph X^r is S^2, the width is $\frac{HW}{S^2}$, and the dimension is C. Then the feature map is obtained by linear mapping Q, K, V $\in R^{S^2 \times \frac{HW}{S^2} \times C}$, where W^q, W^k, $W^v \in R^{C \times C}$ are the projected weights of query, key, value, respectively. The calculation is shown in Eqs. (1), (2) and (3):

$$Q = X^r W^q \tag{1}$$

$$K = X^r W^k \tag{2}$$

$$V = X^r W^v \tag{3}$$

Region-to-region using a directed graph: First construct a directed graph to find the region in which each given region should participate, and calculate the average of Q and K in each region. Thus, get the Q^r, $K^r \in R^{S^2 \times C}$. Second, the adjacency matrix of inter-region correlation and is calculated, as shown in Eq. (4). Finally, the first k connections of each region are retained to trim the correlation graph. The routing index matrix $I_r \in N^{S^2 \times k}$ will stores the indexes of the first k connections row by row. The calculation formula is Eq. (5). I^r where is the index of the first k most relevant regions in row i containing region i.

$$A^r = Q^r (K^r)^T \tag{4}$$

$$I^r = \text{topkIndex}(A^r) \tag{5}$$

Token-to-token attention mechanism: First, using the zone-to-zone routing index matrix, we can compute fine-grained token-to-token attention. Second, for each Query token in zone i, it will focus on k routing zones and will index all key-value pairs as $I^r_{(i,1)}$, $I^r_{(i,2)}, \ldots, I^r_{(i,k)}$. However, it is not easy to do this efficiently because these routing regions are scattered over the feature graph, and the GPU relies on a block-merge memory operation that loads tens of consecutive bytes at a time. Finally, the key-value tensor is first assembled, and then use the attention operation is applied to the assembled key-value pair, as shown in Eq. (6). Here, this paper introduces a local context enhancement term LCE(V), where the function LCE(·) is parameterized with depth-separable convolution, where the convolution kernel size is set to 5.

$$O = \text{Attention}(Q, K^g, V^g) + \text{LCE}(V) \tag{6}$$

3.2 BiFormer Structure

For an input feature graph with width H, height W, and dimension 3. First, the overlapping patch embedding module is used to reduce the spatial resolution of the input feature map, and feature transformation is performed by the BiFormer block to generate a feature map with height, width, and dimension C. In subsequent operations, the number of channels is increased by continuous patch merging, and then N_i connected BiFormer blocks are used to perform transformation operations on the input features. The BiFormer structure is shown in Fig. 2:

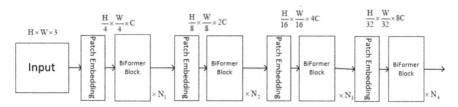

Fig. 2. BiFormer structure

The structure of the BiFormer block is shown in Fig. 3. Note that a 3×3 deep convolution is used to implicitly encode relative positions before entering each BiFormer block. Then, the BRA module and a 2-layer MLP (Multi-Layer Perceptron) module with an expansion rate of e are successively used for cross-position relationship modeling and position-by-position embedding.

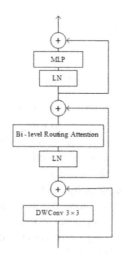

Fig. 3. BiFormer block structure

3.3 BRA-YOLO

The core of BRA-YOLO proposed in this paper is the integration of BiFormer attention mechanism into the Neck layer. This kind of directed graph routing between regions of BiFormer can optimize the related of associated content in the feature graph partitioning, which can not only select more accurate targets in general but also obtain the content of the receptive field in local regions more easily. In addition, because BiFormer only pays attention to some tags in the query adaptation process, it does not distract the attention of other irrelevant tags, which has good performance and high computational efficiency. In addition, BRA-YOLO is difficult to show its advantages for large and medium-sized targets, but it has unexpected effects for small targets. The BRA-YOLO structure diagram is shown in Fig. 4:

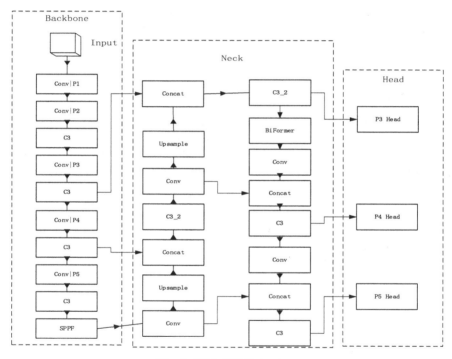

Fig. 4. BRA-YOLO structure

4 Experimental Result

4.1 Experimental Environment

The Pascal visual object class (VOC) dataset is a widely used computer vision dataset for object detection, image segmentation, and image classification. The VOC dataset was created by members of the Computer Vision Research team at the University of

Oxford in the United Kingdom and was used in the PASCAl VOC Challenge. VOC dataset contains a variety of different types of tagged images, and each image has an associated bounding box and the label of the category of objects on the tag. The dataset contains a total of 20 categories, including people, cats, dogs, and cars, with an average of 2.4 objects per image.

VOC2007 contains a total of 9963 images, of which 5011 are in the training set, 4952 are in the test set, each of which is labeled, and a total of 24640 objects are labeled VOC2012 is an evolution of VOC2007, with a total of 11,540 images, of which 5,717 are in the training set, 5,823 are in the verification set, and a total of 27,450 objects are labeled.

In this paper uses the VOC2007+2012 dataset is used for model training and validation. The training set uses VOC2007+2012, and the test set is the test set of VOC2007. The training set contained 16,935 images and the test set contained 4,568 images.

The experiment in this paper is built on the computer system Windows 11, the CPU is AMD Ryzen 7600X 6-Core, the graphics card is Nvidia GeForce RTX3080, and the video memory is 12G. The deep learning framework is Python 3.9.18, torch 1.12.0+CUDA 116.

4.2 Evaluation Index

When detecting a series of models, the performance indexes are as follows: Precision (P), Recall (R), mean Average Precision (mAP), number of parameters (Params), number of floating point operations (FLOPs). The number of parameters (Params) and the number of floating point operations (FLOPs) are used to test whether the model is lightweight. Among them, mAp@0.5% and mAP@0.5:0.95% are used as indicators to evaluate the performance. mAP@0.5% represents the mAP value when the IoU threshold is 0.5, and mAP@0.5:0.95% represents the average mAP value as the step size increases from 0.05 to 95%. The larger the average mAP, the more accurate the model. The equation for calculating the indicators is as follows:

$$P = \frac{TP}{TP + FP} \tag{7}$$

$$R = \frac{TP}{TP + FN} \tag{8}$$

$$AP = \int_0^1 PR \, dr \tag{9}$$

$$mAP = \frac{1}{N} \sum_{i=0}^{n} APi \tag{10}$$

TP is the number of correct judgments; FP is the number of negative cases misjudged; FN is the number of positive cases misjudged.

4.3 Experimental Comparison and Analysis

In this paper, ablation experiments and comparison experiments were used to confirm the effectiveness of the improved model. In the ablation experiment, we conducted comparison experiments between Bi-YOLO and other attention mechanisms to confirm the effect of Bi-YOLO. The ablation experiments are listed in Table 1.

As can be seen from Table 1, after the introduction of BiFormer, although the number of parameters and floating-point operation increased by 0.06% and 2.5%, respectively, the performance of mAP@0.5% increased by 1.303%, mAP@0.5:0.95% increased by 1.389%, and the speed also decreased by 15.742%. It can be seen that Bi-YOLO improves the accuracy of model detection while sacrificing a little speed, which proves the effectiveness of BiFormer to some extent. At the same time, in the experiment, three classical attention mechanisms, CA (Coordinate Attention), CBAM (Convolutional Block Attention), and SimAm, place all three attention on the Neck end. Comparative experiments were done with Bi-YOLO. It can be seen that CA has increased by 0.745% and 0.51% compared to mAP@0.5% and mAP@0.5:0.95% of YOLOv5s, and CBAM has increased by 0.028% and 0.173% compared to mAP@0.5% and mAP@0.5:0.95% of YOLOv5s. SimAm is 0.350% and 0.088% higher than mAP@0.5% and mAP@0.5:0.95% by YOLOv5s. It can be seen that the effect of CA, CBAM, and SimAm on YOLOv5 is not greatly improved, which proves from the side that Bi-YOLO can improve the effect of YOLOv5s. For YOLOv8, the most advanced object detection technology, Bi-YOLO has 3.98% fewer parameters and 9.9% fewer floating-point operations than YOLOv8s, but the speed is similar. However, compared with YOLOv8 lightweight model YOLOv8n, Bi-YOLO has an increase of 0.247% from mAP@0.5%, which can be disguised to show that Bi-YOLO is a relatively balanced algorithm for model lightweight and performance balance.

Table 1. Ablation comparison experiment

Experimental	mAP@0.5%	mAP@0.5:0.95%	Parameter/M	FLOPs/G	Speed/s	Epochs
YOLOV8s	85.99	66.946	11.11	28.5	116.279	200
YOLOv8n	81.516	61.980	3.09	8.1	120.482	200
YOLOv5s	80.472	56.470	7.07	16.1	130.12	200
YOLOv5s+Neck+CA	81.217	56.980	6.74	14.9	115.71	200
YOLOv5s+Neck+CBAM	80.500	56.643	6.74	14.9	114.594	200
YOLOv5s+Neck+SimAM	80.822	56.558	7.06	15.9	130.115	200
BRA-YOLO	81.763	57.79	7.13	18.6	114.378	200
YOLOv5s+Neck3+BiFormer	81.775	57.859	7.13	18.6	114.26	200

4.4 Comparison Experiment of a Lightweight Model

While keeping other experimental environment variables unchanged, the performance of Bi-YOLO was compared with SSD and R-CNN, and the main comparison aspect was lightweight. The structure of the comparison experiment is shown in Table 2.

The SSD uses multi-scale feature maps, large scale feature maps are used to detect small objects, small scale can also be used to detect large objects, using the VGG backbone network to load pre-training weights. Compared with SSD, BRA-YOLO at mAP@0.5% increased by 4.963%, at mAP@0.5:0.95% increased by 10.89%, and the reference count and FLOPs decreased by 19.17% and 44.1%, respectively. It can be seen that Bi-YOLO is superior to SSD in all aspects. For mAP@0.5%, Bi-YOLO can outperform Faster-RCNN by 8.563%. In summary, the BRA-YOLO algorithm can optimize performance well and can be better than some common algorithms.

Table 2. Model comparison experiment

Experimental	mAP@0.5%	mAP@0.5:0.95%	Parameter/M	FLOPs/G	Epochs
SSD	76.8	46.9	26.3	62.7	200
Faster-RCNN	73.2	–	–	–	–
Bi-YOLO	81.763	57.79	7.13	18.6	200

Table 3. Small target comparison experiment

Experimental	Bird	Cow	Horse Cat	Pottedplant	Boat	Chair
YOLOv5s	0.824	0.805	0.875	0.611	0.714	0.622
Bi-YOLO	0.864	0.833	0.893	0.629	0.723	0.648

4.5 Small Target Comparison Experiment

YOLOv5s has some shortcomings on small targets. After improvement, Bi-YOLO has achieved a good detection effect on small targets. In the detection of small targets, the accuracy of birds is improved by 0.040, and other small targets are improved. It can be seen that the optimization and improvement of Bi-YOLO for small target detection by YOLOv5 is effective (Table 3).

5 Conclusion

This paper presents BRA-YOLO, a two-layer routing attentional target detection algorithm based on YOLOv5. In view of the performance deficiencies of YOLOv5s, BiFormer attention mechanism was added to the Neck end based on YOLOv5s to improve the feature fusion capability of Neck, which greatly improved the accuracy at the expense of a small part of the computation. The capability of the model is further improved, and the accuracy of small target detection is optimized. The experiment on VOC2007+2012 data set also shows a good accuracy improvement compared with the latest target detection algorithm YOLOv8n, which has a good performance between balance performance and lightweight.

References

1. Amit, Y., Felzenszwalb, P., Girshick, R.: Object detection. Comput. Vis. Ref. Guide 1–9 (2020)
2. Girshick, R., Donahue, J., Darrell, T., et al.: Region-based convolutional networks for accurate object detection and segmentation. IEEE Trans. Pattern Anal. Mach. Intell. **38**(1), 142–158 (2016)
3. Chen, Y.H., Li, W., Sakaridis, C., et al.: Domain adaptive faster R-CNN for object detection in the wild. In: 2018 IEEE/CVF Conference on Computer Vision and Pattern Recognition. pp. 3339–3348. IEEE, Piscataway (2018)
4. Redmon, J., Divvala, S., Girshick, R., Farhadi, A.: You only look once: unified, real-time object detection. In: 2016 IEEE Conference on Computer Vision and Pattern Recognition (CVPR), pp. 779–788. IEEE, Piscataway (2016)
5. ULTRALYTICS. Yolov5[EB/OL] (2023). https://github.com/ultralytics/yolov5
6. Wu, J., Dong, J., Liu, X., et al.: Lightweight object detection network and its application based on the attention optimization. CAAI Trans. Intell. Syst. **18**(3), 506–516 (2023)
7. Hou, Q., Zhou, D., Feng, J.: Coordinate attention for efficient mobile network design. In: Proceedings of the IEEE/CVF Conference on Computer Vision And Pattern Recognition, pp. 13713–13722 (2021)
8. Hu, J., Shen, L., Albanie, S., et al.: Squeeze-and-excitation networks. IEEE Trans. Patt. Anal. Mach. Intell. **42**(8), 2011–2023 (2020)
9. Chen, S.Z., Zhao, M., Shi, F., Huang, W.: Over-parametric convolution and attention mechanism-fused pleural effusion tumor cell clump segmentation network. J. Image Graph. **28**(10), 3243–3254 (2023)
10. Woo, S., Park, J., Lee, J.Y., et al.: Cbam: convolutional block attention module. In: Proceedings of the European Conference on Computer Vision (ECCV), pp. 3–19 (2018)
11. Fan, H., Zhu, D., Li, Y.: An improved yolov5 marine biological object detection algorithm. In: 2021 2nd International Conference on Artificial Intelligence and Computer Engineering (ICAICE), pp. 29–34. IEEE (2021).
12. Zou, P., Yang, K., Liang, C.: Improved YOLOv5 algorithm for real-time detection of irregular driving behavior. Comput. Eng. Appl. **59**(13), 186–193 (2023)
13. Zhu, L., Wang, X., Ke, Z., Zhang, W., Lau, R.: BiFormer: vision transformer with bi-level routing attention. In: 2023 IEEE/CVF Conference on Computer Vision and Pattern Recognition (CVPR), pp. 10323–10333. Vancouver, BC, Canada (2023)
14. Vaswani, A., Shazeer, N., Parmar, N., et al.L Attention is all you need. Adv. Neural Inform. Process. Syst. **30** (2017)

A Pest Detection Algorithm Based on Improved YOLO

Kangshun Li[1,2(✉)], Shuizhen He[1], and Jiancong Wang[1]

[1] College of Mathematics and Informatics, South China Agricultural University,
Guangzhou 510642, China
likangshun632@sina.com

[2] School of Artificial Intelligence, Dongguan City University, Dongguan 523808, China

Abstract. A pest detection algorithm based on improved YOLO is proposed in this paper to improve the recognition rate of pest detection. Based on the original YOLOv5 model, a new point distance loss function is proposed to reduce redundant computations and improve the model performance. Then an attention module is added to the model to reduce the complexity, keep the performance and improve the detection recognition rate. The experiments on the test set verified the effectiveness of the proposed model. The results show that the mean Average Precision of our proposed model is 95.9% and the mean detection time is 6.1 ms. The proposed model keeps the lightweight features of YOLOv5 with good prospects in smart agriculture applications.

Keywords: YOLO · Loss Function · Intersection over Union · Attention module

1 Introduction

Influenced by pests, a large number of deformed and rotten fruits often appear during the growth of fruits. This seriously restricts the further development of the industry and affects the fruit quality and the economic efficiency of cultivation. Therefore, the primary task of pest control is the pest detection.

At first, insect pests control relied mainly on large-scale and large-dose spraying of pesticides. But this method causes serious environmental pollution, and the pesticide residues will harm consumers' health. In recent years, the development of computer vision technology has made agricultural production gradually intelligent. However, there are still problems such as slow detection speed, low recognition accuracy, large models and difficult deployment in some pest detection methods.

In this paper, we propose a pest detection method based on improved YOLOv5. The proposed method uses a new bounding box regression loss function, called Point Distance Intersection over Union Loss function (PDIoU Loss), to reduce redundant calculations, improve model performance and localization accuracy. Then, Efficient Channel Attention Module (ECA) [1] is added to reduce the model complexity, keep the performance and improve the detection recognition rate. Compared with the YOLOv5s model, the mean Average Precision of the proposed model is improved by 0.97% to 95.9%, and the

© The Author(s), under exclusive license to Springer Nature Singapore Pte Ltd. 2024
K. Li and Y. Liu (Eds.): ISICA 2023, CCIS 2146, pp. 312–325, 2024.
https://doi.org/10.1007/978-981-97-4393-3_26

average detection time is 6.1 ms. The proposed model also keeps the lightweight characteristic of YOLOv5s, which has good application prospect and provides an effective way to preserve farmers' economic benefits.

2   Related Work

In recent years, In order to save resources and protect the environment, people have begun to use various techniques, such as near-infrared light [2], X-ray imaging [3], sound signal [4] and other detection methods to detect pests. With the development of deep learning, some scholars have begun to apply deep learning methods to the field of crops [5, 6].

The Convolutional Neural Network (CNN) removes the manual traces in early image recognition, which improves the detection rate and accuracy [7]. G. Yang et al. [8] implemented tea garden pests recognition by Grubcut algorithm, AlexNet and image significance analysis. H. Li et al. [9] proposed a rapeseed pest detection method based on a deep convolutional neural network to achieve rapid identification and accurate positioning of rapeseed pest target images, with an average accuracy rate of 94.12%. L. Wang et al. [10] achieved the classification and localization of candidate bounding boxes in the real-time detection system for citrus pests, which could detect citrus red spiders and aphids quickly and accurately, with an accuracy of 91% and 89%, respectively.

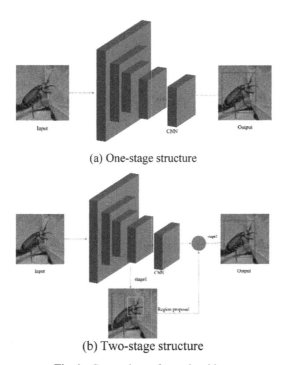

(a) One-stage structure

(b) Two-stage structure

Fig. 1. Comparison of two algorithms

The most representative ones of the current target detection algorithms are mainly of two types. One is the two-stage algorithms based on candidate bounding boxes, such as Regions with CNN features (R-CNN) [11], Fast R-CNN [12], Faster R-CNN [13] and other R-CNN series algorithms. But the detection speed of these methods is slow. The second is the regression-based one-stage detection algorithms, such as Single Shot Multi-Box Detector (SSD) [14] and You Only Look Once (YOLO) series algorithms. Figure 1 shows the comparison between the two methods.

There are five versions of the YOLO series algorithms. YOLOv1 [15] inputs the image to get the localization of prediction boxes, and finds out the classification directly in the output layer. YOLOv2 [16] adopted optimization strategies such as batch normalization, a high score classifier and a prior box to improve the detection speed, accuracy and the number of identified categories. YOLOv3 [17] introduced the Feature Pyramid Network (FPN) [18] and Darknet-53 network, improving the computational speed significantly. YOLOv4 [19] balanced detection accuracy and speed, improving overall performance. YOLOv5 is smaller and faster.

2.1 YOLOv5

YOLOv5 has four structures, YOLOv5s, YOLOv5m, YOLOv5l and YOLOv5x. Among them, the YOLOv5s network has the smallest depth and feature map width in the YOLOv5 series. The latter three are deepened and widened based on YOLOv5s to gradually improve the ability of generalization learning, feature extraction and feature fusion of the network. From Fig. 2, we can see that YOLOv5 is mainly divided into four parts: INPUT, Backbone, Neck and Prediction.

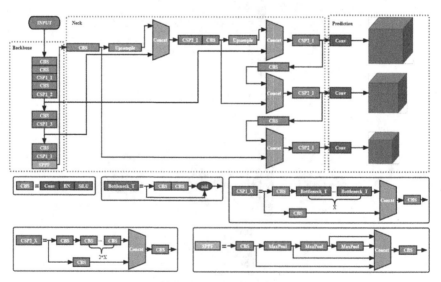

Fig. 2. Schematic diagram of YOLOv5 network structure

2.2 Detection and Recognition Principle of YOLOv5 Algorithm

YOLOv5 has three main tasks on the input part. The first task is the data augmentation based on the mosaic algorithm, which is proposed in the YOLOv4. Its main idea is to combine four pictures into one through random scaling, random cropping and random arrangement, and then calculate the target information of the new picture to expand the online training set. The mosaic not only enriches image backgrounds, adds small targets and improves network robustness, but also reduces training time and memory usage. The second task is adaptive anchor box computation. The difference between the predicted box and the real box is computed in training. Then backpropagation and network parameters updating are performed to compute the best anchor box adaptively. The third task is adaptive picture scaling. In common target detection algorithms, the height and width of the datasets image are usually different. For training convenience, a common method is to scale the input image to a uniform size, and then send them to the detection network. However, after scaling the image to the size of 640×640, YOLOv5 adaptively adds the least black edges to the image, which reduces redundant information and improves the inference speed.

Based on YOLOv4 and the Cross Stage Partial Network (CSPNet) [20], YOLOv5 designed two CSP structures, namely CSP1_X and CSP2_X, where X denotes the number of structures used. These two structures not only enhance the learning ability of CNN but also reduce computational complexity, with a great balance between lightweight and accuracy. CSP1_X is mainly applied to the backbone network part, and CSP2_X is mainly applied to the neck network part.

Referring to Spatial Pyramid Pooling (SPP) [21], the Spatial Pyramid Pooling-Fast (SPPF) is proposed and applied to the backbone part. SPPF merges the feature maps before and after pooling to obtain the refined feature maps, and then performs multiscale fusion to improve the training effect.

Path Aggregation Network (PANet) [22] shortens the information transmission path and improves the feature extraction ability of the neck network. CSP2_X and PANet make it possible to cover the entire feature extraction network with accurate lower layer localization information.

2.3 Bounding Box Regression Loss

The Bounding Box Regression Loss function is important to evaluate the prediction accuracy of the model, which is usually associated with Intersection over Union (IoU). IoU is a criterion for detecting the accuracy of target objects in the dataset. It is the overlap rate between the candidate bounding box and the ground truth bounding box, that is, the ratio of their intersection and union. The early bounding box regression loss function is represented by the Intersection over Union Loss (IoU Loss) function, and its calculation equations are as follows.

$$x_1 = \max(C_{x_1}, G_{x_1}), y_1 = \max(C_{y_1}, G_{y_1}) \tag{1}$$

$$x_2 = \min(C_{x_2}, G_{x_2}), y_2 = \min(C_{y_2}, G_{y_2}) \tag{2}$$

$$C \cap G = \max(0, x_2 - x_1) \times \max(0, y_2 - y_1) \tag{3}$$

$$C \cup G = (C_{x_2} - C_{x_1}) \times (C_{y_2} - C_{y_2}) + (G_{x_2} - G_{x_1}) \times (G_{y_2} - G_{y_2}) - C \cap G \tag{4}$$

$$IoU = \frac{C \cap G}{C \cup G} \tag{5}$$

$$L_{IoU} = 1 - IoU(C, G) \tag{6}$$

where C is the candidate bounding box and G is the ground truth bounding box. (C_x, C_y) and (C_x, C_y) are the coordinates of the upper-left corner and lower-right corner of C respectively. (G_x, G_y) and (G_x, G_y) are the coordinates of the upper-left corner and the lower-right corner of G respectively. (x_1, y_1) and (x_2, y_2) are the coordinates of the upper-left corner and the lower-right corner of the rectangle where C and G intersect, respectively. L_{IoU} is the IoU Loss.

However, the IoU Loss does not consider the distance between the two bounding box. Therefore, YOLOv5 uses Complete Intersection over Union Loss (CIoU Loss) [23] instead of IoU Loss. The CIoU Loss adds a penalty term R_{CIoU} to the IoU Loss and introduces the length-width ratio and normalized distance of the candidate bounding box and the ground truth bounding box. The equations are as follows.

$$R_{CIoU} = \frac{\rho^2(C_{ctr}, G_{ctr})}{d^2} + \alpha v \tag{7}$$

$$L_{CIoU} = L_{IoU} \frac{\rho^2(C_{ctr}, G_{ctr})}{d^2} + \alpha v \tag{8}$$

$$v = \frac{4}{\pi^2} (\arctan \frac{w^{gt}}{h^{gt}} - \arctan \frac{w_2}{h}) \tag{9}$$

$$\alpha = \frac{v}{(1 - IoU) + v} \tag{10}$$

where C_{ctr} and G_{ctr} are the coordinates of the center point of C and G, respectively. L_{CIoU} isthe CIoU Loss $\rho(\cdot)$ is the Euclidean distance and $\rho = \sqrt{(x_2 - x_1)^2 + (y_2 - y_1)^2}$. d is the diagonal length of the smallest box covering C and G. α is the weight function that is used to balance the scale. v is to measure thesimilarity of the aspect ratio between C and G. h^{gt} and w^{gt} represent the height and width of G, respectively. h and w represent the height and width of C, respectively. Figure 3 is a schematic diagram of the CIoU loss.

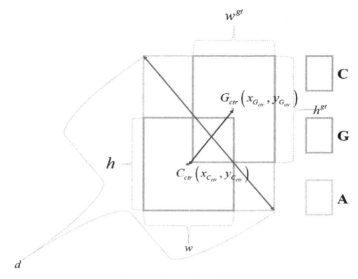

Fig. 3. Schematic of CIoU

2.4 ECA

The Efficient Channel Attention (ECA) module uses a local cross-channel interaction strategy with no dimensionality reduction and a method for adaptively selecting the size of one-dimensional convolutional kernels. It effectively avoids the impact of dimensionality reduction on the channel attention learning effect and achieves performance improvement. Although ECA involves only a few parameters, it has a significant effect gain. The cross-channel interaction of ECA can reduce the complexity of the model with keeping the performance. Figure 4 shows a comparison of various attention modules, and it can be seen that ECA has high accuracy.

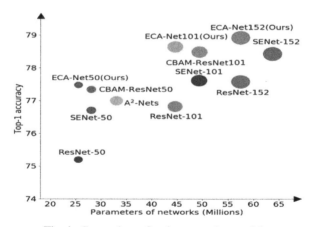

Fig. 4. Comparison of various attention modules

3 Method

3.1 Framework of Proposed Method

In this paper, we make the following improvements based on the YOLOv5s.

(1) A new PDIoU Loss is proposed to be applied to the prediction side to reduce computational consumption and improve model performance and localization accuracy.
(2) ECA is added to the neck network part. ECA makes the convolutional neural network obtain feature maps containing more target information, which reduces model complexity, keeps model performance and improves recognition accuracy.

The algorithm framework proposed in this paper is shown in Fig. 5. The model takes into account the recognition accuracy and detection speed, while retaining the advantage of network lightweight.

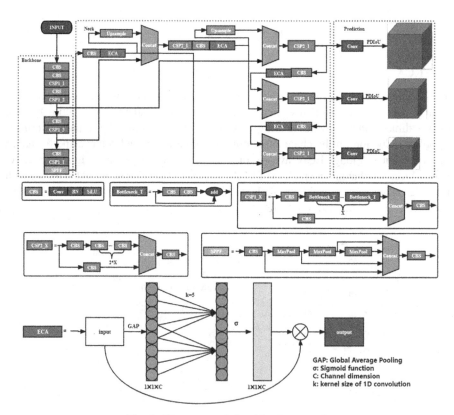

Fig. 5. The proposed algorithm framework

3.2   Point Distance IoU

In this paper, we propose a new PDIoU Loss(L_{PDIoU}) to reduce redundant computations and improve localization accuracy.

Accurate bounding box regression is important for image segmentation, target detection and target tracking. To better represent the position relationship between the candidate bounding box and the ground truth bounding box, we propose PDIoU Loss by using the center points of the above two boxes and the center point of the smallest box covering the two boxes.

The principle of PDIoU Loss is easier. It only needs to add a penalty term R_{PDIoU} to the IoU loss to simplify the calculation of the distance between the candidate bounding box C and the ground truth bounding box G. Let l be the straight line which is determined by the center point of C and the center point of G. Then, the distance d from the center point of the smallest box covering C and G to the straight line l is found by the point-to-straight line distance formula. And schematic diagram of PDIoU. R_{PDIoU} is equal to d^2. Figure 6 shows the schematic diagram of PDIoU

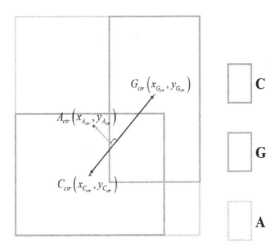

Fig. 6.   Schematic of *PDIoU*

The equation of PDIoU is as follows.

$$a = y_2 - y_1 \tag{11}$$

$$\beta = x_1 - x_2 \tag{12}$$

$$\lambda = x_2 * y_1 - x_1 * y_2 \tag{13}$$

$$R_{PDIoU} = \frac{(\alpha x + \beta x + \lambda)^2}{\alpha^2 + \beta^2} \tag{14}$$

$$PDIoU(C, G) = IoU(C, G) - \eta R_{PDIoU} \qquad (15)$$

$$L_{PDIoU} = 1 - PDIoU \qquad (16)$$

where C is the candidate bounding box. G is the ground truth bounding box. A is the smallest box covering C and G. C_{ctr} and G_{ctr} are the coordinates of the center point of C and G, respectively. A_{ctr} is the coordinate of the center point of A.

The two points C_{ctr} and G_{ctr} are substituted into formulas (11) to (13). A_{ctr} is substituted into formula (14) to obtain R_{PDIoU}, and finally, L_{PDIoU} is obtained by formulas (15) and (16). The main role of η is to balance the gap between the two loss function values. After experimental verification, we take $\eta = 10$.

3.3 ECA in Neck

To reduce the complexity and improve the accuracy of the model, the ECA is added to the four red boxes in Fig. 7.

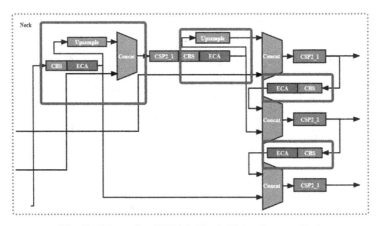

Fig. 7. Schematic of ECA in Neck (Color figure online)

3.4 Algorithm Process

The process of the proposed algorithm is as follows.

(1) The input image enters the SPPF network after a multi-layer convolution in the backbone network part. Then the results before and after pooling are concatenated in the process of 3 times max-pooling. Finally, the dimension and parameters are reduced by standard convolution to improve the computational speed.

(2) In the neck network part, the convolution is performed again, and then the important information of the target object is extracted by the ECA module. Meanwhile,

PANet and FPN perform parameter aggregation at different backbone layers for different detection layers to improve the feature fusion, feature extraction and detection effect. This part reduces the computation of redundant information and improves the detection accuracy and robustness of the model.

(3) PDIoU Loss is added at the prediction side instead of CIoU Loss, which reduces the computations. Moreover, PDIoU not only quickens the overlap of the candidate bounding box and the ground truth bounding box but also improves the performance of the algorithm.

4 Experiments

4.1 Datasets

After manual screening and summary, we collected a total of 850 datasets in the selfmade datasets and related research datasets, which were expanded to 2500 by data augmentation. We used the open source tool labelImg to manually annotate the datasets and saved them in PASCAL VOC format.

The dataset was divided into the training-validation set and test set in the ratio of 8:2. Then the training-validation set was divided into a training set and a validation set in the ratio of 7:3. The number of divided training set, validation set and test set are 1400, 600 and 500, respectively. There are 5 kinds of pests in the datasets, namely, elater, Halyomorpha halys (Hh), Prodenia litura-Adult (PIA), Prodenia litura-Larva (PIL) and thrips, with 500 images for each category.

4.2 Network Structure and Training Parameters

Table 1 shows the parameters of the network model before and after the improvement. Table 2 shows the hyperparameter settings of the proposed network. From Table 1, it can be seen that although the ECA is added to the network, the model adds few parameters and still keeps the characteristics of lightweight.

Table 1. Comparison of network model parameter

Algorithms	Layers	Parameters	GFLOPs
YOLOv5s	270	7.05 M	16
Proposed method	286	7.03 M	15.9

Table 2. Hyperparameters setting

Parameter Name	Parameter Num
epochs	300
batch size	32
image size	640 × 640
mosaic	1

4.3   Evaluation Indicator

In this study, we use Precision, Recall, mean Average Precision (mAP) and mean Detection Time (mDT) as evaluation indicators to evaluate the performance of the proposed network model.

The mAP is the average value of the average precision (AP), which is used as the evaluation indicator for accuracy. The AP can be calculated by the area under the Precision-Recall (P-R) curve.

The mDT is the average detection time of the model, which is used to evaluate whether the model meets the requirement of real-time detection.

4.4   Result and Analysis

The 500 images in the test set were fed into the trained network to examine the recognition effect. The recognition results of our proposed method on the test set are shown in Fig. 8. In Fig. 8(a–e) show the detection results of elater, Hh, PlA, PlL and thrips, respectively. From Fig. 8, it can be seen that the proposed method can identify the target accurately.

(a) elater   (b) Hh   (c) PlA

(d) PlL   (e) thrips

Fig. 8. Examples of experimental results

To verify the effect of the proposed method objectively, we compared the proposed method with mainstream target detection algorithms on the same experimental environment with the same datasets. The experimental results are shown in Table 3.

In terms of accuracy, the mAP of the proposed method is 95.9%, which is 15.27%, 6.5%, 1.82%, 18.81% and 0.97% better than SSD, Faster-RCNN, YOLOv3, YOLOv4 and YOLOv5, respectively.

However, in terms of detection time, the proposed method achieves 6.1ms mDT per image, which is 0.7ms more than the traditional YOLOv5.

In the test experiments, it was found that the proposed model needs to further improve the recognition of small targets.

Table 3. Experimental comparison of several common algorithms

Algorithms	image size	mAP(%)	elater	Hh	PlA	PlL	thrips	mDT(ms)
SSD	300 × 300	83.2	90.6	90.5	90.3	86.6	57.8	18.2
Faster R-CNN	600 × 600	90	96.7	97.8	98.6	93.3	63.6	20
YOLOv3	416 × 416	94.1	98.2	98.7	99	98.8	76	15.7
YOLOv4	416 × 416	80.7	69.6	92.2	97.7	85.3	58.7	8.7
YOLOv5s	640 × 640	94.9	96.7	99	99.5	97.8	81.7	**5.4**
Proposed metheod	640 × 640	**95.9**	97.8	99.2	97.9	95.4	89	6.1

4.5 Ablation Experiments

The proposed method has improved the network structure and loss function of the YOLOv5s model simultaneously. To evaluate the effectiveness, feasibility and optimization of the proposed method, we performed ablation experiments to verify the performance of different parts.

Table 4 shows the results of the ablation experiments on the test set. The experimental statistics were performed using the same test set in the same experimental environment. The results show that each module modification has produced positive optimization. ECA improves mAP to 95.6%. Although the addition of the PDIoU loss function increases the mDT by 0.2 ms, the mAP is improved to 95.7% with an improvement of 0.8%. Finally, though the combination of the two modules increases the detection time by 0.7 ms, the proposed model achieves an optimal recognition accuracy of 95.9%.

Table 4. Ablation data

Algorithms	ECA	PDIoU	mAP(%)	mDT(ms)
YOLOv5s			94.9	**5.4**
Proposed method(1)	✔		95.6	5.8
Proposed method(2)		✔	95.7	5.6
Proposed method	✔	✔	**95.9**	6.1

5 Conclusion

In this paper, we propose a pest detection algorithm based on improved YOLO, which combines the development of smart agriculture and the actual demand of pest detection. We take elater, Hh, PlA, PlL and thrips as the research objects, and add ECA module and PDIoU loss function to the neck end and prediction end of traditional YOLOv5, respectively. The experimental results show that our proposed model increases the detection time by 0.7 ms, but the mAP improves to 95.9%.

Acknowledgment. This work was supported by Key Realm R&D Program of Guangdong Province with Grant No. 2019B020219003, the Key Field Special Project of Guangdong Provincial Department of Education under Grant 2021ZDZX1029.

References

1. Wang, Q., Wu, B., Zhu,P., Li, P., Zuo, W., Hu, Q.: ECA-Net: efficient channel attention for deep convolutional neural networks. In: 2020 IEEE/CVF Conference on Computer Vision and Pattern Recognition (CVPR), pp. 11531–11539 (2020). https://doi.org/10.1109/CVPR42 600.2020.01155
2. Dowell, F.E., Throne, J.E., Wang, D., et al.: Identifying stored-grain insects using near-infrared spectroscopy. J. Econ. Entomol. **92**(1), 165–169 (1999)
3. Wang, H., Gao, X., Chen, T., Chen, Z.: Present situation and prospect of spectrum and image processing technology in crop disease detection. Agric. Mech. Res. **37**(10), 1–7+12 (2015)
4. Vick, K.W., Webb, J.C., Weaver, B.A., et al.: Sound detection of stored-product insects that feed inside kernels of grain. J. Econ. Entomol. **81**(5), 1489–1493 (1988)
5. Ding, W., Taylor, G.: Automatic moth detection from trap images for pest management. Comput. Electron. Agric. **123**, 17–28 (2016)
6. Liu, Z., Gao, J., Yang, G., et al.: Localization and classification of paddy field pests using a saliency map and deep convolutional neural network. Sci. Rep. **6**, 20410 (2016)
7. Qian, K., Li, C., Chen, M., et al.: ship target and key parts detection algorithm based on YOLOv5. Syst. Eng. Electron. 1–14 (2022)
8. Yang, G., Bao, Y., Liu, Z.: Location and recognition of tea garden pests based on image significance analysis and convolutional neural network. Trans. Chin. Soc. Agric. Eng. **33**(06), 156–162 (2017)
9. Hengxiam, L., Chenfeng, L., Meng, Z., Jia, S.: A detecting method for the rape pests based on deep convolutional neural network. J. Hum. Agric. Univ. (Natl. Sci. Edn.) **45**(05), 560–564 (2019)

10. Wang, L., Lan, Y., Liu, Z., et al.: Development and experiment of the portable real-time detection system for citrus pests. Trans. Chin. Soc. Agric. Eng. **37**(09), 282–288 (2021)
11. Girshick, R., Donahue, J., Darrell, T., et al.: Rich feature hierarchies for accurate object detection and semantic segmentation. IEEE Comput. Soc. 580–587 (2013)
12. Girshick, R.: Fast R-CNN. IEEE Comput. Soc. 1440–1448 (2015)
13. Ren, S., He, K., Girshick, R., et al.: Faster R-CNN: towards real-time object detection with region proposal networks. IEEE Trans. Pat. Anal. Mach. Intell. **39**(6), 1137–1149 (2017)
14. Liu, W., Anguelov, D., Erhan, D., et al.: SSD: Single Shot MultiBox Detector. Computer Vision – ECCV 2016, pp. 21–37 (2016)
15. Redmon, J., Divvala, S., Girshick, R., et al.: You only look once: unified, real-time object detection. In: 2016 IEEE Conference on Computer Vision and Pattern Recognition (CVPR), pp. 779–788 (2016)
16. Redmon, J.,Farhadi, A.: YOLO9000: better, faster, stronger. In: 2017 IEEE Conference on Computer Vision and Pattern Recognition (CVPR), pp. 6517–6525 (2017)
17. Redmon, J.,Farhadi, A.: YOLOv3: an incremental improvement. arXiv pre-print server arXiv: 1804.02767v1 (2018)
18. Lin, T.-Y., Dollar, P., Girshick, R., et al.: Feature pyramid networks for object detection. In: 2017 IEEE Conference on Computer Vision and Pattern Recognition (CVPR), 936–944 (2017)
19. Bochkovskiy, A., Wang, C.-Y., Hong, Y.: YOLOv4: optimal speed and accuracy of object detection. arXiv pre-print server arXiv: 2004.10934v1 (2020)
20. Wang, C.-Y., Mark Liao, H.-Y., Wu, Y.-H., et al.: CSPNet: a new backbone that can enhance learningcapability of CNN. In: 2020 IEEE/CVF Conference on Computer Vision and Pattern Recognition Workshops (CVPRW), pp. 1571–1580 (2020)
21. He, K., Zhang, X., Ren, S., et al.: Spatial pyramid pooling in deep convolutional networks for visual recognition. IEEE Trans. Pattern Anal. Mach. Intell. **37**(9), 1904–1916 (2015)
22. Liu, S., Qi, L., Qin, H., et al.: Path aggregation network for instance segmentation. In: 2018 IEEEConference on Computer Vision and Pattern Recognition (CVPR), pp. 8759–8768 (2018)
23. Zheng, Z., Wang, P., Liu, W., et al.: Distance-iou loss: faster and better learning for bounding box regression. In: The Thirty-Fourth AAAI Conference on Artificial Intelligence (AAAI-20), pp. 12993–13000 (2020)

Improving Interactive Differential Evolution for Cartoon Face Image Combination

Bo Tang[1,2], Fei Yu[2(✉)], Qingrong Ou[3], Bang Liang[1,2], and Jian Guan[2]

[1] School of Computer Science, Minnan Normal University, Zhangzhou 363000, China
[2] School of Physics and Information Engineering, Minnan Normal University,
Zhangzhou 363000, China
yufei@whu.edu.cn
[3] School of Electronic Information, Zhangzhou Institute of Technology, Zhangzhou
363000, China

Abstract. In the general cartoon face methods, these methods can generate realistic cartoon face images, but rarely consider the subjectivity of users. In this condition, algorithms lack direct user intervention or the user operation method is not simple enough, so they cannot meet the needs of different users. In this paper, a method of cartoon face generation based on interactive differential evolution is introduced to overcome the deficiencies of general cartoon face methods. In the new proposed oppositional-mutual learning interactive differential evolution algorithm, named OMIDE, the two strategies of dynamic oppositional learning and mutual learning are combined to reduce the number of generations of algorithm. In addition, the user's dynamic selection is introduced to solve the problem that users are not satisfied with some parts of the cartoon face image while others are satisfied in the later stages of the algorithm. The generation of the proposed method is extensively evaluated by comparisons between it and other methods. In addition, the method of cartoon face generation based on interactive differential evolution is confirmed by a set of experiments to be able to satisfy the needs of different users.

Keywords: Cartoon Face · Dynamic Oppositional Learning · Dynamic Selection · Mutual Learning · Interactive Differential Evolution

1 Introduction

Cartoon face images refer to the creation of face portraits in a cartoon style. It not only preserves the personalized facial features and expressions of a subject, but also exhibits a unique artistic style. Its implementation method has attracted the attention of many researchers. The research results are widely used in game production, online communication, virtual reality film, television production, internet privacy protection, and so on.

The task of cartoon face generation is to transform real-face photos into vivid and exquisite cartoons. At present, there are many methods to generate realistic cartoon faces.

© The Author(s), under exclusive license to Springer Nature Singapore Pte Ltd. 2024
K. Li and Y. Liu (Eds.): ISICA 2023, CCIS 2146, pp. 326–339, 2024.
https://doi.org/10.1007/978-981-97-4393-3_27

Liu et al. [1] proposed a new cartoon synthesis method based on the active appearance model. To improve cartoon synthesis performance, the method focuses on high-level semantic features. Zhuang et al. [2] introduced a method for fine-grained face-to-cartoon generation. It includes a multi-branch translation model architecture and a two-stage training process. Zhang et al. [3] used generative adversarial network combined with an attention mechanism to cartoonish human faces and achieved a good balance in the conversion task of style and content. These methods can preserve the characteristics of the original face photos to the greatest extent, but the degree of user intervention in image generation is limited. The cartoon face image generation method based on the component library can overcome this shortcoming by adding more user intervention. Liu et al. [4] presented NatureFace to build a face feature model based on statistical data and combined it with the component library to generate face cartoon images. Nejad et al. [5] adopted edge detection, template matching, and Hermit interpolation to establish the contour of local components, integrate shadow, illumination, and other information, and adjust the final synthesis through the characteristic parameters of the sample. Although methods of components-based cartoon face generation have more user participation, the operation is not simple enough, and users need some relevant knowledge. However, cartoon face generation is subjective, different users have different requirements, and most users do not have relevant knowledge.

To address all the above challenges, this paper proposes a method of participating directly and simplifying operations for users by oppositional-mutual learning interactive differential evolution (OMIDE), which combines OMIDE and cartoon face library to generate cartoon face images. Oppositional-mutual learning (OML) is a hybrid of dynamic opposite learning and mutual learning. Besides, in the later generation, users often are not satisfied with some parts of the individual, but the rest are satisfied, so the user's dynamic selection is introduced. The user's dynamic selection means that the user can manually adjust each component of the cartoon images. In this paper, the interactive differential evolution algorithm (IDE) is applied to cartoon face generation to establish a human-computer cooperation mode. For the entire system, the algorithm is responsible for synthesizing the overall images, and the user is responsible for online evaluation. The evaluation information is fed back to the algorithm to generate face components for evaluation again. After several iterations, images satisfying the user's preferences are finally obtained.

2 Related Work

2.1 Interactive Differential Evolution

Takagi et al. [6] proposed an interactive differential evolution algorithm based on pairing comparison, which is a global optimization technology that introduces human subjective preference into the algorithm [7]. At present, IDE has been applied in many aspects, such as autonomous sensory meridian response sound [8], visual illusion [9], color separation [10] and so on.

IDE is an improvement based on differential evolution algorithm (DE). DE is one of the most powerful stochastic real parameter optimization algorithms. Differential evolution algorithm is mainly composed of four steps: initializing population, crossover, mutation and selection. Figure 1 shows the typical flow of the DE algorithm. The difference between IDE and DE is that user subjective evaluation replaces individual fitness comparison. Details of DE are as follows.

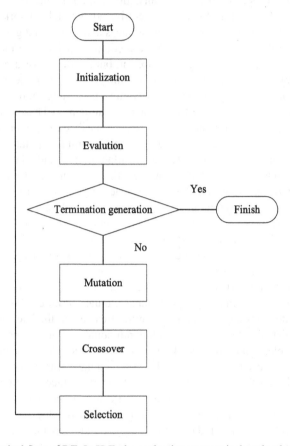

Fig. 1. A typical flow of DE. In IDE, the evaluation process is done by the human user.

Initialization: NP evenly distributed individuals are randomly generated, and each individual is composed of n-dimensional vectors. Its expression can use the following formula:

$$X_i(0) = \left[x_{i,1}(0), x_{i,2}(0), x_{i,3}(0), ..., x_{i,n}(0)\right] \; i = 1, 2, 3, ..., NP \qquad (1)$$

where n is the number of variables, NP is the population size, and 0 is the number of initial evolutionary populations.

The j-dimensional value of the ith individual is taken as follows:

$$X_{i,j}(0) = L_{min} + \text{rand}(0, 1)(L_{max} - L_{min}) \; i = 1, 2, 3, ..., NP, \; j = 1, 2, 3, ..., n \qquad (2)$$

where rand $(0, 1)$ is a random number among $(0, 1)$. L_{max} and L_{min} are the upper and lower bounds of the parametric variable, respectively.

Mutation: in the tth generation, three individuals X_{r1}, X_{r2}, X_{r3} are randomly selected from the population, and the three vectors are different from each other, and the generated mutation vector is:

$$H_i(t) = X_{r1}(t) + F(X_{r2}(t) - X_{r3}(t)) \tag{3}$$

where the parameter F is the scaling factor that controls the difference vector, t is the number of population generation.

Crossover: cross the individual vector with the mutation vector to generate the trial vector:

$$U_{i,j}(t) = \begin{cases} H_{i,j}(t), & \text{if } \text{rand}(0, 1) < CR \text{ or } j = j_{\text{rand}} \\ X_{i,j}(t), & \text{else} \end{cases} \tag{4}$$

where the parameter CR is the crossover probability, j_{rand} is a random integer between 1 and n.

Selection: the trial vector $U_i(t)$ generated in the previous step is compared with the target vector $X_i(t)$. If the fitness value of the trial vector is better than that of the target vector, the target vector is replaced to generate the next generation population. The formula is as follows:

$$X_i(t+1) = \begin{cases} U_i(t), & \text{if } f(U_i(t)) < f(X_i(t)) \\ X_i(t), & \text{else} \end{cases} \tag{5}$$

2.2 Triple Comparison-Based Interactive Differential Evolution

Pei et al. [11] proposed a triple comparison-based interactive differential evolution algorithm (TIDE), which generates the opposition vector of the target vector or the trial vector by an opposition-based learning strategy and finally retains the best individual among the three. Specifically, it is different from pairing comparison, in which users evaluate three candidate schemes and choose the one that best suits their preference.

2.3 Dynamic Opposite Learning

Typically, in evolutionary algorithms, the search starts at a random point in the search space so that after a few iterations it may converge to the optimal solution. Opposition based learning (OBL) is likely to produce individuals closer to the optimal solution than random sampling population, which has been shown to improve the performance of DE algorithm [12–14].

Let there be a solution $x_i = (x_1, x_2, \ldots, x_n)$ such that, where a and b are the lower and upper bound of the solution x_i. An opposite solution in a one-dimensional space can be defined as follows:

$$x_o = a + b - x \tag{6}$$

Similarly, an opposite solution in an n-dimensional space can be defined as follows:

$$x_{o,i} = a + b - x_i \tag{7}$$

Xu et al. [15] presented a dynamic opposite learning (DOL) strategy as a variant of OBL, which has an asymmetric search space between the current point and a dynamic opposite point for improving the global optimization ability. A dynamic opposite solution is defined as in an n-dimensional space and can be defined as follows:

$$x_{d,i} = x_i + m_1\left(m_2 x_{o,i} - x_i\right) \tag{8}$$

where m_1 and m_2 are random numbers among (0, 1) that obey uniform distribution.

2.4 Mutual Learning

Mutual learning (ML) is a strategy for individuals to learn from each other [16]. Individuals with good fitness may get some better information by learning from individuals with poor fitness. Therefore, a mutual learning solution is calculated as follow:

$$x_{ml,i} = x_i + m_3(x_m - x_i) \tag{9}$$

where m_3 is a random number among $(-1, 1)$ that obeys uniform distribution, and x_m is the individual randomly selected from current population. Besides, m_3 is different than [17].

2.5 User Dynamic Selection

In the application of interactive evolutionary algorithm, especially in the later generations of the algorithm, users are not satisfied with some parts of the individual, but with the rest. A user dynamic selection technique is developed in light of this phenomenon. User dynamic selection means that users can manually adjust each feature, and change the unsatisfied features according to their inner needs and preferences to obtain satisfied individuals. Figure 2 is an example of user dynamic selection. The result shown in the blue box is the user operation. The corresponding result is that the hair color changes to red.

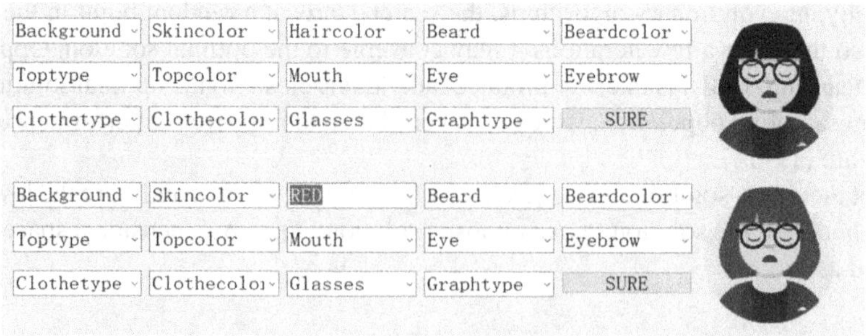

Fig. 2. An example of user dynamic selection. The hair color is adjusted manually by the user.

3 Proposed Method

3.1 Oppositional-Mutual Learning

We present oppositional-mutual learning interactive differential evolution algorithm (OMIDE) for cartoon face image combination. It uses DOL and ML to realize the balance of algorithm exploration and exploitation. DOL has rich exploration ability because of its asymmetric and dynamic search space, but its exploitation ability needs to use weight parameters to adjust the dynamic search space. Using ML instead of weight parameters to enhance exploitation in oppositional-mutual learning [17].

To improve IDE, OML is employed for population initialization and population evolution. In the population initialization, the population P_0 is generated randomly in the search space. In addition, a new initial population OP_0 is generated by (10), and n better individuals are selected from $\{P_0 \cup OP_0\}$. In the initialization, where m_3 is a random number among (-1, 1) that obeys uniform distribution individual, and X_m is select randomly in the population P_0.

$$Hi(t) = \begin{cases} X_i(t) + m_1(m_2(L_{\max} - L_{\min}) - X_i(t)), & \text{if} \quad \text{rand}(0, 1) < J_p \\ X_i(t) + m_3(X_m(t) - X_i(t)), & \text{else} \end{cases} \quad (10)$$

where $L_{\max}$ is the upper bound of each dimension, and it is updated according to the maximum of the current population; $L_{\min}$ is the lower bound of each dimension, and it is updated according to the minimum of the current population; t is the number of population generations.

In the population evolution, applying OML strategy can generate a new solution, which may provide a better solution than the current one. A new population OP_t can be generated by (10), if the selection probability is smaller than the jumping rate J_r. NP better individuals are selected from $\{P_t \cup OP_t\}$. Where m_3 is a random number among (-1, 1) that obeys uniform distribution, and J_p is the probability of controlling the DOL or ML. X_m $(m \neq i)$ is an individual randomly selected from the current population.

3.2 Coding Information

Cartoon face image generation problem belongs to combinatorial optimization prob-lem. In this paper, IDE is used to adjust the eigenvalues of each part of the cartoon face, and each component is coded as a real value. The range of eigenvalues for each component is shown in Table 1. Besides, Fig. 3 shows an example of a vector representation of an individual.

IDE individuals are continuous non-integers. In order to adapt to the cartoon face combination problem, at the initial stage, all individuals are scaled and rounded in a certain proportion. In the process of evolution, after crossover operation, all dimensions of each individual should be scaled in a certain proportion and rounded down.

Table 1. Face components.

Component	Value range
Background	[0,13]
Skin color	[0,7]
Hair color	[0,9]
Beard	[0,5]
Beard color	[0,9]
Top type	[0,31]
Top color	[0,14]
Mouth	[0,11]
Eye	[0,11]
Eyebrow	[0,12]
Clothe type	[0,6]
Clothe color	[0,8]
Glasses	[0,6]
Graph type	[0,10]

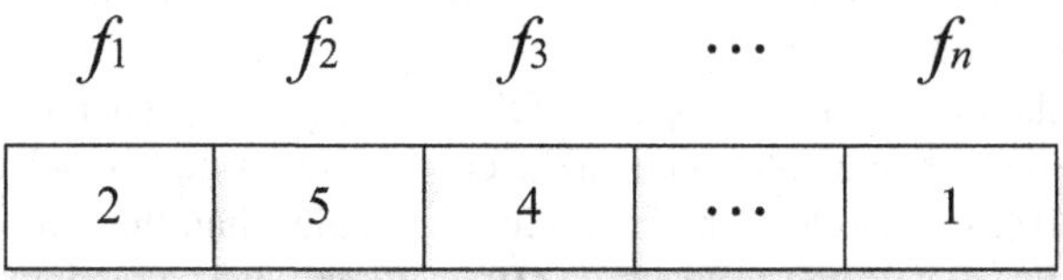

f_1	f_2	f_3	$\cdots$	f_n
2	5	4	$\cdots$	1

Fig. 3. Chromosome example.

3.3 Cartoon Face Image Combination Based on IDE

In the proposed OMIDE, the interaction between the user and DE system is to evalu-ate cartoon face images through interactive interface. Cartoon face images correspond to IDE individuals. Figure 4 shows the pattern of the OMIDE method, which consists of three main elements: interactive interface, user, and DE system. Through the interactive interface, the user evaluates the combination of images from the cartoon face library and selects the most satisfactory one. These images correspond to vectors in IDE, that is, the target vector and the trial vector or the opposite vector. After the user makes a choice, the system retains the best satisfactory solution in each group as the next generation population, and presents the cartoon face image to be evaluated next time through the interactive interface. Repeating these processes will find the best cartoon face image suitable for users' preferences. For oppositional-mutual learning interactive differential evolution with user dynamic selection (OMUIDE), it differs from OMIDE in that the user can manually adjust the dissatisfied part of a cartoon face image in the later generation. Figure 2 is an example. Its details can see Sect. 2.5.

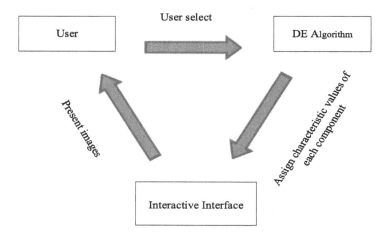

Fig. 4. Pattern of the proposed method.

The user evaluates and selects using the mouse click function. The interactive interface for TIDE is shown in Fig. 5. Three cartoon face images form a group, corresponding to the target vector, trial vector, and opposition vector in TIDE, as shown in the blue box. When the user selects a certain image through the mouse, the corresponding number will be displayed in the text. The default value of the number is 0, that is, the user has not evaluated it. "1" indicates that the user selects the first cartoon face image in a group, "2" indicates that the user selects each group of the second cartoon face image in a group. "3" indicates that the user selects the third cartoon face image in a group. Other methods use the interactive interface in Fig. 6. Two cartoon face images are a group, corresponding to the target vector and trial vector or opposition vector. "1" indicates that the user selects the first cartoon face image in a group. "2" indicates that the user selects the second cartoon face image in a group. In particular, the interface of OMUIDE provides the function for users to dynamically adjust any component of cartoon face images, and the currently selected image is displayed in the upper right corner.

4   Experiment

In order to verify the feasibility of proposed methods, experiments were carried out in the simulated environment and the real human-computer interaction environment. Normalize all characteristic values to the range (0, 10). All parameters are set in advance. The number of individuals in each iteration is $NP = 10$, the jumping rate $J_r = 0.3$, the scaling factor $F = 0.6$, the crossover probability $CR = 0.5$, and $J_p = 0.3$. The settings of F and CR refer to previous studies on DE [18, 19].

Fig. 5. Interactive interface for TIDE.

Fig. 6. Interactive interface for others.

4.1 Interaction Experiment

In order to study the effectiveness of proposed methods, the interaction experiment consists of two experiments. Experiment 1 shows the specific effect of cartoon face images and user evaluation of different methods. Experiment 2 compares the user satisfaction of different methods.

In Experiment 1, users select each group of cartoon face images through the interactive interface. Six college students were invited to experiment alone and told the specific operation. Each user repeated the operation 10 times to find the satisfactory solution of each user. Examples of Cartoon face images satisfying different users are shown in

Fig. 7. Different users expect to generate different face images, so their cartoon face images are also different in shape.

The average iterations of satisfactory solutions obtained by each user using different methods are counted respectively. The final results are shown in Table 2. The experimental results show that by using several algorithms, different users can find a satisfactory solution when the number of generations is less than 30, and different users get different solutions. There were no significant differences between the IDE algorithm and the OIDE algorithm. The TIDE algorithm slightly reduces the number of iterations, while the OMIDE algorithm and OMUIDE algorithm significantly reduce the number of iterations.

(a) User 1   (b) User 2   (c) User 3

(d) User 4   (e) User 5   (f) User 6

Fig. 7. Cartoon face images of different users satisfied.

Experiment 2 is a further experiment based on Experiment 1. Several cartoon face images in Experiment 1 are selected as representatives and evaluated by different users. Because there is no specific fitness function in the IDE and other algorithms, the fitness value of every cartoon face image cannot be obtained. In order to directly understand the effect of the proposed methods, let each user score two representative images as their fitness value [20]. The score range is 1 to 10. The higher the score, the higher the user satisfaction. These two representative images are from generation 4 and generation 14 respectively. These images correspond to the one that the user is satisfied with in the evaluation process.

Figure 8 shows the progress of the average value of the user's subjective score. The image fitness showed an obvious upward trend, and there are significant differences in fitness between the 4th and 14th generation images. TIDE scored the highest scores in

Table 2. Comparison of satisfactory solutions obtained by different methods.

	IDE	OIDE	TIDE	OMIDE	OMUIDE
User 1	17.8	17.4	16.8	15.7	13.4
User 2	18.7	18.6	18.1	16.4	14.1
User 3	15.6	14.7	14.2	14.6	12.4
User 4	18.4	18.2	18.3	16.1	13.5
User 5	22.5	21.1	18.7	18.9	14.9
User 6	19.5	19.4	18.8	16.6	13.7
Average iterations	18.8	18.2	17.5	16.4	13.7

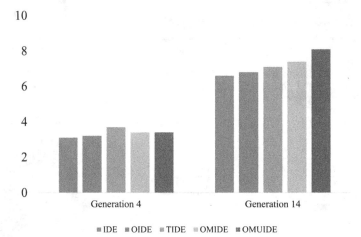

■ IDE ■ OIDE ■ TIDE ■ OMIDE ■ OMUIDE

Fig. 8. Progress of scoring by different methods.

the 4th generation, while OMIDE and OMUIDE scored higher than other methods in the 14th generation. Because OMIDE and OMUIDE use the same strategy in the early generations, their scores were similar in the 4th generation.

It can be seen that OMIDE and OMUIDE have fewer iterations and higher scores, which is because the OML strategy and user dynamic selection work. When OML is applied, the solution space is sampled again, which may produce a solution closer to the user's expectation and remain in the next generation. When user dynamic selection is applied, users can manually change dissatisfied parts to obtain satisfactory individuals. Therefore, it can better satisfy users' preferences than other methods.

4.2 Analysis of Parameter J_p and Dimension N Sensitivity

To further discuss the influence of OML on the performance of the algorithm, three groups of J_p were selected for comparative experiments. DOL is helpful for the diversity of the population and raises exploration ability. On the other hand, ML helps to accelerate convergence and improve exploitation ability. J_p is the probability of controlling the selection of DOL or ML, so it affects the balance between exploration and exploitation and further affects the number of user evaluations. Table 3 shows the average results of iterations under different values of J_p. The results show that J_p is 0.3, the number of iterations is the least, and J_p is 0.7, the number of iterations is the most. The number of iterations is positively correlated with J_p. Therefore, the value of J_p is 0.3 in this paper. Besides, J_p may have different values for different problems.

Similarly, different feature dimensions affect user evaluation, and three groups of n are selected for controlled experiments. Table 4 shows the average results of iterations under different values of n. The deleted components take the same value. The experimental results show that the larger the dimension n is, the more iterations users get satisfactory solutions. Besides, OMUIDE has a great improvement when n takes different values. For other methods, the smaller the dimension n, the smaller the difference between algorithms.

Table 3. The number of iterations of obtaining satisfactory solutions on different values of J_p.

Value of J_p	User 1	User 2	User 3	User 4	User 5	User 6	Average
0.3	15.7	16.4	14.6	16.1	18.9	16.6	16.4
0.5	15.1	17.4	14.9	17.8	20.2	16.0	16.9
0.7	16.9	18.1	15.3	16.6	19.6	20.1	17.8

Table 4. The number of iterations of different methods on different values of n.

Value of n	IDE	OIDE	TIDE	OMIDE	OMUIDE
14	18.8	18.2	17.5	16.4	13.7
10	17.2	16.9	16.2	15.4	12.5
7	15.1	14.6	14.3	13.9	11.5

5 Conclusion

This paper presents oppositional-mutual learning interactive differential evolution algorithm with user dynamic selection to reduce the generation of the algorithm and relieve user fatigue. Aiming at the problem that the general cartoon face methods lack sufficient user operation or that the operation is not simple enough, the proposed method is applied to cartoon face generation. The experimental results show that this method is effective and can get the user satisfactory solution. Compared with the other methods, it has advantages in reducing the number of iterations and meeting the personalized needs of users. It's important to note that the user's dynamic selection has some limits. It's only useful if each portion of the individual is independent and users understand the part that has to be modified. Therefore, in the future, we will further study a strategy with wider applicability to alleviate user fatigue.

References

1. Liu, S., Li, H., Xu, L.: Face cartoon synthesis based on the active appearance model. In: 2012 IEEE 12th International Conference on Computer and Information Technology, pp. 793–797. IEEE, Chengdu, China (2012)
2. Zhuang, N., Yang, C.: Few-shot knowledge transfer for fine-grained cartoon face generation. In: 2021 IEEE International Conference on Multimedia and Expo (ICME), pp. 1–6. IEEE, Shenzhen, China (2021)
3. Zhang, T., Yu, L., Tian, S.: CAMGAN: combining attention mechanism generative adversarial networks for cartoon face style transfer. J. Intell. Fuzzy Syst. 42(3), 1803–2181 (2022)
4. Liu, Y., Su, Y., Shao, Y., Wu, Z., Yang Y.: A face cartoon producer for digital content service. In: Jiang, X., Ma, M.Y., Chen, C.W. (eds.) Mobile Multimedia Processing. WMMP 2008. LNCS, vol. 5960, pp. 188–202. Springer, Berlin, Heidelberg (2010) https://doi.org/10.1007/978-3-642-12349-8_11
5. Nejad, S.S., Balafar, M.A.: Component-based cartoon face generation. Electronics 5(4), 76 (2016)
6. Takagi, H.: Interactive evolutionary computation: fusion of the capabilities of EC optimization and human evaluation. Proc. IEEE 89(9), 1275–1296 (2001)
7. Takagi, H., Pallez, D.: Paired comparison-based interactive differential evolution. In: Proceedings of World Congress on Nature and Biologically Inspired Computing, pp. 375–380. IEEE, Coimbatore, India (2009)
8. Fukumoto, M.: The efficiency of interactive differential evolution on creation of ASMR sounds. In: Tan, Y., Shi, Y. (eds.) Advances in Swarm Intelligence. ICSI 2021. LNCS, vol. 12689, pp. 368–375 Springer, Cham (2021). https://doi.org/10.1007/978-3-030-78743-1_33
9. Mohamad, Z.S., Darvish, A., Rahnamayan, S.: Eyeillusion enhancement using interactive differential evolution. In: 2011 IEEE Symposium on Differential Evolution (SDE), pp. 1–7. IEEE. Paris, France (2011)
10. Mushtaq, H., Rahnamayan, S., Siddiqi, A.: Color separation in forensic image processing using interactive differential evolution. J. Forensic Sci. 60(1), 212–218 (2015)
11. Pei, Y., Takagi, H.: Triple and quadruple comparison-based interactive differential evolution and differential evolution. In: Proceedings of the Twelfth Workshop on Foundations of Genetic Algorithms, pp. 173–182. ACM, Adelaide, SA (2013)
12. Rahnamayan, S., Tizhoosh, H.R., Salama, M.M.A.: Opposition-based differential evolution. IEEE Trans. Evol. Comput. 12(1), 64–79 (2008)
13. Ahandani, M.A., Hosein, A.R.: Opposition-based learning in the shuffled differential evolution algorithm. Soft. Comput. 16(8), 1303–1337 (2012)
14. Choi, T.J., Togelius, J., Cheong, Y.G.: A fast and efficient stochastic opposition-based learning for differential evolution in numerical optimization. Swarm Evol. Comput. 60 (2021)
15. Xu, Y., Yang, Z., Li, X., Kang, H., Yang, X.: Dynamic opposite learning enhanced teaching-learning-based optimization. Knowl. Based Syst. 188, 104966 (2020)
16. Rao, R.V., Savsani, V.J., Vakharia, D.P.: Teaching-learning-based optimization: a novel method for constrained mechanical design optimization problems. Comput. Aided Des. 43(3), 303–315 (2011)
17. Xu, Y., et al.: An enhanced differential evolution algorithm with a new oppositional-mutual learning strategy. Neurocomputing 435, 162–175 (2021)

18. Gamperle, R., Muller, S., Koumoutsakos, P.: A parameter study for differential evolution. In: Proceedings of WSEAS International Conference on Advances in Intelligent Systems, Fuzzy Systems, Evolutionary Computation, pp. 293–29 (2002)
19. Yang, Z., Tang, K.: An over-view of parameter control and adaptation strategies in differential evolution algorithm. CAAI Trans. Intell. Syst. **6**(5), 415–423 (2011)
20. Fukumoto, M., Inoue, M., Imai, J.I.: User's favorite scent design using paired comparison-based Interactive Differential Evolution. IEEE Congress on Evolutionary Computation, pp. 1–6. IEEE, Barcelona, Spain (2010)

Mask Reconstruction Augmentation and Attention Aggregation for Stereo Matching

Zhaokui Li[1][(✉)], Zhongxin Yang[1], Jinen Zhang[1], and Jinrong He[2]

[1] School of Computer Science, Shenyang Aerospace University, Shenyang, China
lzk@sau.edu.cn, {yangzhongxin,zhangjinen}@stu.sau.edu.cn
[2] School of Mathematics and Computer Science, Yan'an University, Yan'an, China
hejinrong@yau.edu

Abstract. Stereo matching is a significant task in computer vision and is widely used for depth prediction. Although deep learning stereo matching networks have achieved impressive results in recent years, the current models are prone to overfitting on synthetic datasets, resulting in poor generalization performance, and the ability of the current cost aggregation modules to perceive global context information is limited. In this paper, we propose a stereo matching framework with mask reconstruction augmentation and attention aggregation (MRA-AA) to address the above issues. Firstly, inspired by the work of applying masked image reconstruction to self-supervised learning, we design a novel data augmentation module suitable for stereo matching to improve the generalization performance of the model, namely mask reconstruction augmentation. Secondly, an attention aggregation module is proposed. The aggregation module can explicitly model the interdependencies between the channels of cost volume and capture the long-range dependence. Therefore, the module can help the model obtain more accurate disparity estimation in thin structures and texture-less regions. Furthermore, the convex upsampling module is used to upsample the cost volume to alleviate the blurred edges issue. Experimental results on Scene Flow and KITTI benchmarks show that our proposed method outperforms some existing methods.

Keywords: Deep Learning · Stereo Matching · Mask Reconstruction Augmentation · Attention Aggregation · Convex Upsampling

1 Introduction

Stereo matching is a hot topic in recent years and is widely applied in tasks such as autonomous driving, 3D scene reconstruction, and robot vision. The goal of stereo matching is to find the corresponding matching points in the given left and right images. The relative distance of the corresponding matching points is called disparity. Using disparity d, we can obtain the depth z according to

© The Author(s), under exclusive license to Springer Nature Singapore Pte Ltd. 2024
K. Li and Y. Liu (Eds.): ISICA 2023, CCIS 2146, pp. 340–353, 2024.
https://doi.org/10.1007/978-981-97-4393-3_28

the formula $z = \frac{fB}{d}$, where f denotes the focal length of the camera and B is the baseline length of the camera. In general, stereo matching networks can be divided into four steps: matching cost computation, cost aggregation, disparity regression, and disparity refinement.

In recent years, a large number of end-to-end stereo matching networks have emerged as a result of the rapid development of convolutional neural networks (CNN), greatly enhancing prediction accuracy. Kendall et al. [12] proposed a geometry and context network (GC-Net), which utilizes 3D convolutional layers to process the cost volume obtained by connecting left and right feature maps. Additionally, they uses a differentiable soft argmin operation to generate the disparity map along the disparity dimension of the cost volume. Chang et al. [2] proposed a pyramid stereo matching network (PSMNet), which introduces the spatial pyramid pooling module to fuse multi-scale information. In addition, a stacked 3D hourglass aggregation network is presented to learn more context information. A group-wise correlation stereo network (GwcNet), proposed by Guo et al. [8], which uses group-wise correlation volume and improves 3D stacked hourglass networks to expand PSMNet [2], achieves good performance. Xu et al. [21] designed a novel cost volume upsampling method and proposed a framework that takes into account high efficiency and high accuracy. In view of the high computational cost of 3D CNN, Gan et al. [5] proposed a depth shift module to model the cost volume along the depth dimension, achieving a balance between computational cost and accuracy.

Despite the remarkable results obtained by the above networks on Scene Flow [15], the generalization performance is still weak. The end-to-end stereo matching models have strong learning ability, are prone to overfitting on synthetic datasets, and cannot fully learn common feature information, such as structural information, between synthetic and real-world datasets, which makes the model perform poorly on real-world datasets. In addition, the previous cost aggregation modules simply stacked 3D CNN to aggregate context information, but the effective receptive field was small and did not fully capture long-range dependence, which is not conducive to disparity estimation in thin structures and texture-less regions.

To solve the above problems, we propose a stereo matching framework with mask reconstruction augmentation and attention aggregation (MRA-AA). Firstly, masked autoencoder (MAE) [9] applied masked images to self-supervised learning and achieved visual representation learning by reconstructing masked pixels. Inspired by MAE, we propose a simple and effective mask reconstruction augmentation module. The module not only regulates the training process, but also helps the feature extractor learn more structural information. Secondly, we propose an attention aggregation module, which introduces Squeeze-and-Excitation (SE) [11] block and Large Kernel Attention (LKA) [7] in the cost aggregation module. SE block can obtain the importance of each channel of the cost volume through learning so as to realize the recalibration of channel-wise features. LKA can help the model perceive global context information. Thirdly, inspired by the convex upsampling module in Recurrent All-Pairs Field Transforms (RAFT) [19], we apply the module to upsample the cost volume. Compared with the previous module of upsampling cost volume, which directly uses bilinear interpolation, the convex upsampling module refers to context information when upsampling the spatial dimension of the cost volume. In this way, the module can alleviate the blurred edges issue. Our main contributions are as follows:

- We propose a mask reconstruction augmentation module for stereo matching, which can improve the generalization performance of the model.
- We propose an attention aggregation module to improve the prediction accuracy of the model in thin structures and large texture-less regions.
- We use convex upsampling module to upsample the cost volume, which can alleviate the blurred edges issue.

2 Related Work

In this section, we briefly describe the relevant work of stereo matching in data augmentation and cost aggregation, including our improvements in these two directions.

2.1 Data Augmentation

In deep learning models, data augmentation is often used to improve generalization performance, which is more common in object detection, semantic segmentation, and image classification. Most of the previous work was to adjust the network structure to make it perform well. However, the impact of data augmentation on the generalization performance of the network is ignored.

Lipson et al. [14] designed a stereo matching network based on the optical flow network RAFT (RAFT-Stereo). Zhao et al. [29] proposed an error aware iterative network (EAI-Stereo). They use the following data augmentation methods: adjusting image saturation, perturbing the image, and randomly stretching the image. Rao et al. [17] summarized data augmentation methods suitable for stereo matching: random cropping, chromatic transformation, random translation transformation, and dataset balancing. The data augmentation methods proposed in their paper enable PSMNet [2] and GC-Net [12] to achieve better generalization performance. The above stereo matching data augmentation methods are easily mastered by CNN and their ability to increase sample diversity is limited. We propose a novel data augmentation module to improve generalization performance, namely mask reconstruction augmentation. See Sect. 3.1 for more details.

2.2 Cost Aggregation

Cost aggregation is a key step in stereo matching, and its purpose is to obtain accurate similarity measure to accurately reflect the correlation between pixels. Much work on stereo matching focuses on cost aggregation. Kendall et al. [12]

tried to use 3D CNN to aggregate the context information of the initial cost volume. The cost aggregation module of PSMNet [2] and GwcNet [8] adopted a stacked 3D hourglass aggregation network. However, the above methods simply stacking 3D CNN cannot significantly improve the ability to aggregate information. CNN can learn local context information better, but it cannot efficiently perceive global context information, which is crucial for stereo matching.

In some work, attention mechanisms are introduced into the cost aggregation module to obtain better aggregation ability. Zhang et al. [27] proposed a 3D channel attention module in the cost aggregation module to recalibrate the cost volume for more correct matching cost. Zhang et al. [28] introduced a 3D attention recording module in the cost aggregation module to obtain robust cost volume. The above methods only focus on the interdependencies between channels. To explicitly model the interdependencies between the channels of cost volume and capture the long-range dependence, we propose an attention aggregation module. Benefiting from the aggregation module, our model can obtain more accurate disparity estimation in thin structures and texture-less regions. See Sect. 3.2 for more details.

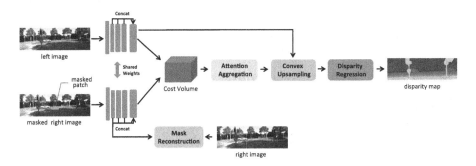

Fig. 1. Overall framework of proposed MRA-AA.

3   Method

Our proposed modules are described in detail in this section, including the mask reconstruction augmentation module, attention aggregation module, and convex upsampling module. We take GwcNet [8] as the basic network, and the above proposed modules are embedded into the basic network to form a novel stereo matching framework. The overall framework is shown in Fig. 1. The left image and the masked right image are extracted by a feature extractor to obtain left and right feature maps. The left and right feature maps are used to construct the cost volume, and then the cost volume is successively passed through an attention aggregation module and a convex upsampling module to obtain a more refined cost volume. Finally, a disparity regression module is used to obtain the disparity map. In addition, the masked right image goes through a mask reconstruction augmentation module to learn structural information.

3.1   Mask Reconstruction Augmentation

Data augmentation is one of the most effective methods to alleviate the overfitting issue of the model, and there are few data augmentation methods suitable
for stereo matching tasks. We propose a simple and efficient data augmentation
module, namely mask reconstruction augmentation, to improve generalization
performance.

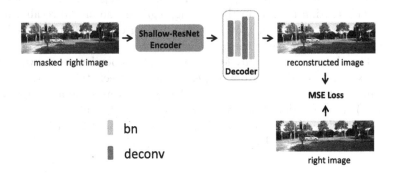

Fig. 2. The structure of mask reconstruction augmentation module.

The whole stereo matching network is divided into two branches, the first branch
is the basic stereo matching network, and the second branch is the mask reconstruction augmentation module. As shown in Fig. 2, the masked image is passed
through the encoder-decoder structure to obtain the reconstructed images.

For stereo matching, masking too many patches in the image brings too much
error information to the cost volume, which is not conducive to the following cost
aggregation. In addition, our model only masks the right image to ensure that the
following convex upsampling module is not disturbed. Combined with the above
points, we only mask a patch in the right image. During the mask reconstruction
augmentation process, the module uses the surrounding pixel information of the
masked patch to reconstruct the masked pixels for visual representation learning.
We use the mean of each image as the masked pixel value of the image. For the
size of the masked patch, we select different size patches for experiments, and
the specific details are described in the experiment section.

Shallow features such as structural information are common features between
synthetic images and real-world images, namely more meaningful generalization
features that are more helpful for generalization performance. The feature extractor is a ResNet [10] network. We choose the shallower network of ResNet as the
encoder structure to force the feature extractor to pay more attention to structural information, which helps to improve the feature extractor's ability to learn
generalization features. The decoder structure is lightweight by stacking several
layers of deconvolution and batch normalization layers.

It is worth noting that the image is masked only during the training phase.
We use the mean square error between the reconstructed image and the original

image as the objective function to train the mask reconstruction augmentation module. The module learns the structural information of the entire image, such as the edges and textures, to recover the masked area. In this way, it alleviates the overfitting of the model to large synthetic datasets and enables a more robust model to be trained.

3.2 Attention Aggregation

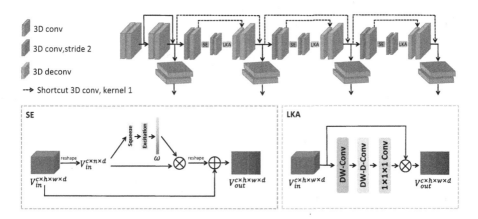

Fig. 3. The structure of attention aggregation module.

The cost volume consists of a group-wise correlation volume and a compact concatenation volume. Cost aggregation is a crucial step in stereo matching. Constructing an efficient cost aggregation module is good for obtaining more accurate matching costs for the cost volume. To obtain a more accurate disparity estimation in thin structures and texture-less regions, we propose an attention aggregation module, as is shown in Fig. 3. The module consists of a pre-hourglass module (the first four 3D convolutions) and three hourglass networks. In the training phase, the cost aggregation outputs four aggregated cost volumes, while in the testing phase, only the final aggregated cost volume is output. The output aggregated cost volume is used for subsequent convex upsampling and disparity regression. In order to explicitly model the interdependencies between the channels of cost volume and capture the long-range dependence, each hour-glass structure introduces SE [11] block and LKA [7] in the encoder and decoder, respectively.

Our SE block structure is shown in the lower left corner of Fig. 3. In the module, we define the input cost volume as $V_{in}^{c \times h \times w \times d}$, each disparity dimension contains a feature map of size $c \times h \times w$. Only focus on the channel dimension in the module, so reshape it as $V_{in}^{c \times n \times d}$, where $n = h \times w$. Specifically, the cost volume passes through Squeeze operation, namely global average pooling layer, to represent the global distribution of channel-wise feature responses. Followed by an Excitation operation with two fully connected layers and a Sigmoid activation function to obtain channel-wise weights. The weights ω are multiplied by V_{in} to create a new cost volume to improve the features that are more useful for the task at hand. After that, the cost volume is reshaped to the original size and then added to the input V_{in} to get the final cost volume V_{out}.

We adaptively introduce LKA [7] in the decoder, showing it in the lower right corner of Fig. 3. For convenience, in this module, we still use V_{in} to represent the input cost volume and V_{out} to represent the output cost volume. Capturing long-range dependence is crucial to help the model perceive global context information. Large kernel convolution can make it, but the huge number of parameters in large kernel convolution brings a computational burden to the model. To reduce computation, a large kernel convolution can be decomposed into three parts: a depth-wise convolution (DW-Conv), a depth-wise dilation convolution (DW-D-Conv), and a channel convolution ($1 \times 1 \times 1$ Conv). LKA consists of the above three parts, and the operations are defined as

$$Attention = F_{CC} \left(F_{DDC} \left(F_{DC} \left(V_{in} \right) \right) \right) \tag{1}$$

$$V_{out} = Attention \otimes V_{in} \tag{2}$$

where F_{DC} represents depth-wise convolution. F_{DDC} represents depth-wise dilation convolution. F_{CC} represents channel convolution. *Attention* denotes attention map, and $\otimes$ denotes the product of elements.

3.3 Convex Upsampling

To obtain a predicted disparity map of the same size as the ground truth, the cost volume often needs to be upsampled, which directly affects the quality of the disparity map. In previous stereo matching networks [2,8], the size of the cost volume was generally recovered by bilinear interpolation upsampling. However, the data-independent upsampling method is not an optimal solution and may lead to the blurred edges issue. Inspired by the module of upsampling flow field in RAFT [19], we apply the module to upsample the spatial dimension of the cost volume.

The size of the cost volume output by the attention aggregation module is $1 \times \frac{1}{4}H \times \frac{1}{4}W \times \frac{1}{4}D$. To recover the cost volume of size $1 \times H \times W \times D$, the spatial dimension of the cost volume is first upsampled using convex upsampling, and then the disparity dimension is upsampled using linear interpolation. Specifically, two convolution layers are used to obtain a weight matrix of size $\frac{1}{4}H \times \frac{1}{4}W \times (4 \times 4 \times 9)$ without Softmax operation. The weight matrix learned in the left feature map contains context information, which can be used to refer to the surrounding pixel information when upsampling. Using it, each matching cost value of the cost volume of size $1 \times H \times W \times \frac{1}{4}D$ to be the convex combination of its 9 neighbors at the coarse resolution, as shown in Fig. 4. After that, linear interpolation is used to upsample the disparity dimension to obtain the final cost volume of size $1 \times H \times W \times D$. The cost volume after upsampling is processed through the disparity regression module to obtain a disparity map. Specifically, the probability volume is first obtained through the Softmax function, and finally, the predicted disparity map is obtained along the disparity dimension using the soft argmin function [12].

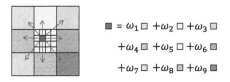

Fig. 4. Illustration of the convex upsampling module.

3.4   Loss Function

The final loss fuction L is the sum of the disparity loss and the masked right image reconstruction loss, The final loss is denoted by

$$L = \sum_{i=0}^{i=3} \lambda_i \cdot Smooth_{L1} \left(\widehat{d}_i - d_{gt} \right) + l_{recon} \tag{3}$$

where λ_i denotes the coefficient of the ith predicted disparity map, $\widehat{d}_i$ is the ith predicted disparity map, and d_{gt} represents the ground-truth disparity map. The smooth L1 loss is expressed as

$$Smooth_{L1}(x) = \begin{cases} 0.5x^2, & if \ |x| < 1 \\ |x| - 0.5, & otherwise \end{cases} \tag{4}$$

the reconstruction loss is denoted by

$$l_{recon} = \frac{1}{N} \sum_{i=1}^{N} (R_r(i) - R_t(i))^2 \tag{5}$$

where R_r represents the reconstructed right image, and R_t represents the true right image. N represents the number of pixels in the entire image.

4   Experiment

4.1   Datasets and Evaluation Metrics

Scene Flow. Scene Flow [15] is a synthetic dataset which contains 35,454 training image pairs and 4,370 testing image pairs in 960 × 540 pixel resolution. It is composed of three synthetic data subsets, namely FlyingThings3D, Driving, Monkaa, all of which provide dense ground-truth disparity maps.

KITTI. KITTI consists of KITTI 2015 [16] and KITTI 2012 [6], and it is a dataset collected on real-world driving scenes with sparse ground-truth disparity maps. KITTI 2015 has 200 training image pairs and 200 testing image pairs. KITTI 2012 has 194 training image pairs and 195 testing image pairs. We also validate generalization performance on KITTI 2015 and KITTI 2012.

Metrics. For Scene Flow, we use the end-point error (EPE) and the percentage of disparity outliers (D1) whose disparity errors are greater than $\max(3px, 0.05d_{gt})$ as the evaluation metrics, where d_{gt} represents the ground-truth disparity. For KITTI 2015, we show the percentage of disparity outliers (D1) in background regions, foreground regions, and all regions. For KITTI 2012, we report the percentage of incorrect pixels for non-occluded (Noc) and all (All) pixels.

4.2   Implementation Details

Our approach is implemented with Pytorch, and all the experiments are carried out with two NVIDIA RTX 3090 GPUs. The batch size is set to 10, and crop size is set to 512×256 to train our network. We use Adam [13] optimizer, with $\beta_1 = 0.9$, $\beta_2 = 0.999$. The coefficients of four predicted disparity maps are set as $\lambda_0 = 0.5$, $\lambda_1 = 0.5$, $\lambda_2 = 0.7$, $\lambda_3 = 1.0$. We set the maximum disparity value to 192.

For Scene Flow [15], we train the model for 25 epochs. The initial learning rate is set to 0.001 decayed by a factor of 2 after epoch 12, 14, 16, 18, 20, 22, and 24. For KITTI, we fine-tune the pre-trained model on Scene Flow for another 350 epochs on KITTI 2015 [16] and KITTI 2012 [6]. The learning rate is 0.001 and it is down-scaled by 10 after 230 epochs.

4.3   Ablation Study

In this section, we mainly validate the effectiveness of the proposed mask reconstruction augmentation (MRA), attention aggregation (AA) and convex upsampling (Cup) on Scene Flow testing datasets and KITTI 2012 testing datasets.

Table 1. Ablation study on Scene Flow [15] and KITTI 2012 [6].

Method	MRA	AA	Cup	Scene Flow		KITTI 2012 3px(%)	
				EPE(px)	D1(%)	Noc	All
GwcNet				0.76	2.71	1.32	1.70
GwcNet-MRA	✓			0.71	2.53	1.29	1.66
GwcNet-AA		✓		0.70	2.44	1.30	1.67
GwcNet-MRA-AA	✓	✓		0.68	2.41	1.28	1.64
GwcNet-MRA-AA-Cup	✓	✓	✓	0.65	2.27	1.27	1.62

We embed proposed modules sequentially into the baseline GwcNet [8]. For Scene Flow, metrics EPE and D1 are adopted. For KITTI 2012, we report the percentage of incorrect pixels whose disparity errors are greater than 3 pixels for non-occluded (Noc) and all (All) pixels. As shown in Table 1, our proposed mask reconstruction augmentation, attention aggregation and convex upsampling all improve the prediction accuracy of the model.

4.4   The Effect of Masked Patches with Different Sizes

We carry out the experiments with masked patches of different sizes, and the results of the experiment are shown in Table 2. We add mask reconstruction augmentation to baseline GwcNet and use Scene Flow testing datasets and KITTI 2012 validation dataset to validate the effect of different patch sizes on model performance. Metrics EPE (px) is adopted in Table 2. As a result of our experiments, we select the size of $0.1H \times 0.1W$.

Table 2. The effect of different sizes of masked patch on Scene Flow and KITTI 2012. **Bold**: Best.

patch size	$0.05H \times 0.05W$	$0.10H \times 0.10W$	$0.15H \times 0.15W$	$0.20H \times 0.20W$
Scene Flow	0.719	**0.717**	0.721	0.739
KITTI 2012	0.640	**0.639**	0.643	0.654

4.5   Comparative Experiments on Scene Flow and KITTI

Scene Flow We validate our model on Scene Flow testing set. We compare with the stereo matching networks in recent years, including GwcNet, guided aggregation network (GANet), learning effective architecture for stereo matching (LEAStereo), correlate-and-excite network (CoEx), and context and geometry interaction stereo matching network (CGI-Stereo). Final model benefits from the proposed modules and achieves the significant EPE of 0.65. As shown in Table 3, our model is highly competitive.

Table 3. Comparison with the state-of-the-art on Scene Flow. **Bold**: Best.

Method	GwcNet [8]	GANet [26]	LEAStereo [4]	CoEx [1]	CGI-Stereo [23]	MRA-AA
EPE(px)	0.98	0.84	0.78	0.68	**0.64**	0.65

KITTI. We validate the performance of our model on real-world datasets KITTI 2012 [6] and KITTI 2015 [16]. As shown in Table 4, we validate the cross-domain generalization performance of the model using the unseen KITTI training sets, which is more challenging because the model is only trained on

Scene Flow [15]. Threshold ($3px$) error rates (%) are adopted in Table 4. We use the training set of KITTI 2015 (200 training image pairs) and the training set of KITTI 2012 (194 training image pairs) to validate the generalization performance of network. We compare with the cross-domain generalization performance of the stereo matching networks in recent years, including hierarchical discrete distribution decomposition network (HD^3), PSMNet, GwcNet, and GANet. As shown in Table 4, Our model is better.

Table 4. Cross-domain generalization performance on KITTI. **Bold**: Best.

Method	HD^3 [25]	PSMNet [2]	GwcNet [8]	GANet [26]	MRA-AA
KITTI 2015 [16]	26.5	16.3	22.7	11.7	**9.4**
KITTI 2012 [6]	23.6	15.1	20.2	10.1	**8.8**

In addition, we also validate the results of the fine-tuning on KITTI datasets, as shown in Table 5. The pre-trained model obtained on Scene Flow [15] is used to fine-tune on KITTI. The results of KITTI 2012 and KITTI 2015 testing sets obtained by our model are uploaded to the evaluation server, and excellent results are achieved. We compare with the stereo matching networks in recent years, including PSMNet, GwcNet, adaptive aggregation network(AANet+), sparse cost volume for stereo matching (SCV-Stereo), multi-dimensional cooperative network (MDCNet), hierarchical iterative tile refinement network (HITNet), and fast attention concatenation volume for stereo matching (Fast-ACVNet+). As shown in Table 5, our model has a competitive advantage. Qualitative comparisons are shown in Fig. 5. The yellow boxes represent the differences in apparent regions. Our model performs better than other stereo matching models in the sky, thin structures, and texture-less regions. In addition, the target edges are clearer. This proves that all our proposed modules play a corresponding role.

Table 5. Comparison with the excellent models on KITTI benchmarks. **Bold**: Best.

Method	KITTI 2015 [16]						KITTI 2012 [6]			
	All pixels			Noc pixels			3px(%)		4px(%)	
	D1-bg	D1-fg	D1-all	D1-bg	D1-fg	D1-all	Noc	All	Noc	All
PSMNet [2]	1.86	4.62	2.32	1.71	4.31	2.14	1.49	1.89	1.12	1.42
GwcNet [8]	1.74	3.93	2.11	1.61	3.49	1.92	1.32	1.70	0.99	1.27
AANet+ [24]	1.65	3.96	2.03	1.49	3.66	1.85	1.55	2.04	1.20	1.58
SCV-Stereo [20]	1.67	3.78	2.02	1.52	3.47	1.84	1.27	1.68	0.97	1.29
MDCNet [3]	1.76	3.68	2.08	1.61	3.26	1.88	1.54	1.97	-	-
HITNet [18]	1.74	**3.20**	1.98	1.54	**2.72**	**1.74**	1.41	1.89	1.14	1.53
Fast-ACVNet+ [22]	1.70	3.53	2.01	-	-	-	1.45	1.85	1.06	1.36
MRA-AA	**1.61**	3.56	**1.93**	**1.47**	3.24	1.76	**1.27**	**1.62**	**0.93**	**1.20**

Fig. 5. Qualitative results on KITTI 2015 [16] and KITTI 2012 [6]. The first two columns are the results on KITTI 2015, and the last two columns are the results on KITTI 2012. The first row is the left images, the second row is the results of PSMNet, the third row is the results of GwcNet, the fourth row is the results of AANet+, the fifth row is the results of SCV-Stereo, the sixth row is the results of HITNet, and the seventh row is the results of our model MRA-AA.

5   Conclusion

In this paper, we propose a stereo matching framework with mask reconstruction augmentation and attention aggregation. While the mask reconstruction augmentation module regulates the training process, it also encourages the feature extractor to learn more structural information that is more helpful for generalization performance. The attention aggregation module not only models the interdependencies between the channels of cost volume, but also captures global context information. In this way, the module can improve the prediction accuracy in thin structures and texture-less regions. In addition, we employ convex upsampling module to recover the size of cost volume, which alleviates the blurred edges issue. In the experiment part, we conduct ablation experiments on Scene Flow and KITTI 2012 to validate the effectiveness of the proposed modules. On KITTI 2015 and KITTI 2012, we carry out cross-domain generalization validation and fine-tuning performance validation, respectively. Benefiting from our proposed modules, excellent results are obtained in the above experiments.

Acknowledgements. This work was supported in part by the National Natural Science Foundation of China(62171295 and 62366053), and the Applied Basic Research Project of Liaoning Province(2023JH2/101300204).

References

1. Bangunharcana, A., Cho, J.W., Lee, S., Kweon, I.S., Kim, K.S., Kim, S.: Correlate-and-excite: Real-time stereo matching via guided cost volume excitation. In: 2021 IEEE/RSJ International Conference on Intelligent Robots and Systems (IROS), pp. 3542–3548. IEEE (2021)
2. Chang, J.R., Chen, Y.S.: Pyramid stereo matching network. In: Proceedings of the IEEE Conference on Computer Vision and Pattern Recognition, pp. 5410–5418 (2018)
3. Chen, W., Jia, X., Wu, M., Liang, Z.: Multi-dimensional cooperative network for stereo matching. IEEE Robot. Autom. Lett. **7**(1), 581–587 (2021)
4. Cheng, X., et al.: Hierarchical neural architecture search for deep stereo matching. Adv. Neural. Inf. Process. Syst. **33**, 22158–22169 (2020)
5. Gan, W., Wu, W., Chen, S., Zhao, Y., Wong, P.K.: Rethinking 3D cost aggregation in stereo matching. Pattern Recogn. Lett. **167**, 75–81 (2023)
6. Geiger, A., Lenz, P., Urtasun, R.: Are we ready for autonomous driving? the kitti vision benchmark suite. In: 2012 IEEE Conference on Computer Vision and Pattern Recognition, pp. 3354–3361. IEEE (2012)
7. Guo, M.H., Lu, C.Z., Liu, Z.N., Cheng, M.M., Hu, S.M.: Visual attention network. Computational Visual Media, pp. 1–20 (2023)
8. Guo, X., Yang, K., Yang, W., Wang, X., Li, H.: Group-wise correlation stereo network. In: Proceedings of the IEEE/CVF Conference on Computer Vision and Pattern Recognition, pp. 3273–3282 (2019)
9. He, K., Chen, X., Xie, S., Li, Y., Dollár, P., Girshick, R.: Masked autoencoders are scalable vision learners. In: Proceedings of the IEEE/CVF Conference on Computer Vision and Pattern Recognition, pp. 16000–16009 (2022)
10. He, K., Zhang, X., Ren, S., Sun, J.: Deep residual learning for image recognition. In: Proceedings of the IEEE Conference on Computer Vision and Pattern Recognition, pp. 770–778 (2016)
11. Hu, J., Shen, L., Sun, G.: Squeeze-and-excitation networks. In: Proceedings of the IEEE Conference on Computer Vision and Pattern Recognition, pp. 7132–7141 (2018)
12. Kendall, A., et al.: End-to-end learning of geometry and context for deep stereo regression. In: Proceedings of the IEEE International Conference on Computer Vision, pp. 66–75 (2017)
13. Kingma, D.P., Ba, J.: Adam: a method for stochastic optimization. arXiv preprint arXiv:1412.6980 (2014)
14. Lipson, L., Teed, Z., Deng, J.: Raft-stereo: multilevel recurrent field transforms for stereo matching. In: 2021 International Conference on 3D Vision (3DV), pp. 218–227. IEEE (2021)
15. Mayer, N., et al.: A large dataset to train convolutional networks for disparity, optical flow, and scene flow estimation. In: Proceedings of the IEEE Conference on Computer Vision and Pattern Recognition, pp. 4040–4048 (2016)
16. Menze, M., Geiger, A.: Object scene flow for autonomous vehicles. In: Proceedings of the IEEE Conference on Computer Vision and Pattern Recognition, pp. 3061–3070 (2015)
17. Rao, Z., Dai, Y., Shen, Z., He, R.: Rethinking training strategy in stereo matching. IEEE Transactions on Neural Networks and Learning Systems (2022)
18. Tankovich, V., Hane, C., Zhang, Y., Kowdle, A., Fanello, S., Bouaziz, S.: Hit-net: hierarchical iterative tile refinement network for real-time stereo matching.

In: Proceedings of the IEEE/CVF Conference on Computer Vision and Pattern Recognition, pp. 14362–14372 (2021)

19. Teed, Z., Deng, J.: RAFT: recurrent all-pairs field transforms for optical flow. In: Vedaldi, A., Bischof, H., Brox, T., Frahm, J.-M. (eds.) Computer Vision – ECCV 2020: 16th European Conference, Glasgow, UK, August 23–28, 2020, Proceedings, Part II, pp. 402–419. Springer International Publishing, Cham (2020). https://doi.org/10.1007/978-3-030-58536-5_24

20. Wang, H., Fan, R., Liu, M.: Scv-stereo: learning stereo matching from a sparse cost volume. In: 2021 IEEE International Conference on Image Processing (ICIP), pp. 3203–3207. IEEE (2021)

21. Xu, B., Xu, Y., Yang, X., Jia, W., Guo, Y.: Bilateral grid learning for stereo matching networks. In: Proceedings of the IEEE/CVF Conference on Computer Vision and Pattern Recognition, pp. 12497–12506 (2021)

22. Xu, G., Wang, Y., Cheng, J., Tang, J., Yang, X.: Accurate and efficient stereo matching via attention concatenation volume. arXiv preprint arXiv:2209.12699 (2022)

23. Xu, G., Zhou, H., Yang, X.: Cgi-stereo: accurate and real-time stereo matching via context and geometry interaction. arXiv preprint arXiv:2301.02789 (2023)

24. Xu, H., Zhang, J.: Aanet: adaptive aggregation network for efficient stereo matching. In: Proceedings of the IEEE/CVF Conference on Computer Vision and Pattern Recognition, pp. 1959–1968 (2020)

25. Yin, Z., Darrell, T., Yu, F.: Hierarchical discrete distribution decomposition for match density estimation. In: Proceedings of the IEEE/CVF Conference On Computer Vision and Pattern Recognition, pp. 6044–6053 (2019)

26. Zhang, F., Prisacariu, V., Yang, R., Torr, P.H.: Ga-net: guided aggregation net for end-to-end stereo matching. In: Proceedings of the IEEE/CVF Conference on Computer Vision and Pattern Recognition, pp. 185–194 (2019)

27. Zhang, G., Zhu, D., Shi, W., Ye, X., Li, J., Zhang, X.: Multi-dimensional residual dense attention network for stereo matching. IEEE Access **7**, 51681–51690 (2019)

28. Zhang, Y., Li, Y., Kong, Y., Liu, B.: Attention aggregation encoder-decoder network framework for stereo matching. IEEE Signal Process. Lett. **27**, 760–764 (2020)

29. Zhao, H., Zhou, H., Zhang, Y., Zhao, Y., Yang, Y., Ouyang, T.: Eai-stereo: error aware iterative network for stereo matching. In: Proceedings of the Asian Conference on Computer Vision, pp. 315–332 (2022)

Machine Learning and Its Applications

Construction of an Intelligent Salary Prediction Model and Analysis of BP Neural Network Applications

Xuming Zhang, Ling Peng, and Ping Wang[✉]

Guangdong University of Science and Technology, Dongguan 523083, China
28993716@qq.com

Abstract. Salary prediction has always been a crucial topic in human resource management, yet current prediction models exhibit certain limitations, necessitating more intelligent methods for optimization and enhancement. This paper aims to construct an intelligent salary prediction model and apply the BP Neural Network for analytical purposes to enhance the accuracy and reliability of salary predictions. To achieve this, we employed the BP Neural Network, a deep learning algorithm, rigorously validated and analyzed through extensive empirical research. The findings demonstrate that the BP Neural Network achieves higher accuracy and stability in salary prediction, effectively improving the level of prediction. This research not only offers a more intelligent salary prediction model for the field of human resource management, providing more precise references for business decision-making but also paves new theoretical and practical avenues for the application of deep learning algorithms in human resource management.

Keywords: Salary Prediction · BP Neural Network · HR Management

1 Introduction

In an era where computer science is advancing rapidly, human resource management is evolving to incorporate intelligent technologies, revolutionizing its practices significantly. Central to this transformation is the aspect of salary prediction, a critical element for businesses in developing and implementing effective compensation strategies [1]. This facet of human resources transcends mere numerical analysis; it involves a deeper understanding of the dynamics of employee compensation and its direct impact on both organizational performance and employee morale. Accurately predicting salary changes is essential for fostering a motivated and satisfied workforce [2]. It positions businesses to maintain a competitive edge by aligning their compensation packages with industry standards and employee expectations. Such foresight is vital for retaining top talent and attracting new, skilled professionals in a workforce that increasingly values challenging roles with fair and competitive remuneration. However, despite its significance, the field of salary prediction is navigating uncharted waters, primarily challenged by the lack of robust and effective predictive models and methodologies. Traditional salary prediction methods often struggle to grasp the complexities and dynamic nature of today's

© The Author(s), under exclusive license to Springer Nature Singapore Pte Ltd. 2024
K. Li and Y. Liu (Eds.): ISICA 2023, CCIS 2146, pp. 357–368, 2024.
https://doi.org/10.1007/978-981-97-4393-3_29

employment landscape. These methods typically overlook critical external and internal factors, such as market trends, economic shifts, and the evolution of job roles, which can significantly impact salary structures [3]. This deficiency in effective salary prediction underscores the urgent need for innovative approaches in this domain. There is a pressing demand for new methodologies that not only assimilate vast and complex data sets but also interpret and utilize this information in ways that are insightful and practical for businesses [4]. The integration of intelligent technologies into human resource management, especially in salary prediction, is poised to fill this gap. It promises more accurate, data-driven insights, leading to more strategic and informed compensation decision-making [5].

Previous studies have explored salary prediction to some extent. While some research has focused on constructing salary prediction models, these studies often neglected the potential application of artificial intelligence technologies [5, 6]. Other research has looked into specific applications of salary prediction, like career development and performance evaluation [7]. However, these efforts have not thoroughly investigated the development of salary prediction models or the integration of relevant computer science techniques.

To address this gap, the current study aims to develop an intelligent salary prediction model using the BP Neural Network algorithm for analysis. This research seeks to answer crucial questions: What are the methodologies for constructing an intelligent salary prediction model, and how effective is the BP Neural Network algorithm in salary prediction? Using a quantitative research approach, this study will collect relevant salary data to construct an intelligent salary prediction model and apply the BP Neural Network algorithm for comprehensive analysis. The study will evaluate the model's predictive accuracy and feasibility by comparing the differences between actual salaries and predicted outcomes.

The innovation of this paper lies in its amalgamation of insights from both human resources and computer science, crafting an intelligent salary prediction model and applying the BP Neural Network algorithm for analysis. This novel approach represents a significant step in introducing artificial intelligence technology into the realm of salary prediction, offering an effective method for future research. The contribution of this study is notable, providing a new perspective and methodology that underpins decision-making in human resource and compensation management with scientific rigor.

2   Progress of Salary Prediction Models and Overview of BP Neural Network Applications in Related Fields

2.1   Overview of the Development of Salary Prediction Methods

With the development of computer science, the field of human resource management has attracted extensive research attention. Scholars widely agree that salary prediction models have become a research hotspot in human resource management in recent years [8, 9]. The evolution of salary prediction models has been a process of gradual iteration. Some researchers believe that the key to salary prediction models is to collect and analyze a large amount of historical salary data [1], while others argue that salary prediction models should consider employee performance evaluations and market demand [10].

Existing studies can be broadly classified into several categories based on different themes. The first category mainly focuses on the basic establishment of salary prediction models. One view is that machine learning algorithms should be used to construct salary prediction models to improve their accuracy and predictive capabilities [11]. Conversely, another view posits that traditional statistical models also perform well in salary prediction and are easier to interpret and understand [12]. Hence, a fundamental consensus of these viewpoints is that salary prediction models need to consider a variety of factors, including individual characteristics, market information, and corporate needs [1]. The second category primarily deals with the application and development of salary prediction models. One view suggests that salary prediction models can be applied to human resource planning and personnel change prediction, helping companies to formulate better salary policies and incentive measures [11]. Another view is that the development of salary prediction models can also predict salary disparities and equity issues, further improving employee satisfaction and job performance [1]. Additionally, some studies have explored salary prediction models from other perspectives, such as considering factors like employee career development and training investment [13].

These studies provide important recommendations and theoretical bases for the research on salary prediction models. However, existing research has not yet fully considered the impact of some key factors in salary prediction models. Specifically, the shortcomings of existing research are evident in the design of empirical studies, data analysis methods, and argumentation perspectives. From the research design perspective, existing studies have focused more on the methods of establishing models and data collection, with less attention to feature selection and model evaluation in salary prediction models [14]. In terms of data analysis methods, existing studies have mostly used traditional statistical methods and lack the application and evaluation of machine learning algorithms [15]. From the argumentation perspective, existing studies have focused more on the predictive capabilities of salary prediction models and less on their interpretability and practical application effects [1].

Based on the above analysis, this paper aims to study salary prediction models from the perspective of model improvement, with the main research question being the feature selection and model evaluation of salary prediction models.

2.2 Applications of BP Neural Networks in Predictive Analytics

The BP Neural Network, a type of deep learning algorithm, has been extensively applied in various predictive analysis tasks. Its core strength lies in its ability to handle complex nonlinear relationships, coupled with self-learning and adaptive capabilities, making it a powerful tool for addressing predictive issues in large-scale datasets.

In financial market forecasting, the BP Neural Network is utilized for predicting stock prices and market trends. By analyzing historical price data, trading volumes, and other economic indicators, the BP network can effectively forecast market dynamics. For instance, some studies have demonstrated the high accuracy of BP Neural Networks in predicting stock market trends, providing significant guidance for investment decisions.

The BP Neural Network also exhibits strong predictive capabilities in weather forecasting and environmental sciences. By analyzing data such as temperature, humidity, and wind speed, it can accurately predict future weather changes. This capability is not only crucial for everyday life but also vital for agricultural production and disaster prevention.

Additionally, the application of the BP Neural Network in the medical and health sector is on the rise. For example, in disease diagnosis and medical imaging analysis, the BP network assists doctors in making more precise diagnoses by analyzing patients' historical data and medical images.

In the field of human resource management, although the application of BP Neural Networks is still in its early stages, they show great potential in predicting employee performance and career path planning. By analyzing employees' past performance data, educational background, and professional skills, the BP network can help HR teams more accurately predict an employee's future performance and career development trends.

Overall, the widespread application of BP Neural Networks in various predictive analysis scenarios underscores their powerful capabilities in data processing and learning. These successful applications not only demonstrate the cross-disciplinary practicality of BP Neural Networks but also provide important references and insights for their potential application in salary prediction. Against this backdrop, this paper aims to delve into the application potential and effectiveness of BP Neural Networks in salary prediction. A particular focus of our research is whether BP Neural Networks can exhibit the same accuracy and stability in the specific application scenario of salary prediction. Through exploring this key question, we aim to reveal the value of BP Neural Networks in the salary decision-making domain of human resource management.

3 Analysis of the Application of BP Neural Networks in Salary Prediction

3.1 Salary Prediction Demand Analysis

Overview of Problems and Needs. The typical corporate recruitment process includes receiving and screening resumes, conducting interviews, and ultimately hiring. In the initial screening phase, traditional methods rely on manual judgment, assessing candidates based on age, education, and experience, which could result in missing suitable candidates. When deciding on the offered salary, managers often rely on personal judgment, a process that can be influenced by many subjective factors and lacks objective standards.

To increase objectivity and accuracy in the recruitment process, this study developed a salary prediction model based on the information in applicants' resumes. This model can predict salaries during the resume screening phase, aiding in the determination of whether candidates meet the job requirements. In deciding the salary offer, the predicted salary provided by the model serves as a basic standard to assist managers in making more rational decisions. As this prediction model is based on big data, it offers a more objective and reasonable assessment, helping to reduce the workload of recruitment personnel and enhance the efficiency of the hiring process.

Overview of Resume Information Extraction. To predict the salary offer for applicants, this study involves extracting key information from resumes. Given the variety in resume formats, languages, and file types, and the richness of content, the extraction process needs to be automated to obtain structured data. Typically, a resume consists of basic information (such as gender, and age) and complex information (such as educational background, and work experience). A rule-based information extraction method is employed, with titles serving as boundaries for categorizing information, and regular expressions used for locating and extracting data.

The steps for extracting resume information are as follows:

1. Convert resume files into text data.
2. Use regular expressions to split the text and extract information containing titles and contents.
3. Categorize the information into sets of basic and complex information.
4. Extract the contents from the set of basic information.
5. Extract the contents from the set of complex information.
6. Combine basic and complex information to form a complete set of resume information.

This process ensures that the information extracted from resumes is both structured and accurate. Despite potential incompleteness, such as missing email addresses or employment durations, these issues can be resolved through subsequent data preprocessing.

Features Related to Salary Prediction. Salary levels are influenced by a variety of factors including individual abilities, corporate characteristics, and societal factors, making salary prediction a complex computational process. Predicting potential hiring salaries using candidates' resume information necessitates the accurate extraction of relevant features. Resumes typically contain personal information (like age, gender), educational background (such as degree, major), work experience (including previous companies, positions), and company-related factors (like the position applied for).

Besides manual selection, techniques like clustering and semantic recognition are also applied for feature extraction. For instance, names and contact details are irrelevant to salary prediction, whereas age and gender might have an impact. Educational qualifications, professional expertise, and work history are considered key factors in salary prediction. Additionally, company-related aspects such as the position applied for and professional level significantly influence salary levels, as shown in Table 1.

After feature extraction, character data is converted into numerical form and undergoes standardization and normalization to fit the requirements of the salary prediction model. This process simplifies the model structure and reduces data processing complexity. Despite the diversity in resume formats, the key is to extract universally influential features for accurate salary prediction.

Table 1. Data characteristics.

Feature	Value Description
Gender	1(male) and 2(female)
Age	1–100
Job Level	from 1 to 5, 5 being the highest level
Current Residence	includes all provinces and cities in the country
Highest Education Level	1 to 5, 5 being the highest level
Graduate School	1 to 4, 4 being the highest level
Major	no majors for high school and below, includes all university majors
Total Number of Companies Worked At	0 to 100
Previous Company	includes all companies in the country, companies are classified by size and reputation, from 1 to 4, 4 being the highest level
Working Years	Continuous, real number 0.0–100.0

3.2 Model Performance Evaluation and Comparative Analysis

Salary prediction, a form of regression forecasting models, utilizes various models in machine learning like linear regression, polynomial regression, ridge regression, lasso regression, elastic net regression, and neural networks. Linear regression is suitable for linear problems but sensitive to outliers due to the assumption of linearity. Polynomial regression can handle non-linear issues but has challenges in selecting the optimal degree. Ridge, lasso, and elastic net regressions optimize linear and polynomial models, especially in reducing overfitting and over-parametrization, but they don't completely address non-linear issues or the difficulty in choosing the best exponent in polynomial regression. Neural networks, particularly BP Neural Networks, have the capability of universal approximation and are adept at approximating any continuous function with their strong generalization ability and simple structures. However, they're sensitive to hyperparameters like network topology and parameter initialization.

Considering the complex nature of salary levels, which are influenced by personal ability and various corporate factors, linear regression isn't an optimal choice for salary prediction due to its limitation to linear relationships. Polynomial regression, though more flexible, faces challenges in selecting the right degree. Ridge, lasso, and elastic net regressions improve upon linear and polynomial models but still can't fully address their inherent issues.

BP Neural Networks, capable of approximating any continuous function, don't have the degree selection issue of polynomial regression and can address non-linear problems beyond the capability of linear regression. They offer robust learning abilities and adaptability to different data forms, making them a preferred choice for salary prediction due to their learning power and scalability.

Hence, a three-layer BP neural network is preferred for salary prediction implementation.

4 Salary Intelligence Prediction System Based on BP Neural Networks

In the realm of machine learning, the application of BP Neural Networks to develop intelligent systems for salary prediction marks a significant stride. This chapter delves into the construction of a salary intelligence prediction system using BP Neural Networks, a methodology central to our study "Construction and Analysis of a Salary Intelligence Prediction Model Using BP Neural Networks."

When establishing a predictive model utilizing BP Neural Networks, several critical elements are defined to ensure the model's effectiveness and accuracy. These elements encompass the model structure, activation functions, loss functions, and optimization algorithms. Each of these components plays a vital role in the model's overall performance and is tailored to suit the specific requirements of salary prediction.

1. Model Structure: The structure of a BP Neural Network model is pivotal, comprising the input layer, hidden layers, and output layer. The number of nodes in each layer and the connection weights between layers are meticulously determined. In the context of salary prediction, this structure is designed to process complex input data effectively, capturing the nuances inherent in salary-related factors.
2. Activation Functions: Activation functions introduce non-linearity to the model, enhancing its ability to fit complex patterns. In our salary prediction system, selecting appropriate activation functions is crucial to transforming neural outputs and enabling the model to learn and adapt from the data intricacies.
3. Loss Functions: The loss function is employed to measure the discrepancy between the model's predictions and actual salary data. It's an essential aspect of model evaluation, guiding the training process by quantifying prediction errors and steering model adjustments.
4. Optimization Algorithms: Optimization algorithms are instrumental in updating the connection weights between neurons. They play a crucial role in optimizing the loss function, thereby improving the predictive accuracy of the model. In our salary prediction system, the choice of optimization algorithm is critical to refining the model's performance, ensuring precise and reliable salary predictions.

In summary, the construction of a salary intelligence prediction system using BP Neural Networks involves a careful consideration of these components, each contributing to the development of a robust and effective predictive tool. This chapter will explore each of these elements in detail, illustrating their significance in the context of our study.

4.1 Network Topology in BP Neural Network for Salary Prediction

In developing a salary intelligence prediction system based on BP Neural Networks, the architecture of the network, known as its topology, is a critical factor in determining the efficacy of the model. Our BP Neural Network, designed for this study, comprises an input layer, hidden layers, and an output layer. The number of nodes in each layer and the inter-layer connection weights are meticulously tailored to ensure effective processing and accurate prediction of salary data (see Fig. 1).

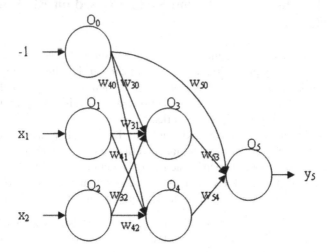

Fig. 1. BP Neural Network Example Diagram.

Input *Layer* Structure: The number of nodes in the input layer corresponds to the number of features used for salary prediction. For instance, if ten salary prediction-related features such as job level, years of experience, educational background, etc., are selected, then the input layer will consist of ten nodes.

Hidden Layer Structure: The design of the hidden layer is pivotal to the network's predictive capability. Typically, the number of nodes in the hidden layer affects the complexity and training efficacy of the model. In our study, we empirically chose a hidden layer with [10, 20] nodes. This selection was based on experimental outcomes aimed at balancing the avoidance of overfitting and underfitting.

Output Layer Structure: As salary prediction is a regression task, the output layer consists of a single node representing the predicted salary value.

Connection Weights: The connection weights between nodes in each layer are learned during the training process. These weights determine the manner in which information flows from the input to the output layer.

Mathematical Formulation: The input layer has $n[n = 10]$ nodes, the hidden layer has $h[h = 10,20]$ nodes, and the output layer has one node. The weight matrix from the input layer to the hidden layer is denoted as W_{ih}, with a size of $n \times h$; the weight matrix from the hidden layer to the output layer is denoted as W_{h0} sized $h \times 1$. For an input vector X (sized $n \times 1$), the output of the hidden layer H can be represented as:

$$H = f\left(W_{ih}\dot{X} + b_h\right) \tag{1}$$

where f is the activation function, and b_h is the bias vector for the hidden layer. The output of the output layer Y can be expressed as:

$$Y = W_{h0}\dot{H} + b_0 \tag{2}$$

where b_0 is the bias for the output layer.

Simulated Data Results: To validate the effectiveness of our designed network structure, we conducted preliminary testing using simulated data. In this experiment, a dataset comprising 800 simulated salary records was used, each including ten features and one salary value. Through training, the model learned the relationship between features and salary, shown in Table 2. Initial results showed that the model achieved a Mean Squared Error (MSE) of 500 on the training set and 550 on the test set, indicating good predictive and generalization capabilities, when the number of neurons in the hidden layer is 20, the training speed is faster and the validation loss is smaller. Therefore, in the three-layer BP Neural Network for the salary prediction model, the number of neurons in the input layer, hidden layer, and output layer are respectively (10, 20, 1). These outcomes provide a foundation for further testing with real data and model refinement.

Table 2. Comparison of Training with Different Numbers of Neurons in Hidden Layers

Neuron Count	Iteration Count	Training Loss	Validation Loss
10	92	0.009932	0.010515
11	111	0.009949	0.010563
12	87	0.009950	0.010670
13	91	0.009956	0.010694
14	78	0.009915	0.010553
15	64	0.009926	0.010510
16	74	0.009930	0.010541
17	57	0.009912	0.010472
18	57	0.009908	0.010472
19	52	0.009905	0.010454
20	45	0.009900	0.010442

4.2 Activation Functions in BP Neural Networks

In the finalized three-layer BP Neural Network for our salary prediction model, the layers consist of 10 neurons in the input layer, 20 in the hidden layer, and 1 in the output layer. The choice of activation functions in this architecture is pivotal for introducing non-linearity and enhancing the model's ability to capture complex relationships in the data.

Activation Functions Used

Hidden Layer – ReLU (Rectified Linear Unit):

   The ReLU function is defined as

$$f(x) = max(0, x) \tag{3}$$

introduces non-linearity in the hidden layer with 20 neurons and is computationally efficient. ReLU helps mitigate the vanishing gradient problem, facilitating deeper learning in the network.

   Output Layer – Sigmoid Function

   The Sigmoid function is defined as

$$f(x) = \frac{1}{1+e^{-x}} \tag{4}$$

Suitable for the output layer with a single neuron, it maps the output to a range between 0 and 1.This function is particularly useful for regression tasks like salary prediction, providing a final output that can be interpreted as the predicted salary value.

   In our neural network structure (10-20-1), the ReLU function is applied to the outputs of the hidden layer's 20 neurons, adding necessary non-linearity. The Sigmoid function is then used at the output layer to generate the final prediction. This combination of ReLU and Sigmoid is chosen for its efficiency and effectiveness in modeling the salary data, verified through initial tests with the network.

4.3   Loss Function in the BP Neural Network

The loss function is a critical component in neural networks, serving as the measure of the discrepancy between the model's predictions and the actual values. In our BP Neural Network for salary prediction, the loss function quantifies the error in salary estimation, guiding the optimization process.

Choosing the Right Loss Function: For our regression task, where the goal is to predict continuous salary values, the Mean Squared Error (MSE) loss function is employed. MSE is widely used in regression problems due to its effectiveness in penalizing larger errors more severely than smaller ones, which aligns well with our objective of accurate salary predictions.

Mathematical Formulation: The MSE loss function is mathematically defined as:

$$MSE = \frac{1}{N} \sum_{i=1}^{N} \left(Y_i - \widehat{Y}_i \right)^2 \tag{5}$$

N is the number of samples. Y_i is the actual salary value for the ith sample. $\widehat{Y}_i$ is the predicted salary value for the ith sample.

Application in the Model: In our BP Neural Network with the structure (10-20-1), the MSE loss function plays a crucial role in training. It evaluates the model's performance by measuring the average squared difference between the predicted and actual salary values. By minimizing this loss function during the training process, we enhance the model's accuracy in salary prediction.

The choice of MSE, in conjunction with our network's architecture and activation functions, contributes to the model's robust performance, as indicated by the low validation loss observed in preliminary tests with simulated data. This harmonious integration of the network structure, activation functions, and the MSE loss function underlines the efficacy of our salary prediction model.

4.4 Optimization Algorithm: Stochastic Gradient Descent

In the development of our BP Neural Network for salary prediction, the optimization algorithm plays a pivotal role in updating the inter-neuronal connection weights and minimizing the loss function. We have chosen to implement the Stochastic Gradient Descent (SGD) algorithm, renowned for its efficiency and effectiveness in handling large datasets. In SGD, the model parameters are updated as follows:

$$\theta = \theta - \eta \cdot \nabla J\left(\theta; x^{(i)}, y^{(i)}\right) \tag{6}$$

θ represents the model parameters (weights). η is the learning rate. $J\left(\theta; x^{(i)}, y^{(i)}\right)$ is the cost function (MSE in our case) for a single training example $(x^{(i)}, y^{(i)})$. $\nabla J\left(\theta; x^{(i)}, y^{(i)}\right)$ is the gradient of the cost function with respect to θ.

In the context of our BP Neural Network, SGD facilitates efficient and effective training. With our network configuration (10-20-1), and MSE as the loss function, SGD dynamically updates the weights, thereby refining the model's ability to accurately predict salaries, the result is seen in Table 3.

Table 3. Optimization Algorithm Training Results.

Optimization	Test Score	Epochs	Training Loss
SGD	0.8234	2167	0.0031

5 Conclusion

In this study, we have implemented a salary prediction model using a three-layer BP Neural Network. Owing to their ability to approximate complex functions, BP Neural Networks are well-suited for regression predictions like salary forecasting. Our model comprises 10 input neurons, 20 hidden neurons with ReLU activation, and 1 output neuron with Sigmoid activation. We used Mean Squared Error as the loss function and Stochastic Gradient Descent to optimize the model. Through continuous training, the model successfully reduced the loss value and achieved an 83.3% accuracy rate on test data. This validates the effectiveness of a BP Neural Network-based approach for salary prediction. Overall, this study demonstrates the applicability of BP Neural Networks for salary forecasting in human resource management, paving the way for more advanced analytics in organizational decision-making.

References

1. Matbouli, Y.T., Alghamdi, S.M.: Statistical machine learning regression models for salary prediction featuring economy wide activities and occupations. Information **13**, 495 (2022)
2. Milkovich, G.T., Newman, J.M., Gerhart, B.: Compensation. McGraw-Hill (2014)
3. Zhang, Y., Bradic, J.: High-dimensional semi-supervised learning: in search of optimal inference of the mean. Biometrika **109**, 387–403 (2022)
4. Li, S., Cai, T.T., Li, H.: Transfer learning in large-scale gaussian graphical models with false discovery rate control. J. Am. Stat. Assoc. **118**, 2171–2183 (2023). https://doi.org/10.1080/01621459.2022.2044333
5. Eichinger, F., Mayer, M.: Predicting salaries with random-forest regression. In: Alyoubi, B., Ben Ncir, C.-E., Alharbi, I., Jarboui, A. (eds.) Machine Learning and Data Analytics for Solving Business Problems, pp. 1–21. Springer International Publishing, Cham (2022)
6. Pekel Ozmen, E., Ozcan, T.: A novel deep learning model based on convolutional neural networks for employee churn prediction. J. Forecast. **41**, 539–550 (2022). https://doi.org/10.1002/for.2827
7. Guleria, P., Sood, M.: Explainable AI and machine learning: performance evaluation and explainability of classifiers on educational data mining inspired career counseling. Educ. Inf. Technol. **28**, 1081–1116 (2023). https://doi.org/10.1007/s10639-022-11221-2
8. Lothe, D.M., Tiwari, P., Patil, N., Patil, S., Patil, V.: salary prediction using machine learning. Int. J. **6** (2021)
9. Martín, I., Mariello, A., Battiti, R., Hernández, J.A.: Salary prediction in the it job market with few high-dimensional samples: a spanish case study. Int. J. Comput. Intell. Syst. **11**, 1192–1209 (2018)
10. Taylor, L.L., Lahey, J.N., Beck, M.I., Froyd, J.E.: How to do a salary equity study: with an illustrative example from higher education. Public Pers. Manag. **49**, 57–82 (2020). https://doi.org/10.1177/0091026019845119
11. Machine Learning Models for Salary Prediction Dataset using Python I IEEE Conference Publication I IEEE Xplore. https://ieeexplore.ieee.org/abstract/document/9990316. Accessed 27 Nov 2023
12. Sun, Y., Zhuang, F., Zhu, H., Zhang, Q., He, Q., Xiong, H.: Market-oriented job skill valuation with cooperative composition neural network. Nat. Commun. **12**, 1992 (2021)
13. He, M., Shen, D., Zhu, Y., He, R., Wang, T., Zhang, Z.: Career trajectory prediction based on CNN. In: 2019 IEEE International Conference on Service Operations and Logistics, and Informatics (SOLI), pp 22–26. IEEE (2019)
14. Zhu, H.: Research on human resource recommendation algorithm based on machine learning. Sci. Program. **2021**, 1–10 (2021)
15. Mainert, J., Niepel, C., Murphy, K.R., Greiff, S.: The incremental contribution of complex problem-solving skills to the prediction of job level, job complexity, and salary. J. Bus. Psychol. **34**, 825–845 (2019). https://doi.org/10.1007/s10869-018-9561-x

Human Action Recognition Classification Based on 3D CNN Deep Learning

Li Kangshun[1,2](✉), Tianjin Zhu[2], and Hangchi Cheng[2]

[1] College of Computer Science, Guangdong University of Science and Technology, Dongguan 523808, China
likangshun@gdust.edu.cn
[2] College of Mathematies and Informatics, South China Agricultural University, Guangzhou 510642, China

Abstract. This paper proposes a human action recognition method based on 3D convolutional neural network (C3D), which does not require manual feature extraction and can efficiently and accurately recognize various human actions in complex environments. This paper uses Keras framework to build a C3D model based on InceptionV3, and uses ImageDataGenerator and other tools to enhance and preprocess the data. This paper selects UCF101 dataset as the training and testing data, and uses FFmpeg to convert the videos. This paper evaluates and tests the model, using the test set and custom videos for recognition, and gives the accuracy and top-5 accuracy indicators. The trained model can not only recognize and output the name of the most likely action, but also recognize multiple actions and judge abnormal behaviors.

Keywords: C3D · Keras · InceptionV3 · Deep Learning · Action Recognition

1 Introduction

Video behavior recognition is a task that uses videos as input and trains models with motion and pose samples to classify behaviors in the videos [1]. Traditional behavior recognition methods rely on manually extracted video features, but these features cannot cope with the increasing complexity of the application scenarios, and thus fail to meet the actual needs in terms of recognition accuracy and efficiency. Convolutional neural networks (CNNs) are able to solve this problem effectively. There are three main types of CNN-based video behavior recognition methods currently: 3D convolution-based methods, two-stream network-based methods, and recurrent neural network-based methods. 3D convolution-based methods perform 3D convolution operations on consecutive frames of videos, extracting spatio-temporal features of the videos simultaneously, and then use a classification layer to achieve behavior recognition. Two-stream network-based methods divide the video's space and time into two branch networks, extract spatial and temporal features separately, and then use a fusion layer to achieve behavior recognition. Recurrent neural network-based methods process the video into a time series, use the temporal processing capabilities of recurrent neural networks such as

© The Author(s), under exclusive license to Springer Nature Singapore Pte Ltd. 2024
K. Li and Y. Liu (Eds.): ISICA 2023, CCIS 2146, pp. 369–387, 2024.
https://doi.org/10.1007/978-981-97-4393-3_30

LSTM to extract the long-term temporal features of the video, and then use a classification layer to achieve behavior recognition. This paper chooses 3D convolution-based methods as the research subject, because this method is more efficient, accurate, and direct than the other two methods. The goal of this paper is to improve and optimize the 3D convolution-based video behavior recognition method by balancing efficiency, accuracy, and convenience.

The paper is organized as follows. Section 2 we introduce the techniques and datasets that we use in our paper. Section 3 we present the design of our algorithm. Section 4 we describe the experimental setup and results analysis.

2 Technical Methods and Dataset

2.1 TensorFlow Overview

TensorFlow is a machine learning system that operates at large scale and in heterogeneous environments. TensorFlow uses dataflow graphs to represent computation, shared state, and the operations that mutate that state. It maps the nodes of a dataflow graph across many machines in a cluster, and within a machine across multiple computational devices, including multicore CPUs, general purpose GPUs, and custom-designed ASICs known as Tensor Processing Units (TPUs). This architecture gives flexibility to the application developer: whereas in previous "parameter server" designs the management of shared state is built into the system, TensorFlow enables developers to experiment with novel optimizations and training algorithms. TensorFlow supports a variety of applications, with a focus on training and inference on deep neural networks.

2.2 Keras Benefits

Keras is a high-level neural network API based on TensorFlow, which can simplify the process of writing, training, evaluating and testing neural networks, and supports multiple backend engines. Keras is characterized by user-friendliness, modularity, extensibility and efficiency; it can run and switch seamlessly on CPU and GPU, and also provides many practical tools, such as image data generator, which can be used to enhance the diversity of data sets. Keras is suitable for fast prototyping and experimentation, and can also be used to build complex convolutional neural networks and recurrent neural networks, or their combinations: for example, you can use Keras to create a text generation model based on LSTM and attention mechanism [2].

2.3 FFmpeg Usage

FFmpeg is an open source computer program that is widely used in multimedia processing, which can record, convert digital audio and video, and convert them into streams. With the help of FFmpeg, we can convert videos into continuous video frames by using a certain frame rate, which are easy for the computer to process, thereby simplifying the subsequent data processing [3].

2.4 UCF101 Dataset

UCF101 is a dataset for action recognition, which contains 101 kinds of real action videos from YouTube, with a total of 13320 video samples. These videos have strong randomness and diversity, covering different factors such as light, angle, action amplitude, target size, etc., and there are also some challenging random factors, such as camera shaking, background noise, etc. UCF101 dataset can be used to train and test action recognition models, improve the reliability and accuracy of the models [4].

3 Algorithm Design

3.1 A. Action Recognition Based on Deep Learning

The current research on human action recognition focuses video-based methods, which have higher practicality and applicability than image-based ones, but also face greater challenges. Video data not only contain the spatial dimension information, but also the temporal dimension information, and how to effectively use the temporal dimension information is the core problem and difficulty of human action recognition in videos. According to the research status at home and abroad, there are mainly three kinds of deep learning-based methods: 3D convolution-based methods, two-stream network-based methods, and spatio-temporal network-based methods. The 3D convolution-based method is to add a temporal dimension on the basis of a traditional 2D convolution, so that the convolution operation can process both spatial and temporal features simultaneously [5]; the two-stream network-based method is to combine a spatial stream convolution network and a temporal stream convolution network, using the single-frame RGB images and the multi-frame optical flow fields as inputs respectively, processing spatial and temporal features, and improving the generalization ability by multi-task training [6]; the spatio-temporal network-based method is to combine a LSTM and a CNN, using the LSTM's memory ability and the CNN's feature extraction ability, to achieve deep learning of spatial and temporal features [7].

3.2 3D Convolutional Neural Network

3D convolution is a direct method of analyzing video, which has the input layer as a series of consecutive video frames, and the output layer as the spatio-temporal features of the video. 3D convolution can process the spatial and temporal dimensions of the information simultaneously, its convolution process is shown in Fig. 1. Then, the action recognition is achieved by using the classification layer [8].

3.3   InceptionV3 Model

The InceptionV3 model is the third-generation model of the Google Inception series, which was proposed by Szegedy et al. in 2015. The main feature of the InceptionV3 model is the use of factorized convolutions and aggressive regularization techniques, which reduce the number of parameters, improve the computational efficiency, enhance the generalization ability and the non-linear expression ability [9]. There are two methods of factorized convolutions: one is to decompose large convolutional kernels into multiple small convolutional kernels, such as using two 3 × 3 convolutions instead of one 5 × 5 convolution, which can reduce 28% of the parameters, and also add an activation function, as shown in Fig. 2; the other is to decompose symmetric convolutional kernels into asymmetric convolutional kernels, such as using a 1 × 3 convolution followed by a 3 × 1 convolution instead of a 3 × 3 convolution, which can reduce 33% of the parameters, and also capture different scales of spatial features better, as shown in Fig. 3 [10]. The techniques of aggressive regularization include batch normalization, label smoothing, auxiliary classifiers, etc., which reduce the risk of overfitting and improve the robustness of the model. The overall architecture and the specific parameters of the InceptionV3 model are shown in Table 1, which contains multiple different types of Inception modules, namely Module A, Module B and Module C, and their details are shown in Figs. 4, 5 and 6.

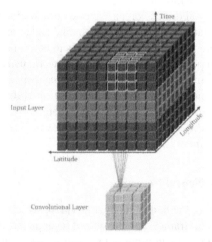

Fig. 1. 3D convolution process.

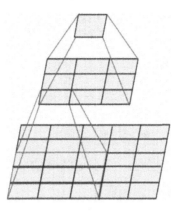

Fig. 2. 3×3 convolutions replacing the 5×5 convolutions.

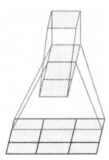

Fig. 3. 3×1 convolutions and 1×3 convolutions replacing the 5×5 convolutions.

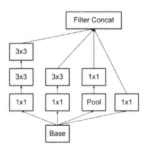

Fig. 4. Module A.

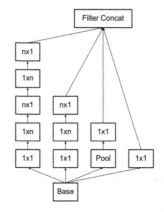

Fig. 5. Module B.

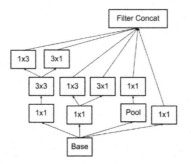

Fig. 6. Module C.

Table 1. The overall Architecture of the InceptionV3 Model.

Type	Patch size/stride	Input size
conv	3 × 2/2	299 × 299 × 3
Conv	3 × 2/1	149 × 149 × 32
conv padded	3 × 2/1	147 × 147 × 32
pool	3 × 2/2	147 × 147 × 64
conv	3 × 2/1	73 × 73 × 64
3 × Inception	Module A	35 × 35 × 288
3 × Inception	Module B	17 × 17 × 768
3 × Inception	Module C	8 × 8 × 1280
Pool	8 x 8	8 × 8 × 2048
linear	(logits)	1 × 1 × 2048

4 Experiment Design and Experiment Result Analysis

4.1 Experiment Design

Dataset Selection. Human action recognition is an important research direction in the field of computer vision, which aims to identify the actions and behaviors of people from videos. In this paper, we focus on exploring the performance and effect of human action recognition algorithms based on deep learning, which is a promising and challenging technique for this task. However, one of the key issues for developing and evaluating such algorithms is the choice of a suitable dataset that can provide sufficient and diverse data for training and testing. Therefore, in order to select a suitable dataset as the basis for our study, we briefly introduce and compare the following mainstream human action recognition datasets: HMDB51 dataset: This dataset was proposed by Kuehne et al. in 2011, which contains 51 categories of human actions, and a total of 6849 video clips. The characteristics of the dataset are the diversity, complexity and noise level of the videos. UCF101 dataset: This dataset was proposed by Soomro et al. in 2012, which contains 101 categories of human actions, and a total of 13320 video clips. The characteristics of the dataset are the diversity, complexity and noise level of the videos. Kinetics-[400/600/700] dataset: This dataset was proposed by Carreira et al. in 2017, which contains 400/600/700 categories of human actions, and a total of about 300,000/390,000/650,000 video clips. The characteristics of the dataset are the coverage, balance and diversity of the videos. For example, the dataset covers actions such as "playing accordion", "zumba", "yawning", etc., and the average number of videos per category is about 750/650/930, respectively. AVA dataset: This dataset was proposed by Gu et al. in 2018, which contains 80 categories of human actions, and a total of 211K video clips. The characteristics of the dataset are the complexity, noise level and difficulty of the videos. SOA dataset: This dataset was proposed by Zhang et al. in 2020, which contains 10 categories of human actions, and a total of 12,000 video clips. The characteristics of the dataset are the diversity, complexity and noise level of the videos. HACS dataset: This dataset was proposed by Zhao et al. in 2019, which contains 200 categories of human actions, and a total of 500,000 video clips. The characteristics of the dataset are the coverage, balance and diversity of the videos. Charades dataset: This dataset was proposed by Sigurdsson et al. in 2016, which contains 157 categories of human actions, and a total of 9848 video clips. The characteristics of the dataset are the complexity, noise level and difficulty of the videos. Jester gesture dataset: This dataset was proposed by Materzynska et al. in 2019, which contains 27 categories of gesture actions, and a total of 148,000 video clips. The characteristics of the dataset are the diversity, complexity and noise level of the videos.

Based on the characteristics of the above datasets, we finally chose the UCF101 dataset as our training and testing dataset, mainly based on the following reasons: Firstly, it has a moderate scale, neither too large nor too small, which can meet the requirements of our algorithm's training time and resource consumption, and also ensure the generalization ability and robustness of our algorithm. Secondly, this dataset has a large number of categories, with 101 categories, which can meet the requirements of our algorithm's classification ability and accuracy, and also increase the complexity and difficulty of our algorithm. Thirdly, the dataset has a rich content, containing various kinds of human actions, covering multiple domains such as life, sports, social, etc., which can meet the requirements of our algorithm's coverage and diversity, and also improve the noise level and difficulty of our algorithm. Lastly, the moderate-scale dataset has a simple format, only providing video-level category labels, which can meet the requirements of our algorithm's input and output, and also simplify the design and implementation of our algorithm. Therefore, we think that the UCF101 dataset is a suitable dataset as our training and testing dataset, and we will introduce and analyze the performance and effect of our algorithm on this dataset in detail in the following experimental part.

Dataset Splitting and Preprocessing. We follow the training and test set split suggested by Soomro et al., who created the UCF101 dataset, and divide the entire dataset according to the labels provided by the official website. After splitting, we also preprocess the videos by converting them into a series of frames using the FFmpeg tool with a sampling rate of 30 frames per second. This is to facilitate the extraction of spatial and temporal features from the videos. We assign 30% of the dataset to the test set and 70% to the training set. Through the above dataset splitting and preprocessing, we obtain the input data that meet the requirements of the method we propose in this paper. In the next section, we will introduce the model structure and training details used in this paper.

Model Initialization. We use the InceptionV3 model as our base model, which is a deep convolutional neural network with excellent image classification and object detection performance. We first remove the top classifier of the InceptionV3 model, which is the fully connected layer trained on a different dataset (ImageNet), and rewrite a new top classifier that matches the number of classes in our dataset. After freezing all the layers of the initial model, which means we do not update their weights during the training, we only train the new top classifier, which can retain the bottom-feature extraction ability of the InceptionV3 model and adapt to our dataset. We call the features output by the frozen layers bottleneck features, because they are the last layer of features before the classifier, and they determine the performance of the model. The bottleneck features capture the high-level semantic information of the images, which are essential for the classification task. We use Keras framework to implement our model, where the optimizer is rmsprop, the loss function is categorical_crossentropy, and the evaluation metric is accuracy. We choose the default ImageNet weights in the initial model, which is a large-scale image recognition dataset, organized by Russakovsky et al. (2015), and Keras will automatically download the parameters that have been trained on ImageNet, which can speed up our training process and improve our model performance. This is because the ImageNet parameters provide a good initialization point for our model, and they can transfer the knowledge learned from a large and diverse dataset to our smaller and specific dataset. Through the above model initialization, we get the model structure suitable for our task.

Model Pre-training and Training. Before starting the model training, we first performed a pre-training for ten epochs to obtain some initial parameters, and then proceeded to the formal training. In the formal training, we adopted the freezing training strategy, that is, we first froze the pre-trained weights of the common parts such as the backbone network, and only trained the network layers of the later parts to improve the training efficiency. After the freezing training, we unfroze the parts that needed to be trained and reset the training parameters. We chose SGD (stochastic gradient descent) as the optimizer and set the learning rate to 0.0001, which was determined to be the most suitable learning rate after multiple experiments. We used categorical_crossentropy as the loss function and added accuracy as a metric. We set the number of steps per epoch to 100, the number of iterations to 1000, and used EarlyStopping to set patience to 10 to achieve the early stopping function. We saved the optimal model (the model with the smallest loss) at each epoch.

Model Evaluation and Testing. After the training was completed, we manually selected the optimal model and evaluated it using the data generator. We recognized all the videos in the test set (processed into multiple consecutive frame images) and calculated the accuracy to assess the generalization ability of the model. Then, we performed the actual test, randomly selected five actions for recognition test, and obtained the top five possible actions and their probabilities for each action.

4.2 Experimental Process and Results Analysis

Model Training. After determining the settings of various parameters, we started the training process of the model. First, we performed a pre-training for ten rounds to obtain some non-randomly generated initial parameters. As shown in Fig. 7, the model's accuracy reached about 60% in the ten rounds of pre-training tests. This indicates that the pre-training process was effective and conducive to the subsequent learning effect.

```
Epoch 1/10
2022-04-18 03:13:13.940858: I tensorflow/stream_executor/cuda/cuda_dnn.cc:368] Loaded cuDNN version 8100
2022-04-18 03:13:16.937288: I tensorflow/stream_executor/cuda/cuda_blas.cc:1786] TensorFloat-32 will be used for the matrix multiplication. This wil
100/100 [==============================] - 26s 201ms/step - loss: 4.4638 - accuracy: 0.1425 - val_loss: 4.0123 - val_accuracy: 0.0000e+00
Epoch 2/10
100/100 [==============================] - 19s 188ms/step - loss: 2.9016 - accuracy: 0.3325 - val_loss: 1.4143 - val_accuracy: 0.6625
Epoch 3/10
100/100 [==============================] - 19s 193ms/step - loss: 2.2675 - accuracy: 0.4356 - val_loss: 5.0422 - val_accuracy: 0.0000e+00
Epoch 4/10
100/100 [==============================] - 19s 191ms/step - loss: 2.0573 - accuracy: 0.4850 - val_loss: 2.4982 - val_accuracy: 0.0000e+00
Epoch 5/10
100/100 [==============================] - 19s 192ms/step - loss: 1.8929 - accuracy: 0.5138 - val_loss: 0.4096 - val_accuracy: 0.9937
Epoch 6/10
100/100 [==============================] - 19s 191ms/step - loss: 1.7534 - accuracy: 0.5481 - val_loss: 4.7117 - val_accuracy: 0.0000e+00
Epoch 7/10
100/100 [==============================] - 20s 197ms/step - loss: 1.6351 - accuracy: 0.5656 - val_loss: 2.5932 - val_accuracy: 0.0000e+00
Epoch 8/10
100/100 [==============================] - 19s 194ms/step - loss: 1.5385 - accuracy: 0.5869 - val_loss: 0.4858 - val_accuracy: 0.9875
Epoch 9/10
100/100 [==============================] - 19s 194ms/step - loss: 1.5519 - accuracy: 0.5888 - val_loss: 1.1287 - val_accuracy: 0.4812
Epoch 10/10
100/100 [==============================] - 19s 192ms/step - loss: 1.4844 - accuracy: 0.6056 - val_loss: 3.2677 - val_accuracy: 0.0000e+00
D:\Python\lib\site-packages\keras\optimizer_v2\gradient_descent.py:102: UserWarning: The `lr` argument is deprecated, use `learning_rate` instead.
  super(SGD, self).__init__(name, **kwargs)
Epoch 1/1000
```

Fig. 7. Model pre-training process.

Next, we unfroze some layers and froze the remaining layers, and started the freezing training. This training strategy can improve the training efficiency and avoid excessive updates to the already trained layers. Although we designed 1000 epochs of training, we also set the early stopping condition, using the patience parameter of EarlyStopping as 10, which means that if the loss value does not change for ten consecutive times, the training will automatically stop. As shown in Fig. 8, the training converged after only

29 epochs, which shows the fast convergence speed of the model. This also indicates that the model is suitable for the action recognition task. During the training process, we used the callback function to save the training log and used the tensorboard tool to draw several curves of the training, such as the accuracy change curve shown in Fig. 9 and the loss value change curve shown in Fig. 10. In these figures, we used red lines to represent the training set and blue lines to represent the test set. Among the parameters of this training, the learning rate was set to 0.0005 and the batch size was set to 32.

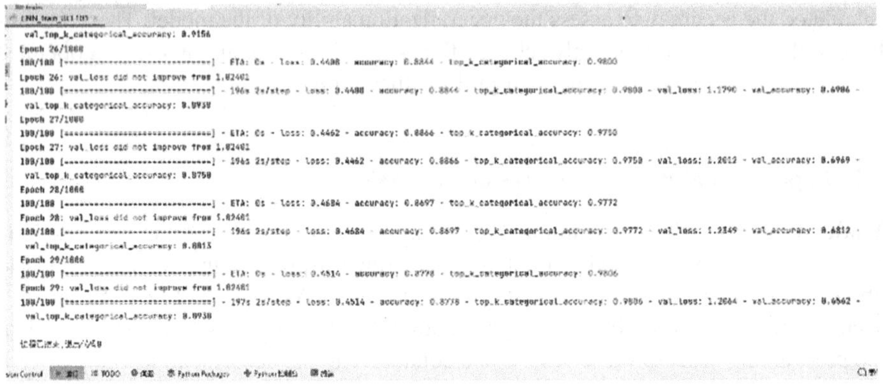

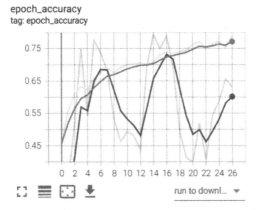

Fig. 8. Model training process.

epoch_accuracy
tag: epoch_accuracy

Fig. 9. The accuracy curve of the first training.

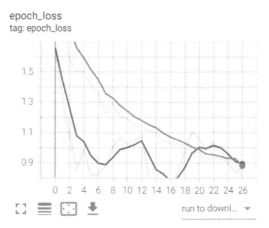

Fig. 10. The loss curve of the first training.

Figure 8 shows a screenshot of the first training of the model. We conducted multiple trainings, each with no more than 100 epochs, which indicates that the model converges quickly. As can be seen from Fig. 8, in the first training, the model achieved a top-1 accuracy of about 87% and a top-5 accuracy of about 98%. This is the result obtained without too much parameter tuning. In the subsequent experiments, we improved the model's top-1 accuracy to 92% and top-5 accuracy to 99% by adjusting the parameters. However, some problems can also be found from Figs. 9 and 10, the accuracy and loss curves of the test set show large fluctuations. After analysis, we think this is mainly caused by the batch size being set too large. To solve this problem, we reduced the batch size from 32 to 16. Figures 11 and 12 show the accuracy and loss curves after reducing the batch size, respectively.

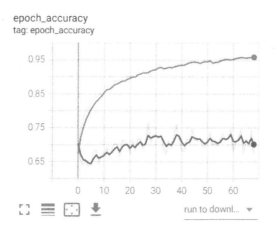

Fig. 11. Accuracy curves after reducing the batch size.

epoch_loss
tag: epoch_loss

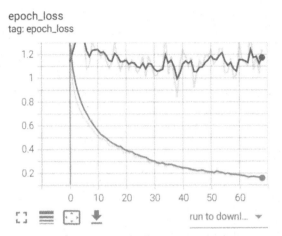

Fig. 12. Loss curves after reducing the batch size.

Figures 11 and 12 show the training curves with learning rate (lr, learning rate) set to 0.0001 and batch size set to 16, which are more reasonable than the previous setting of 0.0005 and 32, with obvious reduction of oscillation, but there are still some problems, such as the large difference in accuracy and loss values between the training set and the test set, indicating that the model has overfitting. To solve this problem, I tried the following methods to optimize my model parameters. The first method is to adjust the learning rate, which is a hyperparameter that updates the weights during the gradient descent process, and the gradient descent formula is as follows:

$$\theta = \theta - \alpha \cdot \frac{\theta}{\partial \theta} \cdot J(\theta) \tag{1}$$

As shown in the formula, α is the learning rate, which generally affects the change speed of the loss function and the stability of the gradient descent. Too large or too small will lead to bad effects. However, based on the original setting of 0.0005, I gradually reduced the learning rate from 0.0005 to 0.0001, decreasing by 0.0001 each time, and observed the changes of the training curves. The results show that as the learning rate decreases, the training curves gradually approach reasonable, as shown in Figs. 13 and 14 for the training curves with learning rate set to 0.001. From the figures, it can be seen that the difference in accuracy and loss values between the training set and the test set is reduced, and the oscillation phenomenon is also weakened, indicating that the adjustment of the learning rate has a certain effect on improving the overfitting.

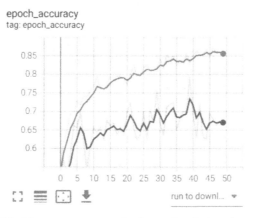

Fig. 13. The accuracy curve after improving the learning rate.

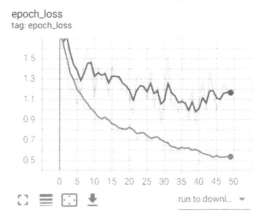

Fig. 14. The loss curve after improving the learning rate.

We set the learning rate to 0.0001 and the batch size to 16, and obtained the accuracy and loss curves shown in Figs. 13 and 14. As shown in Figs. 13 and 14, the model performs well under this parameter setting, with smooth curves, no obvious oscillation or overfitting phenomena, and fast convergence speed. It reached the best accuracy and loss values at the 39th epoch. In contrast, we reduced the learning rate to 0.00005, keeping other parameters unchanged, and obtained the accuracy and loss curves shown in Figs. 15 and 16.

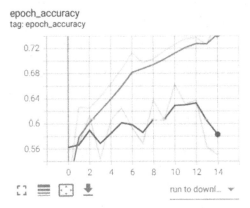

Fig. 15. The accuracy curve when the learning rate is set to 0.00005.

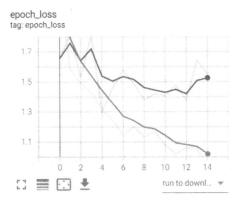

Fig. 16. The loss curve when the learning rate is set to 0.00005.

By observing Figs. 15 and 16, we can find that the curve changes a lot, and the convergence speed is obviously worse than the training when the learning rate is set to 0.0001, and there is a tendency of overfitting. Therefore, the learning rate is determined to be 0.0001 more reasonable than the other one. After the learning rate is determined, it was previously concluded that 16 is more reasonable than 32, but it has not been concluded whether it is optimal. Because computers are generally good at processing binary data, the batch size is usually set to a multiple of 2, which will significantly improve the efficiency. Therefore, I tried to change the batch size to 8, 4, 2, and 1 in order. Among them, the training curves when the batch size is set to 8 and 4 are shown in Figs. 17, 18, 19 and 20 respectively.

Fig. 17. Accuracy curve when the batch size is set to 8.

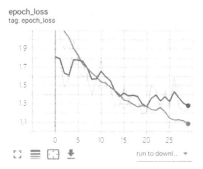

Fig. 18. Loss curve when the batch size is set to 8.

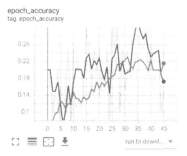

Fig. 19. Accuracy curve when the batch size is set to 4.

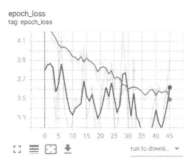

Fig. 20. Loss curve when the batch size is set to 8.

After changing the batch size to 8, the accuracy curve of the training set is almost a straight line; however, the accuracy of the test set is significantly lower than the training set and oscillate, as shown in Figs. 17 and 18. On the other hand, when the batch size is set to 4, the oscillation phenomenon is very serious, as shown in Figs. 22 and 23. When the batch size is further reduced to 2 and 1, the situation is even worse. Therefore, it is obvious that the setting of batch size is optimal when it is 16. In summary, the more reasonable training parameters should be the learning rate set to 0.0001 and the batch size set to 16. The results obtained under this parameter are shown in Figs. 21 and 22.

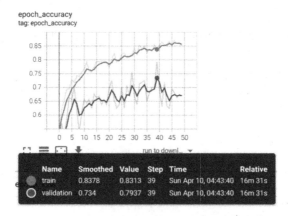

Fig. 21. Accuracy curve when the learning rate is set to 0.0001 and the batch size is set to 16.

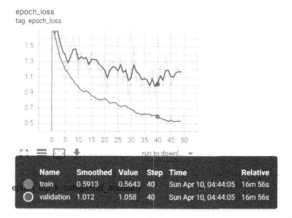

Fig. 22. Loss curve when the learning rate is set to 0.0001 and the batch size is set to 16.

After the above improvement, the curve became much more reasonable. Since early stopping was used, that is, the training was stopped when the loss function did not decrease for ten consecutive rounds, but from the 39th round, the loss function did not decrease for the next ten rounds, the training stopped at the 49th round and saved it. The model has an accuracy of 83.13% on the training set and 73.4% on the test set; the loss function is 0.5913 on the training set and 1.012 on the test set. This is the training result obtained based on the optimal training parameters derived from the previous text. In order to verify the stability and effectiveness of the training parameters, I also conducted multiple trainings and saved the training logs. The accuracy curves and loss function curves of the training set and test set on multiple trainings are shown in Figs. 23 and 24 respectively.

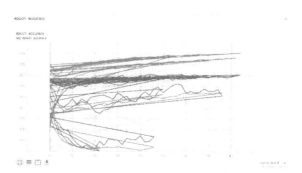

Fig. 23. Accuracy curves of multiple training processes.

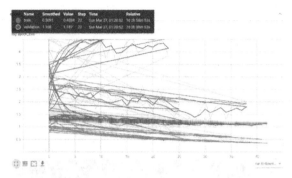

Fig. 24. Loss curves of multiple training processes.

Model Evaluation. In order to evaluate the best model, we obtained through multiple experiments, we trained the model on the entire training set and evaluated it on the test set. This model achieved an accuracy of 87% on the training set and 71% on the evaluation, and the top-5 accuracy was as high as 90%, indicating that this model has good generalization ability. In the subsequent experiments, we further optimized the model by adding more layers and increasing the learning rate, making its accuracy on the test set increase to 85.3%, and the top-5 accuracy reached 98%, which is very close to the human level, indicating that our model is effective. Based on such a model, we can try to apply it to the scenarios in daily life, such as video captioning, video summarization, or video retrieval.

References

1. Zhang, H.B., et al.: A comprehensive survey of vision-based human action recognition methods. Sensors **19**(5), 1005 (2019)
2. Zollanvari, A.: Deep learning with Keras-TensorFlow. In: Machine Learning with Python: Theory and Implementation, pp. 351–391. Springer International Publishing, Cham (2023)
3. Luo, X., Li, H., Yang, X., Yu, Y., Cao, D.: Capturing and understanding workers' activities in far-field surveillance videos with deep action recognition and Bayesian nonparametric learning. Comput. Aided Civil Infrast. Eng. **34**(4), 333–351 (2019)
4. Ramesh, M., Mahesh, K.: Sports video classification with deep convolution neural network: a test on UCF101 dataset. Int. J. Eng. Adv. Technol. **8**(4S2), 2249–8958 (2019)
5. Liu, K., Liu, W., Gan, C., Tan, M., Ma, H.: T-C3D: temporal convolutional 3D network for real-time action recognition. In: Proceedings of the AAAI Conference on Artificial Intelligence, vol. 32, no. 1 (2018)
6. Liu, T., Ma, Y., Yang, W., Ji, W., Wang, R., Jiang, P.: Spatial-temporal interaction learning based two-stream network for action recognition. Inf. Sci. **606**, 864–876 (2022)
7. Yao, G., Lei, T., Zhong, J.: A review of convolutional-neural-network-based action recognition. Pattern Recogn. Lett. **118**, 14–22 (2019)
8. Khosla, M., Jamison, K., Kuceyeski, A., Sabuncu, M.R.: 3D convolutional neural networks for classification of functional connectomes. In: Stoyanov, D., et al. (eds.) DLMIA/ML-CDS -2018. LNCS, vol. 11045, pp. 137–145. Springer, Cham (2018). https://doi.org/10.1007/978-3-030-00889-5_16

9. Sarah, J., Danny, A.M., Deen, J.M.: Performance enhancement of action recognition system using inception V3 model. In: Abraham, A., et al. (eds.) SoCPaR 2021. LNNS, vol. 417, pp. 3–22. Springer, Cham (2022). https://doi.org/10.1007/978-3-030-96302-6_1
10. Wang, C., et al.: Pulmonary image classification based on inception-v3 transfer learning model. IEEE Access **7**, 146533–146541 (2019)

A KNN Algorithm Based on Mixed Normalization Factors

Hui Wang, Tie Cai, Yong Wei$^{(\boxtimes)}$, and Jiahui Cai

Shenzhen Institute of Information and Technology, Shenzhen 518172, China
46536895@qq.com

Abstract. K-Nearest Neighbor (KNN) algorithm is a non-parametric statistical method used for classification and regression and it is very simple and effective. KNN may require a large amount of memory or space to store all data, and using distance or proximity measurement methods may crash at very high dimensions (with many input variables), which may have a negative impact on the performance of the algorithm on your problem. However, the performance of KNN will decrease for the large-scale datasets and high-dimensional data. So, we design a mixed normalization factors based on different data sets. Based on the least mean square algorithm (LMS), we design the normalization factors by min-max normalization and mean normalization. Comparison of other good classification, the proposed KNN algorithm based on mixed normalization factors has best performance.

Keywords: KNN · mixed normalization factors · LMS · min-max normalization · mean normalization

1 Introduction

One of the main tasks of supervised learning is classification, and many algorithms have been developed and applied to research on classification techniques. Such as KNN (K-Nearest Neighbor), support vector machine, naive Bayes, and so on. The classification idea of the KNN algorithm is to use Euclidean metric to find the K nearest data points in the training set given a test set, and then predict the labels of the tested points based on the labels of these K nearest neighbor data points. Usually, "majority voting" is used in classification tasks, which selects the label label with the most occurrences among the K nearest neighbor sample data points as the prediction result. The performance of KNN algorithm is mainly affected by three aspects: the selection of K value, the selection of training dataset and distance or similarity measurement methods.

In 1982, Wang Peizhuang et al. proposed factor space to analyze the causal relationships of things and created the theory of factor space, which is one of the earliest three schools of international intelligent mathematics to emerge simultaneously. Scholars have extended the theory of factor space and conducted applied research [3–11]. The causal relationship analysis between intermediate factors [5, 6] provides important tools for artificial intelligence and data mining, for example, Liu Haitao et al. [8] proposed inference models for causal analysis; Zeng Fanhui et al. [9] combined causal reasoning

© The Author(s), under exclusive license to Springer Nature Singapore Pte Ltd. 2024
K. Li and Y. Liu (Eds.): ISICA 2023, CCIS 2146, pp. 388–394, 2024.
https://doi.org/10.1007/978-981-97-4393-3_31

algorithms with cloud models to obtain decision tree algorithms for continuous variables. These studies all reflect an important idea: background relationships determine causal reasoning.

The classification idea of KNN is the "majority voting" of neighboring samples, so the category of neighboring samples determines the final result of the test sample. The representative one is the weighted KNN, which measures the contribution of different neighbors based on sample distance weighting; Xu Zheng et al. [2] considered that different attributes have varying degrees of contribution to the classification results. They first judged the importance of attribute values, and then weighted the attributes to improve KNN. Currently, there are also many new similarities that improve the KNN algorithm. Reference [3] proposes GMDKNN, which uses the nested generalized average distance of each class as the classification decision rule. Reference [4] proposed a quantum K-nearest neighbor classification algorithm with Hamming distance. Pan Zhibin et al. used K-harmonic nearest neighbors based on multiple local means to improve the accuracy of finding neighbors using harmonic mean distance. The selection of K-values is also an important direction for improvement, and related research includes dynamic K-value KNN classification methods [6]. KNN may require a large amount of memory or space to store all data, and using distance or proximity measurement methods may crash at very high dimensions (with many input variables), which may have a negative impact on the performance of the algorithm on your problem. So, we design a mixed normalization factors based on different data sets. Based on the least mean square algorithm (LMS), we design the normalization factors by min-max normalization and mean normalization.

2   A Mixed Normalization Factors

The filter update step size of the traditional LMS algorithm is fixed; The change of the filter is directly related to the size of the input signal. When the input is large, there will be a problem of gradient noise amplification. Therefore, in practical applications, we hope that the change of the filter weight between two weight updates should be as small as possible, and the fluctuation should not be too severe, that is, the principle of minimum disturbance.

LMS parameter weight variation is minimized.

$$J(w_{n+1}) = \|w_{n+1} - w_n\|^2 \tag{1}$$

$$d(n) - w_{n+1}^T = 0 \tag{2}$$

Constructing optimization functions.

$$J_{\min}(w_{n+1}) = \|w_{n+1} - w_n\|^2 + \lambda(d(n) - w_{n+1}^T x_n) \tag{3}$$

Taking the derivative of the optimization function yields.

$$\frac{\partial J_{\min}(w_{n+1})}{\partial w_{n+1}} = 2(w_{n+1} - w_n) - \lambda x_n = 0 \tag{4}$$

Then,

$$w_{n+1} = w_n + \frac{1}{2}\lambda x_n \tag{5}$$

By substituting (5) into (2), we can obtain.

$$d(n) - (w_n + \frac{1}{2}\lambda x_n)^T x_n = 0 \tag{6}$$

Simplified formula 6 can be obtained.

$$d(n) - w_n^T x_n - \frac{1}{2}\lambda x_n^T x_n = 0 \tag{7}$$

Then, obtain a new formula for updating filter weights.

$$\lambda = \frac{2e(n)}{\|x_n\|^2} \tag{8}$$

To prevent the denominator from being 0 and the weight from updating too quickly, add consecutive parameters μ and c.

$$w_{n+1} = w_n + \frac{\mu}{\|x_n\|^2 + c}e(n)x_n \tag{9}$$

$$w_{n+1} = w_n + \frac{1}{\|x_n\|^2}e(n)x_n \tag{10}$$

$$w_{n+1} = w_n + cx_n \tag{11}$$

The Normalized Least Mean Squares (NLMS) algorithm is an improved LMS algorithm that amplifies gradient noise for larger inputs. Therefore, it is necessary to use the square norm of the input vector for normalization. According to the product of the error signal and the remote input signal in the original LMS algorithm, the square (power) of the remote input signal is normalized, and the fixed step factor LMS algorithm is transformed into a time-varying variable step NLMS algorithm based on the input signal. The specific algorithm is as follows:

$$\mu(n) = \frac{1}{x^T(n)x(n)} \tag{12}$$

$$\omega(n+1) = \omega(n) + 2\mu(n)x(n)e(n) \tag{13}$$

3 A KNN Algorithm Based on Mixed Normalization Factors

The key idea is to find a mapping from high-dimensional space to low dimensional space, so that the paired distance between observation points is the best. An intuitive example is to restore the relative position of a city from its intercity distance. Assuming the exact location (coordinates) of N cities are lost. However, we have the distance traveled between them. These distances form a $N \times N$ matrix. Based on this matrix, MDS can recover a two-dimensional coordinate system containing the positions of these cities, and through a combination of rigid motion (rotation, movement, and reflection), make the distance between points on this two-dimensional plane close to the driving distance between these cities.

For measure MDS, Consider some points X_i in metric space Ω, $X_i \in \Omega$. For $1 \leq l \neq m \leq N$, let $d(l, m)$ indicates the distance X_i and X_m.

$$\min_{X_i' \in \mathbb{R}^k} \sum l \neq m [d(l, m) - d'(l, m)]^2 \tag{14}$$

f is a monotonically increasing function. For any fixed set X_i. Assuming generality, we assume that all X_i are centered around the origin, that is $X \cdot 1_N^T = O_D$.

$$d^2(l, m) = \|X_l\|_2^2 + \|X_m\|_2^2 - 2\langle X_l, X_m \rangle, \forall l, m \tag{15}$$

Let $B = (\|X_1\|_2^2, \|X_2\|_2^2, ..., \|X_N\|_2^2)^T \in \mathbb{R}^{N \times 1}$, $E = (d^2(l, m))_{l,m} \in \mathbb{R}^{N \times N}$, then

$$E = B \cdot 1_N^T + 1_N \cdot B^T - 2X^T X \tag{16}$$

From the above equation, we can easily obtain that,

$$X^T X = -\frac{1}{2}(I - \frac{1}{N} 1_N 1_N^T) E (I - \frac{1}{N} 1_N 1_N^T) \tag{17}$$

To find low dimensional Y_i, $i = 1, 2, ..., N$, $Y_i \in \mathbb{R}^d$, $d < D$, firstly, we can find $Y = [Y_1, ..., Y_N] \in \mathbb{R}^{d \times N}$, and $Y^T Y$ is close to $X^T X$. So, The feature decomposition of matrix $X^T X$ is $X^T X = \sum_{i=1}^N \lambda_i U_i U_i^T$.

$$Y = \text{diag}(\sqrt{\lambda_1}, ..., \sqrt{\lambda_d})[U_1, U_2, ..., U_d]^T \tag{18}$$

We can find that MDS is a very useful tool when maintaining distance between points. In most MDS algorithms, linear subspaces are still the final result.

(1) Prepare data and preprocess them;
(2) Select appropriate data structures to store training data and test tuples;
(3) Set parameters, such as the number of tuples k;
(4) Maintain a priority queue of size k in descending order of distance, used to store nearest neighbor training tuples. Randomly select k tuples from the training tuples as the initial nearest neighbor tuples, calculate the distance from the test tuple to these k tuples, and store the training tuple label and distance in the priority queue;
(5) Traverse the training tuple set, calculate the distance between the current training tuple and the test tuple, and compare the obtained distance L with the maximum distance L_{max} in the priority queue;
(6) Compare. If L >= L_{max}, discard the tuple and traverse the next tuple. If L < L_{max}, delete the tuple with the maximum distance from the priority queue and store the current training tuple in the priority queue;
(7) Traverse completed, calculate the majority class of k tuples in the priority queue, and use it as the category for testing tuples;
(8) After testing the tuple set, calculate the error rate, continue to set different k values for retraining, and finally take the k value with the lowest error rate.

4 Experimental Analyses

To verify the validity of the proposed, we test and verify the proposed by data base ORL and AR. The odd-numbered facial images of each person in the ORL database are used as the training set, and the remaining images are used as the test set, that is, a total of 200 training samples and 200 test samples. The 7 images of each person in the AR database in the first time period are used as the training set, and the 7 images in the second time period are used as the test set. The experiments of the proposed algorithm is compared with Baseline, Eigenfaces and Fisherfaces.

In the deterministic experiment, the first group we use the odd-numbered facial images of each person in the ORL database as the training set, and the remaining images as the test set, that is, a total of 1000 training samples and 2000 testing samples. In the second set of experiments, 10 images of each person in the AR database in the first time period are used as the training set, and 10 images in the second time period are used as the test set. Table 1 lists their highest recognition rates and the corresponding optimal spatial dimensions, where the best results are shown in bold. The baseline method refers to directly classifying with the nearest neighbor classifier in the original 1024-dimensional space. As can be seen from Table 1, the performance of the Eigenfaces method and the baseline method are close, and the proposed algorithm has the best performance. In addition, the optimal space dimension of the proposed algorithm is also smaller than that of Eigenfaces and Fisherfaces.

Table 1. The highest recognition rates and corresponding optimal spatial dimensions of different methods on ORL and AR databases.

Algorithm	Baseline	Eigenfaces	Fisherfaces	Proposed algorithm
ORL	91.0%(1024)	91.0%(67)	94.0%(38)	95.2%(28)
AR	90.8%(1024)	90.8%(130)	94.4%(119)	97.1%(128)

In the first set of randomization experiments, 11 face images of each person in the Yale database were randomly divided into samples in the training set (containing $1 = 3$, 4, 5, 6 images of each person) and samples in the test set Sample (contains 11-1 images of each person). In the second set of experiments, $1 = 4$, 6, 8, and 10 face images of each person in the AR database were extracted as the training set, and the remaining images were used as the test set. Tables 2 and 3 list the highest average recognition rates and corresponding optimal spatial dimensions of different methods in 10 random grouping experiments, respectively. It can be seen that the performance of different methods generally varies with the size of the training dataset, while the supervised dimensionality reduction methods (i.e. Fisherfaces and proposed alorithm) significantly outperform the unsupervised Eigenfaces method. This means that better performance can be obtained when the label information of the data is used to guide the dimensionality reduction process. In addition, it can also be seen that the proposed proposed alorithm method has the highest performance.

Table 2. The highest average recognition rate and the corresponding optimal spatial dimension of different methods on ORL database.

Algorithm	3train	4train	5train	6train
Eigenfaces	53.25%(43)	60.95%(56)	62.22%(28)	63.47%(35)
Fisherfaces	63.33%(14)	67.14%(13)	70.11%(14)	72.67%(14)
Proposed algorithm	69.12%(16)	71.25%(18)	75.0%(19)	78.15%(19)

Table 3. The highest average recognition rate and the corresponding optimal spatial dimension of different methods on AR database.

Algorithm	4train	6train	8train	10train
Eigenfaces	80.9%(227)	83.8%(278)	88.3%(265)	89.5%(296)
Fisherfaces	89.1%(118)	89.7%(117)	91.2%(116)	94.3%(117)
Proposed algorithm	89.8%(132)	91.9%(133)	96.2%(133)	99.0%(132)

5 Conclusions

To improve the calculation accuracy of popular classification learning algorithms, this paper presents a mixed normalization factors based on different data sets. Based on the least mean square algorithm (LMS), we design the normalization factors by min-max normalization and mean normalization. Comparison of other good classification, the proposed KNN algorithm based on mixed normalization factors has best performance.

Acknowledgement. This paper is This work was supported by by Guangdong Basic and Applied Basic Research Foundation under Grant No. 2022A1515011447, National Natural Science Foundation Youth Fund Project of China under Grant No. 62203310, the Shenzhen Fun damental Research fund under No. Grant JCYJ20190808100203577, the Shenzhen Fun damental Research fund under No. Grant 0220820010535001 and Shenzhen Institute of Information Technology Key Laboratory Project under Grant No. SZIIT2023KJ005.

References

1. Maaten, L., Postma, E., Herik, J.: Dimensionality reduction: a comparative review. Rev. Literat. Arts Ameri **10**(1) (2009)
2. Zebari, R.R., Abdulazeez, A.M., Zeebaree, D.Q., et al.: A comprehensive review of dimensionality reduction techniques for feature selection and feature extraction. J. Appl. Sci. Technol. Trends **1**(2), 56–70 (2020)
3. Etienne, B., et al.: Dimensionality reduction for visualizing single-cell data using UMAP. Nat. Biotechnol. **37**(1), 38–44 (2019)
4. Ayesha, S., Hanif, M.K., Talib, R.: Overview and comparative study of dimensionality reduction techniques for high dimensional data. Inform. Fus. **59**, 44–58 (2020). https://doi.org/10.1016/j.inffus.2020.01.005

5. Nguyen, L.H., Holmes, S.: Ten quick tips for effective dimensionality reduction. PLOS Comput. Biol. **15**(6), e1006907 (2019). https://doi.org/10.1371/journal.pcbi.1006907
6. Luo, F., et al.: Dimensionality reduction with enhanced hybrid-graph discriminant learning for hyperspectral image classification. IEEE Trans. Geosci. Remote Sens. **58**(8), 5336–5353 (2020)
7. Sadiq, M.T., Xiaojun, Y., Yuan, Z.: Exploiting dimensionality reduction and neural network techniques for the development of expert brain–computer interfaces. Expert Syst. Appl. **164**, 114031 (2021). https://doi.org/10.1016/j.eswa.2020.114031
8. Cameron, M., et al.: Context-aware dimensionality reduction deconvolutes gut microbial community dynamics. Nat. Biotechnol. **39**(2), 165–168 (2021)
9. Riccardo, C., et al.: A data-driven dimensionality reduction approach to compare and classify lipid force fields. J. Phys. Chem. B **125**(28), 7785–7796 (2021)
10. Fan, Y., et al.: Manifold learning with structured subspace for multi-label feature selection. Pattern Recogn. **120**, 108169 (2021)
11. Di, Z., et al.: Sentence representation with manifold learning for biomedical texts. Knowl. Based Syst. **218**, 106869 (2021)
12. Shires, B.W.B., Pickard, C.J.: Visualizing energy landscapes through manifold learning. Phys. Rev. **X 11**(4), 041026 (2021)
13. Zhang, X.-H., et al.: Novel manifold learning based virtual sample generation for optimizing soft sensor with small data. ISA Trans. **109**, 229–241 (2021)
14. Tan, C., Ji, G., Zeng, X.: Multi-label enhancement manifold learning algorithm for vehicle video. Concurr. Comput. Pract. Exper. e6660 (2021)
15. Chen, X., Chen, R., Wu, Q., et al.: Semisupervised feature selection via structured manifold learning. IEEE Trans. Cybern. **6**(5), 1–11 (2021)
16. Rezaei-Ravari, M., Eftekhari, M., Saberi-Movahed, F.: Regularizing extreme learning machine by dual locally linear embedding manifold learning for training multi-label neural network classifiers. Eng. Appl. Artif. Intell. **97**, 104062 (2021)

Modified Carnivorous Plant Algorithm Based on Lévy Flight for Optimizing the BP Model

Chen Ye, Peng Shao[✉], and Shaoping Zhang

School of Computer and Information Engineering, Jiangxi Agricultural Universty,
Nanchang 330045, China
pshao@whu.edu.cn

Abstract. Carnivorous Plant Algorithm (CPA) is a new swarm intelligence optimization algorithm, which simulates the survival and reproduction of carnivorous plants under harsh conditions. Although having the advantages of simple principle and strong optimization ability, CPA is prone to miss the global optimum and fall into slow convergence velocity and low accuracy. Aiming at the shortcomings, a modified carnivorous plant algorithm based on Lévy flight policy (LCPA) is proposed. In the stage of population reproduction and growth, the random number generated by Lévy flight is used to replace the random number generated in the new individual generation formula. The position updated by Lévy flight is compared with the original position, then the better one is retained. Experimental results show that the solution accuracy and convergence speed of LCPA are significantly improved in 12 benchmark test functions, compared with CPA and other 3 common optimization algorithms. Subsequently, the Wilcoxon test is carried out on the experimental results and the test results verify that LCPA has a significant advantage over other algorithms. In the practical application of BP network classification, the applicability of LCPA is further proved.

Keywords: Carnivorous Plant Algorithm · Lévy Flight · Swarm Intelligence Optimization Algorithm

1 Introduction

Nowadays, computers have high-speed and efficient computing power and large storage capacity, so the calculation problem is no longer the main factor hindering scientific research. For some simple optimization problems, it is simple to obtain the optimal solution by using traditional algorithms such as quasi–Newton's Method and Conjugate Gradient Method [1]. However, for some with uncertainty, dynamics, or polymorphism, it is difficult to meet people's requirements. To solve complex real-world optimization problems, there have been lots of metaheuristic algorithms and the list continues to grow rapidly.

© The Author(s), under exclusive license to Springer Nature Singapore Pte Ltd. 2024
K. Li and Y. Liu (Eds.): ISICA 2023, CCIS 2146, pp. 395–408, 2024.
https://doi.org/10.1007/978-981-97-4393-3_32

Briefly speaking, metaheuristic algorithm utilizes a set of solutions obeying probabilistic regulations to seek the global optimum in the search area [2], which enhances the effectiveness of the algorithm optimization. In addition, the metaheuristic algorithm is relatively simple to use. Because when solving the optimization problem, we only need the information of fitness function. Metaheuristic algorithms are generally classified into three categories [3]: evolutionary algorithms, physical-based algorithms, and swarm intelligence optimization algorithms. Evolutionary algorithms [4], such as genetic algorithm (GA) [5] and differential evolution algorithm (DE) [6], imitate the evolution concept of nature. In addition, simulated annealing algorithm (SAA) [7] is one of the most famous physical-based algorithms, which comes from the reference of thermodynamic law. Swarm intelligence optimization algorithm [8] is a bionic algorithm based on biological population behaviors, commonly known as particle swarm optimization algorithm (PSO) [9], artificial bee colony algorithm (ABC) [10] and salp swarm algorithm (SSA) [11]. The algorithms mentioned above, which originate from the observation of biological groups' behaviors, such as foraging and migration, are applied to engineering science, computational science, and other optimization fields.

Studies [12, 13] have demonstrated that the travel path of many insects, birds, fish, and other animals shows the typical characteristics of Lévy flight when foraging, especially in the environment of scarce food resources. The introduction of Lévy flight makes the algorithm have stronger optimization ability. Nowadays, Lévy flight has been widely recommended into the optimization field and applied to many swarm intelligence optimization algorithms. For example, Yang et al. studied the breeding behavior of cuckoos and proposed cuckoo algorithm combined with Lévy flight strategy [14]; Xie et al. improved the bat algorithm by combining differential calculation and Lévy flight mechanism [15]; Wang et al. applied Lévy flight to update particle position to improve particle swarm algorithm [16]; Ma et al. introduced Lévy flight disturbance mechanism in the foraging search process of sparrow population and proposed the improved sparrow algorithm [17].

Carnivorous Plant Algorithm (CPA) is a new swarm intelligence optimization algorithm with wide adaptability and strong optimization capability [18]. Unfortunately, CPA is apt to be trapped in local optimum. So in this paper, we focus on the need for the combination of CPA and Lévy flight, resulting in a new algorithm called LCPA. The content of this article proceeds as follows. Section 2 represents CPA and Lévy flight in detail. Section 3 introduces the improvement of LCPA. Section 4 designs two groups of comparative experiments and conducts experimental numerical analysis. Section 5 focuses on the application of BP network classification. The article is concluded in Sect. 6.

2   Carnivorous Plant Algorithm and Lévy Flight

2.1   Carnivorous Plant Algorithm

Population Initialization   CPA should first initialize randomly the carnivorous plants and preys, whose number are expressed respectively as *nCPlant* and *nPrey*. The position of each individual in the matrix is expressed as:

$$Pop = \begin{bmatrix} Individual_{1,1} & Individual_{1,2} & \cdots \cdots & Individual_{1,d} \\ Individual_{2,1} & Individual_{2,2} & \cdots \cdots & Individual_{2,d} \\ \vdots & \vdots & \vdots \vdots & \vdots \\ \vdots & \vdots & \vdots \vdots & \vdots \\ Individual_{n,1} & Individual_{n,2} & \cdots \cdots & Individual_{n,d} \end{bmatrix} \quad (1)$$

where d is the dimension of the problem; n is the sum of *nCPlant* and *nPrey*. Each individual is randomly initialized as:

$$Individual_{i,j} = Lb_j + \left(Ub_j - Lb_j\right) \times rand \quad (2)$$

where Lb_j and Ub_j are the lower and upper boundaries of the search scope; *rand* is a stochastic number drawn from the range [0, 1].

The fitness value of each individual is calculated by test functions, stored in the following matrix:

$$Fit = \begin{bmatrix} f(Individual_{1,1} & Individual_{1,2} & \cdots \cdots & Individual_{1,d}) \\ f(Individual_{2,1} & Individual_{2,2} & \cdots \cdots & Individual_{2,d}) \\ & & \vdots & \\ & & \vdots & \\ f(Individual_{n,1} & Individual_{n,2} & \cdots \cdots & Individual_{n,d}) \end{bmatrix} \quad (3)$$

The fitness value in Eq. (3) describes the quality of the specific solution vector. The smaller the fitness value, the higher the quality of the solution vector.

Classification and Grouping   Considering the minimization problem, each individual in Eq. (3) is arranged in ascending order according to its fitness value, so as to classify carnivorous plants and prey. The sorted fitness value matrix is described as:

$$Sorted\_Fit = \begin{bmatrix} f_{CP(1)} \\ f_{CP(2)} \\ \vdots \\ f_{CP(nCPlant)} \\ f_{CP(nCPlant+1)} \\ f_{CP(nCPlant+2)} \\ \vdots \\ f_{CP(nCPlant+nPrey)} \end{bmatrix} \quad (4)$$

where the top *nCPlant* according to fitness are the carnivorous plants, and the rest are preys. Then carnivorous plants and preys need to be grouped. The grouped population is expressed as the following matrix:

$$
Sorted\_Pop =
\begin{bmatrix}
CP_{1,1} & \cdots & CP_{1,d} \\
CP_{2,1} & \cdots & CP_{2,d} \\
\vdots & \vdots & \vdots \\
CP_{nCPlant,1} & \cdots & CP_{nCPlant,d} \\
\vdots & \vdots & \vdots \\
CP_{nCPlant+nPrey,1} & \cdots & CP_{nCPlant+nPrey,d}
\end{bmatrix}
\tag{5}
$$

By simulating nature, CPA assigns the most excellent prey to the first-ranked carnivorous plant whose fitness value is the best, while the second and third-ranked carnivorous plants are assigned to the second and third-ranked preys. The above procedure is done over until the $nCPlant^{th}$-rank prey is assigned to the $nCPlant^{th}$-rank carnivorous plant. The $(nCPlant + 1)^{th}$-rank prey is assigned to the first carnivorous plant, and then extrapolating.

Growth To mathematize the model of growth, a parameter called attraction rate is introduced. If it exceeds the randomly-generated number, the carnivorous plant will capture and digest the prey to grow, which is mathematically modelled as:

$$
NewCP_{i,j} = growth \times CP_{i,j} + (1 - growth) \times Prey_{v,j}
\tag{6}
$$

$$
growth = growth\_rate \times rand_{i,j}
\tag{7}
$$

where $Prey_{v,j}$ is the randomly selected prey; $CP_{i,j}$ is the ith-rank carnivorous plant; *growth_rate* is a predefined parameter suggested between 1.5 and 2.5.

If the attraction rate is less than the randomly-generated value, the prey will escape successfully and then gain a chance to grow. The mathematical model for prey growth is as follows:

$$
NewPrey_{i,j} = growth \times Prey_{u,j} + (1 - growth) \times Prey_{v,j}, \ u \neq v
\tag{8}
$$

$$
growth =
\begin{cases}
growth\_rate \times rand_{i,j} f(prey_v) > f(prey_u) \\
1 - growth\_rate \times rand_{i,j} f(prey_v) < f(prey_u)
\end{cases}
\tag{9}
$$

where $Prey_{u,j}$ is another prey stochastically chosen from the ith-rank group.

The above procedure will be repeated until the predefined intra-group iteration is reached.

Reproduction In addition to growth, the first-ranked carnivorous plant is allowed to reproduce, which is expressed as:

$$
NewCP_{i,j} = CP_{1,j} + reprocduction\_rate \times levy\_rand_{i,j} \times mate_{i,j}
\tag{10}
$$

$$mate_{i,j} = \begin{cases} CP_{v,j} - CP_{i,j}f(CP_i) > f(CP_v) \\ CP_{i,j} - CP_{v,j}f(CP_i) < f(CP_v) \end{cases}, i \neq v \neq 1 \qquad (11)$$

where $CP_{v,j}$ is the randomly selected carnivorous plant; $CP_{1,j}$ is the first-ranked carnivorous plant; *reprocduction_rate* is a predefined value between 2.0 and 3.0.

2.2 Lévy Flight

The Lévy flight is a generalized random walk which obeys the Lévy distribution, and its step length is described by the 'heavy tail' probability distribution during the walk. Compared with the ordinary random walk, Lévy flight's short-range local search is interlinked with the occasional long-range global search.

At present, the Mantegna algorithm is usually used to generate random numbers with Lévy flight characteristics, whose expression is represented as follows:

$$levy\_rand = \frac{\mu}{|v|^{1/\beta}} \qquad (12)$$

where μ and v satisfy normal distribution, defined as:

$$\mu \sim N\left(0, \sigma_\mu^2\right) \qquad (13)$$

$$v \sim N\left(0, \sigma_v^2\right) \qquad (14)$$

where σ_μ and σ_v are calculated as follows:

$$\sigma_\mu = \left\{ \frac{\Gamma(1+\beta)\sin\left(\frac{\pi\beta}{2}\right)}{\Gamma\left[\frac{(1+\beta)}{2}\right]\beta 2^{\frac{(1+\beta)}{2}}} \right\} \qquad (15)$$

$$\sigma_v = 1 \qquad (16)$$

The value of β in the above formulas is usually 1.5.

To prove the superiority of Lévy flight strategy, the Lévy flight and random walk strategy were simulated by MATLAB. The steps in Lévy flight and random walk were both set to 300. The simulation results are shown as Fig. 1.

Seen from Fig. 1 that: Lévy flight strategy can extend the search area and improve the globally-search efficiency due to its special walking mode, namely the combination of long and short steps; on the contrary, the search range of random walk strategy is more intensive, which is more conducive to local optimization in the development stage.

3 Modified LCPA

Normally, the mode shifts of Lévy flight tend to occur in food-deprived environments, while organisms are dominated by random movements in food-rich areas [19]. CPA is to simulate carnivorous plants to prey, grow and reproduce in harsh environments. Considering that the trajectory of flies, ants and other preys has the characteristics of Lévy flight, the Lévy flight policy is introduced to improve CPA.

In this paper, Mantegna method is used to generate random number which obeys the Lévy distribution to replace the ordinary random number.

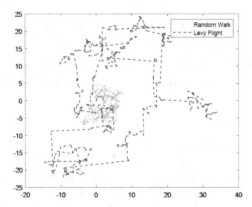

Fig. 1. Lévy flight and random walk strategy simulation.

3.1 Lévy Flight Design in Growth

When carnivorous plants successfully prey, $levy\_rand_{i,j}$ calculated by Eq. (12) is used in place of $rand$ in Eq. (7). The improved growth model of new carnivorous plants is described as:

$$NewCP_{i,j}\prime = growth\prime \times CP_{i,j} + (1 - growth\prime) \times Prey_{v,j} \tag{17}$$

$$growth\prime = growth\_rate \times levy\_rand_{i,j} \tag{18}$$

Similarly, the growth model improved by Lévy flight strategy of preys is as follows:

$$NewPrey_{i,j}\prime = growth\prime \times Prey_{u,j} + (1 - growth\prime) \times Prey_{v,j}, \ u \neq v \tag{19}$$

$$growth\prime = \begin{cases} growth\_rate \times levy\_rand_{i,j} f(prey_v) > f(prey_u) \\ 1 - growth\_rate \times levy\_rand_{i,j} f(prey_v) < f(prey_u) \end{cases} \tag{20}$$

The greedy mechanism should be used to compare the fitness of original position and the improved one, and the position with better fitness will be retained.

$$NewCP_{i,j} = \begin{cases} NewCP_{i,j} f(NewCP_{i,j}\prime) > f(NewCP_{i,j}) \\ NewCP_{i,j}\prime f(NewCP_{i,j}\prime) < f(NewCP_{i,j}) \end{cases} \tag{21}$$

$$NewPrey_{i,j} = \begin{cases} NewPrey_{i,j} f(NewPrey_{i,j}\prime) > f(NewPrey_{i,j}) \\ NewPrey_{i,j}\prime f(NewPrey_{i,j}\prime) < f(NewPrey_{i,j}) \end{cases} \tag{22}$$

In this way, the population is updated by combining the long and short steps of Lévy flight, which can enhance the capacity of the algorithm to leap out of the local optimum and approach the global optimal solution as close as possible.

3.2 Population Reproduction Process of LCPA

The random number subject to Lévy flight distribution is also recommended into the reproduction of the carnivorous plant.

$$NewCP_{i,j'} = CP_{1,j} + reprocduction\_rate \times levy\_rand_{i,j} \times mate_{i,j} \qquad (23)$$

Finally, using Eq. (22), the better solution is retained according to the fitness greedy selection.

4 Simulation Experiment and Results Analysis

4.1 Test Function and Experiment

Twelve benchmark test functions with different mathematical characteristics are selected from reference [20], to prove the performance of the improved algorithm LCPA. F_1-F_7 are unimodal functions and F_8-F_{12} are multimodal functions, which are applied to test the capability of algorithms, including convergence rate and optimization accuracy.

There are 4 kinds of comparison algorithms: CPA, DE, PSO and SSA. To fairly contrast the property of the algorithms, the parameters of all algorithms are set according to the parameters referred in the original paper [6, 9, 11, 18]. The parameters used in each algorithm are generalized as Table 1.

Table 1. Parameter settings of all algorithms.

Algorithm	Parameter Setting
PSO	$population\,size = 30$; $c_1 = c_2 = 2$; $\omega Max = 0.9$; $\omega Min = 0.2$
DE	$population\,size = 30$; $F = 3$; $Cr = 0.8$
SSA	$population\,size = 30$; $c_1 = rand$; $c_2 = rand$
CPA	$nCPlant = 10$; $nPrey = 20$; $group\_iteration = 2$; $attraction\_rate = 0.8$; $growth\_rate = 2$; $reproduction\_rate = 1.8$
LCPA	$nCPlant = 10$; $nPrey = 20$; $group\_iteration = 2$; $attraction\_rate = 0.8$; $growth\_rate = 2$; $reproduction\_rate = 1.8$; $\beta = 1.5$

To avoid accidental errors, for each test function, all algorithms were simulated in 10 independent times, and the average value and standard deviation of the experiment were recorded. When the iterations reached the maximum, which was set to 1000, each experiment was terminated.

4.2 Comparative Analysis of Algorithm Performance

There are two groups of experiments with different dimensions to evaluate the capability of LCPA. Table 2 summarizes the mean value and standard deviation of CPA and other comparison algorithms on F_1-F_{12} with Dim = 30, while Table 3 presents the results with Dim = 50. Figure 2 shows the convergence curve of all algorithms when dealing with the test function (Dim = 30).

As shown in the optimization convergence curve, LCPA performs best in convergence speed in most cases. Especially in F_1, F_3 and F_4, the convergence curves of CPA, SSA, PSO and DE are basically overlapping, where these algorithms are always trapped in local optimums thus leading to slower optimality seeking. Conversely, LCPA combined with Lévy flight do not have the above situation. It is because Lévy flight strategy effectively expands the search range that the algorithm can leap out of local optimum in time in the later period.

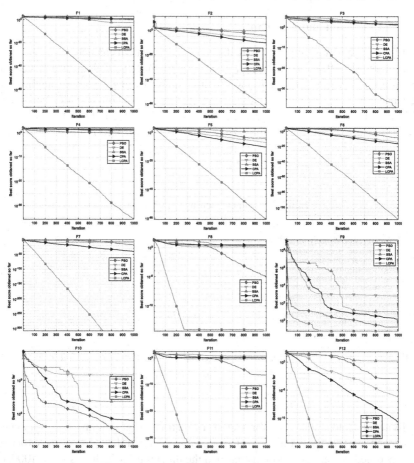

Fig. 2. Convergence curve of all algorithms.

By analyzing experimental data in Table 2 and Table 3, it can be concluded that for all unimodal functions, the optimization accuracy of LCPA is significantly superior to other algorithms. LCPA always finds the optimal solution closest to the theoretical optimum. Besides, the standard deviations of the results in all unimodal functions are

Table 2. Comparison of results on F_1-F_{12} with 30D.

Function	Performance	PSO	DE	SSA	CPA	LCPA
F_1	Mean	2.94e + 00	1.19e + 00	6.86e-03	1.07e + 00	4.58e-75
	Std	1.97e + 00	4.95e-01	5.14e-03	1.91e + 00	7.01e-75
F_2	Mean	3.03e-04	2.68e-08	1.05e + 00	4.54e-11	6.84e-63
	Std	5.43e-04	1.56e-08	9.59e-01	2.75e-11	1.17e-62
F_3	Mean	1.63e + 01	6.94e + 01	4.92e + 02	2.00e + 01	1.98e-40
	Std	7.35e + 00	3.35e + 01	3.38e + 02	1.55e + 01	4.17e-40
F_4	Mean	5.31e-01	2.56e + 01	8.03e + 00	2.19e + 01	1.34e-36
	Std	1.14e-01	9.50e + 00	3.43e + 00	5.88e + 00	1.88e-36
F_5	Mean	2.69e-04	2.08e-07	6.21e + 00	4.56e-10	2.11e-62
	Std	3.65e-04	2.34e-07	9.08e + 00	5.45e-10	4.48e-62
F_6	Mean	5.31e-07	7.63e-09	6.00e-01	9.72e-17	1.53e-115
	Std	1.65e-06	2.41e-08	8.49e-01	1.82e-16	3.72e-115
F_7	Mean	1.28e-15	1.82e-09	6.12e-09	5.58e-07	0
	Std	2.39e-15	4.19e-09	2.30e-09	3.98e-07	0
F_8	Mean	8.83e-05	5.95e-01	2.05e + 00	2.23e + 00	5.15e-15
	Std	1.24e-04	8.09e-01	6.01e-01	1.28e + 00	1.50e-15
F_9	Mean	2.07e-02	2.53e + 01	1.05e + 00	3.47e + 00	2.07e-02
	Std	4.37e-02	7.95e + 01	1.64e + 00	1.64e + 00	4.37e-02
F_{10}	Mean	3.30e-03	1.87e + 02	1.42e + 00	5.23e-01	4.39e-03
	Std	5.31e-03	5.80e + 02	3.23e + 00	1.19e + 00	5.67e-03
F_{11}	Mean	6.33e-02	6.44e-01	7.93e + 00	3.60e + 00	4.54e-02
	Std	1.42e-01	1.13e + 00	3.15e + 00	3.25e + 00	1.44e-01
F_{12}	Mean	9.51e-04	9.58e-08	3.02e + 00	1.12e-11	3.26e-15
	Std	1.60e-03	1.65e-07	1.71e + 00	1.06e-11	1.13e-15

closest to zero, which means LCPA has strong stability. The convergence accuracy of LCPA is much better than that of the original algorithm CPA on all seven single-peak functions, all of which are improved by at least 35 orders of magnitude. As for F_7, the optimal solution found by LCPA has reached zero.

Table 3. Comparison of results on F_1-F_{12} with 50D.

Function	Performance	PSO	DE	SSA	CPA	LCPA
F_1	Mean	6.49e-04	8.18e-05	2.50e-10	8.27e-07	4.01e-93
	Std	6.05e-04	1.34e-04	9.25e-11	2.49e-06	8.23e-93
F_2	Mean	1.36e-01	3.98e-03	4.64e + 00	1.94e-04	1.44e-50
	Std	1.70e-01	3.16e-03	1.04e + 00	4.38e-04	1.86e-50
F_3	Mean	5.86e + 02	5.45e + 03	4.80e + 03	2.62e + 03	4.83e-26
	Std	1.73e + 02	2.07e + 03	1.77e + 03	8.01e + 02	1.32e-25
F_4	Mean	2.28e + 00	4.59e + 01	1.73e + 01	4.46e + 01	2.86e-27
	Std	2.01e-01	2.41e + 01	2.86e + 00	4.22e + 00	3.98e-27
F_5	Mean	2.29e-01	3.12e-01	4.65e + 01	7.07e-04	4.56e-49
	Std	3.97e-01	9.62e-01	3.31e + 01	5.50e-04	9.88e-49
F_6	Mean	2.10e-02	5.96e-04	7.94e + 00	6.82e-06	5.02e-91
	Std	2.34e-02	9.19e-04	7.23e + 00	4.49e-06	8.70e-91
F_7	Mean	3.35e-01	3.79e-08	5.55e-07	1.28e-19	0
	Std	3.12e-01	6.12e-08	3.42e-07	3.02e-19	0
F_8	Mean	1.00e + 00	3.30e + 00	3.48e + 00	3.42e + 00	7.99e-15
	Std	7.17e-01	2.59e + 00	8.18e-01	2.20e + 00	0
F_9	Mean	1.87e-02	1.23e + 04	8.16e + 00	1.81e + 00	6.22e-02
	Std	3.00e-02	3.13e + 04	3.21e + 00	1.77e + 00	1.17e-01
F_{10}	Mean	1.57e-02	2.51e + 04	5.74e + 01	7.92e-01	2.20e-03
	Std	3.23e-02	4.99e + 04	1.43e + 01	1.36e + 00	4.63e-03
F_{11}	Mean	5.00e + 00	4.02e + 00	1.16e + 01	1.07e + 01	4.36e-01
	Std	3.40e + 00	1.59e + 00	5.59e + 00	4.21e + 00	7.05e-01
F_{12}	Mean	7.96e-02	1.12e-03	6.40e + 00	1.76e-05	6.88e-15
	Std	1.46e-01	1.22e-03	2.02e + 00	1.39e-05	3.66e-15

The comparison results on the multimodal functions reveal that LCPA can overcome the attraction of regional extremum to some extent, and ensure the convergence speed while finding a better solution. In the process of optimizing F_8, F_{11} and F_{12}, the optimization accuracy of LCPA perform best. For F_9, overall, PSO has better convergence accuracy and speed, but it is not as stable as LCPA. For F_{10}, PSO performs best in the low-dimensional case of Dim = 30, followed by LCPA, while LCPA performs better than PSO and other algorithms in the higher-dimensional case of Dim = 50.

For the results of the test function with dimension 30, the Wilcoxon test [21] is applied to prove whether there is a significant difference between LCPA and the comparison algorithm. It can be seen from Table 4 that the p-values are almost less than 0.05, which means that the superiority of the algorithm proposed in this paper is significant compared with other algorithms (Table 4).

Table 4. p-value of Wilcoxon statistical test results (Dim = 30).

Dataset	LCPA vs PSO	LCPA vs DE	LCPA vs SSA	LCPA vs CPA
F_1	1.83e-04	1.83e-04	1.83e-04	1.83e-04
F_2	1.83e-04	1.83e-04	1.83e-04	1.83e-04
F_3	1.83e-04	1.83e-04	1.83e-04	1.83e-04
F_4	1.83e-04	1.83e-04	1.83e-04	1.83e-04
F_5	1.83e-04	1.83e-04	1.83e-04	1.83e-04
F_6	1.83e-04	1.83e-04	1.83e-04	1.83e-04
F_7	6.39e-05	6.39e-05	6.39e-05	6.39e-05
F_8	1.10e-04	1.10e-04	1.10e-04	1.09e-04
F_9	8.86e-03	2.03e-03	1.10e-04	1.19e-03
F_{10}	9.93e-02	4.04e-03	1.46e-03	2.37e-02
F_{11}	1.33e-03	7.56e-04	8.74e-05	8.74e-05
F_{12}	1.83e-04	1.83e-04	1.83e-04	1.83e-04

5 Practical Applications of the Modified Algorithm

The proposal and improvement of the optimization algorithm are finally used to solve practical application problems. This chapter describes how to apply the modified algorithm LCPA to BP neural network classification problem, which is under the simulation experiments using MATLAB R2021a. The parameters of the algorithms involved are the same as those set in Sect. 4.1, and each algorithm runs independently 30 times.

5.1 Neural Network Training Based on LCPA

Back-Propagation (BP) Network [22] is a concept proposed by scientists led by Rumelhart and McClelland in 1986. It is a multi-layer feedforward network with back-propagation training according to the errors. BP network has been extensively implemented in model classification, image processing, function approximation and time series prediction due to high nonlinearity and strong generalization ability. Structurally, BP network is composed of input layer, hidden layer, and output layer. The hidden layer, sandwiched between the input layer and the output layer, can be one layer or multiple layers. Figure 3 shows the three-layer BP neural network topology.

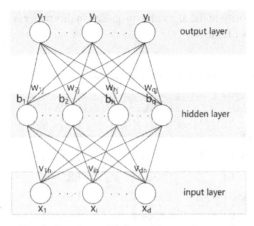

Fig. 3. Topology of 3-layer BP neural network.

Essentially, the BP algorithm takes the square of the network error as the objective function, and constantly adjusts the weights and thresholds of the network through back propagation, so that the objective function decreases along the negative gradient direction to reach the minimum value. However, when the problem is more complex or difficult to obtain gradient information, BP neural network may be helpless, which leads to slow convergence speed and local minimization problems. For the purpose of avoiding the above situation, this paper introduces the carnivorous plant algorithm based on Lévy flight to continuously optimize the weight and threshold of BP network, to minimize the sum of squared errors of the network. BP model and LCPA algorithm are combined to form LCPA-BP model. In the LCPA-BP model, the position vector of the individual in the LCPA represents the connection weights and thresholds of the neural network; the dimension corresponds to the number of ownership values and thresholds; the sum of squared errors of the network is regarded as the fitness function in the LCPA algorithm. The smaller the value of the fitness function, the more excellent the search performance of the algorithm. When population iteration times reach the maximum or the error precision finds the optimal one, the optimization process will be terminated and the best connection weights and thresholds in the BP neural network will be output.

The mathematical model of mean square error for all training cases is as follows:

$$F = \frac{1}{N} \sum_{t=1}^{N} \sum_{j=1}^{C} \left(y_{i,j} - t_{i,j}\right)^2 \tag{24}$$

where N is the total number of training samples; C is the number of output nodes; $y_{i,j}$ is the actual output of the network; $t_{i,j}$ is the expected output of the network.

In order to certify the superiority of LCPA-BP model, three popular data sets from UCI machine learning and Intelligent Systems Center, including Iris, Wine, and Wine Quality, are selected for comparative experiments. The characteristics of the above three data sets are shown in Table 5. Table 6 shows the comparison results of the classification accuracy of the LCPA-BP model and the original BP on each data set.

Table 5. Characters of UCI datasets.

Dataset	Number of samples	Number of features
Iris	150	4
Breast Tissue	106	9
Wine	178	13

Table 6. Classification accuracy of each data set.

Dataset	Model	Classification accuracy
Iris	LCPA-BP	100%
	BP	95.2%
Breast Tissue	LCPA-BP	80%
	BP	57.7%
Wine	LCPA-BP	100%
	BP	94%

It is obvious that the classification accuracy of the LCPA-BP model is enormously higher than that of the BP model. For both Iris and Wine datasets, the classification accuracy of the LCPA-BP model is as high as 100%. For Breast Tissue dataset, the classification accuracy is improved by more than 20% compared with the original model.

6 Conclusion

This paper improves the carnivorous plant algorithm using Lévy flight strategy and proposes an improved algorithm LCPA. The introduction of Lévy flight strategy can enhance the diversity of population and improve the global exploration ability of the algorithm to a certain extent. For the purpose of verifying the effectiveness of the modified algorithm, 12 benchmark functions are utilized to demonstrate the performance of LCPA. The comparison results indicate that, the proposed algorithm LCPA has faster convergence rate, higher accuracy, and more excellent global search ability, especially in resolving high-dimensional continuous optimization problems. Meanwhile, the superiority of LCPA is well proved in the Wilcoxon test. Finally, the practical application of BP neural network classification proves the feasibility of LCPA, not just limited to theoretical basis. It is believed that this algorithm can perform well in other real-world problems in the future.

Acknowledgement. This work is supported by the National Natural Science Foundation of China (No. 71863018 and No. 71403112), Jiangxi Provincial Social Science Planning Project (No. 21GL12) and Technology Plan Projects of Jiangxi Provincial education Department (No. GJJ200424).

References

1. Wang, X., Hu, H., Liang, Y., Zhou, L.: On the mathematical models and applications of swarm intelligent optimization algorithms. Arch. Comput. Methods Eng. **29**(6), 3815–3842 (2022)
2. Dokeroglu, T., Sevinc, E., Kucukyilmaz, T., Cosar, A.: A survey on new generation metaheuristic algorithms. Comput. Ind. Eng. **137**, 106040 (2019)
3. Hare, W., Nutini, J., Tesfamariam, S.: A survey of non-gradient optimization methods in structural engineering. Adv. Eng. Softw. **59**, 19–28 (2013)
4. Alba, E., Tomassini, M.: Parallelism and evolutionary algorithms. IEEE Trans. Evol. Comput. **6**(5), 443–462 (2002)
5. Mirjalili, S.: Genetic algorithm. Evolutionary Algorithms and Neural Networks: Theory Appl., 43-55 (2019)
6. Pant, M., Zaheer, H., Garcia-Hernandez, L., Abraham, A.: Differential evolution: a review of more than two decades of research. Eng. Appl. Artif. Intell. **90**, 103479 (2020)
7. Suman, B., Kumar, P.: A survey of simulated annealing as a tool for single and multiobjective optimization. J. Oper. Res. Soc. **57**, 1143–1160 (2006)
8. Beheshti, Z., Shamsuddin, S.M.H.: A review of population-based meta-heuristic algorithms. Int. J. Adv. Soft Comput. Appl **5**(1), 1–35 (2013)
9. Gad, A.G.: Particle swarm optimization algorithm and its applications: a systematic review. Arch. Comput. Methods Eng. **29**(5), 2531–2561 (2022)
10. Yurtkuran, A., Emel, E.: An adaptive artificial bee colony algorithm for global optimization. Appl. Math. Comput. **271**, 1004–1023 (2015)
11. Mirjalili, S., Gandomi, A.H., Mirjalili, S.Z., Saremi, S., Faris, H., Mirjalili, S.M.: Salp swarm algorithm: A bio-inspired optimizer for engineering design problems. Adv. Eng. Softw. **114**, 163–191 (2017)
12. Pavlyukevich, I.: Lévy flights, non-local search and simulated annealing. J. Comput. Phys. **226**(2), 1830–1844 (2007). https://doi.org/10.1016/j.jcp.2007.06.008
13. Barthelemy, P., Bertolotti, J., Wiersma, D.S.: A Lévy flight for light. Nature **453**(7194), 495–498 (2008)
14. Yang, X. S., Deb, S.: Cuckoo search via Lévy flights. In 2009 World congress on nature & biologically inspired computing (NaBIC), pp. 210–214. IEEE (2009, December)
15. Xie, J., Zhou, Y., Chen, H.: A novel bat algorithm based on differential operator and Lévy flights trajectory. Computational intelligence and neuroscience, Vol 2013(1), p .453812(2013)
16. Qingxi, W., Xiaobo, G.: Particle swarm optimization algorithm based on Levy flight. Appl. Res. Comput. **33**(9), 2588–2591 (2016)
17. Ma, W., Zhu, X.: Sparrow search algorithm based on Levy flight disturbance strategy. J. Appl. Sci. **40**(1), 116–130 (2022)
18. Ong, K.M., Ong, P., Sia, C.K.: A carnivorous plant algorithm for solving global optimization problems. Appl. Soft Comput. **98**, 106833 (2021)
19. Emary, E., Zawbaa, H.M., Sharawi, M.: Impact of Lèvy flight on modern meta-heuristic optimizers. Appl. Soft Comput. **75**, 775–789 (2019)
20. Jamil, M., Yang, X.S.: A literature survey of benchmark functions for global optimisation problems. Int. J. Math. Model. Numer. Optimisation **4**(2), 150–194 (2013)
21. Smida, Z., Cucala, L., Gannoun, A., Durif, G.: A Wilcoxon-Mann-Whitney spatial scan statistic for functional data. Comput. Stat. Data Anal. **167**, 107378 (2022)
22. Xie, R., Wang, X., Li, Y., Zhao, K.: Research and application on improved BP neural network algorithm. In: 2010 5th IEEE Conference on Industrial Electronics and Applications, pp. 1462–1466 IEEE (2010, June)

CR-IFSSL: Imbalanced Federated Semi-Supervised Learning with Class Rebalancing

Yutong Xie[1], Haiyan Liang[1], Xianmin Wang[1(✉)], Jing Li[1(✉)], Ziyu Cheng[1], Siming Huang[1], Feng Liu[2], and Li Guo[3]

[1] Guangzhou University, Guangzhou, Guangdong, China
{xianmin,lijing}@gzhu.edu.cn
[2] Luoding Polytechnic, Ximengang, Luoding, Guangdong, China
[3] Zhongshan Road Primary School, Ganzhou, Jiangxi, China

Abstract. Federated semi-supervised learning is a popular research area known for its ability to preserve data privacy, utilize client data efficiently, and reduce the need for a large number of manually labeled samples. However, most existing studies focus on scenarios where client data is independently and identically distributed (IID). In practical applications, client data often deviates from the IID assumption, posing significant challenges for existing federated semi-supervised learning algorithms. In this paper, we proposed a novel approach called Class Rebalancing Imbalanced Federated Semi-Supervised Learning (CR-IFSSL) designed specifically to address the class imbalance issue in non-IID settings, thereby expanding the applicability of federated semi-supervised learning. The primary principle of our method is to conduct local class rebalance training for each individual client. Specifically, to handle the problem of class imbalance, each client's contribution is weighted according to its estimated class distribution when selecting pseudo-label samples. This ensures that samples from minority classes are chosen more frequently, thereby mitigating the class imbalance issue. Furthermore, our method gradually adjusts the alignment strength in the selftrained predictive data distribution to fine-tune the model'sperformance. Experimental results demonstrate that CR-IFSSL outperforms existing federated semi-supervised learning algorithms when dealing with class imbalance data, thereby offering valuable insights for real-world data applications.

KeywordS: federated semi-supervised learning · class imbalance · class rebalancing · pseudo-labeled samples

1 Introduction

Traditional machine learning methods usually need to train the entire dataset on a centralized server. This process involves the transfer of large amounts of sensitive data and therefore introduces significant privacy vulnerabilities. In contrast, Federated Learning (FL) [1–3] takes a decentralized approach, distributing the training process to local

© The Author(s), under exclusive license to Springer Nature Singapore Pte Ltd. 2024
K. Li and Y. Liu (Eds.): ISICA 2023, CCIS 2146, pp. 409–419, 2024.
https://doi.org/10.1007/978-981-97-4393-3_33

devices. This innovative approach involves model training on independent datasets distributed across different locations or devices. [4, 5] It only shares model parameter updates with a central server, eliminating the need to exchange raw data directly and providing strong data privacy protection [6–8]. This distributed training mode not only reduces the data transmission cost and delay, but also enables global model optimization without affecting the privacy of individual data.

One common limitation of existing Federated Learning (FL) methods is that they primarily focus on supervised learning settings, assuming that locally private data are fully labeled. However, in practical scenarios, it's unlikely that all data examples will come with complex annotations. For instance, consider the context of federated learning on a device, where users may be reluctant to invest time and effort in manually labeling their data. To address this issue, several studies have introduced the concept of Semi-Supervised Federated Learning (FSSL) [9–11]. Semi-supervised Federated Learning extends the application of semi-supervised learning methods to federated learning scenarios, aiming to combine the advantages of both techniques to tackle real-world problems more effectively.

Existing Federated Semi-Supervised Learning (FSSL) methods [12] typically assume that the training data follows the independent and identically distributed (IID) condition. However, in real-world scenarios, data often deviates from this assumption and exhibits class imbalance, which is a common manifestation of non-IID data distribution. For instance, consider an animal classification dataset where there are numerous images of common animals like dogs and cats, but very few samples of rare animals like snow leopards. As the rarity of a category increases, the cost associated with collecting more samples for that category increases exponentially. This non-IID data distribution poses significant challenges for the field of federated semi-supervised learning. In essence, the presence of class imbalance in real-world datasets makes it difficult to apply traditional FSSL methods directly. Addressing this issue is essential for developing more effective and practical federated semi-supervised learning techniques that can handle the complexities of real data distributions.

In response to the challenge posed by the inability of existing Federated SemiSupervised Learning (FSSL) methods to handle imbalanced class data [13, 14] in non-IID scenarios, we proposed a novel semi-supervised federated training framework called CR-IFSSL, illustrated in Fig. 1. This framework addresses this issue through two main strategies. Firstly, we improve the reliability and accuracy of online pseudolabeling by progressive distribution alignment, while selecting samples from minority classes more frequently according to their estimated class distribution, achieving class rebalancing and reducing the overhead of manual data processing. Secondly, we adopt a random selection approach for local training using subsets of local clients, coupled with a distance-reweighted model aggregation method to generate a sub-consensus model. This iterative process is repeated multiple times, making the federated learning model more robust. This framework enables the application of class-imbalanced data to FSSL. Experimental results demonstrate that the proposed method significantly enhances modeling outcomes in comparison to existing FSSL approaches, offering a promising solution for challenging real-world scenarios.

The main contribution of our work can be summarized as follows:

- We proposed a progressive distribution alignment approach to improve the reliability and accuracy of online pseudo-labeling. This approach enhances the pseudo-label generation process and achieves class rebalanceing, resulting in improved model performance and a reduction in the manual data processing overhead, particularly in scenarios involving imbalanced class data under non-IID conditions.
- We present a novel distance-reweighted model aggregation method to create a sub-consensus model. This approach involves local training on randomly selected subsets of local clients and employs distance-based reweighting during the model aggregation process. Through iterative repetitions of this procedure, the federated learning model gains increased robustness.

2 Related Work

2.1 Semi-Supervised Learning

Recent years have observed a significant advancement of SSL [15]. While deep learning has achieved remarkable success in tasks like image and speech recognition, it often requires a large number of parameters to learn useful abstractions. This parameter-rich nature increases the risk of overfitting, where the model becomes too specialized to the training data and performs poorly on unseen examples.Semi-supervised learning, as an approach in machine learning, aims to address the limitations of deep learning models by utilizing both labeled and unlabeled data during training. By incorporating both labeled and unlabeled data, these models can learn more robust and generalized representations, leading to improved performance on various tasks.

2.2 Federated Semi-Supervised Learning

Federated Learning [16], a decentralized approach in machine learning, has gained significant attention in recent years. It allows multiple participants to collaboratively train a shared model without directly sharing their private data. This approach addresses privacy concerns and data security while enabling the development of powerful models.

In traditional machine learning, data from different sources are centralized for training. However, in scenarios where data is sensitive or distributed across multiple devices or organizations, centralization may not be feasible or desirable due to privacy regulations or proprietary concerns. This is where federated learning [3, 17, 18] comes into play.

In federated learning, each participant (such as a user's device or an organization's server) locally trains a model using their own private data. The models are then aggregated, typically by a central server, to create a global model that encapsulates the knowledge from all participants. During this process, only model updates or gradients are exchanged, ensuring that the raw data remains secure and private on individual devices.

Privacy and security are maintained through various techniques such as encryption and differential privacy. Encryption ensures that data remains protected during transmission, and differential privacy adds noise to the shared gradients to prevent the exposure of individual data points. These mechanisms allow participants to contribute their data without compromising its confidentiality.

2.3 Class Imbalanced Semi-Supervised Learning

Class imbalance refers to the situation where the number of samples in different classes is significantly unbalanced, resulting in the performance degradation of machine learning models. SSL methods aim to leverage both labeled and unlabeled data to improve model performance, but they may suffer from class imbalance. This post addresses the class imbalance problem in machine learning, specifically in the context of semi-supervised learning (SSL). In this paper, we propose a method called Suppressed Consistency Loss (SCL) that works robustly in the context of class imbalanced semi-supervised learning (CISSL). In this paper, we analyze the existing SSL methods in CISSL, and propose a solution SCL. Experiments show that SCL performs well in CISSL environment compared with the existing SSL and Class Imbalance Learning (CIL) methods.

3 Methodology

Our proposed method(CR-IFSSL) is outlined in Fig. 1. Specifically, there are two stages: 1) Fully-unlabeled clients are trained locally through class rebalancing, selecting samples from a few classes as pseudo-labeled samples, and using progressive distribution alignment to improve the reliability and accuracy of online pseudo-labels. 2) A distance reweighted model aggregation approach was used to generate subconsensus models. By repeating this process several times, a set of subconsensus models is obtained and these models are aggregated into a global model.

3.1 Problem Setup

Our study first establishes the problem of class imbalance in federated semisupervised learning in the class imbalanced case. In the class imbalanced semisupervised federated environment, fully-labeled and fully-unlabeled clients were considered, denoted as $\{C^1,\ldots,C^m\}$ and $\{C^{m+1},\ldots,C^{m+n}\}$, respectively. Each fullylabeled client has a local dataset defined as

$$D^l = \left\{ \left(X_i^l, Y_i^l \right) \right\}_{i=1}^{N^l} \tag{1}$$

while the dataset D^l of fully-unlabeled clients is defined as $\left\{ \left(X_i^u \right) \right\}_{i=1}^{N^u}$. The number of training examples in X of class 1 is denoted as N_i, i.e. $\sum_{i=1}^{n} N_i = N$. We assume that the classes are sorted in descending order,i.e. $N_1 > N_2 \geq \ldots \geq N_n$. For an L-class classification task, our goal is for each client's classifier f to generalize well on class-balanced test data after the first stage, and to obtain a better global model θ_{glob} in the second stage by simultaneously utilizing labeled and unlabeled data in a distributed scheme.

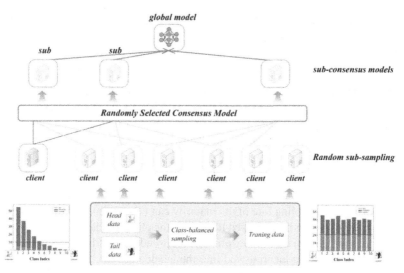

Fig. 1. An overview of our proposed CR-IFSSL. Firstly, we perform classbalanced sampling on each client for local training to address class imbalance issues. Next, we conduct multiple rounds of random sub-sampling, selecting samples from different clients for training. This helps expand the labeled dataset and improves the model's generalization ability. After the sub-sampling process, we aggregate the obtained sub-models through a consensus mechanism. This involves merging and averaging the model parameters to generate the aggregated model. Finally, we aggregate the global model once again to obtain the final integrated model. By employing this approach, we can leverage the data from various.

3.2   Local Training by Class Rebalancing

Training models on imbalanced data tends to have certain advantages in recall for the majority class but may be lacking in precision for the minority class. To address this issue, in the first stage, a self-training framework with class rebalancing was employed, using the high precision of the minority classes to alleviate the decrease in the recall rate. This approach effectively tackles class imbalance, unlike traditional classifiers that exhibit biases due to imbalanced class distributions. Our proposed CR-IFSSL achieves class re-balancing by adjusting sample weights, avoiding further biases caused by online pseudo-labels generated from unlabeled data and mitigating the impact of class imbalance.

Self-training is an iterative approach widely used to handle datasets with incomplete labels [19]. Conventional methods involve training a teacher model on labeled data and using this model to generate pseudo-labels for unlabeled data. However,to solve the class imbalance problem, our method combines fullylabeled and fully-unlabeled client data, especially in fully-unlabeled client data, adhering to the principle of class rebalance: The unlabeled samples of the less frequent category l were selected as the specific subset of the pseudo-label set.

Since the data distribution of the fully-labeled and fully-unlabeled clientsamples follows the class imbalanced distribution law of realistic data, this problem can be solved by estimating the category distribution from the fully labeled client dataset.

For local training on labeled clients, the labeled client loss Ls is the crossentropy between the model prediction and the label:

$$L_s = y_i log(\hat{y}_i) \tag{2}$$

For local training on unlabeled clients, different from the traditional selftraining method by filtering the prediction using a score threshold [20], we adopt an adaptive sampling rate adjustment strategy in the pseudo-label set $\hat{S}$ gradually including unlabeled samples predicted as class "l" at a fine-tuning rate. This rate μl is calculated using a formula

$$\mu_l = \left(\frac{N_{L+1-l}}{N1}\right)^{\alpha} \tag{3}$$

In this way, the sampling rate of the majority can be reduced, and the sampling rate of the minority class can be improved to achieve the effect of class balance.

After each iteration of the training process, the pseudo-labeled samples obtained from the unlabeled dataset are incorporated into the labeled dataset, followed by retraining the local model. Formally, the unlabeled client loss Lu can be formulated as:

$$L_u = p_i \, log(\hat{p}_i) \tag{4}$$

where $\hat{p}_i$ is the prediction of local data from the local model, is the pseudolabel obtained after screening.

3.3 Aggregate A Global Model

In the second stage, our method uses random subsampling to obtain multiple subconsensus models. Specifically, at the beginning of each training round, M independent random subsamples were performed on K clients in order to obtain a set of randomly selected subsets. Then, a subset of each client is initialized using the global model and local training is performed for each client. Next, the global model is aggregated using the distance-based model aggregation (DMA) method [20], which involves reweighting the model by its distance to obtain the global model.

For each subset, we initially calculate the intra-subset averaged model θavg :

$$K_{total} = \sum_{i=1}^{K} K_i \text{ and } \theta_{avg} = \sum_{i=1}^{K} \frac{N_i}{N_{total}} \theta_i \tag{5}$$

where Ki represents the amount of data for the i-th client, θi stands its local model.DMA method scales ωi as follows:

$$\omega_i = \frac{K_i}{K_{total}} exp\left(-\beta \cdot \frac{\|\theta_i - \theta_{avg}\|_2}{K_i}\right), \tag{6}$$

$$\text{and } \overline{\omega}_i = \frac{\omega_i}{\sum_j \omega_j} \tag{7}$$

In the above equation, β is a hyper-parameter and $\|\theta_i - \theta_{avg}\|_2$ denotes the L2 norm of the model gradient between the local model and the intra-subset averaged model.

Then normalize the intra-subset model weight to [0,1]. A set of sub-consensus models is obtained, and were present their equal-weighted average as the final global model θglob:

$$\theta_{glob}^{t+1} = \frac{1}{M} \sum_{m=0}^{M-1} \theta_{sub}^m \tag{8}$$

In the given equation, θ_{sub}^m represents the mth sub-consensus model. Subsequently, the $t + 1^{th}$ synchronization round is performed with θ_{glob}^{t+1} as the initialization.

4 Experiments

To demonstrate the effectiveness and robustness of our proposed method, we conducted experiments on two benchmark datasets and further evaluated the CR-IFSSL under different ratios of unlabeled data. To simulate the FL (Federated Learning) setting, we randomly divided the training set into 10 different subsets, representing 10 local clients. We evaluated the model performance on each client using 10%, 20%, 30%, 40%, and 50% labeled data.

The CIFAR-100 dataset contains 100 image categories, each with 600 RGB images of 32 * 32 size, which is more challenging, while the SVHN dataset contains a large number of real-world color images rather than synthetic images, closer to practical application scenarios and more meaningful to apply in practice. And to better simulate the phenomenon of class imbalance, we used the CIFAR-100-LT and SVHN-LT datasets to conduct the experiments. During training on SVHN-LT and CIFAR-100-LT datasets, we followed the approach described in [5] (referencing RESC) and employed a simple CNN as the backbone for feature extraction. The CNN consisted of two 5×5 convolutional layers, a 2×2 max pooling layer, and two fully connected layers. Additionally, to ensure fair comparison across methods, we used the same classification network on each client. This was done in the context of federated learning settings.

We used the Dirichlet distribution (γ), where all two benchmark had $\gamma = 0.8$, as suggested by previous methods [21–23]. This strategy resulted in varying classes and sample quantities among the clients, meaning not all clients had samples from all classes. We reported the average results from three independent runs and extensively evaluated the classification performance using the AUC and ACC metrics. In both the RSCFed method and our proposed CR-IFSSL, we utilized a class-imbalanced dataset. As shown in Fig. 2, when the ratio of labeled to unlabeled CIFAR100-LT data was 4:6, the performance improved by 4%.Subsequently, we transitioned to employing the SVHN-LT dataset, where we observed a 1% enhancement in accuracy with an equal split of labeled and unlabeled clients (5:5 ratio), as depicted in Fig. 3.

4.1 Implementation Details

Firstly, we set the imbalance degree to 0.01 and used an exponential distribution to generate sample quantities. Based on the obtained sample quantities for each class, we performed sampling on the original datasets to create class-imbalanced datasets, namely SVHN-LT and CIFAR-100-LT. Next, we set the learning rates for labeled clients and

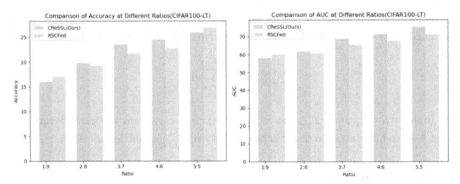

Fig. 2. The AUC and ACC results of the two methods are compared at different ratios on the CIFAR100-LT dataset.

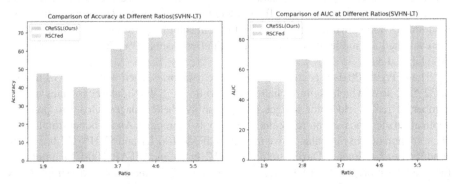

Fig. 3. The AUC and ACC results of the two methods were compared across different ratios using the SVHN-LT dataset.

unlabeled clients of all methods on SVHN-LT and CIFAR-100-LT as 0.03 and 0.021, respectively, with a batch size of 64. The training process consisted of two stages: the first 500 epochs involved training each client with class re-balancing and L2 regularization strength (weight decay) set to 0.0005. After that, the models were aggregated into a global model, and these two steps were repeated. For the next 500 epochs, we set the number of sub-sampling operations (M) to 3 and the number of local clients used in each sub-sampling operation (K) to 5.

4.2   Comparison with State-of-the-Arts

We conducted a comparison with the recent federated semi-supervised learning method RSCFed. RSCFed utilizes a distance-weighted model aggregation strategy to aggregate sub-consensus models into a global model, aiming to achieve consensus. We systematically studied the performance of both methods under different proportions of labeled and unlabeled clients, focusing mainly on accuracy and recall as key evaluation metrics. Through comprehensive analysis and comparative experiments, we aim to gain a deeper understanding of the effectiveness and robustness of these methods in the federated learning framework and provide scientific insights for improving and optimizing federated semisupervised learning.

Table 1 and Table 2 present the quantitative results of our approach and state-of-the-art methods on two benchmark datasets across five different proportions. Upon observation, it is evident that CR-IFSSL achieves the highest AUC in all scenarios, and it demonstrates superior performance in terms of accuracy in mostcases. This indicates that the classifiers trained using our method possess higher true positive rates and lower false positive rates at different thresholds, enabling better discrimination between positive and negative instances. One significant contributing factor is that we perform local class re-balancing training for each client prior to aggregation.

Table 1. Comparison of RSCFed and CR-IFSSL(Ours) on CIFAR100-LT

Labeled: Unlabeled	AUC(RSCFed)	ACC(RSCFed)	AUC(Ours)	ACC(Ours)
1:9	0.60	0.17	0.581	0.16
2:8	0.606	0.192	0.617	0.198
3:7	0.654	0.217	0.687	0.235
4:6	0.676	0.228	0.713	0.245
5:5	0.712	0.269	0.755	0.259

Table 2. Comparison of RSCFed and CR-IFSSL(Ours) on SVHN-LT

Labeled: Unlabeled	AUC(RSCFed)	ACC(RSCFed)	AUC(Ours)	ACC(Ours)
1:9	0.519	0.465	0.524	0.478
2:8	0.661	0.398	0.668	0.404
3:7	0.847	0.711	0.858	0.61
4:6	0.868	0.721	0.877	0.673
5:5	0.884	0.715	0.891	0.725

5 Conclusion

In this paper, we address a significant but often neglected concern concerning local clients in the context of federated semi-supervised learning, with a particular emphasis on class-imbalanced non-IID datasets. Our main contribution of our work involves conducting local class rebalance training for each individual client. Following this, we aggregate the global models through a combination of random sampling and distance-weighted model aggregation techniques. The results of our experiments on two class-imbalanced datasets demonstrate that our approach consistently outperforms state-of-the-art methods in the majority of cases. These findings provide strong empirical evidence for the effectiveness of our proposed methodology.

Acknowledgment. This work was supported by the National Natural Science Foundation of China (No. 62072127, No. 62002076), Natural Science Foundation of Guangdong Province (No. 2023A1515011774), Project 6142111180404 supported by CNKLSTISS.

References

1. McMahan, B., Moore, E., Ramage, D., Hampson, S., Arcas, B.A.: Communication-efficient learning of deep networks from decentralized data. In: Artificial Intelligence and Statistics, pp. 1273–1282 (2017)
2. Kairouz, P., et al.: Advances and open problems in federated learning. Found. Trends® Mach. Learn. **14**(1–2), 1–210 (2021)
3. Yang, Q., Liu, Y., Chen, T., Tong, Y.: Federated machine learning: concept and applications. ACM Trans. Intell. Syst. Technol. (TIST) **10**(2), 1–19 (2019)
4. Liu, Y., et al.: Fedvision: an online visual object detection platform powered by federated learning. In: Proceedings of the AAAI Conference on Artificial Intelligence, vol. 34, no. 08, pp. 13172–13179 (2020)
5. Li, Q., He, B., Song, D.: Model-contrastive federated learning. In: Proceedings of the IEEE/CVF Conference on Computer Vision and Pattern Recognition, pp. 10713–10722 (2021)
6. Kaissis, G.A., Makowski, M.R., Rückert, D., Braren, R.F.: Secure, privacy-preserving and federated machine learning in medical imaging. Nat. Mach. Intell. **2**(6), 305–311 (2020)
7. Kumar, R., et al.: Blockchain-federated-learning and deep learning models for covid-19 detection using ct imaging. IEEE Sensors J. **21**(14), 16301–16314 (2021)
8. Yang, D., et al.: Federated semi-supervised learning for COVID region segmentation in chest CT using multi-national data from China, Italy, Japan. Med. Image Anal. **70**, 101992 (2021)
9. Jeong, W., Yoon, J., Yang, E., Hwang, S.J.: Federated semi-supervised learning with inter-client consistency & disjoint learning. arXiv preprint arXiv:2006.12097 (2020)
10. Yurochkin, M., Agarwal, M., Ghosh, S., Greenewald, K., Hoang, N., Khazaeni, Y.: Bayesian nonparametric federated learning of neural networks. In: International Conference on Machine Learning, pp. 7252–7261 (2019)
11. Wang, H., Yurochkin, M., Sun, Y., Papailiopoulos, D., Khazaeni, Y.: Federated learning with matched averaging. arXiv preprint arXiv:2002.06440 (2020)
12. Krizhevsky, A., Sutskever, I., Hinton, G.E.: Imagenet classification with deep convolutional neural networks. Adv. Neural Inf. Process. Syst. **25** (2012)
13. Cui, Y., Jia, M., Lin, T.-Y., Song, Y., Belongie, S.: Classbalanced loss based on effective number of samples. In: Proceedings of the IEEE/CVF Conference on Computer Vision and Pattern Recognition, pp. 9268–9277 (2019)
14. Jamal, M.A., Brown, M., Yang, M.-H., Wang, L., Gong, B.: Rethinking class-balanced methods for long-tailed visual recognition from a domain adaptation perspective. In: Proceedings of the IEEE/CVF Conference on Computer Vision and Pattern Recognition, pp. 7610–7619 (2020)
15. Lee, D.-H., et al.: Pseudo-label: the simple and efficient semi-supervised learning method for deep neural networks. In: Workshop on Challenges in Representation Learning, ICML, vol. 3, no. 2, p. 896. Atlanta (2013)

16. Huang, D., Li, J., Chen, W., Huang, J., Chai, Z., Li, G.: Divide and adapt: active domain adaptation via customized learning. In: Proceedings of the IEEE/CVF Conference on Computer Vision and Pattern Recognition, pp. 7651–7660 (2023)

17. Kone˘cn'y, J., McMahan, H.B., Yu, F.X., Richt'arik, P., Suresh, A.T., Bacon, D.: Federated learning: Strategies for improving communication efficiency. arXiv preprint arXiv:1610.05492 (2016)

18. Zhao, Y., Li, M., Lai, L., Suda, N., Civin, D., Chandra, V.: Federated learning with non-IID data. arXiv preprint arXiv:1806.00582 (2018)

19. Yarowsky, D.: Unsupervised word sense disambiguation rivaling supervised methods. In: 33rd Annual Meeting of the Association for Computational Linguistics, pp. 189–196 (1995)

20. Liang, X., Lin, Y., Fu, H., Zhu, L., Li, X.: RSCFED: random sampling consensus federated semi-supervised learning. In: Proceedings of the IEEE/CVF Conference on Computer Vision and Pattern Recognition, pp. 10154–10163 (2022)

21. Wang, H., Yurochkin, M., Sun, Y., Papailiopoulos, D., Khazaeni, Y.: Federated learning with matched averaging. In: International Conference on Learning Representations (2020). https://openreview.net/forum?id=BkluqlSFDS

22. Yurochkin, M., Agarwal, M., Ghosh, S., Greenewald, K., Hoang, N., Khazaeni, Y.: Bayesian nonparametric federated learning of neural networks. In: International Conference on Machine Learning, pp. 7252–7261. PMLR (2019)

23. Li, T., Sahu, A.K., Zaheer, M., Sanjabi, M., Talwalkar, A., Smith, V.: Federated optimization in heterogeneous networks. Proc. Mach. Learn. Syst. **2**, 429–450 (2020)

Kernel Fence GAN: Unsupervised Anomaly Detection Model Based on Kernel Function

Lu Niu[1,2] and Shaobo Li[1,3(✉)]

[1] Chengdu Institute of Computer Applications, Chinese Academy of Sciences, Chengdu 610041, China
lishaobo@gzu.edu.cn
[2] School of Computer Science and Technology, University of Chinese Academy of Sciences, Beijing 100049, China
[3] State Key Laboratory of Public Big Data, Guizhou University, Guiyang 550025, China

Abstract. Data imbalance and difficult construction of high-dimensional data cause the low accuracy of anomaly detection. This paper proposes an unsupervised anomaly detection model Kernel Fence GAN to increase the accuracy. In this adversarial generative network, the generator's loss function consists of three items: the encirclement loss limits the generated samples to lie at the decision boundary of normal samples, the central dispersion loss maximizes the range of the decision boundary, and the enhanced dispersion loss uses kernel function to nonlinearly map the generated samples to RKHS space, which can maximize sample differences and avoid outliers. The purpose of training the discriminator is to accurately identify whether the input sample is a normal sample in the original dataset or an abnormal sample generated by the generator. In this paper, validation experiments are performed on two datasets: on two-dimensional synthetic dataset, the experiment result shows that the model has faster convergence rate than Fence GAN. On the KDD99 dataset, compared with traditional and advanced deep learning methods, both precision and recall have been correspondingly improved.

Keywords: Generative Adversarial Network (GAN) · kernel function · anomaly detection · Reproducing Kernel Hilbert Space (RKHS)

1 Introduction

Anomaly detection [1] is a challenging research problem in supervised and semi-supervised learning. The common anomaly detection method is to estimate the density of normal samples and then identify samples that do not conform to the normal density distribution. Anomaly refers to sample points in low-density areas. However, as the data dimension becomes higher, the data becomes sparse, and it is impossible to judge whether the samples are abnormal by low density. This phenomenon is often called the high-dimensional curse, and the difficulty of constructing high-dimensional data will lead to the low accuracy of anomaly detection. Anomaly detection has been widely used in various fields: medical diagnosis [2], industrial manufacturing, automatic driving and so on.

© The Author(s), under exclusive license to Springer Nature Singapore Pte Ltd. 2024
K. Li and Y. Liu (Eds.): ISICA 2023, CCIS 2146, pp. 420–428, 2024.
https://doi.org/10.1007/978-981-97-4393-3_34

For low-dimensional data, the commonly used method for anomaly detection is to observe normal samples to determine classification boundaries, including one-class SVMs [3] and kernel density estimation [4], etc. However, these methods exhibit limited efficacy when handling high-dimensional data. Existing anomaly detection models primarily focus on reducing data dimensionality. Yet, this reduction often leads to diminished information content, potentially discarding crucial data information. For example, methods based on AutoEncoder frameworks, such as Variational AutoEncoder [5], aim to derive a low-dimensional latent space from input samples through nonlinear mappings between encoders and decoders.

Generative adversarial network (GAN) [6] is implicit generative model that consist of a generator and a discriminator. When solely normal samples are present in the anomaly detection training process, it falls under the category of one-class classification. In the training phase of the generated model, only normal samples are used. Consequently, the algorithm used to generate the model can naturally extend to anomaly detection. Schlegl et al. [7] were the first to propose the use of generative adversarial network for anomaly detection in medical images. However, their approach required time-consuming gradient descent computations for each test sample, leading to expensive calculation times. ALAD [8], building upon the cyclic consistency and stability of bi-directional GANs, introduced a method that calculates the reconstruction errors of samples to assess anomalies' abnormality. Nguyen et al. [9] employed a multi-hypothesis network to comprehend the intricate distribution of high-dimensional data. This network comprised an encoder and multiple decoders.

Drawing from the aforementioned considerations, a novel approach harnessing kernel function is introduced to train Fence GAN [10] specifically tailored for anomaly detection, termed Kernel Fence GAN. In order to validate the accuracy and robustness of this proposed model, experiments were conducted on both 2D synthetic datasets and the KDD99 datasets. The obtained results outperformed those achieved by traditional and advanced deep learning methods, including Fence GAN.

2 Background

2.1 GAN (Generative Adversarial Network)

In the original GAN proposed by Goodfellow et al., a finite set of true samples $\{x_i\}_{i=1}^n \subset \mathcal{X} \subset R^d, d \in \mathbb{Z}^+, i = 1, 2, \ldots, N$ is given, these samples are sampled from an unknown density distribution $q(x)$. Generator in GAN seek to map from a simple density distribution $p_z(z)$ (such as a normal distribution, etc.) to a complex density distribution $p(\theta; z)$. In the training process, $p(\theta; z)$ is as close as possible to the real sample distribution $q(x)$, the generator $G_\theta(z)$ is realized by the multi-layer perceptron and θ is the corresponding network parameter. The discriminator $D_\phi(x')$ also consists of a multilayer perceptron, which ϕ is the corresponding network parameter, and the output value of the discriminator is 0 or 1, determining whether the sample originates from the generator or the real samples. Their respective loss functions are:

$$L_{G_\theta} = \frac{1}{N} \sum_{i=1}^{N} \log\big(1 - D_\phi(G_\theta(z_i))\big) \tag{1}$$

$$L_{D_\phi} = \frac{1}{N} \sum_{i=1}^{N} \left[\log(D_\phi(x_i)) + \log(1 - D_\phi(G_\theta(z_i))) \right] \tag{2}$$

2.2 Fence GAN

The primary objective of Fence GAN is to modify the loss function of the original GAN's generator. This modification aims not only to emulate the density distribution pattern of normal sample data but also to generate samples positioned at the boundary of this distribution. Samples situated at the extremities of the normal sample distribution serves the purpose of aiding in the identification of abnormal samples. In the training of original GAN framework, the discriminator typically assigns 0 to real samples and 1 to generated samples. In the context of anomaly detection, the labeling convention is reversed, wherein normal samples are labeled as 1 and abnormal samples as 0. The specific formulations for the loss functions governing the generator and discriminator in Fence GAN are detailed as follows:

$$L_{G_\theta} = \frac{1}{N} \sum_{i=1}^{N} \log\left| \alpha - D_\phi(G_\phi(z_i)) \right| + \beta \times \frac{1}{\frac{1}{N} \sum_{i=1}^{N} (\|G_\theta(z_i) - \mu\|_2)} \tag{3}$$

$$L_{D_\phi} = \frac{1}{N} \sum_{i=1}^{N} \left[-\log(D_\phi(x_i)) - \gamma \log(1 - D_\phi(G_\theta(z_i))) \right] \tag{4}$$

$$\mu = \frac{1}{N} \sum_{i=1}^{N} G_\theta(z_i)$$

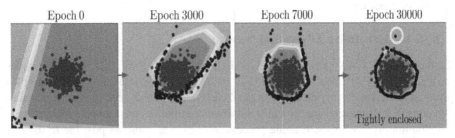

Fig. 1. The result of Fence GAN on a unimodal normal distribution in two dimensions.

In (3), two hyperparameters: α and β, are present. The training objective of Fence GAN is no longer $p(\theta; z)$ equal to $q(x)$, but to generate samples distributed on the edge of $\mathcal{X}$, denoated $\delta\mathcal{X}$, $\alpha \in (0, 1)$ as the discriminator score values of these edge samples. Only when a sample is generated precisely at the edge of $\mathcal{X}$ can the generator's loss function attain a low value. The minimum point for the generator's loss function is reached when the discriminator score equals α. As widely acknowledged, generative adversarial networks are difficult to train [11], often prone to mode collapse. Fence GAN introduces a solution to mitigate mode collapse by incorporating a dispersion loss into the generator's loss function as a regularization term. This addition aims to ensure that the generated sample points comprehensively cover the entirety of $\delta\mathcal{X}$, rather than being confined to a small portion. The dispersion loss is measured by the L_2 distance from the generated samples to their central point. Figure 1 shows the results of 30,000 iterations on a unimodal normal distribution in two dimensions.

3 Kernel Fence GAN

To enhance the classification accuracy of Fence GAN when dealing with high-dimensional and intricate datasets, we have introduced a novel anomaly detection method that integrates GAN with a kernel function. This kernel function plays a pivotal role in mapping the input samples to the Reproducing Kernel Hilbert Spaces (RKHS), facilitating the measurement of distances between samples.

Kernel function $k(x, x') = \langle \phi(x), \phi(x') \rangle$ nonlinearly map samples from the original space to the RKHS [12] of higher or potentially infinite dimensions. RKHS represents a rigorously defined inner product space exhibiting properties of separability, completeness, and the crucial reproducing property. The RBF (Radial Basis Function) stands as the most widely utilized kernel function, also referred to as the Gaussian kernel function. The formulation of the RBF kernel function is as follows:

$$RBF(x, x') = \exp\left(-\frac{1}{2\sigma^2}\|x - x'\|^2\right) \tag{5}$$

In the above formula, replace x' with μ, μ is defined in formula (4):

$$RBF(x, \mu) = \exp\left(-\frac{1}{2\sigma^2}\|x - \mu\|^2\right) \tag{6}$$

Kernel function and GAN [13] have found widespread application in anomaly detection tasks, but predominantly in contexts involving time series data or neural networks of greater complexity than the original GAN framework. Our proposition introduces an algorithm that amalgamates a kernel function with Fence GAN, aiming to detect anomalies within unsupervised data. This integration aims to circumvent the generation of abnormal samples and mitigate the uneven distribution observed in the generated samples depicted in Fig. 1. To achieve this, we augment the regularization term $RBF(x, \mu)$ to the right side of the formula (5). We call this adversarial generation algorithm Kernel Fence GAN. Consequently, the modified loss function for the generator and discriminator within the Kernel Fence GAN is expressed as follows:

$$L_{G_\theta}^{Kernel\ FGAN} = \frac{1}{N} \sum_{i=1}^{N} \log|\alpha - D_\phi(G_\phi(z_i))|$$

$$+\beta \times \frac{1}{\frac{1}{N} \sum_{i=1}^{N} (\|G_\theta(z_i) - \mu\|_2)} + \delta \times \frac{1}{N} \sum_{i=1}^{N} (RBF(G_\theta(z_i), \mu)) \quad (7)$$

$$L_{D_\phi}^{Kernel\ FGAN} = \frac{1}{N} \sum_{i=1}^{N} \left[-\log(D_\phi(x_i)) - \gamma \log(1 - D_\phi(G_\theta(z_i))) \right] \quad (8)$$

Among, β, $\delta \in R^+$. Formula (7) is the generator's loss, three components constitute the right side of the equation. The first component, termed the encirclement loss, limits the generated samples locate on the decision boundary of the normal samples. The second component, referred to as the central dispersion loss, and the third, termed the enhanced dispersion loss, collectively aim to avert model collapse and address diversity limitations. The central dispersion loss specifically dictates that generated points should not only reside on the decision boundary but also maximize their distance from the center point. Incorporating the second dispersion loss decelerates convergence, and even at Epoch 7000 (as depicted in Fig. 1), many blue points persist beyond the boundary of the normal distribution. Notably, certain sample points deviate significantly from the center point, manifesting as outliers, such as the blue sample point indicated by the yellow circle in Fig. 1. Therefore, it is necessary to add the third regularization term to measure the distance between the generated points and the center point in the Reproducing Kernel Hilbert Spaces.

The reduction in weight assigned to the center dispersion loss diminishes the emphasis on maximizing the distance between the generated sample and the center point. Consequently, the occurrence of outliers significantly distant from the center point is mitigated, and the generated samples tend to cluster around the right decision boundary. Introducing the third enhanced dispersion loss, designed to augment sample diversity, further ensures a more even distribution of the generated blue samples across the decision boundary. It's essential to note that the effectiveness of the enhanced dispersion loss is contingent upon the foundation laid by the center dispersion loss. If solely relying on the enhanced dispersion loss, the generated samples fail to maximize their distance from the center point within the original space. Consequently, a smaller, self-imposed "boundary" is identified, deviating from the precise delineation of the normal sample boundary. Hence, the central dispersion hyperparameter β is larger than the enhanced dispersion hyperparameter δ. Equation (8) denotes the discriminator loss function within the Kernel Fence GAN framework. Similar to Fence GAN, this function discriminates between generated and original normal samples, placing heightened emphasis on the classification of normal samples.

4 Experiment

Our experimental methodology aligns with the principles of Fence GAN. Initially, we undertake the training phase of Kernel Fence GAN using a two-dimensional synthetic dataset, generating samples gradually conforming to the decision boundaries of normal samples. Subsequently, we conduct anomaly detection tasks on the KDD99 dataset to evaluate the accuracy and robustness of Kernel Fence GAN. This evaluation includes comparative analysis against other prevalent anomaly detection methodologies.

4.1 2D Synthetic Dataset

We verified the convergence process of Kernel Fence GAN using a two-dimensional artificial dataset derived from random sampling of a two-dimensional unimodal normal distribution, depicted as red sample points in Fig. 2. In the visual representation, the color gradient of the shaded background corresponds to the discriminator score, where higher scores are denoted by red and lower scores by blue. As elucidated in Sect. 2.2, a score of 1 is assigned to normal samples, while abnormal samples receive a score of 0. For the training of this two-dimensional dataset, both the generator and discriminator are implemented using relatively straightforward three-layer fully connected neural networks. The loss function for both the generator and discriminator comprises the summation involving the statistical metric on the right side of the equality in formulas (7) and (8).

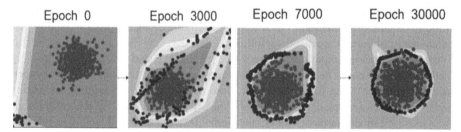

Fig. 2. The result of Kernel Fence GAN on a unimodal normal distribution in two dimensions with central dispersion loss

The hyperparameter σ necessitates manual configuration. In our experiments, we adapted the σ value based on the number of input samples and the dimension of the sample features. Four hyperparameters, namely $\alpha, \beta, \delta, \gamma$, require specification within the loss function. Figure 2 displays the results obtained after 0, 3000, 7000, and 30000 iterations of Kernel Fence GAN applied to the two-dimensional unimodal normal distribution dataset. Hyperparameter values: $\alpha = 0.5, \beta = 10, \delta = 0.3, \gamma = 0.1$. Comparing the outcomes of Epoch 3000 and Epoch 7000 with those depicted in Fig. 1, it is evident that the blue generated samples, within Fig. 2, exhibit increased dispersion along the identified decision boundary. Notably, at Epoch 7000 in Fig. 2, the blue sample points present a more accurate delineation of the decision boundary with wider coverage compared to Epoch 7000 in Fig. 1. Hence, the convergence rate of Kernel Fence GAN is notably faster than that of Fence GAN. Upon contrasting the results of 30,000 iterations between Fig. 1 and Fig. 2, it is apparent that the blue abnormal sample points, illustrated by the yellow circle in Fig. 1, are absent in Fig. 2. This disparity can be attributed to the reduction in weight applied to the central dispersion loss within the generator's loss function of Kernel Fence GAN. This adjustment ensures that generated samples do not deviate excessively from the center point.

4.2   KDD99 Dataset

In order to substantiate the superiority of our proposed Kernel Fence GAN over alternative algorithms, we conducted experimental comparisons using the KDD99 dataset. This dataset comprises 5 million records, presenting considerable scale wherein the original dataset. It offers a 10% partition for training purposes alongside a distinct testing subset. The training subset encompasses a singular normal identity type and encompasses 22 attack types. Our experimental methodology aligns with the procedural framework delineated in [14]. During the training phase of the Kernel Fence GAN, solely the attack type data is introduced, with the primary objective of discerning normal type data. We employ 50% of the attack type data as the training dataset, while utilizing the remaining 50% of the attack type data in conjunction with the normal type data for constituting the test set.

Table 1 illustrates the average outcomes derived from the initial 10 iterations of Kernel Fence GAN applied to the KDD99 dataset. Both the generator and discriminator architectures employ simplistic three-layer fully connected neural networks. The optimization is facilitated by Adam for the generator and SGD for the discriminator, while maintaining a consistent batch size of 256 throughout the process. The hyperparameter configurations are set as follows: $\alpha = 0.1, \delta = 0.01, \beta = 30, \gamma = 0.1$. The table provides a comparative analysis where traditional and advanced deep learning methodologies serve as benchmarks. Our proposed method exhibits a notably competitive performance when juxtaposed with these advanced methodologies, showcasing higher precision and recall rates. Notably, the F1 score attained by the ensemble f-AnoGAN [15] model surpasses that of the Kernel Fence GAN, attributed to its intricate network architecture comprising a group of generators and discriminators. Similarly, our model demonstrates superior performance over Fence GAN, showcasing a 2.2% enhancement in accuracy and a 2.16% improvement in recall rates.

Table 1. Results of anomaly detection on the KDD99 dataset.

Model	Precision	Recall	F1
OC-SVM	0.7457	0.8523	0.7954
DSEBM-r	0.8521	0.6472	0.7328
DSEBM-e	0.8619	0.6446	0.7399
DAGMM-NVI	0.9290	0.9447	0.9268
DAGMM	0.9297	0.9442	0.9269
AnoGANfm	$0.88 \pm 3 \times 10^{-2}$	$0.83 \pm 3 \times 10^{-2}$	$0.89 \pm 3 \times 10^{-2}$
AnoGANsigmoid	0.8 ± 0.01	0.8 ± 0.01	0.8 ± 0.01
Efficient-GANfm	0.9 ± 0.01	$0.95 \pm 3 \times 10^{-2}$	$0.91 \pm 7 \times 10^{-2}$
Efficient-GANsigmoid	$0.92 \pm 7 \times 10^{-2}$	$0.96 \pm 1 \times 10^{-2}$	$0.94 \pm 4 \times 10^{-2}$
Fence GAN	$0.954 \pm 9 \times 10^{-3}$	$0.969 \pm 9 \times 10^{-3}$	$0.95 \pm 2 \times 10^{-2}$
EGBAD	0.920	0.958	0.939
f-AnoGAN	0.935	0.986	0.960
EGBAD^{en}	0.972	0.960	0.966
f − AnoGAN^{en}	0.967	0.990	**0.979**
Kernel Fence GAN	$0.975 \pm 7 \times 10^{-3}$	$0.995 \pm 3 \times 10^{-3}$	$0.962 \pm 2 \times 10^{-3}$

5 Conclusions

In this research paper, we introduce Kernel Fence GAN, a generative adversarial model specifically tailored for unsupervised anomaly detection. By integrating a kernel function into the generator loss function of the Fence GAN framework, we achieve a nonlinear mapping of generated samples from the original sample space to RKHS. This novel addition of the kernel function aims to maximize the discrepancy between the generated samples and a central point, effectively mitigating mode collapse issues while enhancing the diversity of generated samples. Furthermore, our approach reduces the impact of center dispersion loss on the generated samples, leading to improved accuracy in defining the decision boundary and effectively minimizing the occurrence of abnormal generated samples. Through extensive evaluations conducted on two datasets, Kernel Fence GAN consistently outperforms other existing models, establishing its superior performance in anomaly detection tasks. Designing and training stable GAN based on kernel functions present considerable challenges due to the dependency on selecting appropriate kernel functions and parameter settings. Moving forward, our future research endeavors will focus intensively on addressing these challenges associated with anomaly detection.

Acknowledgement. This work was supported by the National Natural Science Foundation of China, grant number "52275480".

References

1. Chandola, V., Banerjee, A., Kumar, V.: Anomaly detection: a survey. ACM Comput. Surv. **41**(3) (2009)
2. Fernando, T., Gammulle, H., Denman, S., et al.: Deep learning for medical anomaly detection–a survey. ACM Comput. Surv. (CSUR) **54**(7), 1–37 (2021)
3. Schölkopf, B., Williamson, R., Smola, A., Shawe-Taylor, J., Platt, J.: Support vector method for novelty detection. In: Proceedings of the 12th International Conference on Neural Information Processing Systems, pp. 582–588 (1999)
4. Nicolau, M., Mcdermott, J., et al.: One-class classification for anomaly detection with kernel density estimation and genetic programming. In: European Conference on Genetic Programming, pp. 3–18. Springer (2016)
5. Kingma, D.P., Welling, M.: Auto-Encoding Variational Bayes. arXiv.org (2014)
6. Goodfellow, I., Pouget-Abadie, J., Mirza, M., et al.: Generative Adversarial Nets. Neural Information Processing Systems. MIT Press (2014)
7. Schlegl, T., Seeböck, P., Waldstein, S.M., et al.: Unsupervised anomaly detection with generative adversarial networks to guide marker discovery. In: Proceedings of the 2017International Conference on Information Processing in Medical Imaging, LNCS, vol. 10265, pp. 146–157. Springer, Cham (2017)
8. Zenati, H., Romain, M., Foo, C.S., et al.: Adversarially learned anomaly detection. In: 2018 IEEE International Conference on Data Mining (ICDM). IEEE (2018)
9. Nguyen, D.T., Lou, Z., Klar, M., et al.: Anomaly detection with multiple-hypotheses predictions. In: International Conference on Machine Learning. PMLR (2019)
10. Ngo, C.P., Winarto, A.A., Li, C., et al.: Fence GAN: Towards Better Anomaly Detection (2019)
11. Bińkowski, M., Sutherland, D.J., Arbel, M., et al.: Demystifying MMD GANs. arXiv.org (2018)
12. Shawe-Taylor, J., Cristianini, N.: Kernel Methods for Pattern Analysis. Publications of the American statistical association (2004). https://doi.org/10.1198/jasa.2006.s153
13. Lee, C.K., Cheon, Y.J., Hwang, W.Y.: Studies on the GAN-based anomaly detection methods for the time series data. IEEE Access **PP**(99), 1 (2021)
14. Zenati, H., Foo, C.S., Lecouat, B., et al.: Efficient GAN-based anomaly detection (2018). https://doi.org/10.48550/arXiv.1802.06222
15. Han, X., Chen, X., Liu, L.P.: GAN Ensemble for Anomaly Detection (2020). https://doi.org/10.48550/arXiv.2012.07988

A News Recommendation Approach Based on the Fusion of Attention Mechanism and User's Long and Short Term Preferences

Yi Xiong[✉]

College of Mathematics and Information, South China Agricultural University, Guangzhou, Guangdong, China
1245237977@qq.com

Abstract. Nowadays, more and more news readers are reading news online and they have access to millions of news articles from multiple sources. To help users find correct and relevant content, news recommendation systems have been developed to alleviate the problem of information overload and to recommend news items that may be of interest to news readers. Compared to traditional recommendation domains such as product recommendation and video recommendation, in the news recommendation domain, the update speed of information is very fast, so when recommending news content to users, it is necessary to consider the deeper characteristics of the user's representation and the recommendation speed of the news, which in turn improves the accuracy of the news recommendation user satisfaction, and increases the user's stickiness. Therefore, this paper proposes a news recommendation method that integrates the attention mechanism with the user's short-term and long-term preferences, which breaks through the problem that the news feature extraction does not make full use of the full text and the modeling of the user's interest is not granular enough and timely, and at the same time, while achieving a high level of accuracy, it also meets the requirements of high timeliness.

Keywords: recommender systems · deep learning · attention mechanisms

1 Introduction

With the development of the Internet, all kinds of information such as news information grows exponentially, and in the face of massive data, how to help users quickly get the news they are interested in is very important. As a key technology to effectively solve the information overload problem, recommender system is one of the research hotspots in industry and academia [1]. Based on this, news recommender systems have emerged.

As an important field in recommender systems, news recommender systems aim to help users select news of interest, which is important for improving the reading experience [2]. Most of the traditional news recommendation methods rely on manual feature engineering to construct user behavior feature matrices and news feature matrices. However, these methods rely heavily on manual feature engineering, which requires a

© The Author(s), under exclusive license to Springer Nature Singapore Pte Ltd. 2024
K. Li and Y. Liu (Eds.): ISICA 2023, CCIS 2146, pp. 429–442, 2024.
https://doi.org/10.1007/978-981-97-4393-3_35

large number of researchers with specialized domain knowledge to complete, and the cost is very huge. News recommendation combined with deep learning constructs user models differently from traditional recommendation models and handles timeliness and other news recommendation related issues in a more advanced way. Compared with traditional recommendation fields such as product recommendation and video recommendation, in the news recommendation field, the updating speed of information is very fast, so when recommending news content to users, it is necessary to consider the deeper characteristics of the user's representation and the recommendation speed of the news to improve the accuracy of the news recommendation and user satisfaction, and to increase the user's stickiness. The research on news recommendation based on deep learning is more and has achieved better results, but there are still some problems in the accuracy and efficiency of news recommendation.

Currently, many content-based deep learning news recommendation methods have been proposed, and the core idea is to obtain user and news representations through a user encoder and a news encoder for recommendation. Wang et al. learn news representations from the headlines of news articles through a convolutional neural network, and then based on the similarity between the candidate news and the history of news browsing of each user, they learn to obtain the user from the news representation representation [3].Wu et al. proposed to use a personalized word-level and news-level attention network to learn news and user representations from the headlines of news, which argues that users do not pay the same attention to different words in news headlines. This personalized attention network uses the embedding vectors of the user's IDs to generate the query vectors. Wu et al. also used a multi-head self-attention network to model news headlines and user's history of news browsing [4].

In fact, the above news recommendation methods only learn news representations and user representations from part of the textual information of the news or individual features of the user, ignoring the fact that the user clicks on the news which is affected by multiple factors, and the convolutional neural network is unable to obtain the long-distance textual relationship between the words of the headlines. In addition to this, existing news recommendation models are also deficient in more granular content feature mining of news, often ignoring word-level interactions between different clicks on news from the same user, which contain rich and detailed cues to infer user interests, and ignoring word-level behavioral interactions, which may lead to poorer user modeling. Moreover, users need an efficient news recommendation process, and the efficiency requirement for user short-term preference feature extraction is even more so, while the efficiency of the existing short-term preference feature extraction approach is not efficient enough, and the efficiency of recommending news to users needs to be improved while ensuring accuracy. In these cases, the semantic feature extraction of news is not precise enough and the user interest modeling is not accurate enough, which leads to the news recommended to the user is not timely and precise enough.

Therefore, this paper proposes an approach to improve news encoder using multi-head attention network, i.e., learning the relationship between long distance between words and words in different types of news information, improving the news feature representation, and fusing it into a user encoder that combines users' long term and short term interest representations as a way of improving the results of news recommendation.

In the news encoder, a multi-head attention network is used and combined with an attention mechanism for feature extraction of three types of textual modal information, namely, headline, category and subcategory, focusing on the long-distance interaction relationship among the words therein, and an attention mechanism is added because different categories of information may have different amounts of information; and the user encoder is used to learn the user's representation by combining the user's long-term and short-term preferences. By using Fastformer users' fine-grained behavioral features for fast feature extraction, the final user short-term preference behavior vectors are obtained, which are then connected with the user long-term preference behavior vectors obtained through the user's ID information as the user representation. Through a large number of experiments on real news datasets, it is proved that the method in this paper can effectively improve the accuracy of news recommendation and other indicators.

2 Prepare Work

2.1 Related Work

News recommendation is an important task in the field of data mining and has been extensively studied in recent years. Popular news reading platforms can generate thousands of news every day, and due to time constraints, it is impossible for users to read all the news [5]. Therefore, personalizing news recommendations to users' interests is an important task for news reading platforms [4]. Some existing traditional approaches employ artificial feature engineering for news and user representation learning. For example, Liu et al. proposed to represent news with topic categories and interest features predicted by Bayesian models, and to represent users with click distribution features of news categories [6]. Son et al. extracted topic and location features from Wikipedia pages as news representations, and proposed ELSA, a location-based explicit semantic analysis model for news recommendation [7]. Lei et al. use an implicit Dirichlet distribution model to generate topic distribution features as news representations, while sessions are represented by the topic distribution of news viewed in the session, and user representations are constructed based on time-weighted session representations [8]. However, these methods not only rely on manual feature engineering, but also require a lot of domain knowledge and effort to process news data.

In recent years, news recommendation methods based on deep learning have been proposed one after another, and many of them are unable to distinguish between users' long-term and short-term hobbies. The LSTUR model proposed by An expresses long-term and short-term interests by combining news text features and the time period of users' historical browsing behavior, in which the feature extraction of news headlines employs convolutional neural networks with the attention mechanism [9], but it only learns the news representation from news headlines, and learning the feature representation of news in this way is not sufficient.

Since news data has categories, subject categories, summaries and other information in addition to headlines, in order to be able to use them as a unified learning representation to extract the deep features of news information, this paper adopts a multi-head attention network combined with the attention mechanism to combine different categories of information to learn the news representation and improve the user representation.

In addition, it learns user short-term preference from user's news browsing through GRU network In the future when the amount of data and model volume gradually increase, the disadvantage of its inability to perform parallel computation may lead to a reduction in the speed of news recommendation, which reduces user stickiness. Therefore, in this paper, we first connect the user's behavioral data of clicking news as a long document, transfer the user feature extraction task to the news feature extraction task, and then add Fastformer after the news embedding layer module because it can quickly learn the word-level interactions between the news as a representation of the user's short-term interests to improve the efficiency of user feature extraction [10].

Extensive experiments on real news datasets show that the method in this paper is able to better learn news and user representations and improve the accuracy of news recommendation compared to existing popular methods [11].

2.2 Methodological Design

News recommender system is designed to help users select the news they may be interested in from a huge amount of news, the core idea is to get the news representation based on the news information such as title, category, etc., and at the same time to get the user representation based on the user information such as the history of browsing the news, and finally to calculate the probability of the user clicking on the news through the news representation and the user representation [12], and then push a number of news with the highest clicking probability to the user, the main process is shown in Fig. 1.

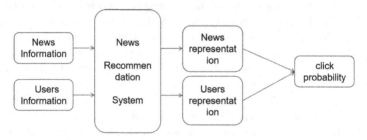

Fig. 1. Main flow of news recommendation

From Fig. 1, it can be seen that the key to achieve excellent recommendation effect of the news recommendation system lies in how to utilize the existing information to get a more comprehensive and detailed news representation and user representation [13].

In popular news reading platforms, the news displayed in front of the user's eyes generally include the title, category and subcategory of the three types of text-modal information, the user clicks on a piece of news may be attracted by the title of the content, or it may be interested in categories like sports or subcategories like basketball, so both the title, category and subcategory are very important for the news recommender model to learn to get the news representation are all very important. In addition, for users, their recent historical browsing behavior undoubtedly best represents their short-term interest, however, the user's interest changes over time, for example, in the long term, a user has not browsed the soccer news many times, but it is likely that he/she will read the news of soccer matches during the soccer World Cup, and if only from the history of browsing behavior to represent the user there is a possibility that the The same user representation occurs for two different users. It is necessary to model users from both a short-term, changing perspective and a long-term, fixed perspective.

Therefore, this paper proposes a news recommendation method based on the fusion of the attention mechanism and the user's long and short-term preferences, aiming to obtain a comprehensive and detailed news representation through the learning of three types of textual modal information, namely, news title, category and subcategory, and at the same time, adding the user's ID information in the modeling of the user using the user's history of news browsing to construct a news recommendation model, which is divided into a news encoder and a user encoder.

3   Model Structure

The news recommendation model based on the fusion of attention mechanism and user's long and short term preferences is mainly composed of three modules: news encoder, user encoder and click probability prediction. The main role of the news encoder is to get the news represented using vectors, while the main role of the user encoder is to get the user represented using vectors, and finally the click probability prediction converts the news representation and the user representation into the probability that the user clicks on the news, and the basic structure of the model is shown in Fig. 2.

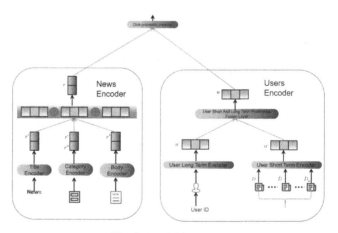

Fig. 2. Model Structure

3.1 News Encoder

The main work of the news encoder is to model the textual modal information of the news, i.e., the three types of textual modal information, i.e., headline, category and subcategory of the news are transformed into the final news vector, which is used as the news representation. The news encoder includes headline encoder, category encoder, subcategory encoder and pooling layer, and its architecture is shown in Fig. 3.

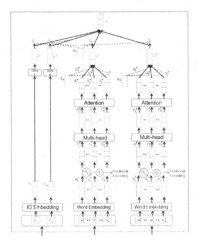

Fig. 3. News Encoder

The headline encoder is divided into four layers, the first is the Word Embedding layer (Word Embedding), the second is the Position Embedding layer (Position Embedding), the third is the Multi-head Self-attention Network layer (Multi-head Self-attention Network), and the fourth is the Attention Network layer (Attention). The main work of the headline encoder is to process the textual information which is the news headline, i.e., input the words of the news headline, output the headline vector and use it as the headline representation.

The word embedding layer converts news headlines from a sequence of words to a sequence of low-dimensional embedding vectors. First, a word sequence $[w_1^t, w_2^t, \ldots, w_m^t]$ of news headlines is input to the word embedding layer, where w_m^t is the mth word of the news headline. Then, it is converted by the word embedding layer to a low-dimensional word vector sequence, where is the first low-dimensional word vector $[e_1^t, e_2^t, \ldots, e_m^t]$ of the news headline, and is used as input to the next layer.

The positional embedding layer fuses trigonometric positional vectors with a sequence of low-dimensional word vectors so that the vectors in the sequence of low-dimensional word vectors have relatively positional sequential features. Typically, trigonometric positional vectors use trigonometric functions (e.g., sine or cosine functions) to encode and represent features in the sequence. The position of a word in an utterance is very important, and the same word arranged in different ways may not have the same semantic meaning, e.g., the words "I", "look", and "you" may express "I look" and "you". For example, the words "I", "look" and "you" can express two meanings: "I look at you" and "you look at me", and the difference only lies in who is in the verb in the words "I" and "you"? The only difference is who precedes or follows the verb "look" in the words "I" and "you", so the inclusion of word order information helps to learn the title semantic information more accurately.

Therefore, the use of trigonometric position vectors can help the positional embedding layer better understand the structure and order of news headlines, thus improving the accuracy and personalization of the recommendation algorithm. Trigonometric position vectors and low-dimensional word vector sequences have the same number of words, i.e., the same number of dimensions d, and the position vectors are calculated by the following formula:

$$PE_{(pos,2i)} = \sin(pos/10000^{\frac{2i}{d}}) \tag{1}$$

$$PE_{(pos,2i+1)} = \cos(pos/10000^{\frac{2i}{d}}) \tag{2}$$

where pos denotes the word position and i denotes the dimension, the sequence of low-dimensional word vectors $[e_1^t, e_2^t, \ldots, e_m^t]$ input to the positional embedding layer is converted to $[e_1^p, e_2^p, \ldots, e_m^p]$ by adding the trigonometric positional vectors, where it is the mth low-dimensional word vector of the news headline after adding the positional information and is outputted as the input to the next layer.

The multi-head self-attention network layer can learn the long-distance relationship between words, thus further enriching the semantic information of headlines. Therefore, a multi-head attention network is used to convert a sequence of low-dimensional word vectors into a multi-head self-attention vector, where the ith word in the kth head self-attention network is denoted as

$$\alpha_{i,j}^k = \frac{\exp(e_i^T Q_k^w e_j)}{\sum_{m=1}^{M} \exp(e_i^T Q_k^w e_m)} \tag{3}$$

$$h_{i,k}^w = V_k^w \left(\sum_{j=1}^{M} \alpha_{i,j}^k e_j \right) \tag{4}$$

where Q_k^w and V_k^w are the network parameters of the kth head's self-attention, $\alpha_{i,j}^k$ denotes the weight value of the interaction relationship between the ith word and the jth word, e_i and e_j represent the low-dimensional word vectors of the ith and jth words, respectively, and ultimately the output of the ith word, after passing through the head's self-attention network is the splicing of the representations of the outputs of the respective head's self-attention networks, where is the representation of the The output of the ith word after passing through the head self-attention network, $h_{i,h}^w$, and the sequence of low-dimensional word vectors input to the multi-head self-attention network layer is finally converted into a sequence of multi-head self-attention vectors, $h_i^w = [h_{i,1}^w; h_{i,2}^w; \ldots; h_{i,h}^w]$, where h_m^p is the mth multi-head self-attention vector of the news headline obtained through the multi-head self-attention network;

In a news sentence, different words may have different levels of importance to the news sentence, i.e., they may have different levels of importance to the representation of the sentence. For example, in the news statement "Today, the NBA team Lakers beat the Cavaliers by a large margin," it is clear that the word "win" is more important than the word "today" in understanding this news sentence. The word "win" is more important and informative than the word "today". Since the attention network can focus on the importance of different words in the text statement and assign different weights to them. Therefore, the attention network layer is used to select the important words in the news headlines, specifically, that is, to select the important word sequences for the output multiple attention vector sequences of the previous layer, in which the formula for calculating the attention weight α_i^w of the ith word in the news headline is:

$$\alpha_i^w = q_w^T \tanh\left(V_w \times h_i^w + v_w\right) \tag{5}$$

$$\alpha_i^w = \frac{\exp\left(\alpha_i^w\right)}{\sum_{j=1}^{M} \exp\left(\alpha_j^w\right)} \tag{6}$$

where V_w, v_w and are the attention network parameters, α_j^w represents the attention weight of the j th word, and the final representation of the news headline r^t is the weighted sum of the multi-head self-attention vector h_i^w and the attention weight α_i^w, with the following formula:

$$r^t = \sum_{i=1}^{M} \alpha_i^w h_i^w \tag{7}$$

In addition to the title of a news article, which represents the news to a certain extent, the categories and subcategories of the news article are also informative. For example, on many news sites, news articles are labeled with their corresponding subject categories, such as "sports" and "music" and their corresponding subcategories "basketball" and "jazz," and so on. "Jazz" and so on. If a news reader often reads news in the "music" category, it can be assumed that the user has a strong preference for news in the "music" category. Therefore, the feature information of news categories and subcategories are extracted and integrated into the learning of news features. The category and subcategory encoders are similar, with the first layer being the ID embedding layer (Category Embedding) and the second layer being the fully connected layer (Dense). The IDs of the categories and subcategories of the input news are feature extracted as vector representations of the categories and subcategories.

In addition to the title and category information of the news, the body of the news contains a lot of information about the news article, which is very important and representative for the feature learning of the news. The main work of the body encoder is to process this information of the body of the news, extract the feature information as the feature representation of the body of the news, and incorporate it into the feature information of the news, which is similar to the structure of the headline encoder, so it will not be repeated.

3.2 User Encoder

The main role of the user encoder is to learn the user's characteristics, i.e., user's interest preferences, from the user's historical news browsing. Usually, user's preference is categorized into user's short-term preference and user's long-term preference. Among them, the user's long-term preference, the user's short-term preference. Therefore, the user encoder consists of two, the user short-term preference encoder and the user long-term preference encoder.

The user short-term preference encoder is divided into three layers, in which the first layer is the user history behavior transformation layer and the second layer is the Fastformer layer. The main function of the user short-term preference encoder is to extract features from the user's historical behavior and convert it into a vector of user's historical behavior as the user's short-term preference representation, the specific structure is shown in Fig. 4:

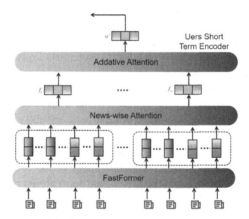

Fig. 4. User Short Term Encoder

First, the user's historical behavior sequence $[h_1, h_2, \ldots h_k]$ is converted into a long user document, which is used as input for the next layer.

Second, Fastformer enables efficient global context modeling of long sequential data. Long user documents are used as input in which word-level behavioral interactions within and between news can be captured to understand user interests at a fine-grained level. Fastformer can be used to quickly learn word-level interactions between news, and modeling fine-grained behavioral interactions at the word level in a user's history of clicking on news is used to efficiently and accurately model a user's short-term preferences, which first generalizes the news sequence input into a global embedding vector, and then interacts each embedded news information vector with the global embedding vector with the following formula:

$$q = Att(q_1, q2, \cdots qi, \cdots, q_L), q_i = W_q h_i \tag{8}$$

$$k = Att(q * k_1, \cdots, q * k_L), k_i = W_k h_i \tag{9}$$

$$\hat{h}_i = W_o(k * v_i), v_i = W_v h_i \tag{10}$$

where h_i and $\hat{h}_i$ denote the input and output of the ith news message in the embedded news message, q_i represents the ith embedded news message vector, k_i represents the ith embedded news message key, k represents the global embedded news message key, v_i represents the value of the ith embedded news message, $*$ denotes the product of the elements, and $Att()$ denotes the additive attention network, which is the reason why the Fastformer can reduce the time complexity from quadratic complexity to linear complexity by linearly converting the learned global matrix to global vector through additive attention network, W_q, W_k, W_v, and W_o represent the global training parameters; then, connect the outputs of the different attention headers and construct a unified context of the ith embedded news message representing g_i, as the input of the next layer.

Since among the news read by a user, different news have different news information and a user may read more than one news, additive attention mechanism is used in the pooling layer to perform the aggregation operation. First, the aggregation operation is performed on the output of the previous layer, the contextual representation of the embedded news information. Finally, the news representations are then pooled and aggregated through a layer of additive attention mechanism layer to construct the user's short-term preference representation.

The main role of the user long term preference encoder is to obtain the long term preference representation of the user, which is divided into two layers, where the first layer is the ID embedding layer and the second layer is the full connectivity layer.

Since user preferences often contain both short-term and long-term preferences of users, and the contribution of long-term and short-term user preferences to user interests may be different. Therefore, in order to better integrate the two interests of the user, the user's short-term preference and the user's long-term preference are directly spliced together to obtain the complete preference representation of the user through the full connectivity layer.

3.3 Model Training

The role of the news click probability prediction module is to predict the probability of a user clicking on a candidate news item [14]. Usually, due to the simplicity and efficiency of inner product computation, the probability of a user clicking on a news is obtained using an inner product operation on the user representation vector and the news representation vector. Inspired by [15] and [16], the model is trained using random negative sample sampling, specifically, K negative samples are randomly selected for training while training on each positive sample data. Then, the click probability of $\hat{y}^+$ positive samples of news and the click probability of K negative samples of news are combined to serve as the news click probability prediction. Finally, the softmax function is used to normalize the function of the click probability prediction, and the final result is expressed.

4 Experiments

4.1 Dataset

The dataset used in this paper is the publicly available Microsoft News Dataset MIND-small (Microsoft News Dataset-small) [17]. The source of the MIND-small dataset is the user behavior logs from Microsoft News, which are extracted from 50,000 users and their behavior logs are recorded, and is the news recommended benchmark dataset. The information about MIND-small is shown in Table 1:

Table 1. Statistics of the dataset in our experiments.

	MIND-small
Users	50000
News	51282
Samples	5843454
Positive Samples	236344
Negative Samples	5607110

4.2 Performance Evalution

We evaluate the performance of our approach by comparing it with several baseline methods, The results of comparing different methods are summarized in Table 2.

Table 2. The performance of different methods on news recommendation

Model	AUC	MRR	NDCG@5	NDCG@10
DeepFM	0.5612	0.2215	0.2411	0.3116
DKN	0.5898	0.2333	0.2653	0.3301
NAML	0.6096	0.2586	0.2898	0.3592
FIM	0.6133	0.2611	0.3056	0.3611
NRMS	0.6196	0.2635	0.3121	0.3689
LSTUR	0.6285	0.2706	0.3201	0.3772
FLSTUR-NR	**0.6379**	**0.2827**	**0.3336**	**0.3878**

From the experimental results, it can be seen that FLSTUR-NR model in AUC, MRR, NDCG@5 and NDCG@10 reaches 0.6379, 0.2827, 0.3336, 0.3878, the recommendation effect are due to the effect of other news recommendation algorithms, therefore, it is summarized as follows:

(1) Using techniques such as neural nets can be able to learn the representation of users and news from data better than traditional manual feature extraction techniques.
(2) When modeling the user, the method of combining the user's long and short-term preferences can obtain a better representation of the user and improve the effect of news recommendation
(3) The use of multi-attention mechanism to extract long-distance feature associations between words can obtain better news representation and improve the accuracy of recommendation.

4.3 Ablation Experiment

In order to verify the effectiveness of user modeling by combining the user's long and short-term preferences, experiments were conducted using either of the user's short-term preference module and the user's long-term preference module, respectively, and the results of the experiments are shown in Fig. 5:

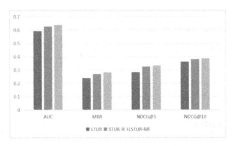

Fig. 5. The effectiveness of incorporating long-tern user representations (LTUR) and short-term user representations (STUR).

As a result, it can be found that combining both short-term user preferences and long-term user preferences is useful for the model to capture interests that are not used by the user, and that short-term user preferences are better modeled than long-term user preferences, where the introduction of long-term user preferences may introduce more noise leading to performance degradation.

4.4 Effect of Important Parameters on Experimental Results

During model training, an epoch represents the process of training the entire dataset once. However, in actual model training, only one training is often not enough, so the epoch value needs to be adjusted to perform multiple training. In the process of deep learning training, increasing the epoch value can avoid overfitting or underfitting problems, making the model get a better fit on the training. Therefore, it is very important to choose the appropriate epoch value to train the model, this paper further explores the effect of different epoch values on the FLSTUR-NR model and the results are shown in Fig. 6:

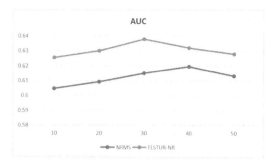

Fig. 6. Impact of different epochs on modeling effectiveness

The epoch values were set to 10, 20, 30, 40 and 50 for training. As can be seen from the figure, the model is most effective when epoch is equal to 30. When epoch is less than 30, the model is improving and the model is in an underfitting state. When epoch is greater than 30, the effect of the model begins to decline, and the model is in an overfitting state.

5 Conclusion and Future Work

This chapter proposes a text modality-based news recommendation model FLSTUR-NR and its structure, which contains three structures: news encoder, user encoder and click prediction probability encoder. Among them, the news encoder contains a title encoder, a category encoder, a body encoder, and an attention pooling layer, which performs feature extraction on the title, category, and body of the news, respectively, and then fuses the different information of the news through the attention pooling layer according to different weights to obtain the final news representation. The user encoder consists of a short-term preference encoder and a long-term preference encoder, where the user's historical clicking behavior sequence is converted into a long document sequence by linking operations, and then the long document is subjected to fast and fine-grained word-level feature extraction by Fastformer as a short-term preference representation; and the user ID is embedded as a long-term preference representation. Finally, the user's short and long term preferences are fused as the final representation of the user. Extensive experiments on a real-world dataset collected by MSN News show that our approach can effectively improve the performance of news recommendation.

References

1. An, M., Wu, F., Wu, C., et al.: Neural news recommendation with long-and short-term user representations. In: Proceedings of the 57th Annual Meeting of the Association for Computational Linguistics, pp. 336–345 (2019)
2. Blei, D.M., Ng, A.Y., Jordan, M.I.: Latent Dirichlet allocation. J. Mach. Learn. Res. **3**, 993–1022 (2003)
3. Cheng, H.T., Koc, L., Harmsen, J., et al.: Wide & deep learning for recommender systems. In: Proceedings of the 1st Workshop on Deep Learning for Recommender Systems, pp. 7–10 (2016)

4. Hu, J., Shen, L., Sun, G.: Squeeze-and-excitation networks. In: Proceedings of the IEEE Conference on Computer Vision and Pattern Recognition, pp. 7132–7141 (2018)

5. Guo, H., Tang, R., Ye, Y., et al.: DeepFM: a factorization-machine based neural network for CTR prediction. arXiv preprint arXiv:1703.04247 (2017)

6. Hu, Y., Qiu, Z., Wu, X.: Denoising Neural Network for News Recommendation with Positive and Negative

7. IJntema, W., Goossen, F., Frasincar, F., et al.: Ontology-based news recommendation. In: Proceedings of the 2010 EDBT/ICDT Workshops, pp. 1–6 (2010)

8. Kalman, D.: A singularly valuable decomposition: the SVD of a matrix. Coll. Math. J. **27**(1), 2–23 (1996)

9. Kuo, F.F., Shan, M.K., Lee, S.Y.: Background music recommendation for video based on multimodal latent semantic analysis. In: 2013 IEEE International Conference on Multimedia and Expo (ICME), pp. 1–6. IEEE (2013)

10. Li, L., Wang, D., Li, T., et al.: Scene: a scalable two-stage personalized news recommendation system. In: Proceedings of the 34th International ACM SIGIR Conference on Research and Development in Information Retrieval, pp. 125–134 (2011)

11. Li, L., Zheng, L., Yang, F., et al.: Modeling and broadening temporal user interest in personalized news recommendation. Expert Syst. Appl. **41**(7), 3168–3177 (2014)

12. Lian, J., Zhang, F., Xie, X., et al.: Towards Better Representation Learning for Personalized News Recommendation: a Multi-Channel Deep Fusion Approach. IJCAI. 2018, pp. 3805–3811

13. Liu, J., Dolan, P., Pedersen, E.R.: Personalized news recommendation based on click behavior. In: Proceedings of the 15th International Conference on Intelligent User Interfaces, pp. 31–40 (2010)

14. Lommatzsch, A., Kille, B., Hopfgartner, F., et al.: Newsreel multimedia at mediaeval 2018: news recommendation with image and text content. In: Working Notes Proceedings of the MediaEval 2018 Workshop. CEUR-WS (2018)

15. Lu H, Zhang M, Ma S. Between clicks and satisfaction: Study on multi-phase user preferences and satisfaction for online news reading[C]//The 41st International ACM SIGIR Conference on Research & Development in Information Retrieval. 2018: 435–444

16. Ning X, Karypis G. Slim: Sparse linear methods for top-n recommender systems[C]//2011 IEEE 11th international conference on data mining. IEEE, 2011: 497–506

17. Phelan O, McCarthy K, Smyth B, et al. Terms of a feather: content-based news discovery and recommendation using Twitter[J]. Information Retrieval (ECIR-11), 2010, 18: 21

18. Qi T, Wu F, Wu C, et al. Personalized news recommendation with knowledge-aware interactive matching[C]//Proceedings of the 44th International ACM SIGIR Conference on Research and Development in Information Retrieval. 2021: 61–70

19. Vaswani A, Shazeer N, Parmar N, et al. Attention is all you need[J]. Advances in neural information processing systems, 2017, 30

Research of the Three-Dimensional Spatial Orientation for Non-visible Area Based on RSSI

Huabei Nie[1], Jianqiao Shen[1]([✉]), Haihua Zhu[2], Ani Dong[1], Yongcai Zhang[1], and Yi Niu[1]

[1] College of Artificial Intelligence, DongGuan City College, Dongguan 523419, China
331096487@qq.com
[2] College of Computer and Information Technology, Nanyang Normal University, Nanyang 473061, China

Abstract. This paper research the stereo space location of invisible areas based on RSSI. Introduce the basic knowledge of RSSI and spatial positioning, and analyze the related existing technologies. On the basis of distance correction, this correction model converts the RSSI value of the unknown Wi-Fi direct connected device into a number which delegates the distance. Then constructs the relative spatial position by combination the geometric principle, so as to compute the location of the node. The unknown Wi-Fi direct connected device receives information from the three known Wi-Fi direct connected devices. The positioning application of this ranging correction model is widely used.

Keywords: Stereoscopic Space · Positioning · RSSI · Invisible Region · Ranging Correction Model

1 Introduction

Wi-Fi devices can connect anytime, anywhere without a Wi-Fi access point. The future will be an era of full devices interconnection. We can use smart phones, tablets and other devices in the future, and all those devices in the future will not be limited by the routing, and realize the interconnection of mobile phones and cars! This is a strong foundation for the future development of the Internet of Things [1, 2].

Wireless positioning technology is a communication technology that carries out positioning and tracking by detecting the positioning parameters (such as transmission time, phase angle, angle of arrival, amplitude, etc.) of wireless signals during trans-mission. Its positioning accuracy depends on the positioning algorithm [3]. Common wireless location technologies include Wi-Fi, RFID, Bluetooth, etc. With the widely deployment of Wi-Fi wireless access points and the fact that Wi-Fi modules have became the basic configuration of intelligent terminals, users can quickly build a Wi-Fi wireless positioning platform composed of wireless access points and intelligent handheld terminal devices [4–7].

Wi-Fi Direct is a set of software protocols, which is different from the traditional P2P mode. That is, Wi-Fi devices can not pass through the wireless network base station [8–12]. Wi-Fi Direct positioning is a positioning method, which uses a spatial positioning

© The Author(s), under exclusive license to Springer Nature Singapore Pte Ltd. 2024
K. Li and Y. Liu (Eds.): ISICA 2023, CCIS 2146, pp. 443–450, 2024.
https://doi.org/10.1007/978-981-97-4393-3_36

technology based on the RSSI. Compared to the GPS method, Wi-Fi Direct positioning has the advantages of low cost, easy to use, etc., and can be applied in most of applications. The positioning technology based on RSSI is more suitable for the current market demand of location based services (LBS). Although easy to implement, there are some difficulties. Because the using of the radio wave transmission loss model, The positioning technology based on received signal intensity indication (RSSI) [13] is vulnerable to magnetic fields and due to Inaccurate positioning results which caused by environmental noise such as obstacles.

2 Description of the Problem

For short-distance data transmission, we often turn on Bluetooth or Wi-Fi Direct function. It is assumed that student D's phone open Wi-Fi Direct function and debug into silent mode and close shock. When student D opens the closet, thrust the phone in his closet, then D flips clothes, which makes the phone hidden. At the same time, there are his classmates A, B, and C in the dormitory. The three students (A, B and C) can receive the Wi-Fi Direct signals of the students D's smart-phone.

2.1 Ranging Principle Based on RSSI

The relation [16] of transmitted power (expressed by PT) and the received power (expressed by PR) of a radio signal can be represented by the formula (1). In the formula (1), d delegates the distance between transmitter and receiver, n is the propagation factor, also is called attenuation factor, and n's size (value) depends on the wireless signal propagation environment. Such as formula (1):

$$P_R = P_T / d^n \tag{1}$$

In order to apply the description of power gain (dB), we take the logarithm on both sides in the formula (1), and multiply a constant 10 at the same time, then obtain the formula (2):

$$10 \lg P_R = 10 \lg P_T - 10n \lg d \tag{2}$$

Transmit power PT unchanged, when the n's value is smaller, the signal attenuation in the communication process is smaller too, the signal can be transmitted over greater distances.

2.2 Analysis and Comparison of Common Simulation Models

Establishing the relationship between RSSI and spatial location is the key to RSSI based positioning. Based on the basic theory of RSSI, here gives the comparison of common models. Corresponding simulations were carried out on those models, including free space infinite electric propagation path loss model simulation, logarithmic normal distribution model simulation, logarithmic distance path loss model, wall attenuation factor simulation model and CHAN propagation model.

The selection of propagation model is very important for the calculation of pathloss. Usually, different types of propagation models will produce different path losses. The common propagation models are simulated, and different propagation models are studied, or different propagation models are combined, so that the parameters of different loss which affecting the distance and each model can be intuitively displayed. Only after comparison can the transmission model which closest to the real situation be clearly selected.

3   Proposed Scheme

3.1   Mending for Distance Measurement Based on RSSI

In practical application, RSSI will be influenced by many factors. In the indoor environment, the interference is mainly composed of the wall, ground reflection, barrier effect and so on, which lead to the increasing attenuation of wireless signal, especially, the wireless signal can be attenuated sharply in the metal barrier, even blocked.

The intensity of the measurement signal is converted to a distance based on the RSSI model, the key step to be resolved is the values of the parameters A and n should be relatively accurate. Environment largely influences the value of A and n. Along with the surrounding environment changes, the values of the parameters A and n will be changed. After a lot of tests, the values range of n and A in different environments are obtained, as shown in Table 1.

Table 1. The values range of N and A in different environments

Environments	A (dB)	N
Park	(32.7–36.0)	(2.7–3.4)
Stair, balcony	(33.5–44.2)	(1.4–2.4)
Big yard	(34.2–39.0)	(2.8–3.8)
barrier	(34.5–38.2)	(4.6–5.1)
Stone walls	(37.2–41.5)	(3.3–3.7)
greensward	(33.2–36.4)	(3.0–3.9)
Beach	(37.5–40.8)	(3.8–4.6)

Both A and n values need to be refined according to the specific environment which can make the values of A and n more suitable for indoor environmental parameters. Then there are the following formula (3)–(6) for estimating:

$$n = \hat{n} = \sum_{i=1}^{n}(p_i - \bar{p})RSSI_i / \sum_{i=1}^{n}(p_i - \bar{p})^2 \tag{3}$$

$$A = \overline{RSSI} - n\bar{p} \tag{4}$$

where in

$$\bar{p} = \frac{1}{n} \sum_{i=1}^{n} p_i \tag{5}$$

$$\overline{RSSI} = \frac{1}{n} \sum_{i-1}^{n} RSSI_i \tag{6}$$

Thus, the specificity of RSSI data is improved. The n can be obtained in two ways: the first one is from the continuous transmission time domain, the cost is the measurement cycle of this method will be longer; The second method is uses different power transmission modes to transmit chips, which send information and corresponding processing with different power, in order to reduce the error of distance measurement, but the cost of such a situation is high.

3.2 3D Spatial Orientation

There are 3 known reference point acquisition signals from the current group, we correct the parameter of the distance measurement model, which translate the strength indicator of received signal into a distance value. The spatial localization algorithm realizes 3D orientation of the known reference points and the unknown point with all measurement distances. The spatial location process based on RSSI shown in Fig. 1.

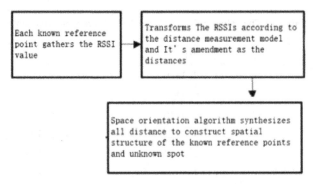

Fig. 1. Diagram of spatial location process based on RSSI

The known Wi-Fi direct-attached device is related to the unknown Wi-Fi direct attached devices by the correction method. The Known Wi-Fi direct-attached devices received information from the unknown Wi-Fi direct-attached equipment, so it collects signal strength RSSI of a column data from Wi-Fi direct-attached equipment ID. When information group which are received from the unknown Wi-Fi direct-attached equipment exceed a certain threshold (experiment group is 100), we can calculate out the RSSI value of undetermind Wi-Fi direct-attached equipment. Then we can calculate out the distance factor d by the path loss index and attenuation parameters, which is the distance between the known Wi-Fi direct-attached devices and undetermind Wi-Fi direct-attached devices. All the known Wi-Fi direct-attacheddevices use this method to calculate out d1, d2, ..., dn.

The reference point A is the center of the two concentric sphere, the two concentric spherical radius are dAB and dAC. dAB indicates the distance translated by RSSI of the known reference point A (receiving terminal) from the reference point B (transmitting terminal) based on a distance measurement model. It can be seen that the reference point B is in the sphere (A point is the sphere's center, the radius is dAB). Similarly, dAC indicates the distance translated by the received signal strength indicator (RSSI) of the known reference point A (receiving terminal) from the reference point C (transmitting terminal) based on a distance measurement model. It can be seen that the reference point C is in the sphere (point A is center of the sphere, the radius is dAC). The reference point B is the center of the two concentric sphere, the two concentric spherical radius are dAB and dBC. The reference point C is the center of the two concentric sphere, the two concentric spherical radius are dBC and dAC. For ease of understanding, all the spheres by the reference points A, B, C are projected on a plane. RSSI of three reference points shown in Fig. 2.

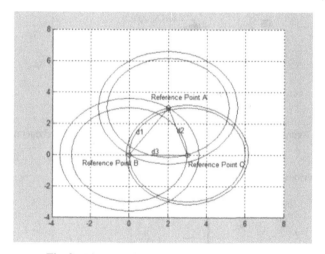

Fig. 2. Diagram of RSSI of three reference points

3.3 Calculation of Network Approximate Positioning Error

Firstly, we need to obtain the correction based on RSSI distance measurement from the reference node (beacon node) to each adjacent node. Then, we use the maximum likelihood estimation method (or trilateral measurement method) to estimate the relative coordinate position, and finally compare the relative coordinate position with the actual coordinate position [17].

The reference node (beacon node), $M(x0, y0)$, obtains its own coordinates through information such as the coordinates and correction distance of its neighboring reference nodes (beacon nodes), M(x1, y1), M(x2, y2), M(xk, yk), and then compares it with the actual coordinates to obtain its own coordinate error.

The formula for calculating the coordinate error of the reference node (beacon node) is:

$$\begin{cases} \Delta x_i = \hat{x} - x_i \\ \Delta y_i = \hat{y} - y_i \end{cases} \tag{7}$$

$(\hat{x}_i, \hat{y}_i)$ and (xi, yi) are the calculated and actual coordinates of the reference node (beacon node), respectively.

Similarly, other reference nodes (beacon nodes) can also use this method to obtain their coordinate errors. An example diagram is shown in Fig. 3.

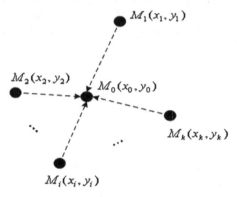

Fig. 3. Example diagram of self positioning of reference nodes (beacon nodes)

The approximate positioning error of the defined network is:

$$\begin{cases} \overline{\Delta x_\iota} = \frac{1}{m} \sum_{i=1}^{m} \Delta x_i \\ \overline{\Delta y_\iota} = \frac{1}{m} \sum_{i=1}^{m} \Delta y_i \end{cases} \tag{8}$$

Among them, Δx_i, Δy_i (coordinate error of reference node (beacon node) M1), m = k + 1.

Network approximate positioning is commonly used to correct the positioning error of unknown nodes, which is essentially the average of the position errors of all reference nodes (beacon nodes). The corrected result of the unknown location node is:

$$\begin{cases} x = \hat{x} + \overline{\Delta x} \\ y = \hat{y} + \overline{\Delta y} \end{cases} \tag{9}$$

$(\hat{x}, \hat{y})$ is an unknown position node, obtain its own coordinate error through the above method.

4   Simulation of 3D Spatial Direction

Add 3 known reference points to the group through the Wi-Fi Direct device point. For example, the known reference point A is converted to the distance d1 through the RSSI model, from which we can know that the unknown point should be on the sphere with reference point A as the center and d1 as the radius. Therefore, three spheres can be constructed from three different reference points according to the distance converted by RSSI. Through 3D space positioning algorithm, we can calculate the values of each reference point. The algorithm outlines the spatial structure based on this point according to the corresponding distance data. And then combines the prediction direction of the unknown points to predict the spatial position of the unknown reference point relative to the three known reference points, as shown in Fig. 4.

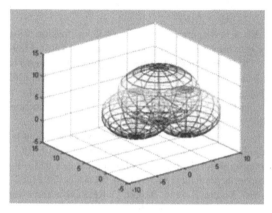

Fig. 4. Spatial schematic diagram of RSSI construction receiving three known reference points from an unknown point

5   Conclusions

In this paper, we review the relevant literature to summarize the status of localization algorithms, focus on analyzing and comparing several important ranging methods, and give the performance evaluation criteria of localization algorithms in wireless sensor networks. Based on the basic theory of RSSI, the RSSI ranging and positioning process is given. The mathematical or empirical model of radio wave propagation based on RSSI is analyzed, and the error of RSSI algorithm is analyzed. The Gaussian model with good correction effect of RSSI measurement value, the correction based on moving average method and the correction based on ranging error estimation are emphatically analyzed. Finally, several models are simulated, analyzed and compared.

Acknowledgment. This work is supported by:
  1. The Key Field Special Project of Guangdong Provincial Department of Education with No. 2021ZDZX1029.

2. The 2021 Key scientific research platforms and projects in ordinary colleges and universities in Guangdong Province, ID: 2021ZDZX3019.

References

1. Li, K., Wang, J., Jalil, H., Wang, H.: A fast and lightweight detection algorithm for passion fruit pests based on improved YOLOv5. Comput. Electron. Agric. **204**, 107534 (2023)
2. Guo, Z., et al.: Perpendicular intersection: locating wireless sensors with mobile beacon. IEEE Trans. Veh. Technol. **59**(7), 3501–3509 (2010)
3. Wang, W., Li, K., Jalil, H., Wang, H.: An improved estimation of distribution algorithm for multi-objective optimization problems with mixed-variable. Neural Comput. Appl. (2022)
4. Ouyang, R., Wong, A., Lea, C.: Received signal strength-based wireless localization via semidefinite programming: noncooperative and cooperative schemes. IEEE Trans. Veh. Technol. **59**(3), 1307–1318 (2010)
5. Wang, G., Yang, K.: A new approach to sensor node localization using RSS measurements in wireless sensor networks. IEEE Trans. Wirel. Commun. **10**(5), 1389–1395 (2011)
6. Zhang, L., Li, K., Qi, Y.: Person re-identification with multi-features based on evolutionary algorithm. IEEE Trans. Emerg. Top. Comput. Intell. **6**(3), 505–518 (2022)
7. Li, K., Fu, X., Wang, F., Jalil, H.: A dynamic population reduction differential evolution algorithm combining linear and nonlinear strategy piecewise functions. Concurr. Comput. **34**(6) (2022). https://doi.org/10.1002/cpe.6773
8. Liu, H., Darabi, H., Banerjee, P., Liu, J.: Survey of wireless indoor positioning techniques and systems. IEEE Trans. Syst. Man Cybern. Part C Appl. **37**(6), 1067–1080 (2007)
9. Cai, Y., et al.: Framework for the marriage of a fluorescence and colorimetric biosensor to detect organophosphorus pesticides. Anal. Chem. **93**, 7275–7282 (2021). ISSN 1520-6882
10. Li, K., Wang, Z., Yao, X., Liu, J., Fang, H., Lei, Y.: Recidivism early warning model based on rough sets and the improved k-prototype clustering algorithm and a back propagation neural network. J. Ambient Intell. Human. Comput. 1–14 (2021). ISSN 1868-5137
11. Wang, J., Ghosh, R.K., Das, S.K.: A survey on sensor localization. J. Control Theory Appl. **8**(1), 2–11 (2010)
12. Ng, J.K.-Y., Lam, K.-Y., Cheng, Q.J., Shum, K.C.Y.: An effective signal strength-based wireless location estimation system for tracking indoor mobile users. J. Comput. Syst. Sci. **79**, 1005–1016 (2013)
13. Muthukrishnan, K., van der Zwaag, B.J., Havinga, P.: Inferring motion and location using WLAN RSSI. In: Fuller, R., Koutsoukos, X.D. (eds.) MELT 2009. LNCS, vol. 5801, pp. 163–182. Springer, Heidelberg (2009). https://doi.org/10.1007/978-3-642-04385-7_12
14. Goldsmith, A.: Wireless Communications. Cambridge University Press, Cambridge (2005)
15. Bahi, P., Padmanabhan, V., RADAR: an in-building RF based user location and tracking system. In: Proceedings of IEEE INFOCOM, vol. 2, pp. 775–784 (2000)
16. Yu, N., Wan, J., Ma, W.: Sampling based 3D localization algorithm for wireless sensor networks. J. Beijing Univ. Posts Telecomm. **31**(3), 13–18 (2008)
17. Crow, B.P., Widjaja, I., Kim, J.G., et al.: IEEE802.11 wireless local area networks. IEEE Commun. Mag. **35**(9), 116–126 (1997)
18. Mao, G., Fidan, B., et al.: Wireless sensor network localization techniques. Comput. Netw. **51**(10), 2529–2553 (2007)

Collaborative Filtering Recommendation Algorithm Based on Improved KMEANS

Xuesong Zhou[✉], Changrui Li, and Jia Shi

CSSC Intelligent Technology Shanghai Co. LTD., Shanghai, China
zhouxuesong1985@163.com

Abstract. The collaborative filtering recommendation algorithm looks for the nearest neighbor set with the target user or item according to the historical track of the user's subscription or browsing of the item, and then predicts the score of the target object according to the score of the user in the nearest neighbor set of the item, and finally recommends the top several items to the user. This algorithm has a high degree of recognition in both academia and industry, but it also has some problems such as cold startup, data sparsity and poor scalability. For this reason, researchers at home and abroad have proposed many methods to improve the existing technology, among which how to accurately mine the user's interest preference is the key to improve the collaborative filtering based recommendation method. Aiming at the defects and problems still existing in the existing collaborative filtering based personalized recommendation methods, through in-depth and systematic research, this paper first adopts semantic similarity computing technology to analyze the similarity of users' comments on projects, and takes the similarity as the similarity of users' preferences in user comments. The method of mixing the similarity with Pearson correlation coefficient and heuristic similarity is designed respectively. At the same time, matrix decomposition technology is used to fill the sparse score matrix, and then the improved KMEANS algorithm is integrated to build the project cluster model on the filled score matrix, and finally complete the project-based collaborative filtering recommendation on the candidate set determined by the cluster model. Experimental results show that the method proposed in this paper improves the precision of collaborative filtering.

Keywords: Recommendation system · User preference · Collaborative filtering · Recommended accuracy · K-MEANS

1 Introduction

Recommendation systems analyze user behavior to suggest information, products, or services aligned with their interests and preferences, without requiring explicit queries. They've been impactful in movies, music, video, and e-commerce. Recommendation systems are categorized into content-based and collaborative filtering-based systems. Typically, they predict user ratings for unreviewed items. Collaborative filtering, focusing on user ratings or usage records rather than delving into user or item characteristics, offers practicality and has seen extensive research and application in both academia and industry.

© The Author(s), under exclusive license to Springer Nature Singapore Pte Ltd. 2024
K. Li and Y. Liu (Eds.): ISICA 2023, CCIS 2146, pp. 451–462, 2024.
https://doi.org/10.1007/978-981-97-4393-3_37

Recommendation systems, continuously a research focus, integrate disciplines like AI, machine learning, data mining, and natural language processing, expanding their application scope. Collaborative filtering, using others' ratings and history, offers personalized services like rating predictions or product suggestions. It surpasses content-based methods by filtering items with challenging content, catering to user-specific needs and interests, and providing unexpected recommendations. Consequently, this paper primarily explores personalized recommendation methods based on collaborative filtering.

2 Research Status

With Web2.0, user-generated online reviews provide clearer insights into preferences and reasons for liking or disliking items, enhancing recommendation systems' accuracy in discerning user interests. This also addresses issues of data sparsity and cold starts. In the last decade, collaborative filtering methods based on user reviews have advanced significantly, propelled by natural language processing technologies that extract semantic and emotional information from comments, further evolving collaborative filtering recommendation methods. In a recent review, literature [1] introduced and analyzed the current recommendation system based on user reviews in detail, pointing out that text analysis technology and concept mining technology can extract a variety of valuable information for the recommendation system from user reviews, and the use of such information extracted from user reviews can effectively alleviate data sparsity and cold start problems. Literature [2] designed a hybrid recommendation system based on similarity based on user comments, and the author found that user ratings alone could not reflect user similarity well. In order to study the usefulness of user reviews and their impact on sales volume, literature [3] combines the econometric analysis method with the subjective analysis technology in text mining technology to solve the problem of customers selecting the best reviews and judging product quality based on reviews to a certain extent, and also helps product producers to obtain reviews that affect customer sales volume and check the content of reviews. Literature [4] studied the problem of mining users' views on topics by using used reviews and forming users' overall judgment on products based on the degree of users' preference for different topics, and proposed a new probability score regression model. Literature [5] combines the recommendation algorithm of nearest-neighbor collaborative filtering based on review content with the matrix decomposition collaborative filtering algorithm based on the review-item scoring matrix. In literature [6, 7], it is assumed that all words in comment statements come from the same topic. An ASUM algorithm, extending the LDA method and utilizing SLDA's simultaneous aspect and emotion information, is proposed for deriving aspecte-motion pairs. Experiments demonstrate that both SLDA and ASUM enhance comment understanding accuracy.

The classic collaborative filtering recommendation method believes that the user, project and the interaction between them in the recommendation system are static and unchanged, but due to the influence of factors such as user, project and context, users and projects may be changing all the time. Literature [8] classifies the influence of time effect on recommendation system into two categories: On this basis, literature [9] elaborated these two types of time effects in detail, arguing that time-sensitive recommendation systems explicitly model time. Literature [10] designed an algorithm for calculating the time weights of different items, aiming at the problem that traditional recommendation systems do not consider the change of users' interests over time. The algorithm increases the weight of recent user ratings and decreases the weight of older user ratings, and introduces this time weight into the project-based collaborative filtering recommendation method.

The evolution of users' interests, reflected in their changing preferences, is crucial for analyzing user behavior, particularly in geolocation-based recommendation systems. Advancements in wireless communication, positioning technologies, and mobile internet have made spatio-temporal trajectory data a key aspect of location-based recommendation system research. In literature [11], the basic principles and main research contents of spatiotemporal trajectory data research are introduced in detail, and various technologies used for spatiotemporal trajectory data calculation are systematically displayed, which is an important reference book for trajectory calculation. In addition to the use of trajectory data in location-based recommendation systems, in recent years, people have also begun to study the impact of time series on recommendation systems in traditional recommendation systems. Literature [12] extracts the timing rules from the purchase sequence of users, and provides recommendation services for users by using the fragment-based collaborative filtering recommendation algorithm. The traditional recommendation system believes that users' interests and preferences are independent. However, with the development of web2.0 and social networks, the relationship between users has attracted the attention of sociology, psychology, computer and other fields. Various studies have found that users' interests and preferences are not independent, but influenced by friends or others in the society. At present, the recommendation system mainly studies the social relationship including trust relationship, attention relationship and user reputation. Literature [13] holds that although users have their own unique tastes and hobbies, they are also more susceptible to the influence of their trusted friends. The final decision of users is made by users according to their own preferences and considering the influence of friends. A method for predicting user scores is designed based on the scores of users' trusted friends on projects and the trust between users and friends. Taking the user similarity based on user rating as the implicit social relationship between users, and the trust relationship between users as the explicit social relationship, and applying these two social relationships between users to the probability matrix decomposition through fusion weights, a recommendation method using the explicit and implicit social relations is proposed. The experiment shows that, the recommendation method using both user preferences and friend preferences has better experimental results than the matrix decomposition method relying solely on rating and the recommendation method relying solely on user preferences or friend preferences.

3 Improved KMEANS Algorithm

In the KMEANS algorithm, the distance between two samples is calculated using the Euler distance of formula 1.

$$d_{ij} = \sqrt{\sum_{p=1}^{k} (x_{ip} - x_{jp})^2} \tag{1}$$

dij is the distance between samples i and j, x_{ip} is the value of the i sample p dimension, and x_{jp} is the value of the j sample p dimension. Before calculation, the original data should be normalized and centralized. In this paper, x_{ij} is processed according to formula 2.

$$\widetilde{x_{\iota k}} = \frac{x_{ik} - \overline{x_k}}{\sqrt{\sigma_k^2}} \tag{2}$$

x_{ik}' is the value after the normalization of x_{ik}, x_{ik} is the k-dimensional value of sample i, and $\overline{x_k}$ is the k-dimensional average. σ_k^2 is the variance of k dimensions. Given the data sample set $D = x1, x2, ..., x_m$, KMEANS wants to divide D into K clusters $C = c1, c2, ..., c_k$ with no intersection between the clusters. The goal according to formula 3 is to minimize the sum of squared errors.

$$SSE = \sum_{i=1}^{k} \sum_{x \in c_i} \|X - U_i\|^2 \tag{3}$$

$$U_i = \frac{1}{C_i} \Sigma_{X \in C_i} x \tag{4}$$

where, K is the number of clusters, X is the sample, Ui is the center point of the KTH cluster, and the larger SSE is, the higher the degree of sample aggregation is. The basic idea is as follows: when k is less than the real cluster number, SSE will decrease greatly because the increase of k will greatly increase the degree of aggregation of each cluster, and when k reaches the real cluster number, the return on the degree of aggregation obtained by increasing k will rapidly decrease, so SSE will decrease sharply, and then tend to be flat with the continuous increase of k value. The point that flattens out first is the appropriate value of K.

As shown in Fig. 1, KMEANS first selects K center points, then calculates the distance between all remaining samples and the center point, selects the center point closest to him, and adds this center point. When all sample points are added to the corresponding center point, recalculates the center point, recirculates the distance between all remaining sample points and the center point, and selects the nearest center point to add, and so on. Until the center point no longer changes, the iteration ends.

KMEANS algorithm has been favored by all walks of life for more than 50 years since its birth. Because of the advantages of agglomeration, the main factors can be attributed to the following aspects:

(1) The implementation efficiency of the algorithm is high, the principle is easy to understand, and it is easy to carry out engineering practice in various fields.

(2) The algorithm has excellent time complexity O(ndkt) when processing data samples, where k represents the number of clusters, d represents the data dimension, t represents the number of iterations, and n is the number of data samples. It can be seen from the formula of time complexity that the parameters in the formula have a linear relationship with the time complexity, so this algorithm is often used as an initialization process.

(3) In the division of data samples, if the classes of the samples to be clustered have obvious differences, the clusters belonging to the same category will be very dense and the clustering quality will be very good when the square error criterion is used for discrimination.

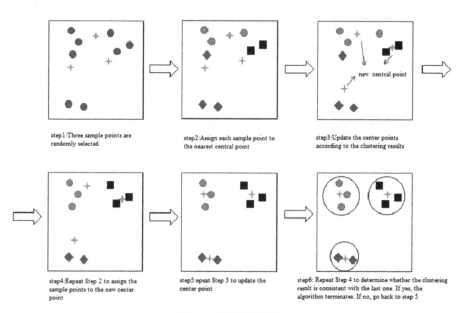

Fig. 1. KMEANS idea

(4) According to relevant literature studies, KMEANS algorithm has good scalability. For example, Dai Tao et al. proposed a parallel KMEANS document clustering algorithm based on CUDA in literature. He Peipei et al. proposed a KMEANS design scheme based on cloud environment in literature and other related literature. Both experiments show that the parallelized KMEANS algorithm has excellent speed ratio. The K value must be given manually, and prior knowledge is required. The clustering result depends on the random selection of the initial cluster center, and may converge to the local optimal solution rather than the global optimal. It is greatly affected by noise and outliers. Ideas for improvement:

KMEANS can be improved by optimizing the selection of random centers, and the specific ideas are as follows:

(1) The initial clustering center is determined according to the distribution density of samples, the clustering center is selected by establishing KD-tree, and the initial clustering center is selected by using the adaptive method of individual contour coefficient. A sample is selected as the first clustering center by the three compromise methods.
(2) The choice of the NTH center is related to the first n-1 centers:
(3) calculates the sum of distances of all samples to the existing center, and selects the distance and largest sample as the new clustering center.
(4) until K centers are selected

The idea behind this process is to spread the initial cluster centers as widely as possible. Fig. 2 and Fig. 3 show the comparison between the original KMEANS effect and the improved KMEANS effect.

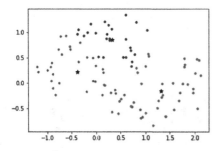

Fig. 2. KMEANS effect picture   **Fig. 3.** Improved KMEANS renderings

The biggest disadvantage of KMEANS is that it is greatly affected by outliers or noise. This is because the way KMEANS updates the center is to calculate the mean value vector in the cluster. Outliers will greatly affect the mean value of an attribute column, resulting in deviation of the center point. Ideas for improvement: Try to replace the center point with a non-center point, and if the total cost is reduced after substitution, do the real substitution. In this way, the center point of the update is always some sample, and in terms of total cost, the center point does not deviate from the cluster. When the number of samples belonging to a category is too small, the category is removed, and when the number of samples belonging to a category is too large and the degree of dispersion is large, the category is divided into two sub-categories. Fig. 4 and Fig. 5 are the comparison between the further improved effect diagram and the original KMEANS effect diagram.

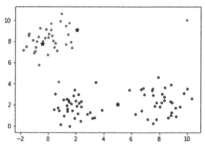 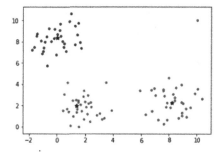

Fig. 4. KMEANS effect picture   **Fig. 5.** Improved KMEANS renderings

4 Collaborative Filtering Recommendation Algorithm Based on the Improved KMEANS Algorithm

Collaborative filtering algorithm is the focus of this paper, and it is also a successful recommendation method. Its main idea is to dig out an internal relationship between users and projects, actively find a set of similar users and projects according to the historical track of users, and then recommend items from this set that target users have preferences, but have not made substantial ratings obviously, which can improve the accuracy of recommendation. The implementation of the general collaborative filtering recommendation algorithm consists of three steps: (1) Obtaining the historical score matrix between users and projects; (2) Adopt appropriate similarity method to measure the association value between users or items and form the group with larger value into the nearest neighbor set; (3) Recommendation is made according to the predicted score of the target user towards the recommended object. Collaborative filtering recommendation belongs to a family of algorithms, which can be divided into two categories through refinement: memory-based collaborative filtering and model-based collaborative filtering. The memory-based collaborative filtering can be divided into user-based and project-based collaborative filtering according to the different similarity objects. Model-based collaborative filtering mainly constructs a recommendation model for R through some data mining algorithms, and finally generates recommendations for target users.

4.1 Similarity Measurement in Recommendation System

In collaborative filtering, a family recommendation algorithm, no matter what kind of classification method is adopted, finding similar users or item sets is a key step, but the similarity measurement method involved in this step is its core. Therefore, two common similarity measures are introduced below.

(1) Cosine similarity

This measurement method is the same as that introduced in the clustering technology section of Sect. 2, but in the recommendation algorithm, this method is described as follows:

The scoring matrix $R(m, n)$ involved in this calculation is generally regarded as a vector on M-dimensional user space and N-dimensional item space, and the correlation degree between users or items is determined according to the cosine value of the Angle between the scoring vectors, whose expression is as follows (5) :

$$w_{ij} = cos\theta_{ij} = \frac{\sum_{k=1}^{n} x_{ki}x_{kj}}{\sqrt{\sum_{k=1}^{n} x_{ki}^2 \sum_{k=1}^{n} x_{kj}^2}} i, j = 1, 2, \cdots p \tag{5}$$

x_{ki} represents user $k's$ rating of item i; x_{kj} represents user $k's$ rating of item j. Represents the set of users who rate item i in the scoring matrix R, ki represents the set of users who rate item i in the scoring matrix R, kj represents the set of users who rate item j in the scoring matrix R. $Kinki \cap kj$

$$w_{ij} = cos\theta_{ij} = \frac{\sum_{I=1}^{n} x_{ki}x_{vi}}{\sqrt{\sum_{i=1}^{n} x_{ki}^2 \sum_{i=1}^{n} x_{vi}^2}} k, v = 1, 2, \cdots n \tag{6}$$

xki represents user $k's$ score on item$i$, and xvi represents user $v's$ score on item i. ki represents the set of scoring items of user $u$in the scoring matrix R, vi represents the set of scoring items of user v in the scoring matrix R, $i\ eki\ Uvi$.

(2) Modified cosine similarity

The biggest difference between the modified cosine similarity method and the cosine similarity method is that it considers the individual differences existing in the score and is an improvement to the cosine similarity method. The improved expression of formula 5 is as follows:

$$w_{ij} = cos\theta_{ij} = \frac{\sum_{k=1}^{n} (x_{ki} - \overline{x}_i)(x_{kj} - \overline{x}_j)}{\sqrt{\sum_{k=1}^{n} (x_{ki} - \overline{x}_i)^2 \sum_{k=1}^{n} (x_{kj} - \overline{x}_j)^2}} i, j = 1, 2, \cdots, p \tag{7}$$

The improved expression of formula 6 is as follows:

$$w_{ij} = cos\theta_{ij} = \frac{\sum_{I=1}^{n} (x_{ki} - \overline{x}_k)(x_{vi} - \overline{x}_v)}{\sqrt{\sum_{i=1}^{n} (x_{ki} - \overline{x}_k)^2 \sum_{i=1}^{n} (x_{vi} - \overline{x}_v)^2}} k, v = 1, 2, \cdots n \tag{8}$$

The collaborative filtering recommendation algorithm based on cluster improvement in this paper includes three main processes, namely: matrix decomposition data preprocessing, improved KMEANS algorithm to build a cluster model, and prediction of user ratings to form recommendations. The entire process is done in Hadoop cluster mode, where 1R and 2R represent the training and test sets divided by the user-project matrix, respectively.

4.2   Matrix Decomposition Data Preprocessing

As for the above recommendation of collaborative filtering based on singular value (SVD) decomposition, this method can decompose a high-dimensional sparse matrix into two lower-order matrices, but this method has some shortcomings: (1) Before decomposing the original matrix, it is necessary to take certain methods to complete the sparse matrix, and the completed matrix will become quite dense, resulting in a very large storage space; (2) SVD is very slow to calculate the score matrix with more than 1000 dimensions, and the actual experimental data are all tens of millions of data. For this reason, some scholars have proposed to use alternate least squares (ALS) as matrix decomposition to avoid the shortcomings of SVD algorithm. Therefore, this paper adopts the ALS algorithm to fill the high-dimensional sparse matrix as the first stage of improving the clustering collaborative filtering algorithm.

In the ALS matrix decomposition algorithm, it is necessary to decompose the scoring matrix $R(m, n)$ and find a low-rank matrix X to approximate the matrix $R(m, n)$, so that $X = UVT$, where U and V are eigenmatrices, and the objective function expression of the algorithm is 9:

$$L(U, V) = \sum_{ij} (R_{ij} - U_i V_j T)^2 + \gamma (||U_i||2 + ||V_j||^2) \tag{9}$$

Where γ is the regularization factor and Rij represents the score value of the original matrix. The distributed design of ALS algorithm under the MapReduce framework is described as follows: Input: user score matrix $R1$, feature dimension d, maximum number of iterations k and regularization factor

Output: Complete new scoring matrix R1 '.

The improved KMEANS algorithm is used to cluster the obtained item-scoring matrix R1'. Its specific idea is to form an item sample set by taking the scores of all users on the same item as a data sample. By calculating the minimum variance of each item in the item set, the threshold value set according to the mean value and the item sample with the minimum variance are used as the new clustering center to divide the cluster, so as to obtain the initial clustering center of KMEANS algorithm to cluster the item. High-quality initial clustering center can reduce the number of iterations of KMEANS algorithm, that is, reduce the computational amount of project clustering, thus improving the efficiency of building clustering model.

5   Experiment and Analysis

The experimental Hadoop platform in this paper consists of one master node and four slave nodes. In the first part, the improved improved KMEANS algorithm uses the data set in the open UCI machine learning library for experimental analysis to verify the accuracy and convergence of the algorithm, and compares it with the traditional K-means algorithm. At the same time, in order to verify the advantages of the improved KMEANS algorithm in processing large-scale data, the acceleration ratio and expansion rate are analyzed. The second part is mainly to improve the experimental analysis of KMEANS clustering collaborative filtering recommendation algorithm on MovieLens data set.

Table 1 below lists the attributes, number of samples, and number of categories of the selected samples in the UCI set. These standard data sets are used to accurately measure the clustering quality of the proposed algorithm.

Table 1. Experimental test data set

Data set	Sample number	Attribute number	Class number
Iris	200	4	3
Wine	800	13	3
lonshpere	1000	34	2
Glass	500	9	6
Wdbc	900	30	2
Soybean-small	1000	35	4

5.1 Experimental Results and Analysis

(1) Improve the comparison of clustering quality between KMEANS and traditional KMEANS algorithm.

This paper uses the following data sets to test the optimized algorithm on a single machine. In order to ensure the reliability of the test results, each sample set is tested 10 times under each algorithm, and the average value of the 10 experiments is taken as the result. The results are shown in Table 2 below.

Table 2. Experimental results of sample data set

Data set	Classical KMEANS			Improved KMEANS		
	accuracy rate (%)	Iteration times	error sum of squares	accuracy rate	Iteration times	error sum of squares
Iris	75.35	9	94.66	89.33	6	82.13
Wine	86.24	7	6.55 + E06	90.27	5	5.62 + E06
lonshpere	72.69	9	6.81 + E06	85.10	7	5.39 + E06
Glass	78.31	12	6.31 + E06	83.47	9	5.71 + E06
Wdbc	86.17	14	6.15 + E06	91.44	10	5.26 + E06
Soybean-small	66.25	16	6.37 + E06	80.11	9	5.82 + E02

As can be seen from Table 2: The clustering accuracy of the proposed method is improved on the selected data set, and the error square and iteration times are reduced, with the improvement intervals ranging from 4.2% to 27.32%, and the reduction intervals ranging from 4.70% to 34.00% and 16.67% to 42.86%, respectively. It can be seen that the text variance can obtain better clustering quality and faster convergence speed.

(2) Improved algorithm acceleration ratio and expansion rate analysis

An important indicator to verify the parallel capability of an algorithm is the acceleration ratio, the acceleration ratio $S = Ts/Tp$, Ts is the time required for a single node to run, Tp is the time required for p (p = 1, 2, 3, 4) nodes to run. According to the experimental results in Table 2, the improved KMEANS algorithm has achieved good results. Therefore, this paper only constructs a larger dataset for the selected dataset Iris. The sizes of 25M (1,541,250 pieces), 50M (3,082,500 pieces), 100M (6,165,000 pieces) and 200M (1,2330,000 pieces) for processing and analysis respectively, and the resulting acceleration is shown in Fig. 6:

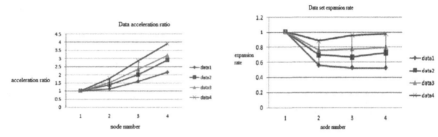

Fig. 6. Iris dataset speedup test results   **Fig. 7.** Iris dataset speedup test results

As can be seen from Fig. 6, the improved KMEANS algorithm has a good acceleration ratio on data sets of different sizes after parallelization. When the amount of data increases, the acceleration ratio shows a linear increase. This shows that the improved algorithm design in this paper is reasonable, especially after the Map stage of KMEANS algorithm, a Combine operation is designed, which greatly reduces the consumption of system network and disk IO, and reduces the execution pressure in the Reduce stage, so the communication cost is very small.

Under normal circumstances, as the cluster nodes continue to increase, the acceleration ratio cannot be infinite upward. Therefore, it is often insufficient to consider the superiority of parallel algorithms only from the acceleration ratio, and the acceleration ratio can not accurately reflect the utilization of clusters. Next, this paper uses the expansion rate to verify the use of the cluster. The formula $H = P/T(T$ represents the acceleration ratio of the algorithm, P represents the number of nodes) is used. The measured results are shown in Fig. 7.

It is not difficult to see from Fig. 7 that the cluster expansion rate of data3 and data4 begins to rise at node 2, while the increase of data2 cluster expansion rate is delayed to node 3. For the data set data1, the cluster expansion rate keeps decreasing, which does not reflect the advantage of cluster. However, it can be obviously seen that when data nodes and data volume increase, the cluster expansion rate of large data volume is always greater than that of small data volume. It is shown that the expansion rate of cluster is closely related to the data size, which indicates that the algorithm is suitable for the processing of large data.

6 Summarize

With the development of new technologies such as big data, cloud computing and artificial intelligence, as well as the Internet of Things, human beings are already in a world of information overload. The recommendation system technology to solve the current problem has also begun to rely on these new technologies to be studied in full swing, and new achievements and applications continue to emerge. This paper focuses on the shortcomings of traditional KMEANS clustering algorithm and collaborative filtering algorithm, proposes solutions and achieves certain results. However, these studies still have some limitations and need to be further explored.

References

1. Chen, Y., Dai, Y., Han, X., et al.: Dig users' intentions via attention flow network for personalized recommendation. Inf. Sci. **547**, 1122–1135 (2021). https://doi.org/10.1016/j.ins.2020.09.007
2. Zhang, Q., Ren, F.: Double Bayesian pairwise learning for one-class collaborative filtering. Knowl.-Based Syst. **229**, 107339 (2021). https://doi.org/10.1016/j.knosys.2021.107339
3. Amer, A.A., Abdalla, H.I., Nguyen, L.: Enhancing recommendation systems performance using highly-effective similarity measures. Knowl.-Based Syst. **217**(4), 106842 (2021). https://doi.org/10.1016/j.knosys.2021.106842
4. Yang, E., Huang, Y., Liang, F., et al.: FCMF: federated collective matrix factorization for heterogeneous collaborative filtering. Knowl.-Based Syst. **220**(1/2), 106946 (2021). https://doi.org/10.1016/j.knosys.2021.106946
5. Zhou, M.Y., Xu, R.Q., Wang, Z.M., et al.: A generic Bayesian-based framework for enhancing top-N recommender algorithms. Inf. Sci. **580**(1) (2021). https://doi.org/10.1016/j.ins.2021.08.048
6. Wang, P., Wang, Y., Zhang, L.Y., et al.: An effective and efficient fuzzy approach for managing natural noise in recommender systems. Inf. Sci. (4) (2021). https://doi.org/10.1016/j.ins.2021.05.002
7. Zhao, J., Li, H., Qu, L., et al.: DCFGAN: an adversarial deep reinforcement learning framework with improved negative sampling for session-based recommender systems. Inf. Sci. **596**, 222–235 (2022)
8. Ai, J., Cai, Y., Su, Z., et al.: Predicting user-item links in recommender systems based on similarity-network resource allocation. Chaos Solitons Fractals **158** (2022)
9. Ortega, A.F., Lara-Cabrera, R., González-Prieto, A., et al.: Providing reliability in recommender systems through Bernoulli matrix factorization. Inf. Sci. **553**, 110–128 (2021). https://doi.org/10.1016/J.INS.2020.12.001
10. Tang, H., Zhao, G., Bu, X., et al.: Dynamic evolution of multi-graph based collaborative filtering for recommendation systems. Knowl.-Based Syst. **228**, 107251 (2021). https://doi.org/10.1016/j.knosys.2021.107251
11. Wang, Y., Deng, J., Gao, J., et al.: A hybrid user similarity model for collaborative filtering. Inf. Sci. 102–118 (2017). https://doi.org/10.1016/j.ins.2017.08.008
12. Wang, Z., Wei, X., Pan, J.: Research on IRP of perishable products based on mobile data sharing environment. Int. J. Cogn. Inform. Nat. Intell. (IJCINI) **15** (2021). https://doi.org/10.4018/IJCINI.20210401.oa10
13. Wang, Z., Liu, X., Zhang, W., et al.: The statistical analysis in the era of big data. Int. J. Model. Identification Control (2022)

Emotion Analysis of Weibo Based on Long Short-Term Memory Neural Network

Li Kangshun[1,2(✉)], Weicong Chen[1], and Yishu Lei[1]

[1] College of Mathematics and Informatics, South China Agricultural University,
Guangzhou 510642, China
likangshun@dgcu.edu.cn

[2] School of Artificial Intelligence, Dongguan City University, Dongguan 523808, China

Abstract. This paper proposes a sentiment analysis method based on the Long Short-Term Memory (LSTM) neural network, which achieves sentiment polarity classification of user comments on Weibo (a Chinese microblogging platform). The results are visualized, providing assistance in monitoring public opinion on Weibo. The study primarily employs both unidirectional and bidirectional LSTM models and compares their performance under different conditions through parallel repeated experiments. The model that performs better in the experiments is selected as the deep learning model. Additionally, Naive Bayes classifier, XGBoost classifier, and Support Vector Machine classifier are trained, and their optimal parameters are determined as machine learning models. The trained deep learning model and machine learning models are combined to form an ensemble model. The model is evaluated and tested using a test set and user comments on a specified topic, providing accuracy and top-5 accuracy metrics.

Keywords: LSTM · Emotion Recognition · Deep Learning · Integrated Model

1 Introduction

Currently, there are three main methods for sentiment analysis: sentiment lexicon based sentiment analysis, machine learning based sentiment analysis, and deep learning based sentiment analysis [1]. The method based on sentiment semantic analysis mainly utilizes a pre organized dictionary set, and judges and classifies all vocabulary in the text according to the prescribed rules of semantics, grammar, etc. to obtain the sentiment tendencies of all vocabulary that make up the sentence, and then estimates and deduces the sentiment tendencies of the entire text. The sentiment analysis method based on machine learning belongs to supervised learning, which relies on manually classifying and labeling the corpus to obtain a large number of corpus models. Then, suitable algorithms are selected to train classifiers, and new data is used to train models to obtain predicted results. Finally, the probability of belonging to each text sentiment category is further calculated [2]. The emotional analysis of text plays an increasingly important role in our Internet life. It can not only deepen our understanding of the text, but also apply the analysis results to such fields as public opinion monitoring, fraud information

© The Author(s), under exclusive license to Springer Nature Singapore Pte Ltd. 2024
K. Li and Y. Liu (Eds.): ISICA 2023, CCIS 2146, pp. 463–470, 2024.
https://doi.org/10.1007/978-981-97-4393-3_38

filtering, etc., which is of great significance [3]. The article constructs a sentiment analysis neural network model based on LSTM. Encode and implement a sentiment analysis model based on long and short-term memory neural networks. The preprocessed data is vectorized into embedding vectors using Word2Vec technology, and then entered into the sentiment model for sentiment classification. MicroBlog is a type of microblog that allows users to easily post or obtain messages online. Typically, users can post text or image content with a word count range of up to 140 through their Weibo mobile app and other clients [4]. The sentiment polarity of user comments is analyzed, and the experiment shows that the sentiment analysis model based on long and short-term memory neural networks constructed in this paper has high accuracy in user comments on Sina Weibo.

2   Technical Methods and Dataset

Deep learning is an important branch of machine learning algorithms, essentially a neural network that mimics the behavior and cognitive patterns of the human brain. Deep learning can analyze and interpret data by simulating the human brain, and can process diverse data such as text and images [5]. With the development of computer hardware, deep learning has rapidly developed while meeting the requirements of computing power, and has achieved considerable results in various fields. Currently, deep neural networks applied in text sentiment analysis mainly include recurrent recurrent neural networks and convolutional neural networks, both of which have good effects in the field of text mining applications.

2.1   Recurrent Neural Networks (RNN) Model

The structure of recurrent neural networks is different from traditional feedforward neural networks. In the RNN sequence, the output content of the current neuron is related to the output content of the previous neuron, which is also the origin of the name RNN recurrent neural network. The input of the hidden layer consists of two parts, namely the output of the input layer and the output of the previous hidden layer, which gives RNN the characteristic of recording past information and displaying it in the current output calculation.

As shown in Fig. 1, RNN is vividly understood as copying the same neural network multiple times, and in these neuron sequences, each neural module will pass its own "memory" data to the next neural module. The memory of RNN is essentially related to its neuron sequence. Based on this characteristic, RNN has been widely used in multiple fields of natural language processing, such as language modeling, and has achieved good results. However, traditional RNNs have a fatal feature, which is that they are prone to a situation called gradient vanishing during the training process. The introduction of the LSTM (long term memory) model has largely overcome the shortcomings of traditional RNNs regarding gradient vanishing and introduced the concept of gates.

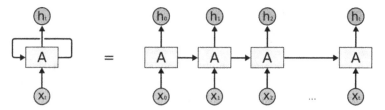

Fig. 1. Structure diagram of recurrent neural network model.

2.2 Support Vector Machine Algorithm

Support Vector Machine (SVM) was first proposed by foreign researchers in 1995. It is a machine learning algorithm that uses methods such as interval, duality, and kernel techniques. The generalization learning ability of SVM is superior to some traditional statistical methods [6]. Its advantage is that it can solve the problem of insufficient sample size and play a good role in high-dimensional and nonlinear pattern recognition. Support Vector Machine can be extended to other types of machine learning research. Currently, SVM has been widely used in fields such as speech recognition, remote sensing image analysis, fault recognition, and prediction [7].

2.3 Long Short Term Memory Neural Network Model

The LSTM model is actually a special RNN model, which introduces the concept of a forget gate layer, which allows LSTM to selectively forget some relatively useless information, reducing the computational complexity of the model, while also possessing the advantages of traditional RNN's memory data features. The architecture of the LSTM neural network model is roughly shown in Fig. 2. The LSTM neural network will filter the information of the cell state through a forget gate, achieving the function of "forgetting". The forget gate will read and output a value range of [0, 1] to assign any number in the cell state. Its encoding meaning is 0 to discard the information completely, and 1 to retain the information completely.

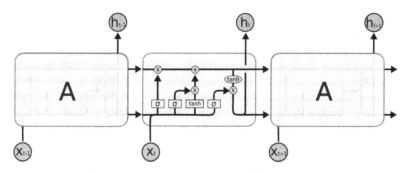

Fig. 2. LSTM neural network model architecture.

3 Algorithm Design

3.1 XGBoost

XGBoost is a model that integrates various tree models and is an improved boosting algorithm of GBDT. Compared with GBDT algorithm, XGBoost has made improvements in error approximation and numerical optimization. XGBoost uses K trees to predict the sum of values obtained from samples as the system's prediction result, and its defined prediction function is as follows (1):

$$\hat{y}_i = \sum_{t=1}^{K} f_t(x_i), f_t \in F \tag{1}$$

Among them, represents the predicted values of samples, K represents the number of trees, F is the set space of CART trees, and X_i represents the feature vectors of i data points. XGBoost can converge faster on the training set and effectively improve training speed by performing second-order Taylor expansion on the cost function, using first and second derivatives [8].

3.2 Naive Bayes

Naive Bayesian inference is the assumption of conditional independence of conditional probability distributions based on Bayesian inference. From this, the expression of the naive Bayesian classifier can be obtained. Due to the assumption of independence between different independent variables and normality of continuous variables as prerequisites, the computational accuracy of the algorithm will be affected to some extent. Its naive Bayesian algorithm is (2):

$$\tilde{y} = argmax_{c \in Y} P(c) \prod_{i=1}^{d} P(x_i|c) \tag{2}$$

4 Experiment Design and Experiment Result Analysis

4.1 Experiment Design

Classifiers trained through different deep learning algorithms or machine learning algorithms may have different effects on different datasets [9]. This paper draws on the idea of ensemble learning and proposes a hybrid sentiment analysis model based on deep learning and machine learning. The optimal model is obtained through multiple parallel repeated experiments on Naive Bayes algorithm, Support Vector Machine algorithm, XGBoost algorithm, and LSTM algorithm, combine these models into an ensemble model, where the predicted output of the ensemble model is the mode of the combined classifier for predicting the sentiment polarity of Weibo user comments. For example, the Naive Bayes algorithm predicts the sentiment polarity of sentences as negative, the support vector machine predicts as positive, XGBoost predicts as positive, and LSTM predicts as positive, So the final output of the integrated model is that the sentiment polarity of the sentence is positive, that is, the comment is biased towards positive comments [10]. The integrated model algorithm flowchart proposed in this paper is shown in Fig. 3.

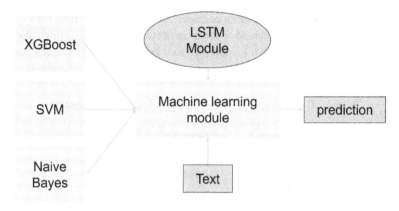

Fig. 3. The integrated model algorithm flowchart.

The experimental process steps designed in this paper are as follows:

A. Use web scraping techniques to crawl user comments on Sina Weibo and save them to a CSV document. Manually clean and label the comment text data in the document, assigning a label of 0 for negative comments and 1 for positive comments. Split the labeled text data into training and validation sets.
B. Perform tokenization, stop word removal, and other preprocessing operations on the training and validation set text data to achieve initial data preprocessing.
C. Train a Word2Vec algorithm on the text data to obtain a word vector model. Replace all words in the text with vectors from the trained Word2Vec space.
D. Build a sentiment analysis model based on Long Short-Term Memory (LSTM) neural network. Use the training set vectors as input to the model. Record the accuracy and loss values of the training set for each training epoch. Test the model on the validation set to obtain accuracy and AUC metrics.
E. Visualize the recorded accuracy, loss values, and AUC metrics using line charts.
F. Use web scraping techniques to crawl user comment text data on a specified topic on Sina Weibo and save it to a CSV document. Perform tokenization, stop word removal, and other operations on the text data. Convert the feature words into word vectors in the Word2Vec model obtained in step 3. Input the transformed data into the trained LSTM model to obtain the sentiment polarity of the user comments. Visualize the results accordingly.

The preprocessed text data is input into the Word2Vec model for training. Considering factors such as the size of the training samples, the Word2Vec model parameters used in this paper are mainly as follows: vector size = 64, min_count = 1, epochs = 1000. By searching for synonyms of specified text feature words, test the effectiveness of the trained Word2Vec model: find "beautiful" and print the possible synonyms and their probabilities in descending order as shown in Table 1 (excerpt); Search for "uncomfortable" and print the possible synonyms and their probabilities in descending order as shown in Table 2 (excerpt):

It can be seen that Word2Vec has a considerable effect on searching for synonyms. Each feature word in each segmented sentence in the text data is transformed into a word vector in Word2Vec space, which is input as an embedding vector into the constructed LSTM neural network.

Table 1. The synonym for "beautiful" in partially trained Word2Vec.

Heading level	Example
'lovely'	0.50734943151474
'beautiful'	0.49486386775970
'great'	0.49229389429092
'love'	0.48042699694633
'handsome'	0.46954420208930

Table 2. The synonym for "Uncomfortable" in partially trained Word2Vec.

Heading level	Example
'Uncomfortable'	0.61197310686111
'beautiful'	0.59778428077697
'great'	0.57840931415557
'love'	0.56028634309768
'handsome'	0.53017503023147

4.2 Experimental Process and Results Analysis

The LSTM neural network models designed in this paper mainly include unidirectional LSTM models and bidirectional LSTM models. The next experiment will set the training epoch to $10 \wedge 2$, with each epoch having $10 \wedge 2$ batches. The CELoss function will be selected, and the learning rate, optimizer type, and optimizer weight will be changed to compare the efficiency of the LSTM model and the Bi LSTM model. Through the comparison of the twelve experimental groups mentioned above in Table 3, it can be seen that Bi LSTM is generally more efficient on the validation set than the unidirectional LSTM model, and there are many factors that affect the training efficiency of the model. For example, in the 9th and 10th groups, a learning rate of 5e−5 was used, and the optimizer did not set weight_decay, resulting in a slight overfitting phenomenon. Based on the comparison of experimental results from various groups, it can be concluded that when Adam is selected as the optimizer with a learning rate of 5e−5 and a weight_decay of 0.01, the Bi LSTM model has the highest accuracy on the validation set. In addition, although the Bi LSTM model has slightly higher accuracy in the validation set than the LSTM model, the training speed of the LSTM model is better than that of the Bi LSTM model.

Table 3. Twelve sets of experiments on LSTM model and Bi LSTM model.

model	T_accuracy	V_accuracy	Loss	V_AUC
LSTM	78.7%	82.8%	0.48827	0.828
Bi-LSTM	83.1%	85.6%	0.39310	0.856
LSTM	92.7%	84.6%	0.23301	0.846
Bi-LSTM	98.2%	85.8%	0.07366	0.858
LSTM	78.2%	83.0%	0.47980	0.830
Bi-LSTM	85.1%	86.6%	0.36999	0.866
LSTM	76.6%	79.0%	0.52441	0.790
Bi-LSTM	79.8%	83.8%	0.46300	0.838
LSTM	94.8%	83.6%	0.17646	0.836
Bi-LSTM	97.4%	82.0%	0.09137	0.820
LSTM	88.4%	85.6%	0.29126	0.856
Bi-LSTM	92.7%	87.0%	0.33003	0.870

5 Conclusion

This paper designs and implements a mixed sentiment analysis model based on long short-term memory neural network, which basically classifies the sentiment polarity of Weibo user comments and visualizes the results, playing a certain auxiliary role in the public opinion supervision of Weibo user comments. The main work can be summarized as using the obtained Weibo user comment corpus for category annotation, dividing the dataset into training and validation sets, and performing certain preprocessing work. The preprocessed dataset is used to train the model; Conduct multiple parallel repeated experiments on the model under different conditions to obtain various evaluation indicators of the model; And this paper adopts the idea of ensemble modeling to propose a new sentiment classification model, which combines classifiers trained by different algorithms in machine learning and deep learning to obtain a text sentiment analysis model with stronger classification performance; Using the LSTM based hybrid sentiment analysis model implemented in this paper, sentiment polarity analysis is performed on crawled Weibo user comments to obtain subjective tendencies of user comments, providing a certain auxiliary role for public opinion supervision.

Acknowledgement. This work is supported by the Key Field Special Project of Guangdong Provincial Department of Education with No. 2021ZDZX1029.

References

1. Liu, Y., Lu, X., Deng, K., Ruan, K., Liu, J.: Method for constructing an emotional dictionary in the field of photography. Comput. Eng. Des. **40**(10), 3037–3042 (2019). https://doi.org/10.16208/j.issn1000-7024.2019.10.051

2. Shang, Y., Zhao, Y.: sentiment analysis and implementation of online comments based on machine learning. J. Dali Univ. **6**(12), 80–86 (2021)

3. Huang, S.: Research on Sentiment Analysis Based Comment Mining System. Nanjing University of Posts and Telecommunications (2021). https://doi.org/10.27251/d.cnki.gnjdc.2021-000084

4. Wang, C., Zhang, H., Mo, X., Yang, W.: A review of sentiment analysis on Weibo. Comput. Eng. Sci. **44**(01), 165–175 (2022)

5. Yang, K., Zhao, M.. Emotional analysis technology based on deep learning. Inf. Commun. (08), 99–101 (2020)

6. Blei, D.M., Ng, A.Y., Jordan, M.I.: Latent dirichlet allocation. J. Mach. Learn. Res. (2003)

7. Wu, Q., Fu, Y.: Overview of support vector machine feature selection methods. J. Xi'an Univ. Posts Telecommun. **25**(05), 16–21 (2020). https://doi.org/10.13682/j.issn.2095-6533.2020.05.003

8. Zhao, B., Liu, B.: Estimation of ground PM_ (2.5) concentration based on stacking. Environ. Eng. **38**(02), 153–159 (2020). https://doi.org/10.13205/j.hjgc.202002022

9. Wang, Z.: Research on personalized emotions of intelligent service robots using BP neural network model in IT service desk. Adhesive **48**(12), 79–82 (2021)

10. Chai, Y.: Implementation research of library book recommendation system based on Word2vec. Electron. Des. Eng. **30**(02), 7–10+15 (2022). https://doi.org/10.14022/j.issn1674-6236.2022.02.02

Author Index

C

Cai, Jiahui I-388
Cai, Tie I-388, II-49, II-135
Cai, Xingjuan I-277
Cao, Jiale I-73, II-337
Chai, Lu II-196
Che, Yongdie II-372
Chen, Bing II-196
Chen, Feiyun II-459
Chen, Weicong I-463
Chen, Xuhang I-266
Chen, Yan I-223, II-28
Chen, Yiheng II-476
Chen, Yongxian II-113
Chen, Yu I-137
Chen, Zeming II-135
Chen, Zhiqiang II-103, II-384
Cheng, Hangchi I-369
Cheng, Huabin I-137
Cheng, Shi II-407
Cheng, Ziyu I-409
Chu, Yih Bing I-231
Cui, Dandan I-165
Cui, Xiaojun I-18, II-170, II-187
Cui, Zhihua I-277

D

Damian, Maria Amelia E. II-3
Diao, Zhenya I-152
Dong, Ani I-443
Dong, Qianqian I-178

F

Fang, Wanhan II-196
Feng, Tian II-127

G

Gao, Zihang I-18, II-170, II-187
Gong, Miao II-476
Guan, Jian I-152, I-326
Guan, Jing I-207

Guo, Li I-409
Guo, Xuan I-292

H

He, Dan II-159
He, Fufa I-178
He, Jinfeng II-103
He, Jinrong I-340
He, Kejin II-287
He, Shuizhen I-312
He, Yongqiang I-277
Hu, Kun II-270
Hu, Min I-125
Huang, Peiquan I-3
Huang, Siming I-409
Huang, Weidong I-292, I-302, II-225
Huang, Xing I-302

J

Jalil, Hassan I-49
Ji, Dong I-165
Ji, Qi II-430
Jiang, Chengyu II-214
Jiang, Tian I-231
Jiang, Zifeng II-39
Jin, Xiao II-56
Jing, Furong I-88

K

Kang, Lanlan I-101, I-192
Kangshun, Li I-49, I-369, I-463, II-15
Kong, Yuyan I-3

L

Lai, Luyan II-351, II-372
Lai, Tao II-360
Lai, Yu I-101
Lei, Jiawei II-459
Lei, Yishu I-463
Li, Changrui I-451
Li, Gouqiong II-476

© The Editor(s) (if applicable) and The Author(s), under exclusive license
to Springer Nature Singapore Pte Ltd. 2024
K. Li and Y. Liu (Eds.): ISICA 2023, CCIS 2146, pp. 471–473, 2024.
https://doi.org/10.1007/978-981-97-4393-3

Li, Huade   II-337, II-394
Li, Jiahao   II-225
Li, Jiawang   II-142
Li, Jing   I-409
Li, Kangshun   I-39, I-266, I-312, II-39,
    II-127, II-360
Li, Mingxing   II-430
Li, Shaobo   I-420
Li, Wei   I-113, I-178
Li, Yuanxiang   II-71
Li, Zhaokui   I-340
Li, Zhengying   I-223
Liang, Bang   I-326
Liang, Haiyan   I-409
Liang, Zhixun   II-71
Liao, Futao   I-125
Lin, Chanjuan   I-26
Lin, Kexin   I-113
Liu, Feng   I-409
Liu, Hui   I-88
Liu, Yi   II-240
Liu, Yue   II-142
Liu, Zitu   II-142
Lu, Huyuan   II-384
Lu, Jintao   II-56

M
Ma, Sha   II-214, II-287
Ma, Xinyu   II-476
Meng, Wen   II-278

N
Nie, Huabei   I-247, I-443
Niu, Lu   I-420
Niu, Yi   I-443

O
Ou, Qingrong   I-326
Ou, Yangcong   I-207
Ou, Yuqi   I-113

P
Peng, Hongxing   II-28
Peng, Ling   I-357

Q
Qian, Haijun   I-247
Qiu, Wenbin   I-223
Qiu, Zhenzhen   I-3

S
Shang, Jingtong   II-270
Shao, Peng   I-395
Shen, Chao   II-430
Shen, Jianqiao   I-247, I-443
Shi, Dehao   II-28
Shi, Jia   I-451
Shi, Yunying   I-62
Su, Hongwei   II-3
Sun, Jian   I-137
Sun, Yuming   II-179, II-208, II-262

T
Tan, Xixian   II-419
Tang, Bo   I-326
Teng, Zi   II-71
Tian, Wenjie   II-142

W
Wang, Hao   II-214, II-287
Wang, Haoliang   II-270
Wang, Hua   II-179, II-208, II-262, II-297,
    II-319, II-442
Wang, Hui   I-125, I-388, II-49, II-135
Wang, Jian'ou   I-26
Wang, Jiancong   I-312
Wang, Juan   I-3
Wang, Junjie   II-15
Wang, Lei   II-450
Wang, Lili   II-270
Wang, Lingwei   II-319, II-442
Wang, Ping   I-357, II-95, II-326
Wang, Shuai   I-125
Wang, Wenjun   I-125
Wang, Xianmin   I-409
Wang, Xiaofeng   I-26
Wang, Yong   II-196
Wang, Yuanbing   II-253, II-309
Wang, Zhiyong   I-62
Wei, Bo   II-56
Wei, Yong   I-388, II-135
Wei, Yunshan   II-113, II-419
Wen, Peihua   II-459
Wu, Jian   I-137

X
Xiang, Lu   I-254
Xiao, Dong   I-125
Xiao, Hongyu   I-18, II-170, II-187
Xie, Mingchen   I-266

Xie, Yutong   I-409
Xiong, Yi   I-429
Xu, Qiner   II-384
Xu, Rui   II-337, II-394
Xu, Yiying   II-240
Xu, Zexin   II-337, II-394

Y

Yan, Yuangao   II-419
Yang, Lei   I-73, II-337, II-394
Yang, Ming   I-207
Yang, Shuxin   II-225
Yang, Zhongxin   I-340
Yao, Jintao   I-3
Yao, Wanyi   II-459
Ye, Chen   I-395
Yi, Yunfei   I-62, II-71
Yu, Chao   II-278
Yu, Fei   I-152, I-326
Yu, Haili   II-240
Yu, Jingkun   II-196
Yuan, Jia   I-254

Z

Zang, Yanhui   I-88
Zeng, Jia   II-459
Zeng, Sanyou   II-476
Zeng, Zhaolian   II-459

Zha, Wentao   II-56
Zhan, Guangsheng   II-407
Zhang, Dongbo   I-254
Zhang, Jiayu   I-39
Zhang, Jibo   I-254
Zhang, Jinbao   II-407
Zhang, Jinen   I-340
Zhang, Jingbo   I-277
Zhang, Jun   II-253, II-309
Zhang, Qing   II-476
Zhang, Shaoping   I-395
Zhang, Shaowei   I-125
Zhang, Xuming   I-357, II-95, II-326
Zhang, Yanjun   I-277
Zhang, Yongcai   I-231, I-443
Zhang, Yuanye   I-73
Zhao, Fei   II-476
Zhao, Jingwen   II-103
Zhao, Nannan   I-18, II-170, II-187
Zhong, Beixin   II-384
Zhong, Yi   I-192
Zhou, Xuesong   I-451
Zhu, Fuyu   II-297
Zhu, Haihua   I-443
Zhu, Tianjin   I-369
Zhu, Wenbin   II-15
Zhu, Zhanyang   I-137
Zong, Xuanyi   II-103

Printed in the United States
by Baker & Taylor Publisher Services